T0361811

INTRODUCTION TO PARTIAL DIFFERENTIAL EQUATIONS FOR SCIENTISTS AND ENGINEERS USING MATHEMATICA

INTRODUCTION TO PARTIAL DIFFERENTIAL EQUATIONS FOR SCIENTISTS AND ENGINEERS USING MATHEMATICA

Kuzman Adzievski • Abul Hasan Siddiqi

CRC Press
Taylor & Francis Group
Boca Raton London New York

CRC Press is an imprint of the
Taylor & Francis Group, an **informa** business

A CHAPMAN & HALL BOOK

CRC Press
Taylor & Francis Group
6000 Broken Sound Parkway NW, Suite 300
Boca Raton, FL 33487-2742

First issued in hardback 2019

© 2014 by Taylor & Francis Group, LLC
CRC Press is an imprint of Taylor & Francis Group, an Informa business

No claim to original U.S. Government works

ISBN-13: 978-1-4665-1056-2 (hbk)

Visit the Taylor & Francis Web site at
http://www.taylorandfrancis.com

and the CRC Press Web site at
http://www.crcpress.com

CONTENTS

v

PREFACE

There is a wide class of physical, real world problems, such as distribution of heat, vibrating string, blowing winds, movement of traffic, stock market prices and brain activity, which can be represented by partial differential equations. The goal of this book is to provide an introduction of phenomena and process represented by partial differential equations and their solutions, both analytical and numerical. In addition, we introduce those mathematical concepts and ideas, namely, Fourier series, integral transforms, Sturm–Liouville problems, necessary for proper understanding of the underlying foundation and technical details.

This book is written for a one- or two-semester course in partial differential equations. It was developed specifically for students in engineering and physical sciences. A knowledge of multi-variable calculus and ordinary differential equations is assumed. Results from power series, uniform convergence, complex analysis, ordinary differential equations and vector calculus that are used here are summarized in the appendices of the book.

The text introduces systematically all basic mathematical concepts and ideas that are needed for construction of analytic and numerical solutions of partial differential equations. The material covers all the elements that are encountered in theory and applications of partial differential equations which are taught in any standard university.

We have attempted to keep the writing style simple and clear, with no sacrifice of rigor. On many occasions, students are introduced to proofs. Where proofs are beyond the mathematical background of students, a short bibliography is provided for those who wish to pursue a given topic further. Throughout the text, the illustrations, numerous solved examples and projects have been chosen to make the exposition as clear as possible.

This textbook can be used with or without the use of computer software. We have chosen to use the Wolfram computer software Mathematica because of its easy introduction to computer computations and its excellent graphics capabilities. The use of this software allows students, among other things, to do fast and accurate computations, modeling, experimentation and visualization.

The organization of the book is as follows:

The book is divided into 8 chapters and 8 appendices. Chapters 1, 2 and 3 lay out the foundation for the solution methods of partial differential equations, developed in later chapters.

Chapter 1 is devoted to basic results of Fourier series, which is nothing but a linear combination of trigonometric functions. Visualization of Fourier series of a function is presented using Mathematica. The subject matter of this chapter is very useful for solutions of partial deferential equations presented in Chapters 5 to 8. Topics of Chapter 1 include Fourier series of periodic functions and their convergence, integration and differentiation, Fourier Sine and Fourier Cosine series and computation of partial sums of a Fourier series and their visualization by Mathematica.

Integral transforms are discussed in Chapter 2. Fourier transforms and their basic properties are discussed in this chapter. The Laplace transform and its elementary properties as well as its differentiation and integration are introduced. The inversion formula and convolution properties of the Laplace and Fourier transform, Heaviside step and the Dirac Delta functions are presented. Applications of the Laplace and Fourier transforms to initial value problems are indicated in this chapter. Parseval's identity and the Riemann-Lebesgue Lemma for Fourier transforms are also discussed.

The concept of the Fourier series, discussed in Chapter 1, is expanded in Chapter 3. Regular and singular Sturm–Liouville Problems and their properties are introduced and discussed in this chapter. The Legendre and Bessel differential equations and their solutions, the Legendre polynomials and Bessel functions, respectively, are discussed in the major part of this chapter.

Chapter 4 provides fundamental concepts, ideas and terminology related to partial differential equations. In addition, in this chapter there is an introduction to partial differential equations of the first order and linear partial differential equations of the second order, along with the classification of linear partial differential equations of the second order. The partial differential equations known as the heat equation, the wave equation, the Laplace and Poisson equations and the transport equation are derived in this chapter.

Analytical (exact) solutions of the wave, heat, Laplace and Poisson equations are discussed, respectively, in Chapters 5, 6 and 7.

Chapter 5 constitutes d'Alambert's method, separation of variable method of the wave equation on rectangular and circular domains. Solutions of the wave equation applying the Laplace and the Fourier transforms are obtained.

Chapter 6 is mainly devoted to the solution of the heat equation using Fourier and Laplace transforms. The multi dimensional heat equation on rectangular and circular domains is also studied. Projects using Mathematica for simulation of the heat equation are presented at the end of the chapter.

In Chapter 7 we discuss the Laplace and Poisson equations on different rectangular and circular domains. Applications of the Fourier and Hankel transform to the Laplace equation on unbounded domains are presented. Examples of applications of Mathematica to the Laplace and Poisson equations are given.

In Chapter 8 we study finite difference methods for elliptic, parabolic and hyperbolic partial differential equations. In recent years these methods have become an important tool in applied mathematics and engineering. A systematic review of linear algebra, iterative methods and finite difference is presented in this chapter. This will facilitate the proper understanding of numerical solutions of the above mentioned equations. Applications using Mathematica are provided.

Projects using Mathematica are included throughout the textbook.

Tables of the Laplace and Fourier transforms, for ready reference, are given in Appendix A and Appendix B, respectively.

Introduction and basics of Mathematica are given in Appendix H.

Answers to most of the exercises are provided.

An Index of Symbols used in the textbook is included.

In order to provide easy access to page numbers an index is included.

ACKNOWLEDGMENTS

The idea for writing this book was initiated while both authors were visiting Sultan Qaboos University in 2009. We thank Dr. M. A-Lawatia, Head of the Department of Mathematics & Statistics at Sultan Qaboos University, who provided a congenial environment for this project.

The first author is grateful for the useful comments of many of his colleagues at South Carolina State University, in particular: Sam McDonald, Ikhalfani Solan, Jean-Michelet Jean Michel and Guttalu Viswanath.

Many students of the first author who have taken applied mathematics checked and tested the problems in the book and their suggestions are appreciated.

The first author is especially grateful for the continued support and encouragement of his wife, Sally Adzievski, while writing this book, as well as for her useful linguistic advice.

Special thanks also are due to Slavica Grdanovska, a graduate research assistant at the University of Maryland, for checking the examples and answers to the exercises for accuracy in several chapters of the book.

The second author is indebted to his wife, Dr. Azra Siddiqi, for her encouragement. The second author would like to thank his colleagues, particularly Professor P. Manchanda of Guru Nanak Dev University, Amritsar India. Appreciation is also given to six research scholars of Sharda University, NCR, India, working under the supervision of the second author, who have read the manuscript of the book carefully and have made valuable comments.

Special thanks are due to Ms. Marsha Pronin, project coordinator, Taylor & Francis Group, Boca Raton, Florida, for her help in the final stages of manuscript preparation.

We would like to thank also Mr. Shashi Kumar from the Help Desk of Taylor & Francis for help in formatting the whole manuscript.

We acknowledge Ms. Michele Dimont, project editor, Taylor & Francis Group, Boca Raton, Florida, for editing the entire manuscript.

We acknowledge the sincere efforts of Ms. Aastha Sharma, acquiring editor, CRC Press, Taylor & Francis Group, Delhi office, whose constant persuasion enabled us to complete the book in a timely manner.

This book was typeset by $\mathcal{AMS}$-TEX, the TEX macro system of the American Mathematical Society.

Kuzman Adzievski

Abul Hasan Siddiqi

CHAPTER 1

FOURIER SERIES

The purpose of this chapter is to acquaint students with some of the most important aspects of the theory and applications of Fourier series.

Fourier analysis is a branch of mathematics that was invented to solve some partial differential equations modeling certain physical problems. The history of the subject of Fourier series begins with d'Alembert (1747) and Euler (1748) in their analysis of the oscillations of a violin string. The mathematical theory of such vibrations, under certain simplified physical assumptions, comes down to the problem of solving a particular class of partial differential equations. Their ideas were further advanced by D. Bernoulli (1753) and Lagrange (1759). Fourier's contributions begin in 1807 with his study of the problem of heat flow presented to the Académie des Sciences. He made a serious attempt to show that "arbitrary" function f of period T can be expressed as an infinite linear combination of the trigonometric functions sine and cosine of the same period T:

$$f(t) = \sum_{n=0}^{\infty} \left[a_n \cos \left(\frac{2n\pi t}{T} \right) + b_n \sin \left(2n\pi \, tT \right) \right].$$

Fourier's attempt later turned out to be incorrect. Dirichlet (1829), Riemann (1867) and Lebesgue (1902) and later many other mathematicians made important contributions to the subject of Fourier series.

Fourier analysis is a powerful tool for many problems, and particularly for solving various differential equations arising in science and engineering. Applications of Fourier series in physics, chemistry and engineering are enormous and almost endless: from the analysis of vibrations of the air in wind tunnels or the description of the heat distribution in solids, to electrocardiography, to magnetic resonance imaging, to the propagation of light, to the oscillations of the ocean tides and meteorology. Fourier analysis lies also at the heart of signal and image processing, including audio, speech, images, videos, radio transmissions, seismic data, and so on. Many modern technological advances, including television, music, CDs, DVDs video movies, computer graphics and image processing, are, in one way or another, founded upon the many results of Fourier analysis. Furthermore, a fairly large part of pure mathematics was invented in connection with the development of Fourier series. This remarkable range of applications qualifies Fourier's discovery as one of the most important in all of mathematics.

1.1 Fourier Series of Periodic Functions.

In this chapter we will develop properly the basic theory of Fourier analysis, and in the following chapter, a number of important extensions. Then we will be in a position for our main task—solving partial differential equations.

A Fourier series is an expansion of a periodic function on the real line $\mathbb{R}$ in terms of trigonometric functions. It can also be used to give expansions of functions defined on a finite interval in terms of trigonometric functions on that interval. In contrast to Taylor series, which can only be used to represent functions which have many derivatives, Fourier series can be used to represent functions which are continuous as well as those that are discontinuous. The two theories of Fourier series and Taylor series are profoundly different. A power series either converges everywhere, or on an interval centered at some point c, or nowhere except at the point c. On the other hand, a Fourier series can converge on very complicated and strange sets. Second, when a power series converges, it converges to an analytic function, which is infinitely differentiable, and whose derivatives are represented again by power series. On the other hand, Fourier series may converge, not only to periodic continuous functions, but also to a wide variety of discontinuous functions.

After reviewing periodic, even and odd functions, in this section we will focus on representing a function by Fourier series.

Definition 1.1.1. Let $T > 0$. A function $f : \mathbb{R} \to \mathbb{R}$ is called *T-periodic* if

$$f(x + T) = f(x)$$

for all $x \in \mathbb{R}$.

Remark. The number T is called a *period* of f. If f is non-constant, the smallest number T with the above property is called *the fundamental period* or simply *period* of f.

Classical examples of periodic functions are $\sin x$, $\cos x$ and other trigonometric functions. The functions $\sin x$ and $\cos x$ have period 2π, while $\tan x$ and $\cot x$ have period π.

Periodic functions appear in many physical situations, such as the oscillations of a spring, the motion of the planets about the sun, the rotation of the earth about its axis, the motion of a pendulum and musical sounds.

Next we prove an important property concerning integration of periodic functions.

Theorem 1.1.1. *Suppose that f is T-periodic. Then for any real number a, we have*

$$\int_0^T f(x)\,dx = \int_a^{a+T} f(x)\,dx.$$

Proof. Define the function F by

$$F(x) = \int\limits_{x}^{x+T} f(t)\, dt.$$

By the fundamental theorem of calculus, $F'(x) = f(x+T) - f(x) = 0$ since f is T-periodic. Hence, F is a constant function. In particular, $F(0) = F(a)$, which implies the theorem. ∎

Theorem 1.1.2. *The following result holds.*
(a) If $f_1, f_2, \ldots f_n$ are all T-periodic functions, then $c_1 f_1 + c_2 f_2 + \ldots + c_n f_n$ is also T-periodic.

(b) If f and g are T-periodic functions so is their product fg.

(c) If f and g are T-periodic functions so is their quotient f/g, provided $g \neq 0$.

(d) If f is T-periodic and $a > 0$, then $f(ax)$ has period $\frac{T}{a}$.

(e) If f is T-periodic and g is any function, then the composition $g \circ f$ is also T-periodic. (It is understood that f and g are such that the composition $g \circ f$ is well defined.)

Proof. We prove only part (d) of the theorem, leaving the other parts as an exercise. Let f be a T-periodic function and let $a > 0$. Then

$$f(a(x + \frac{T}{a})) = f(ax + T) = f(ax).$$

Therefore $f(ax)$ has period $\frac{T}{a}$. ∎

The following result is required in the study of convergence of Fourier series.

Lemma 1.1.1. *If f is any function defined on $-\pi < x \leq \pi$, then there is a unique 2π periodic function $\tilde{f}$, known as the 2π periodic extension of f, that satisfies $\tilde{f}(x) = f(x)$ for all $-\pi < x \leq \pi$.*

Proof. For a given $x \in \mathbb{R}$, there is a unique integer n so that $(2n-1)\pi < x \leq (2n+1)\pi$. Define $\tilde{f}$ by

$$\tilde{f}(x) = f(x - 2n\pi).$$

Note that if $-\pi < x \leq \pi$, then $n = 0$ and hence $\tilde{f}(x) = f(x)$. The proof that the resulting function $\tilde{f}$ is 2π periodic is left as an exercise. ∎

Definition 1.1.2. Let f be a function defined on an interval I (finite or infinite) centered at the origin $x = 0$.

1. f is said to be *even* if $f(-x) = f(x)$ for every x in I.

2. f is said to be *odd* if $f(-x) = -f(x)$ for every x in I.

Remark. The graph of an even function is symmetric with respect to the y-axis, and the graph of an odd function is symmetric with respect to the origin. For example, 5, x^2, x^4, $\cos x$ are even functions, while x, x^3, $\sin x$ are odd functions.

The proof of the following lemma is left as an exercise.

Lemma 1.1.2. *The following results hold.*

1. *The sum of even functions is an even function.*

2. *The sum of odd functions is an odd function.*

3. *The product of even functions and the product of two odd functions is an even function.*

4. *The product of an even and an odd function is an odd function.*

Since integrable functions will play an important role throughout this section we need the following definition.

Definition 1.1.3. A function $f : \mathbb{R} \to \mathbb{R}$ is said to be *Riemann integrable* on an interval $[a, b]$ (finite or infinite) if

$$\int_a^b |f(x)| dx < \infty.$$

Lemma 1.1.3. *If f is an even and integrable function on $[-a, a]$, then*

$$\int_{-a}^a f(x) \, dx = 2 \int_0^a f(x) \, dx.$$

If f is an odd and integrable function on the interval $[-a, a]$, then

$$\int_{-a}^a f(x) \, dx = 0.$$

Proof. We will prove the above result only for even functions f, leaving the case when f is an odd function as an exercise. Assume that f is an even function. Then

$$\int_{-a}^{a} f(x)\, dx = \int_{-a}^{0} f(x)\, dx + \int_{0}^{a} f(x)\, dx = \int_{-L}^{0} f(-x)\, dx + \int_{0}^{a} f(x)\, dx$$

$$= -\int_{a}^{0} f(t)\, dt + \int_{0}^{a} f(x)\, dx = \int_{0}^{a} f(t)\, dt + \int_{0}^{a} f(x)\, dx$$

$$= 2\int_{0}^{a} f(x)\, dx. \qquad \blacksquare$$

Suppose that a given function $f : \mathbb{R} \to \mathbb{R}$ is Riemann integrable on the interval $[-L, L]$. We wish to expand the function f in a (trigonometric) series

$$(1.1.1) \qquad f(x) = \frac{a_0}{2} + \sum_{n=1}^{\infty} \left[a_n \cos\left(\frac{n\pi}{L} x\right) + b_n \sin\left(\frac{n\pi}{L} x\right) \right].$$

Each term of the above series has period $2L$, so if the sum of the series exists, it will be a function of period $2L$. With this expansion there are three fundamental questions to be addressed:

(a) What values do the coefficients a_0, a_n, b_n have?

(b) If the appropriate values are assigned to the coefficients, does the series converge at some some points $x \in \mathbb{R}$?

(c) If the trigonometric series does converge at a point x, does it actually represent the given function $f(x)$?

The answer to the first question can be found easily by using the following important result from calculus, known as the *orthogonality property*. If m and n are any integers, then the following is true.

$$\int_{-L}^{L} \sin\left(\frac{m\pi}{L} x\right) \cos\left(\frac{n\pi}{L} x\right) dx = 0,$$

$$\int_{-L}^{L} \cos\left(\frac{m\pi}{L} x\right) \cos\left(\frac{n\pi}{L} x\right) dx = \begin{cases} 0, & m \neq n \\ L, & n = m \neq 0, \end{cases}$$

$$\int_{-L}^{L} \sin\left(\frac{m\pi}{L} x\right) \sin\left(\frac{n\pi}{L} x\right) dx = \begin{cases} 0, & m \neq n \\ L, & n = m \neq 0. \end{cases}$$

The above orthogonality relations suggest that we multiply both sides of the equation (1.1.1) by $\cos\left(\frac{n\pi}{L}x\right)$ and $\sin\left(\frac{n\pi}{L}x\right)$, respectively. If for a moment we ignore convergence issues (if term-by-term integration of the above series is allowed), we find

(1.1.2)
$$\begin{cases} a_n = \dfrac{1}{L}\displaystyle\int_{-L}^{L} f(x)\cos\left(\dfrac{n\pi}{L}x\right)dx, & n = 0, 1, 2, \ldots \\[2em] b_n = \dfrac{1}{L}\displaystyle\int_{-L}^{L} f(x)\sin\left(\dfrac{n\pi}{L}x\right)dx, & n = 1, 2, \ldots. \end{cases}$$

Therefore, if the proposed equality (1.1.1) holds, then the coefficients a_n and b_n must be chosen according to the formula (1.1.2).

In general, the answers to both questions (b) and (c) is "no." Ever since Fourier's time, enormous literature has been published addressing these two questions. We will see later in this chapter that the convergence or divergence of the Fourier series at a particular point depends only on the behavior of the function in an arbitrarily small neighborhood of the point.

The above discussion naturally leads us to the following fundamental definition.

Definition 1.1.4. Let $f : \mathbb{R} \to \mathbb{R}$ be a $2L$ periodic function which is integrable on $[-L, L]$. Let the Fourier coefficients a_n and b_n be defined by the above formulas (1.1.2). The infinite series

(1.1.3) $$S f(x) = \frac{a_0}{2} + \sum_{n=1}^{\infty}\left[a_n\cos\left(\frac{n\pi}{L}x\right) + b_n\sin\left(\frac{n\pi}{L}x\right)\right]$$

is called the Fourier series of the function f.

Notice that even the coefficients a_n and b_n are well defined numbers, there is no guarantee that the resulting Fourier series converges, and even if it converges, there is no guarantee that it converges to the original function f. For these reasons, we use the symbol $S f(x)$ instead of $f(x)$ when writing a Fourier series.

The notion introduced in the next definition will be very useful when we discuss the key issue of convergence of a Fourier series.

Definition 1.1.5. let N be a natural number. The Nth *partial sum* of the Fourier series of a function f is the trigonometric polynomial

$$S_N f(x) = \frac{a_0}{2} + \sum_{n=1}^{N}\left[a_n\cos\left(\frac{n\pi}{L}x\right) + b_n\sin\left(\frac{n\pi}{L}x\right)\right],$$

where a_n and b_n are the Fourier coefficients of the function f.

A useful observation for computational purposes is the following result.

Lemma 1.1.4. *If f is an even function, then*

$$a_n = \frac{2}{L} \int_0^L f(x) \cos \left(\frac{n\pi}{L} x\right) dx, \quad n = 0, 1, 2, \ldots \ \text{and} \ b_n = 0, \ n = 1, 2, \ldots$$

If f is an odd function, then

$$b_n = \frac{2}{L} \int_0^L f(x) \sin \left(\frac{n\pi}{L} x\right) dx, \quad n = 1, 2, \ldots \ \text{and} \ a_n = 0, \ n = 0, 1, \ldots$$

Proof. It easily follows from Lemma 1.1.1 and Lemma 1.1.3. ∎

Remark. For normalization reasons, in the first Fourier coefficient $\frac{a_0}{2}$ we have the factor $\frac{1}{2}$. Also notice that this first coefficient $\frac{a_0}{2}$ is nothing but the *mean (average)* of the function f on the interval $[-L, L]$.

In the next section we will show that every periodic function of x satisfying certain very general conditions can be represented in the form (1.1.1), that is, as a Fourier series.

Now let us take several examples.

Example 1.1.1. Let $f : \mathbb{R} \to \mathbb{R}$ be the 2π-periodic function which on the interval $[-\pi, \pi]$ is defined by

$$f(x) = \begin{cases} 0, & -\pi \le x < 0 \\ 1, & 0 \le x < \pi. \end{cases}$$

Find the Fourier series of the function f.

Solution. Using the formulas (1.1.2) for the Fourier coefficients in Definition 1.1.4 we have

$$a_0 = \frac{1}{\pi} \int_{-\pi}^{\pi} f(x) \, dx = \frac{1}{\pi} \int_{-\pi}^{0} 0 \, dx + \frac{1}{\pi} \int_{0}^{\pi} 1 \, dx = 1.$$

For $n = 1, 2, \ldots$ we have

$$a_n = \frac{1}{\pi} \int_{-\pi}^{\pi} f(x) \cos nx \, dx$$

$$= \frac{1}{\pi} \int_{-\pi}^{0} 0 \cos nx) \, dx + \frac{1}{\pi} \int_{0}^{\pi} 1 \cos nx \, dx$$

$$= 0 + \frac{1}{n\pi} \big[\sin(n\pi) - \sin(0) \big] = 0;$$

and

$$b_n = \frac{1}{\pi} \int_{-\pi}^{\pi} f(x) \sin nx \, dx$$

$$= \frac{1}{\pi} \int_{-\pi}^{0} 0 \sin nx \, dx + \frac{1}{\pi} \int_{0}^{\pi} 1 \sin nx \, dx$$

$$= 0 - \frac{1}{n\pi} \Big[\cos n\pi - \cos 0 \Big] = \frac{1 - (-1)^n}{n\pi}.$$

Therefore the Fourier series of the function is given by

$$S f(x) = \frac{1}{2} + \frac{2}{\pi} \sum_{n=1}^{\infty} \frac{1}{2n - 1} \sin (2n - 1) \frac{x}{2}.$$

Figure 1.1.1 shows the graphs of the function f, together with the partial sums $S_N f(x)$ of the function f taking $N = 1$ and $N = 3$ terms.

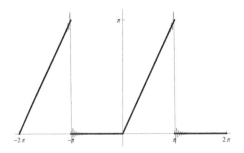

Figure 1.1.1

Figure 1.1.2 shows the graphs of the function f, together with the partial sums of the function f taking $N = 6$ and $N = 14$ terms.

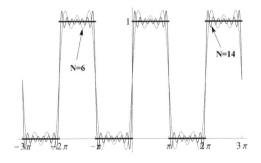

Figure 1.1.2

From these graphs we see that, as N increases, the partial sums $S_N f(x)$ become better approximations to the function f. It appears that the graphs of $S_N f(x)$, are approaching the graph of $f(x)$, except at $x = 0$ or where x is an integer. In other words, it looks like f is equal to the sum of its Fourier series except at the points where f is discontinuous.

Example 1.1.2. Let $f : \mathbb{R} \to \mathbb{R}$ be the 2π-periodic function which on $[-\pi, \pi)$ is defined by

$$f(x) = \begin{cases} -1, & -\pi \le x < 0 \\ 1, & 0 \le x < \pi. \end{cases}$$

Find the Fourier series of the function f.

Solution. The function is an odd function, hence by Lemma 1.1.1 its Fourier cosine coefficients a_n are all zero and for $n = 1, 2, \cdots$ we have

$$b_n = \frac{2}{\pi} \int_0^\pi f(x) \sin(nx)\, dx = \frac{2}{\pi} \int_0^\pi 1 \sin nx\, dx$$

$$= \frac{2}{n\pi} \left[-\cos n\pi + 1 \right] = \frac{2}{n\pi} \left[-(-1)^n + 1 \right].$$

Therefore the Fourier series of $f(x)$ is

$$S f(x) = \sum_{n=1}^{\infty} b_n \sin nx = \sum_{n=1}^{\infty} b_{2n-1} \sin (2n-1)x$$

$$= \frac{4}{\pi} \sum_{n=1}^{\infty} \frac{1}{2n-1} \sin (2n-1)x.$$

Figure 1.1.3 shows the graphs of the function f, together with the partial sums

$$S_N f(x) = \frac{4}{\pi} \sum_{n=1}^{N} \frac{1}{2n-1} \sin \big((2n-1)x\big)$$

of the function f, taking, respectively, $N = 1$ and $N = 3$ terms.

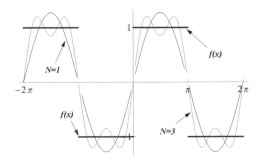

Figure 1.1.3

Figure 1.1.4 shows the graphs of the function f, together with the partial sums of the function f, taking, respectively, $N = 10$ and $N = 25$ terms.

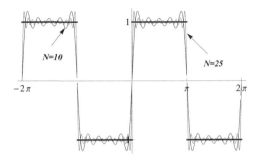

Figure 1.1.4

Figure 1.1.5 shows the graphs of the function f, together with the partial sum $S_N f$ of the function f, taking $N = 50$ terms.

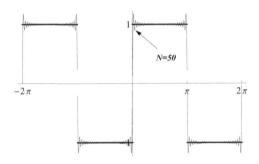

Figure 1.1.5

And again we see that, as N increases, $S_N f(x)$ becomes a better approximation to the function f. It appears that the graphs of $S_N f(x)$ are approaching the graph of $f(x)$, except at $x = 0$ or where x is an integer multiple of π. In other words, it looks as if f is equal to the sum of its Fourier series except at the points where f is discontinuous.

Example 1.1.3. Let $f : \mathbb{R} \to \mathbb{R}$ be the 2π-periodic function defined by $f(x) = |\sin(x)|$, $x \in [-\pi, \pi]$. Find the Fourier series of the function f.

Solution. This function is obviously an even function, hence by Lemma 1.1.1 its Fourier cosine coefficients b_n are all zero. For $n = 1$ we have that

$$a_1 = \frac{2}{\pi} \int\limits_0^\pi f(x) \cos x \, dx = \frac{2}{\pi} \int\limits_0^\pi \sin x \cos nx \, dx = 0.$$

For $n = 0, 2, 3, \ldots$ we have

$$a_n = \frac{2}{\pi} \int_0^\pi f(x) \cos nx \, dx = \frac{2}{\pi} \int_0^\pi \sin x \cos nx \, dx$$

$$= \frac{2}{\pi} \int_0^\pi \frac{1}{2} \left[\sin (n+1)x + \sin (n-1)x \right] dx$$

$$= \frac{1}{\pi} \left[-\frac{\cos (n+1)\pi}{n+1} - \frac{\cos (n1)\pi}{n-1} + \frac{1}{n+1} - \frac{1}{n-1} \right]$$

$$= \begin{cases} 0, & n \text{ odd} \\ \frac{1}{\pi}\left[\frac{2}{n+1} - \frac{2}{n-1} \right], & n \text{ even} \end{cases}$$

$$= \begin{cases} 0, & n \text{ odd} \\ -\frac{4}{(n^2-1)\pi}, & n \text{ even}. \end{cases}$$

Therefore the Fourier series of $f(x)$ is

$$S f(x) = \frac{a_0}{2} + \sum_{n=1}^\infty a_n \cos nx = \frac{a+0}{2} + \sum_{n=1}^\infty a_{2n} \cos 2nx$$

$$= \frac{2}{\pi} - \frac{4}{\pi} \sum_{n=1}^\infty \frac{1}{4n^2 - 1} \cos 2nx.$$

Figure 1.1.6 shows the graphs of the function f, together with the partial sums $S_N f$ of the function f, taking, respectively, $N = 1$ and $N = 3$ terms.

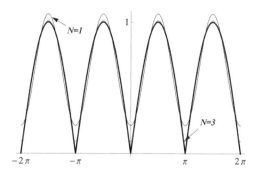

Figure 1.1.6

Figure 1.1.7 shows the graphs of the function f, together with the partial sums $S_N f$ of the function f, taking, respectively, $N = 10$ and $N = 30$ terms.

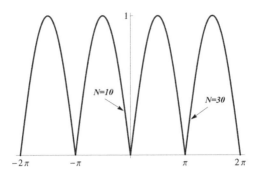

Figure 1.1.7

Remark. A useful observation for computing the Fourier coefficients is the following: instead of integrating from $-L$ to L in (1.1.2), we can integrate over any interval of length $2L$. This follows from Theorem 1.1.1 and the fact that the integrands are all $2L$-periodic. (See Exercise 1 of this section.)

Example 1.1.4. Let $f : \mathbb{R} \to \mathbb{R}$ be the 2π-periodic function defined on $[-\pi, \pi]$ by

$$f(x) = \begin{cases} 0, & -\pi \le x < 0 \\ x, & 0 \le x < \pi. \end{cases}$$

Find the Fourier series of the function f.

Solution. The Fourier coefficients are

$$a_0 = \frac{1}{\pi} \int_{-\pi}^{\pi} f(x)\,dx = \int_{-\pi}^{0} 0\,dx + \int_{0}^{\pi} x\,dx = \frac{1}{\pi}\left(0 + \frac{\pi^2}{2}\right) = \frac{\pi}{2}.$$

For $n = 1, 2, \ldots$ by the integration by parts formula we have

$$a_n = \frac{1}{\pi} \int_{\pi}^{\pi} f(x)\cos nx\,dx = \frac{1}{\pi}\int_{\pi}^{0} 0\cos nx\,dx + \frac{1}{\pi}\int_{0}^{\pi} x\cos nx\,dx$$

$$= 0 + \frac{1}{\pi}\left[\frac{x\sin nx}{n} + \frac{\cos nx}{n^2}\right]\Bigg|_{x=0}^{x=\pi} = \frac{1}{n^2\pi}[(-1)^n - 1].$$

Similarly, for $n = 1, 2, \ldots$ we obtain

$$b_n = \frac{(-1)^{n+1}}{n}.$$

Therefore the Fourier series of $f(x)$ is

$$S f(x) = \frac{a_0}{2} + \sum_{n=1}^{\infty} a_{2n-1}\cos(2n-1)x + \sum_{n=1}^{\infty} b_n \sin nx$$

$$= \frac{\pi}{4} - \frac{2}{\pi}\sum_{n=1}^{\infty} \frac{1}{(2n-1)^2}\cos(2n-1)x + \frac{2}{\pi}\sum_{n=1}^{\infty} \frac{(-1)^n}{n}\sin nx.$$

Figure 1.1.8 shows the graphs of the function f, together with the partial sums $S_N f$ of the function f, taking, respectively, $N = 1$ and $N = 3$ terms.

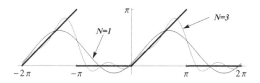

Figure 1.1.8

Figure 1.1.9 shows the graphs of the function f, together with the partial sums $S_N f$ of the function f, taking, respectively, $N = 10$ and $N = 30$ terms.

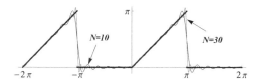

Figure 1.1.9

Figure 1.1.10 shows the graphs of the function f together with the partial sum $S_N f$ of the function f, taking $N = 50$ terms.

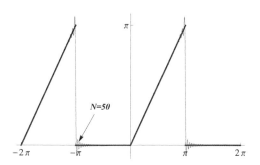

Figure 1.1.10

Example 1.1.5. Let $f : \mathbb{R} \to \mathbb{R}$ be the 2π-periodic function defined on $[-\pi, \pi]$ by

$$f(x) = \begin{cases} \frac{1}{2}\pi + x, & -\pi \leq x < 0 \\ \frac{1}{2}\pi - x, & 0 \leq x < \pi. \end{cases}$$

Find the Fourier series of the function f.

Solution. The Fourier coefficients are

$$a_0 = \frac{1}{\pi} \int_{-\pi}^{\pi} f(x)\, dx = \frac{1}{\pi} \int_{-\pi}^{0} (\frac{1}{2}\pi + x)\, dx + \frac{1}{\pi} \int_{0}^{\pi} (\frac{1}{2}\pi - x)\, dx$$

$$= \frac{1}{\pi}(0 + 0) = 0.$$

For $n = 1, 2, \ldots$ by the integration by parts formula we have

$$a_n = \frac{1}{\pi} \int_{-\pi}^{\pi} f(x) \cos nx\, dx = \frac{1}{\pi} \left\{ \int_{-\pi}^{0} (\frac{\pi}{2} + x) \cos nx\, dx + \int_{0}^{\pi} (\frac{\pi}{2} - x) \cos nx\, dx \right\}$$

$$= \frac{1}{\pi} \left\{ \left[(\frac{1}{2} + x)\frac{\sin nx}{n} + \frac{\cos nx}{n^2} \right]_{x=-\pi}^{x=0} + \left[(\frac{1}{2} - x)\frac{\sin nx}{n} + \frac{\cos nx}{n^2} \right]_{x=0}^{x=\pi} \right\}$$

$$= \frac{2}{n^2 \pi}(1 - \cos n\pi).$$

The computation of the coefficients b'_ns is like that of the a'_ns, and we find that $b_n = 0$ for $n = 1, 2, \ldots$. Hence the Fourier series of $f(x)$ is

$$S f(x) = \frac{4}{\pi} \sum_{n=1}^{\infty} \frac{\cos (2n-1)\pi}{(2n-1)^2}.$$

Figure 1.1.11 shows the graphs of the function f, together with the partial sums $S_N f$ of the function f, taking, respectively, $N = 1$ and $N = 3$ terms.

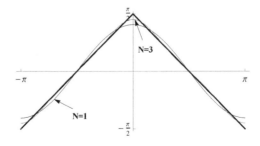

Figure 1.1.11

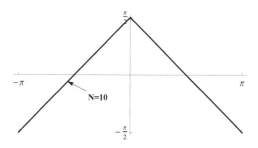

Figure 1.1.12

Figure 1.1.12 shows the graphs of the function f, together with the partial sum $S_N f$ of the function f, taking $N = 10$ terms.

So far we have discussed Fourier series whose terms are the trigonometric functions sine and cosine. An alternative, and more convenient, approach to Fourier series is to use complex exponentials. There are several reasons for doing this. One of the reasons is that this is a more compact form, but the main reason is that this complex form of a Fourier series will allow us in the next chapter to introduce important extensions—the *Fourier transforms.*

Let $f : \mathbb{R} \to \mathbb{R}$ be a $2L$ periodic function which is integrable on $[-L, L]$. First let us introduce the variable

$$\omega_n = \frac{n\pi}{L}.$$

In view of the formulas

$$\cos \alpha = \frac{e^{i\alpha} + e^{-i\alpha}}{2}, \quad \sin \alpha = \frac{e^{i\alpha} - e^{-i\alpha}}{2i},$$

easily obtained from the Euler's formula

$$e^{i\alpha} = \cos \alpha + i \sin \alpha,$$

the Fourier series (1.1.3) can be written in the following form:

(1.1.4)
$$S f(x) = \sum_{n=-\infty}^{\infty} c_n e^{i\omega_n x},$$

where

(1.1.5)
$$c_n = \frac{1}{2L} \int_{-L}^{L} f(x) e^{-i\omega_n x}\, dx, \; n = 0, \pm 1, \pm 2, \dots.$$

This is the *complex form* of the Fourier series of the function f on the interval $[-L, L]$.

Remark. Notice that even though the complex Fourier coefficients c_n are generally complex numbers, the summation (1.1.4) always gives a real valued function $S f(x)$. Also notice that the Fourier coefficient c_0 is the mean of the function f on the interval $[-L, L]$.

We need to emphasize that the real Fourier series (1.1.3) and the complex Fourier series (1.1.4) are just two different ways of writing the same series.

Using the relations $a_n = c_n + c_{-n}$ and $b_n = i(c_n - c_{-n})$ we can change the complex form of the Fourier series (1.1.4) to the real form (1.1.3), and vice versa. Also note that for an even function all coefficients c_n will be real numbers, and for an odd function are purely imaginary.

If m and n are any integers, then from the formulas

$$\int_{-L}^{L} e^{i\omega_n x} e^{i\omega_m x} \, dx = \begin{cases} 0, & m \neq n \\ \frac{1}{2L}, & m = n \end{cases}$$

it follows that the following set of functions

$$(1.1.6) \qquad \left\{ 1, e^{i\omega_1 x}, e^{-i\omega_1 x}, e^{i\omega_2 x}, e^{-i\omega_2 x}, \cdots, e^{i\omega_n x}, e^{i\omega_n x}, \ldots \right\}$$

has the orthogonality property.

Example 1.1.6. Let $f : \mathbb{R} \to \mathbb{R}$ be the 2π-periodic function defined on $[-\pi, \pi]$ by

$$f(x) = \begin{cases} -1, & -\pi \leq x < 0 \\ 1, & 1 \leq x < \pi. \end{cases}$$

Find the complex Fourier series of the function f.

Solution. In this problem, we have $L = \pi$ and $\omega_n = n$. Therefore,

$$c_0 = \frac{1}{2\pi} \int_{-\pi}^{\pi} f(x) \, dx = \frac{1}{2\pi} \int_{-\pi}^{0} -1 \, dx + \frac{1}{2\pi} \int_{0}^{\pi} 1 \, dx = 0.$$

For $n \neq 0$ we have

$$c_n = \frac{1}{2\pi} \int_{-\pi}^{\pi} f(x) e^{-inx} \, dx = \frac{1}{2\pi} \int_{-\pi}^{0} -1 e^{-inx} \, dx + \frac{1}{2\pi} \int_{0}^{\pi} 1 e^{-inx} \, dx$$

$$= \frac{1}{2n\pi} e^{-inx} \Big|_{x=-\pi}^{x=0} - \frac{1}{2n\pi} e^{-inx} \Big|_{x=0}^{x=\pi}$$

$$= -\frac{i}{2n\pi}(1 - e^{in\pi}) + \frac{i}{2n\pi}(e^{-in\pi} - 1) = \frac{i}{n\pi}(\cos(n\pi) - 1)$$

$$= \frac{i}{n\pi}((-1)^n - 1).$$

Therefore, the complex Fourier series is

$$S f(x) = -\frac{2i}{\pi} \sum_{n=-\infty}^{\infty} \frac{e^{(2n-1)ix}}{2n-1}.$$

Exercises for Section 1.1.

1. Let f be a function defined on the interval $[-a, a]$, $a > 0$. Show that:

   (a) If f is even, then $\int_{-a}^{a} f(x)dx = 2\int_{0}^{a} f(x)dx$.

   (b) If f is odd, then $\int_{-a}^{a} f(x)dx = 0$.

2. Let f be a function whose domain is the interval $[-L, L]$. Show that f can be expressed as a sum of an even and an odd function.

3. Let f be an L-periodic function and a any number. Show that

$$\int_{0}^{L} f(x)dx = \int_{a}^{a+L} f(x)dx.$$

4. Show that the following results are true.

   (a) If $f_1, f_2, \ldots f_n$ are L-periodic functions and $c_1, c_2, \ldots, c_n$ are any constants, then $c_1 f_1 + c_2 f_2 + \ldots + c_n f_n$ is also L-periodic.

   (b) If f and g are L-periodic functions, so is their product fg.

   (c) If f and g are L-periodic functions, so is their quotient f/g.

   (d) If f is L-periodic and $a > 0$, then $f(\frac{x}{a})$ has period aL and $f(ax)$ has period $\frac{L}{a}$.

   (e) If f is L-periodic and g is any function (not necessarily periodic), then the composition $g \circ f$ is also L-periodic.

5. Find the Fourier series of the following functions (the functions are all understood to be periodic):

   (a) $f(x) = x$, $-\pi < x < \pi$.

(b) $f(x) = |\sin(x)|$, $-\pi < x < \pi$.

(c) $f(x) = \pi - x$, $0 < x < 2\pi$.

(d) $f(x) = x(\pi - |x|)$, $-\pi < x < \pi$.

(e) $f(x) = e^{ax}$, $0 < x < 2\pi$.

(f) $f(x) = e^{ax}$, $-\pi < x < \pi$.

(g) $f(x) = \begin{cases} x, & -\frac{3}{2} \le x < \frac{1}{2} \\ 1 - x, & \frac{1}{2} \le x \le \frac{3}{2}. \end{cases}$

(h) $f(x) = \begin{cases} 0, & -2 \le x < 0 \\ x, & 0 \le x \le 1 \\ 1, & 1 < x \le 2. \end{cases}$

6. Complete the proof of Lemma 1.1.

7. Show that the Fourier series (1.1.3) can be written in the following form:

$$S\,f(x) = \frac{1}{2}a_0 + \sum_{n=1}^{\infty} A_n \sin\left(\frac{n\pi}{L}x + \alpha\right),$$

where

$$A_n = \sqrt{a_n^2 + b_n^2}, \quad \tan \alpha = \frac{a_n}{b_n}.$$

8. Find the complex Fourier series for the following functions (the functions are all understood to be periodic):

(a) $f(x) = |x|$, $-\pi < x < \pi$.

(b) $f(x) = x$, $0 < x < 2$.

(c) $f(x) = \begin{cases} 0, & -\frac{\pi}{2} < x < 0 \\ 1, & 0 < x < \frac{\pi}{2}. \end{cases}$

9. Show that if a is any real number, then the Fourier series of the function $f(x) = e^{ax}$, $0 < x < 2\pi$, is

$$S\,(f,x) = \frac{e^{2a\pi} - 1}{2\pi} \sum_{n=-\infty}^{\infty} \frac{1}{a - in} e^{inx}.$$

10. Show that if a is any real number, then the Fourier series of the function $f(x) = e^{ax}$, $-\pi < x < \pi$, is

$$S(f,x) = \frac{e^{a\pi} - e^{-a\pi}}{4\pi} \sum_{n=-\infty}^{\infty} \frac{(-1)^n}{a - in} e^{inx}.$$

1.2 Convergence of Fourier Series.

In this section we will show that the Fourier series $Sf(x)$ of a periodic function f converges, under certain reasonably general conditions, on the function f. For convenience only we will consider the interval $[-\pi, \pi]$ and the function $f : \mathbb{R} \to \mathbb{R}$ is 2π-periodic. We begin this section by deriving an important estimate on the Fourier coefficients.

Theorem 1.2.1 (Bessel's Inequality). *If $f : \mathbb{R} \to \mathbb{R}$ is a 2π-periodic and Riemann integrable function on the interval $[-\pi, \pi]$ and the complex Fourier coefficients c_k are defined by (1.1.5), then*

$$\sum_{-\infty}^{\infty} |c_k|^2 \le \frac{1}{2\pi} \int_{-\pi}^{\pi} |f(x)|^2 \, dx.$$

Proof. Let n be any positive integer number. Using $|z|^2 = z\bar{z}$ for any complex number z ($\bar{z}$ is the complex conjugate of z), we have

$$0 \le \left| f(x) - \sum_{k=-n}^{n} c_k e^{ikx} \right|^2 = \left(f(x) - \sum_{k=-n}^{n} c_k e^{ikx} \right) \overline{\left(f(x) - \sum_{k=-n}^{n} c_k e^{ikx} \right)}$$

$$= |f(x)|^2 - \sum_{k=-n}^{n} \left[c_k f(x) e^{ikx} + \overline{c_k} f(x) e^{-ikx} \right] + \sum_{k=-n}^{n} \sum_{j=-n}^{n} c_k \overline{c_j} e^{ikx} e^{-jx}.$$

Dividing both sides of the above equation by 2π, integrating over $[-\pi, \pi]$ and taking into account formula (1.1.5) for the complex Fourier coefficients c_j, the orthogonality property of the set (1.1.6) implies

$$0 \le \frac{1}{2\pi} \int_{-\pi}^{\pi} |f(x)|^2 \, dx - \sum_{k=-n}^{n} \left[c_k \overline{c_k} + \overline{c_k} c_k \right] + \sum_{k=-n}^{n} c_k \overline{c_k}$$

$$= \frac{1}{2\pi} \int_{-\pi}^{\pi} |f(x)|^2 \, dx - \sum_{k=-n}^{n} \left[c_k \overline{c_k} + \overline{c_k} c_k \right] + \sum_{k=-n}^{n} |c_k|^2$$

$$= \frac{1}{2\pi} \int_{-\pi}^{\pi} |f(x)|^2 \, dx - \sum_{k=-n}^{n} |c_k|^2.$$

Letting $n \to \infty$ in the above inequality we obtain the result.   ∎

Remark. Based on the relations for the complex Fourier coefficients c_n and the real Fourier coefficients a_n and b_n, Bessel's inequality can also be stated in terms of a_n and b_n:

$$\frac{1}{4}a_0^2 + \frac{1}{2}\sum_{n=1}^{\infty}\left(a_n^2 + b_n^2\right) \leq \frac{1}{2\pi}\int_{-\pi}^{\pi} |f(x)|^2 \, dx.$$

Later in this chapter we will see that Bessel's inequality is actually an equality. But for now, this inequality implies that the series

$$\sum_{n=-\infty}^{\infty} |c_n|^2, \quad \sum_{n=0}^{\infty} a_n^2 \text{ and } \sum_{n=1}^{\infty} b_n^2$$

are all convergent, and as a consequence we obtain the following important result about the Fourier coefficients:

Corollary 1.2.1 (Riemann–Lebsgue Lemma). *If f is a 2π-periodic and Riemann integrable function on the interval $[-\pi, \pi]$ and c_n (a_n and b_n) are the Fourier coefficients of f, then*

$$\lim_{n\to\infty} c_n = 0, \quad \lim_{n\to\infty} a_n = \lim_{n\to\infty} b_n = 0.$$

The following terminology is needed in order to state and prove our convergence results.

Definition 1.2.1. A function f is called *piecewise continuous* on an interval $[a, b]$ if it is continuous everywhere on that interval except at finitely many points $c_1, c_2, \ldots, c_n \in [a, b]$ and the left-hand and right-hand limits

$$f(c_j^-) = \lim_{\substack{h\to 0 \\ h>0}} f(c_j - h) \text{ and } f(c_j^+) = \lim_{\substack{h\to 0 \\ h>0}} f(c_j + h), \quad j = 1, 2, \ldots, n$$

of f exist and they are finite.

A function f is called *piecewise smooth* on an interval $[a, b]$ if f and its first derivative f' are piecewise continuous functions on $[a, b]$.

A function f is called *piecewise continuous (piecewise smooth)* on $\mathbb{R}$ if it is piecewise continuous (piecewise smooth) on every bounded interval.

Representative graphs of piecewise continuous and piecewise smooth functions are provided in the next several examples.

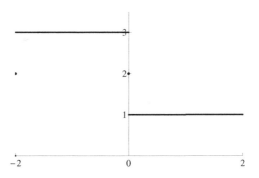

Figure 1.2.1

Example 1.2.1. The function $f : [-2, 2] \to \mathbb{R}$ defined by

$$f(x) = \begin{cases} 3, & -2 < x < 0 \\ 1, & 0 < x \le 2 \\ 2, & x = -2, 0 \end{cases}$$

is piecewise continuous on $[-2, 2]$ since f is discontinuous only at the points $c = -2, 0$. See Figure 1.2.1.

Example 1.2.2. The function $f : [-2, 6] \to \mathbb{R}$ defined by

$$f(x) = \begin{cases} -x, & -2 \le x \le 0 \\ x^2, & 0 \le x \le 1 \end{cases}$$

is piecewise smooth on $[-2, 1]$ since it is continuous on that interval, and it is not differentiable only at the point $c = 0$. See Figure 1.2.2.

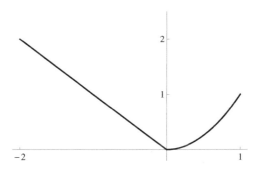

Figure 1.2.2

Example 1.2.3. The function $f : [-4, 4] \to \mathbb{R}$ defined by

$$
f(x) = \begin{cases}
-x, & -4 \le x < 1 \\
-1, & 1 < x \le 3 \\
x - 4, & 3 < x \le 4
\end{cases}
$$

is piecewise smooth on $[-4, 4]$ since it is continuous on that interval, and it is not differentiable at the points $c = 1, 3$. See Figure 1.2.3.

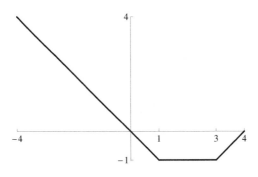

Figure 1.2.3

Example 1.2.4. The function f defined by $f(x) = \sqrt[3]{x}$ (see Figure 1.2.4.) is piecewise continuous but not piecewise smooth on any interval containing $x = 0$, since

$$
f'(0-) = f'(0+) = \infty.
$$

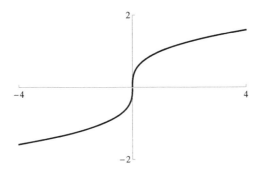

Figure 1.2.4

Before addressing the question of convergence of Fourier series we address the following question of *uniqueness of Fourier series*:

If two functions f and g have the same Fourier coefficients:

$$(1.2.1) \qquad \int_{-\pi}^{\pi} f(x)e^{-inx}\,dx = \int_{-\pi}^{\pi} g(x)e^{-inx}\,dx, \qquad n = 0, \pm 1, \pm 2, \ldots,$$

are the functions necessarily identical? In other words, is a function *uniquely determined* by its Fourier coefficients? The answer is affirmative and for its proof the interested reader can consult the book by G. B. Folland [6].

Theorem 1.2.2 (Uniqueness Theorem). *Let f and g be piecewise continuous on the interval $-\pi \le x \le \pi$ and satisfy (1.2.1), i.e., the two functions have the same Fourier coefficients. Then $f(x) = g(x)$ for all $x \in [-\pi, \pi]$ except perhaps at points of discontinuity.*

Let us recall the definition of a partial sum of a Fourier series. If N is a natural number, then we defined the Nth *partial sum* of the Fourier series of a 2π-periodic function f by

$$(1.2.2) \qquad S_N f(x) = \sum_{n=-N}^{N} c_n e^{inx} = \frac{1}{2}a_0 + \sum_{n=1}^{N} \left[a_n \cos nx + b_n \sin nx \right].$$

If we substitute c_n from formula (1.1.5) into (1.2.2) we have

$$S_N f(x) = \sum_{n=-N}^{N} \left(\frac{1}{2\pi} \int_{-\pi}^{\pi} e^{-iny} f(y)\,dy \right) e^{inx} = \int_{-\pi}^{\pi} \left(\frac{1}{2\pi} \sum_{n=-N}^{N} e^{in(x-y)} \right) f(y)\,dy.$$

Definition 1.2.2. If N is a natural number, then the function

$$(1.2.3) \qquad D_N(x) = \frac{1}{2\pi} \sum_{n=-N}^{N} e^{inx}.$$

is called the Nth *Dirichlet kernel.*

Notice that $D_N(x)$ is a 2π-periodic function. By a change of variable and the periodicity of the function $D_N(x-y)f(y)$ it follows that

$$(1.2.4) \qquad S_N f(x) = \int_{-\pi}^{\pi} D_N(x-y)f(y)\,dy = \int_{-\pi}^{\pi} D_N(y)f(x-y)\,dy.$$

We discuss below some properties of the kernel $D_N(x)$ which plays a crucial role in obtaining convergence results for Fourier series.

Lemma 1.2.1. *The Dirichlet Kernel D_N satisfies the following properties:*

(a)
$$\int_{-\pi}^{\pi} D_N(x)\,dx = 1,$$

(b)
$$D_N(x) = \frac{1}{2\pi}\frac{\sin\left(N+\frac{1}{2}\right)x}{\sin\left(\frac{x}{2}\right)}.$$

Proof. To prove (a) we use the definition of $D_N(x)$. From $(1.2.3)$ we have

$$D_N(x) = \frac{1}{2\pi} + \frac{1}{\pi}\sum_{n=1}^{N}\cos nx,$$

and therefore

$$\int_{-\pi}^{\pi} D_N(x)\,dx = \left[\frac{1}{2\pi}x + \frac{1}{\pi}\sum_{n=1}^{N}\frac{\sin nx}{n}\right]_{x=-\pi}^{x=\pi} = 1.$$

Using the geometric sum formula, from $(1.2.3)$ for $x \neq 0$ we have

$$D_N(x) = \frac{1}{2\pi}e^{-iNx}\sum_{n=0}^{2N}e^{inx} = \frac{1}{2\pi}e^{-iNx}\frac{e^{i(2N+1)x}-1}{e^{ix}-1}$$

$$= \frac{1}{2\pi}\frac{e^{i(N+1)x}-e^{-iNx}}{e^{ix}-1} = \frac{1}{2\pi}\frac{e^{i(N+\frac{1}{2})x}-e^{-i(N+\frac{1}{2})x}}{e^{i\frac{x}{2}}-e^{-i\frac{x}{2}}}$$

$$= \frac{1}{2\pi}\frac{\sin\left(N+\frac{1}{2}\right)x}{\sin\left(\frac{x}{2}\right)},$$

which establishes (b).   ∎

From (a) in Lemma 1.2.2. it follows that

$$(1.2.5) \qquad \int_{-\pi}^{0} D_N(x)\,dx = \int_{0}^{\pi} D_N(x)\,dx = \frac{1}{2}.$$

Plots of the kernel $D_N(x)$ for several values of N are presented in Figure 1.2.5.

Now we are ready to state and prove the main convergence theorem.

Theorem 1.2.3. *If f is a 2π-periodic and piecewise smooth function on the real line $\mathbb{R}$, then*

$$\lim_{N\to\infty} S_N f(x) = \frac{f(x^+)+f(x^-)}{2}.$$

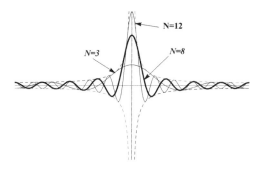

Figure 1.2.5

at every point x. Hence

$$\lim_{N \to \infty} S_N f(x) = f(x)$$

at every point x of continuity of the function f.

Proof. Using the integral representation (1.2.4) for $S_N f$, by Lemma 1.2.1 and formula (1.2.5) for the Dirichlet kernel $D_N(y)$, we have

$$S_N f(x) - \frac{f(x^-) + f(x^+)}{2} = \int_{-\pi}^{\pi} D_N(y) f(x-y) \, dy - \frac{f(x^-) + f(x^+)}{2}$$

$$= \int_{\pi}^{0} D_N(y) \big[f(x-y) - f(x^+) \big] \, dy + \int_{0}^{\pi} D_N(y) \big[f(x-y) - f(x^-) \big] \, dy$$

$$= \int_{0}^{\pi} D_N(y) \big[f(x-y) - f(x^-) \big] \, dy + \int_{\pi}^{0} D_N(y) \big[f(x+y) - f(x^+) \big] \, dy$$

$$= \int_{0}^{\pi} D_N(y) \big[f(x-y) - f(x^-) + f(x+y) - f(x^+) \big] \, dy$$

$$= \frac{1}{2\pi} \int_{0}^{\pi} \frac{\sin \left(N + \frac{1}{2} \right) y}{\sin \left(\frac{y}{2} \right)} \big[f(x-y) - f(x^-) + f(x+y) - f(x^+) \big] \, dy$$

$$= \frac{1}{\pi} \int_{0}^{\pi} \big[\sin \left(N + \frac{1}{2} \right) y \big] \left(\frac{\frac{y}{2}}{\sin \left(\frac{y}{2} \right)} \right) \big[\frac{f(x-y) - f(x^-)}{y} \, dy$$

$$+ \frac{1}{\pi} \int_{0}^{\pi} \big[\sin \left(N + \frac{1}{2} \right) y \big] \left(\frac{\frac{y}{2}}{\sin \left(\frac{y}{2} \right)} \right) \frac{f(x+y) - f(x^+)}{y} \big] \, dy.$$

For a fixed value x define the function $F(y)$ by

$$F(y) = \frac{\frac{y}{2}}{\sin\left(\frac{y}{2}\right)} \left[\frac{f(x-y) - f(x^-)}{y} + \frac{f(x+y) - f(x^+)}{y} \right].$$

It is easy to see that $F(y)$ is an odd function of y on $[-\pi, \pi]$. We claim that for each fixed value of x this function is piecewise continuous on $[-\pi, \pi]$. Indeed, we need to check the behavior of $F(y)$ only at the point $y = 0$. Since f is a piecewise smooth function we have

$$\lim_{y \to 0+} \frac{\frac{y}{2}}{\sin\left(\frac{y}{2}\right)} \left[\frac{f(x-y) - f(x^-)}{y} + \frac{f(x+y) - f(x^+)}{y} \right] = f'(x^+) + f'(x^-).$$

Therefore,

$$S_N f(x) - \frac{1}{2}[f(x^-) + f(x^+)] = \frac{1}{\pi} \int_0^\pi F(y) \sin\left(N + \frac{1}{2}\right) y \, dy$$

$$= \frac{1}{\pi} \int_0^\pi \left\{ F(y) \cos\left(\frac{y}{2}\right) \right\} \sin(Ny) \, dy + \frac{1}{\pi} \int_0^\pi \left\{ F(y) \sin\left(\frac{y}{2}\right) \right\} \cos(Ny) \, dy$$

$$= B_N + A_N,$$

where B_N, A_N are the Nth Fourier coefficients of the piecewise continuous functions

$$\frac{1}{2} F(y) \cos\left(\frac{y}{2}\right), \qquad \frac{1}{2} F(y) \sin\left(\frac{y}{2}\right),$$

respectively. From Corollary 1.2.1 (Riemann–Lebesgue Lemma) it follows that $B_N \to 0$ and $A_N \to 0$ as $N \to \infty$. Therefore

$$S_N f(x) - \frac{f(x^-) + f(x+)}{2} = B_N + A_N \to 0, \text{ as } N \to \infty. \qquad \blacksquare$$

This result, besides its theoretical significance, provides a useful method for finding the sums of certain numerical series.

Example 1.2.5. Let $f : \mathbb{R} \to \mathbb{R}$ be the 2π-periodic function which on $[-\pi, \pi]$ is defined by

$$f(x) = \begin{cases} 0, & -\pi \le x < 0 \\ \sin x, & 0 \le x \le \pi. \end{cases}$$

Find the Fourier series of f and find the sum of the series for $x = k\pi$, $k \in \mathbb{Z}$. Plot f and several partial sums of the Fourier series of f.

Solution. The Fourier series of f is given by

$$S f(x) = \frac{1}{\pi} + \frac{1}{2} \sin x - \frac{2}{\pi} \sum_{n=1}^{\infty} \frac{1}{4n^2 - 1} \cos(2nx).$$

Since f is continuous on $\mathbb{R}$ by Theorem 1.2.3 we have

$$f(x) = \frac{1}{\pi} + \frac{1}{2} \sin x - \frac{2}{\pi} \sum_{n=1}^{\infty} \frac{1}{4n^2 - 1} \cos(2nx)$$

for every $x \in \mathbb{R}$. In particular, for $x = k\pi$, $k \in \mathbb{Z}$ we have

$$f(k\pi) = 0 = \frac{1}{\pi} - \frac{2}{\pi} \sum_{n=1}^{\infty} \frac{1}{4n^2 - 1}$$

and after rearrangement

$$\sum_{n=1}^{\infty} \frac{1}{4n^2 - 1} = \frac{1}{2}.$$

In Figure 1.2.6 we plot f and the partial sums $S_N f$ of f for $N = 1$ and $N = 4$.

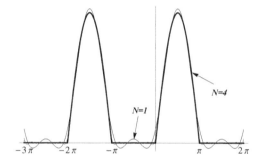

Figure 1.2.6

In Figure 1.2.7 we plot f and the partial sum $S_N f$ of f for $N = 10$.

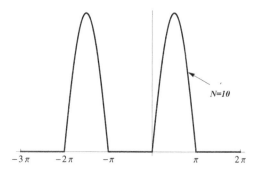

Figure 1.2.7

Example 1.2.6. Show that

$$\sum_{n=1}^{\infty} \frac{1}{4n^2 - 1} = \frac{1}{2}.$$

Solution. The Fourier series of the 2π-periodic function $f(x) = |\sin(x)|$, $x \in [-\pi, \pi]$ is

$$S f(x) = \frac{2}{\pi} - \frac{4}{\pi} \sum_{n=1}^{\infty} \frac{1}{4n^2 - 1} \cos 2nx.$$

Since $x = 0$ is a point of continuity of f the Fourier series of f at $x = 0$ converges to $f(0) = 0$ and thus

$$0 = \frac{1}{2} - \frac{4}{\pi} \sum_{n-1}^{\infty} \frac{1}{4n^2 - 1}.$$

Example 1.2.7. Examine the behavior of the partial sums of the 2π-periodic function f, which on the interval $(-\pi, \pi)$ is given by $f(x) = e^x$.

Solution. The complex Fourier coefficients c_n of the function f are

$$c_n = \frac{1}{\pi} \int_{-\pi}^{\pi} e^{-inx} e^x \, dx = \frac{(e^\pi - e^{-\pi})(-1)^n}{2(1 - ni)\pi}.$$

Using the relations between the complex Fourier coefficients and the real Fourier coefficients we find the real Fourier series of f is

$$S f(x) = \frac{e^\pi - e^{-\pi}}{2\pi} + \frac{e^\pi - e^{-\pi}}{\pi} \sum_{n=1}^{\infty} \frac{(-1)^n}{n^2 + 1} \left[\cos nx - n \sin nx \right].$$

In Figure 1.2.8 we compare the graphs of $S_N f(x)$ for $N = 1, 3$ with the graph of the function f.

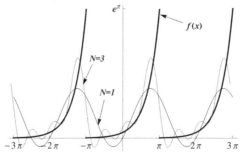

Figure 1.2.8

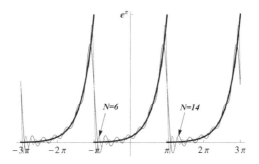

Figure 1.2.9

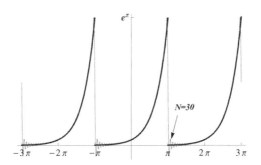

Figure 1.2.10

In Figure 1.2.9 we compare the graphs of $S_N f(x)$ for $N = 6$ and $N = 14$ with the graph of the function f.

In Figure 1.2.10 we compare the graph of $S_N f(x)$ for $N = 30$ with the graph of the function f.

In Figure 1.2.11 we compare the graph of $S_N f(x)$ for $N = 90$ with the graph of the function f in the small interval $(49/50\pi, 51/50\pi)$ around the point $x = \pi$.

If we examine Figure 1.2.8, Figure 1.2.9, Figure 1.2.10 and Figure 1.2.11 more closely we notice the following:

1. The partial sums $S_N f(x)$ converge "nicely" to $f(x)$ as $N \to \infty$ at all points x where f is continuous.

2. The partial sums $S_N f(x)$ exhibit some strange behavior near the points x where f is discontinuous (has a jump) as N increases. We can see in each case that the functions $S_N f(x)$ have noticeable *overshoot* just to the right (and left) of $x = 0$ (with a similar behavior near the points $x = \pm\pi,\ \pm 2\pi$). We also see that these overshoots remain narrower and narrower and their magnitudes remain fairly large even

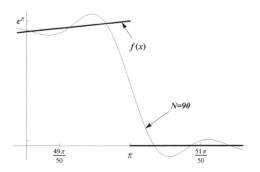

Figure 1.2.11

as N is larger and larger. This curious behavior is known as the *Gibbs phenomenon* and it is a manifestation of the non-uniform convergence of the Fourier series that will be discussed later.

Example 1.2.8. Discuss the Gibbs phenomenon for the 2π-periodic function which on $[-\pi, \pi]$ is defined by

$$f(x) = \begin{cases} 0, & x = -\pi \\ -1, & -\pi < x < 0 \\ 0, & x = 0 \\ 1, & 0 < x < \pi \\ 0, & x = \pi. \end{cases}$$

Solution. The Fourier series of this odd function is

$$S\,f(x) = \frac{4}{\pi} \sum_{n=1}^{\infty} \frac{1}{2n-1} \sin(2n-1)x.$$

In Figure 1.2.12 we plotted the function f and the Nth partial sums

$$S_N\,f(x) = \frac{4}{\pi} \sum_{n=1}^{N} \frac{1}{2n-1} \sin(2n-1)x$$

for $N = 1$ and $N = 3$, respectively.

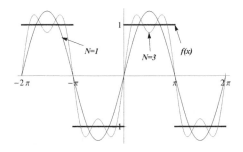

Figure 1.2.12

In Figure 1.2.13 we plotted the function f and the Nth partial sums for $N = 10$ and $N = 25$, respectively.

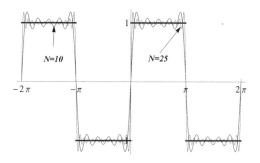

Figure 1.2.13

In Figure 1.2.14 we plotted the function f and the Nth partial sum for $N = 50$.

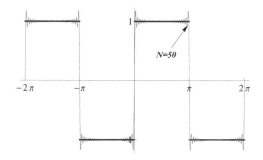

Figure 1.2.14

In Figure 1.2.15 we plotted the function f and the Nth partial sum for $N = 150$.

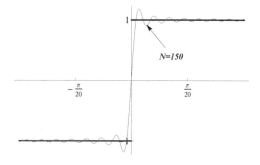

Figure 1.2.15

Let us now examine analytically the Gibbs phenomenon for this function. Since f is continuous at the point $x = 0$ the Fourier series at $x = 0$ converges to $f(0) = 0$. From the continuity of f we have for small positive $x < \pi$, $S_N f(x)$ converges to 1; and for small negative $x > -\pi$, $S_N f(x)$ converges to -1. We are especially interested in the behavior of $S_N f(x)$ in a small neighborhood of 0 as $N \to \infty$. Since both f and $S_N f$ are odd functions, it suffices to consider the case when x is positive and small.

First, we find the relative extrema of $S_N f$. From

$$S_N' f(x) = \frac{4}{\pi} \sum_{n=1}^{N} \cos(2n-1)x$$

and the summation formula (b) in Lemma 1.2.1 we have

$$S_N' f(x) = \frac{2}{\pi} \frac{\sin 2Nx}{\sin x}.$$

From the last expression it follows S_N has relative extrema in the interval $(0, \pi)$ at the points $x = \frac{n}{2N}\pi$, $n = 1, 2, \ldots, 2N-1$. It is easy to show that at the point $x = \frac{\pi}{2N}$ the function $S_N f$ has the first maximum. The value of this maximum is

$$S_N f\left(\frac{\pi}{2N}\right) = \frac{4}{\pi} \sum_{n=1}^{N} \frac{1}{2n-1} \sin\left(\frac{2n-1}{2N}\pi\right).$$

Further, notice that the above sum can be made to look like a Riemann sum for the integral

$$\frac{2}{\pi} \int_0^{\pi} \frac{\sin x}{x} dx.$$

Indeed, consider the continuous function

$$\frac{2}{\pi} \frac{\sin x}{x}.$$

If we partition the segment $[0, \pi]$ into $2N$ equal sub-intervals each of length

$$\Delta x = \frac{\pi}{2N},$$

then

$$\lim_{N \to \infty} S_N\left(\frac{\pi}{2N}\right) = \lim_{N \to \infty} \frac{4}{\pi} \sum_{n=1}^{N} \frac{1}{2n-1} \sin(2n-1)\frac{\pi}{2N}$$

$$= \frac{2}{\pi} \lim_{N \to \infty} \sum_{n=1}^{N} \frac{\sin(2n-1)\frac{\pi}{2N}}{\frac{2n-1}{2N}\pi} \cdot \frac{\pi}{2N}$$

$$= \frac{2}{\pi} \int_0^{\pi} \frac{\sin x}{x} dx,$$

that is,

$$\lim_{N\to\infty} S_N\left(\frac{\pi}{2N}\right) = \frac{2}{\pi}\int_0^\pi \frac{\sin x}{x}\,dx.$$

The last integral is non-elementary, but can be easily approximated. Using a Maclaurin series we have

$$\frac{2}{\pi}\int_0^\pi \frac{\sin x}{x}\,dx = 2 - \frac{\pi^2}{9} + \frac{\pi^4}{300} - \frac{\pi^6}{17640} + \cdots,$$

and so up to three accurate decimal places we have

$$\frac{2}{\pi}\int_0^\pi \frac{\sin x}{x}\,dx \approx 1.179.$$

Therefore

(1.2.6) $$\lim_{N\to\infty} S_N\left(\frac{\pi}{2N}\right) \approx 1.179.$$

Equation (1.2.6) shows that the maximum values of the partial sums $S_N f$ are always greater than 1 near the discontinuity of the function f at $x = 0$, no matter how many terms we choose to include in the partial sums. But 1 is the maximum value of the original function that is being represented by the Fourier series. It follows, therefore, that near the discontinuity, the maximum difference between the values of the partial sums and the function itself (sometimes called the *overshoot* or the *bump*) remains finite as $N \to \infty$. This overshoot is approximately 0.18 in this example.

Exercises for Section 1.2.

1. Which of the following functions are continuous, piecewise continuous, or piecewise smooth on the interval $[-\pi, \pi]$?

   (a) $f(x) = \csc x$.

   (b) $f(x) = \sqrt[3]{\sin x}$.

   (c) $f(x) = \sqrt[4]{\sin^5 x}$.

   (d) $f(x) = \begin{cases} -\cos x, & -\pi \le x \le 0 \\ \cos x, & 0 < x \le \pi. \end{cases}$

(e) $f(x) = \begin{cases} \sin 2x & -\pi \leq x \leq 0 \\ \sin x, & 0 < x \leq \pi. \end{cases}$

2. How smooth are the following functions? That is, how many derivatives can you guarantee them to have?

(a) $f(x) = \displaystyle\sum_{n=-\infty}^{\infty} \frac{1}{n^{4.2} + 2n^6 - 1} e^{inx}$.

(b) $f(x) = \displaystyle\sum_{n=-\infty}^{\infty} \frac{1}{n + \frac{1}{2}} \cos nx$.

(c) $f(x) = \displaystyle\sum_{n=-\infty}^{\infty} \frac{1}{2^n} \cos(2^n x)$.

3. Let f be the 2π-periodic function which on the interval $[-\pi, \pi]$ is given by
$$f(x) = \begin{cases} 0, & -\pi < x \leq 0 \\ \sin x, & 0 < x \leq \pi. \end{cases}$$

(a) Show that the Fourier series of f is given by

$$\frac{1}{\pi} + \frac{1}{2} \sin x - \frac{2}{\pi} \sum_{n=1}^{\infty} \frac{1}{4n^2 - 1} \cos 2nx, \quad x \in \mathbb{R}.$$

(b) Find the Fourier series at $x = k\pi, \ k \in \mathbb{Z}$.

(c) Find the sum of the series

$$\sum_{n=1}^{\infty} \frac{1}{4n^2 - 1}.$$

(d) Plot the graph of f and the Fourier partial sums $S_3 f$, $S_{10} f$, $S_{30} f$ and $S_{100} f$ on the interval $[-2\pi, 2\pi]$.

4. Let f be the 2π-periodic function, which on the interval $[-\pi, \pi]$ is given by
$$f(x) = \begin{cases} x, & 0 < x \leq \pi \\ 0, & \pi < x \leq 2\pi. \end{cases}$$

(a) Show that the Fourier series of f is given by

$$\frac{\pi}{4} - \frac{2}{\pi} \sum_{n=1}^{\infty} \frac{\cos (2n-1)x}{(2n-1)^2} + \sum_{n=1}^{\infty} \frac{(-1)^{n-1}}{n} \sin nx, \quad x \in \mathbb{R}.$$

(b) Find the Fourier series at $x = \pi$.

(c) Plot the graph of f and the Fourier partial sums $S_3 f$, $S_{10} f$, $S_{30} f$ and $S_{100} f$ on the interval $[-2\pi, 2\pi]$.

(d) Examine the Gibbs phenomenon near the point $x = \pi$.

5. Let f be the 2π-periodic function which on the interval $(0, 2\pi]$ is given by

$$f(x) = \frac{\pi - x}{2}.$$

(a) Show that the Fourier series of f is given by

$$\sum_{n=1}^{\infty} \frac{1}{n} \sin nx, \ x \in \mathbb{R}.$$

(b) Find the sums of the Fourier series at $x = 0$ and $x = 2\pi$.

(c) Plot the graph of f and the Fourier partial sums $S_3 f$, $S_{10} f$, $S_{30} f$ and $S_{100} f$ on the interval $[-2\pi, 2\pi]$.

(d) Examine the Gibbs phenomenon near the point $x = 0$.

6. Using the Fourier expansion of the function $f(x) = x(\pi - |x|)$, $-\pi < x < \pi$, and choosing a suitable value of x, derive the following:

$$\sum_{n=1}^{\infty} \frac{(-1)^{n+1}}{(2n - 1)^3} = \frac{\pi^3}{32}.$$

7. Let f be the 2π-periodic function which on the interval $[-\pi, \pi]$ is given by

$$f(x) = \begin{cases} \frac{\pi}{2} - \frac{x}{2} - \frac{x^2}{4\pi}, & -\pi < x \leq 0 \\ \frac{\pi}{2} + \frac{x}{2} - \frac{x^2}{4\pi}, & 0 < x \leq \pi. \end{cases}$$

(a) Show that the Fourier series of f is given by

$$\frac{2\pi}{3} - \frac{1}{\pi} \sum_{n=1}^{\infty} \frac{\cos nx}{n^2}, \ x \in \mathbb{R}.$$

(b) Find the Fourier series at $x = \pi$.

(c) Plot the graph of f and the Fourier partial sums $S_3 f$, $S_{10} f$, $S_{30} f$ and $S_{100} f$ on the interval $[-2\pi, 2\pi]$.

8. Show that each of the following Fourier expansions is valid in the indicated range.

(a) If $0 \leq x \leq 2\pi$, then

$$\frac{x^2}{2} - \pi x = -\frac{\pi^2}{3} + 2 \sum_{n=1}^{\infty} \frac{\cos nx}{n^2}.$$

(b) If $0 < x < 2\pi$, then

$$x^2 = \frac{4\pi^2}{3} + 4 \sum_{n=1}^{\infty} \left[\frac{1}{n^2} \cos nx - \frac{\pi}{n} \sin nx \right].$$

(c) If $-\pi \leq x \leq \pi$, then

$$x^2 = \frac{\pi^2}{3} + 4 \sum_{n=1}^{\infty} \frac{(-1)^n}{n^2} \cos nx.$$

9. Let α be any positive number that is not an integer. Define $f(x) = \sin(\alpha x)$ for $x \in (-\pi, \pi)$. Let $f(-\pi) - f(\pi) = 0$ and extend f to be a 2π-periodic function.

(a) Plot the graph of f when $0 < \alpha < 1$ and when $1 < \alpha < 2$.

(b) Show that the Fourier series of f is

$$S(f, x) = -\frac{2 \sin(\alpha \pi)}{\pi} \sum_{n=1}^{\infty} \frac{(-1)^{n+1}}{\alpha^2 - n^2} \sin nx.$$

(c) Identify all points x in $\mathbb{R}$ at which the series converges.

(d) Prove that the Fourier series converges uniformly to f on any interval $[a, b]$ that does not contain an odd multiple of π.

(e) Take $x = \pi/2$ and $y = \alpha/2$. Use the trigonometric formula $\sin 2\pi y = 2 \sin \pi y \cos \pi y$ in order to show that

$$\pi \sec \pi y = -4 \sum_{n=1}^{\infty} \frac{(-1)^{n+1}}{4y^2 - (2n-1)^2}.$$

10. Let a be a non-zero constant. Show that

$$e^{ax} = \frac{e^{a\pi} - e^{-a\pi}}{2a\pi} + \frac{e^{a\pi} - e^{-a\pi}}{\pi} \sum_{n=1}^{\infty} \frac{(-1)^n}{n^2 + a^2} \left(a \cos nx - n \sin nx \right)$$

for every x, $-\pi < x < \pi$.

11. Let a be any number which is not an integer. Show that the following is true

(a) For any $-\pi \leq x \leq \pi$ we have

$$\cos ax = \frac{\sin a\pi}{a\pi} + \frac{2\sin a\pi}{\pi} \sum_{n=1}^{\infty} \frac{(-1)^{n+1}a}{n^2 - a^2} \cos nx$$

and

(b) For any $-\pi < x < \pi$ we have

$$\sin ax = \frac{2\sin a\pi}{\pi} \sum_{n=1}^{\infty} \frac{(-1)^{n+1}n}{n^2 - a^2} \sin nx.$$

What happens if a is an integer?

===

1.3 Integration and Differentiation of Fourier Series.

In this section we will study the questions of integration and differentiation of Fourier series. If a series of functions converges "nicely" to some function f, then we expect to be able to integrate and differentiate the series term by term and the resulting series should converge to the integral and derivative of the given function f. For example, integration and differentiation of a power series is always valid inside the interval of convergence. In many situations, both operations of integration and differentiation term-by-term of Fourier series lead to valid results, and are quite useful for constructing new Fourier series of more complicated functions. However, in all these situations, the question is, to what do the resulting series converge?

From calculus we know that integration is a smoothing operation, that is, the resulting function is always smoother than the original function. Therefore, we would expect that we would be able to integrate Fourier series without any difficulties. There is, however, one problem: the integral of a periodic function is not necessarily periodic. For example, the constant function 1 is certainly periodic, but its integral is not. On the other hand, integrals of all sine and cosine functions appearing in the Fourier series are periodic. Therefore, only the constant term might cause some difficulty when we try to integrate a Fourier series. Recall that functions which have no constant terms in their Fourier series expansions are exactly those functions which have zero means. Now, we will show that only functions which have zero means remain periodic upon integration. The results discussed below for 2π-periodic functions defined on $[-\pi, \pi]$ can be easily extended to $2L$ periodic functions defined on any interval $[-L, L]$.

Lemma 1.3.1. *If f is a 2π-periodic function, then its indefinite integral*

$$F(x) = \int\limits_0^x f(t)\, dt$$

is 2π-periodic if and only if

$$c_0 = \int\limits_{-\pi}^{\pi} f(x)\, dx = 0.$$

Proof. Assume that the function F is 2π-periodic, i.e., let $F(x+2\pi) = F(x)$ for every x. If we take $x = -\pi$, then it follows $F(\pi) = F(-\pi)$. Therefore

$$\int\limits_0^{\pi} f(t)\, dt = \int\limits_0^{-\pi} f(t)\, dt.$$

If in the right-hand side integral we introduce a new variable t by $u = t+2\pi$, then

$$\int\limits_0^{\pi} f(t)\, dt = \int\limits_{2\pi}^{\pi} f(u - 2\pi)\, du.$$

Since f is 2π-periodic, we have that

$$\int\limits_0^{\pi} f(t)\, dt = \int\limits_{2\pi}^{\pi} f(u)\, du,$$

and so

$$\int\limits_0^{\pi} f(t)\, dt - \int\limits_{2\pi}^{\pi} f(u)\, du = 0.$$

Therefore

$$\int\limits_0^{\pi} f(t)\, dt + \int\limits_{\pi}^{2\pi} f(t)\, dt = 0,$$

that is,

$$\int\limits_0^{2\pi} f(t)\, dt = 0.$$

From the last equation, again by the 2π-periodicity of f, it follows

$$\int\limits_{-\pi}^{\pi} f(t)\, dyt = 0$$

as required.

The proof of the other direction is left as an exercise. ∎

Using the above lemma we easily have the following result.

Theorem 1.3.1. *Let f be a 2π-periodic, piecewise continuous function with complex Fourier coefficients c_n (real Fourier coefficients a_n and b_n) and let*

$$F(x) = \int_0^x f(y)\, dy.$$

If $c_0 = a_0 = 0$, then for all x,

$$F(x) = C_0 + \sum_{\substack{n=-\infty \\ n\neq 0}}^{\infty} \frac{c_n}{in} e^{inx} = \frac{A_0}{2} + \sum_{n=1}^{\infty} \left[-\frac{b_n}{n} \cos nx + \frac{a_n}{n} \sin nx \right],$$

where

$$C_0 = \frac{A_0}{2} = \frac{1}{2\pi} \int_{-\pi}^{\pi} F(x)\, dx$$

is the average of the function $F(x)$ on the interval $[-\pi, \pi]$.

Proof. Integrate term-by-term the Fourier series and use Lemma 1.3.1.   ■

If $c_0 \neq 0$, considering the function $F(x) - c_0 x$, from the previous theorem we have the following result.

Theorem 1.3.2. *Let f be a 2π-periodic, piecewise continuous function with Fourier coefficients c_n and let*

$$F(x) = \int_0^x f(y)\, dy.$$

Then for all x we have that

$$F(x) = c_0 x + C_0 + \sum_{\substack{n=-\infty \\ n\neq 0}}^{\infty} \frac{c_n}{in} e^{inx},$$

where

$$C_0 = \frac{1}{2\pi} \int_{-\pi}^{\pi} F(x)\, dx$$

is the average of F on $[-\pi, \pi]$.

Notice that the last series is not a Fourier series because it contains the term $c_0 x$.

Next we present two theorems regarding differentiation of Fourier series. First we need the following result.

Theorem 1.3.3. *Let c_n be the complex Fourier coefficients (or the real coefficients a_n and b_n) of a 2π-periodic, continuous and piecewise smooth function f and let c_n' $(a_n'$ and $b_n')$ be the Fourier coefficients of the first derivative f' of f. Then*

$$c_n' = inc_n, \quad a_n' = nb_n; \quad b_n' = -na_n.$$

Proof. The fundamental theorem of calculus

$$\int_a^b f'(x)\,dx = f(b) - f(a)$$

is valid not only for functions f which are continuously differentiable on the interval $[a, b]$, but also for functions f which are continuous and piecewise smooth on $[a, b]$.

Applying the integration by parts formula and the fundamental theorem of calculus we have

$$c_n' = \frac{1}{2\pi} \int_{-\pi}^{\pi} f'(x)e^{-inx}\,dx = \frac{1}{2\pi}f(x)e^{-inx}\Big|_{-\pi}^{\pi} - \frac{1}{2\pi}\int_{-\pi}^{\pi} f(x)(-ine^{-inx})\,dx$$

$$= \frac{1}{2\pi}\big[f(\pi)e^{-\pi ni} - f(-\pi)e^{\pi ni}\big] + \frac{in}{2\pi}\int_{-\pi}^{\pi} f(x)e^{-inx}\,dx$$

$$= \frac{1}{2\pi}\big[f(\pi)(-1)^n - f(-\pi)(-1)^n\big] + \frac{in}{2\pi}\int_{-\pi}^{\pi} f(x)e^{-inx}\,dx$$

$$= 0 + inc_n = inc_n.$$

A similar proof works for $a_n' = nb_n$, $b_n' = -na_n$. ■

Theorem 1.3.4. *Let f be a 2π-periodic, continuous and piecewise smooth function, with piecewise smooth derivative f'. If c_n $(a_n$ and $b_n)$ are the Fourier coefficients of f, then the Fourier series $S\,f'$ of f' is given by*

$$S\,f'(x) = \sum_{n=-\infty}^{\infty} inc_n e^{inx} = \sum_{n=1}^{\infty} \big(nb_n \cos nx - na_n \sin nx\big)$$

and converges to $f'(x)$ for each x where f' is continuous. If f' is not continuous at x, then the series converges to $\frac{1}{2}\big[f'(x-) + f'(x+)\big]$.

Proof. The result follows by combining the previous theorem and the Fourier Convergence Theorem. ■

Let us illustrate the above results with several examples.

Example 1.3.1. Integrate term-by-term the Fourier series of the 2π-periodic function $f : \mathbb{R} \to \mathbb{R}$ which for $x \in [-\pi, \pi]$ is given by

$$f(x) = \pi - |x|$$

and obtain a new Fourier series.

Solution. Since the given 2π-periodic function $f : \mathbb{R} \to \mathbb{R}$ is continuous on $\mathbb{R}$, its Fourier series $Sf(x)$ converges to $f(x)$ for every x. Since f is even, its Fourier coefficients b_n are all zero. By computation we find $a_0 = 0$ and

$$a_{2n-1} = \frac{8}{\pi} \frac{1}{(2n-1)^2} \quad \text{and} \quad a_{2n} = 0 \quad \text{for} \quad n = 1, 2, \dots.$$

Therefore the Fourier series of f is given by

$$\pi - 2|x| = \frac{8}{\pi} \sum_{n=1}^{\infty} \frac{1}{(2n-1)^2} \cos(2n-1)x.$$

Since $a_0 = 0$, by Theorem 1.3.1, integrating both sides of the above equation we obtain

$$x(\pi - |x|) = \frac{4}{\pi} \sum_{n=1}^{\infty} \frac{1}{(2n-1)^3} \sin(2n-1)x.$$

Example 1.3.2. Differentiating term-by-term the Fourier series of the 2π-periodic function $f : \mathbb{R} \to \mathbb{R}$, which on the interval $[-\pi, \pi]$ is given by

$$f(x) = \begin{cases} -x \sin x, & -\pi < x < 0 \\ x \sin x, & 0 < x < \pi \end{cases}$$

obtain a new Fourier series and find the sum of the series

$$\sum_{n=1}^{\infty} \frac{n^2}{(2n+1)^2(2n-1)^2}.$$

Solution. The given 2π-periodic, odd function f is continuous and piecewise smooth, with Fourier series

$$f(x) = \frac{\pi}{2} \sin x - \frac{16}{\pi} \sum_{n=1}^{\infty} \frac{n}{(2n-1)^2(2n+1)^2} \sin 2nx.$$

By direct calculations we have that

$$f'(x) = \begin{cases} \sin x + x \cos x, & -\pi < x < 0 \\ -\sin x - x \cos x, & 0 < x < \pi. \end{cases}$$

Since
$$\lim_{x\to 0^+} f'(x) = \lim_{x\to 0^-} f'(x) = 0$$
and
$$\lim_{x\to \pi^-} f'(x) = \lim_{x\to -\pi^+} f'(x) = -\pi$$

the continuation of the function f' on the real $\mathbb{R}$ is continuous. Therefore, term-by-term differentiation of the Fourier series of f gives us

$$f'(x) = \frac{\pi}{2} \cos x - \frac{32}{\pi} \sum_{n=1}^{\infty} \frac{n^2}{(2n-1)^2(2n+1)^2} \cos 2nx.$$

If we rearrange the terms of the Fourier series of f', for $x \in [0, \pi]$ we obtain

$$\sum_{n=1}^{\infty} \frac{n^2}{(4n^2-1)^2} \cos 2nx = \frac{\pi^2}{64} \cos x - \frac{\pi}{32} f'(x)$$

$$= \frac{\pi^2}{64} \cos x - \frac{\pi}{32} \sin x - \frac{\pi}{32} \cos x.$$

Inserting $x = 0$ in the last series we have that

$$\sum_{n=1}^{\infty} \frac{n^2}{(4n^2-1)^2} = \frac{\pi^2}{64}.$$

Example 1.3.3. Let $f : \mathbb{R} \to \mathbb{R}$ be the function given by

$$f(x) = \frac{1}{5 - 3\cos x}, \quad x \in \mathbb{R}.$$

Show that

$$f(x) = \frac{1}{4} + \frac{1}{2} \sum_{n=1}^{\infty} \frac{1}{3^n} \cos nx, \ x \in \mathbb{R}.$$

Further, if the function $g : \mathbb{R} \to \mathbb{R}$ is given by

$$g(x) = \frac{\sin x}{5 - 3\cos x}, \quad x \in \mathbb{R},$$

show that

$$g(x) = \frac{2}{3} \sum_{n=1}^{\infty} \frac{1}{3^n} \sin nx, \ x \in \mathbb{R}.$$

Finally, find the sum of the series

$$\sum_{n=1}^{\infty} \frac{1}{3^n} \cos nx.$$

Solution. Since

$$\left|\frac{e^{ix}}{3}\right| = \frac{1}{3} < 1 \quad \text{for every } x \in \mathbb{R},$$

by the complex geometric sum formula we obtain

$$\sum_{n=1}^{\infty} \frac{e^{inx}}{3^n} = \sum_{n=1}^{\infty} \left(\frac{e^{ix}}{3}\right)^n = \frac{e^{ix}}{3} \cdot \frac{1}{1 - \frac{e^{ix}}{3}} = \frac{e^{ix}}{3 - e^{ix}} \cdot \frac{3 - e^{-ix}}{3 - e^{-ix}}$$

$$\cdot \frac{3e^{ix} - 1}{9 - 6\cos x + 1} = \frac{1}{2} \cdot \frac{3\cos x - 1}{5 - 3\cos x} + i\frac{1}{2} \cdot \frac{3\sin x}{5 - 3\cos x}.$$

Hence

$$\frac{1}{4} + \frac{1}{2} \sum_{n=1}^{\infty} \frac{1}{3^n} \cos nx = \frac{1}{4} + \frac{1}{2} Re\left\{\sum_{n=1}^{\infty} \frac{1}{3^n} e^{inx}\right\} = \frac{1}{4} + \frac{1}{2} \cdot \frac{1}{2} \cdot \frac{3\cos x - 1}{5 - 3\cos x}$$

$$= \frac{1}{5 - 3\cos x} = f(x),$$

and

$$\frac{2}{3} \sum_{n=1}^{\infty} \frac{1}{3^n} \sin nx = \frac{2}{3} Im\left\{\sum_{n=1}^{\infty} \frac{1}{3^n} e^{inx}\right\} = \frac{2}{3} \cdot \frac{1}{2} \cdot \frac{3\sin x}{5 - 3\cos x} = g(x).$$

By termwise integration we obtain

$$\int_0^x g(t)\, dt = \int_0^x \frac{\sin t}{5 - 3\cos t}\, dt = \frac{1}{3}\ln(5 - 3\cos x) - \frac{1}{3}\ln 2$$

$$= \frac{2}{3} \sum_{n=1}^{\infty} \frac{1}{3^n} \int_0^x \sin nx\, dx = -\frac{2}{3} \sum_{n=1}^{\infty} \frac{1}{n3^n}(\cos nx - 1),$$

hence by a rearrangement

$$\sum_{n=1}^{\infty} \frac{1}{n3^n} \cos nx = \sum_{n=1}^{\infty} \frac{1}{n3^n} + \frac{1}{2} - \frac{1}{2}\ln 2 - \frac{1}{2}\ln(5 - 3\cos x)$$

$$= \ln\frac{3}{2} + \frac{1}{2} - \frac{1}{2}\ln 2\frac{1}{2}\ln(5 - 3\cos x)$$

$$= \ln 3 - \frac{1}{2}\ln 2 - \frac{1}{2}\ln(5 - 3\cos x).$$

In the above we used the following result:

$$\sum_{n=1}^{\infty} \frac{1}{n3^n} = \ln\frac{3}{2},$$

which easily follows if we take $x = \frac{1}{3}$ in the Maclaurin series

$$-\ln(1-x) = \sum_{n=1}^{\infty} \frac{x^n}{n}, \quad |x| < 1.$$

The Fourier convergence theorem gives conditions under which the Fourier series of a function f converges point-wise to f. Working with infinite series can be a delicate matter and we have to be careful. Since a uniform convergent series can be integrated term by term, it would be much better if we had absolute and uniform convergence. Let us recall these definitions and a few related results.

Definition 1.3.1. An infinite series of functions f_n

$$\sum_{n=1}^{\infty} f_n(x)$$

converges *absolutely* on a set $S \subseteq \mathbb{R}$ if the series

$$\sum_{n=1}^{\infty} |f_n(x)|$$

converges for every $x \in S$.

Definition 1.3.2. An infinite series of functions f_n

$$\sum_{n=1}^{\infty} f_n(x)$$

converges *uniformly* on a set $S \subseteq \mathbb{R}$ to a function f on S if the sequence $(S_N f)$ of the partial sums

$$S_N f(x) = \sum_{n=1}^{N} f_n(x)$$

converges uniformly to f on S, i.e., for every $\epsilon > 0$, there exists an integer $N = N(\epsilon)$ such that

$$|S_n f(x) - f(x)| < \epsilon$$

for all $x \in S$ and all $n \geq N$.

Remark. The reason for the term *"uniform convergence"* is that the integer N depends only on ϵ and not on the choice of the point $x \in S$.

Important consequences of uniform convergence are the following:

Theorem 1.3.5. *If an infinite series of continuous functions $f_n(x)$ on a set $S \subseteq \mathbb{R}$ converges uniformly on S to a function f, then f is also continuous on S.*

Theorem 1.3.6. *If an infinite series of functions $f_n(x)$ on a set $S \subseteq \mathbb{R}$ converges uniformly on S to a function f, then we can integrate the series term by term and the resulting integrated series*

$$\int_a^x \left(\sum_{n=1}^{\infty} f_n(y) \right) dy = \sum_{n=1}^{\infty} \int_a^x f_n(y)\, dy = \int_a^x f(y)\, dy$$

is also uniformly convergent on S.

Theorem 1.3.7. *If an infinite series of differentiable functions $f_n(x)$ on a set $S \subseteq \mathbb{R}$ is such that the series*

$$\sum_{n=1}^{\infty} f_n'(x)$$

converges uniformly on S, then the series

$$\sum_{n=1}^{\infty} f_n(x)$$

is also uniformly convergent on S and

$$\left(\sum_{n=1}^{\infty} f_n(x) \right)' = \sum_{n=1}^{\infty} f_n'(x), \quad \text{for every } x \in S.$$

A useful criterion for absolute and uniform convergence is the following test:

Theorem 1.3.8 (Weierstrass M-test). *If there are positive constants M_n such that*

$$|f_n(x)| \leq M_n \text{ for } n = 1, 2, \ldots \text{ and every } x \in S,$$

and the series

$$\sum_{n=1}^{\infty} M_n$$

converges, then the series

$$\sum_{n=1}^{\infty} f_n(x)$$

converges absolutely and uniformly on S.

The notion of uniform convergence will help us to investigate the already mentioned *"nice"* Fourier coefficients in the question what are "nice" Fourier coefficients? How quickly the Fourier coefficients decay to zero determines how many terms of the Fourier series we need take in order to get a better approximation. If the Fourier series decay more quickly, then the Fourier coefficients are "nicer." The next theorem shows precisely how the rate of convergence of Fourier coefficients (series) is related to the smoothness of the function.

Theorem 1.3.9. *Let f be a 2π-periodic function on $\mathbb{R}$ and let $k \in \mathbb{N}$. If f and its first $k - 1$ derivatives $f', \ldots, f^{(k-1)}$ are all continuous on $\mathbb{R}$ and the kth derivative $f^{(k)}$ is piecewise continuous on $\mathbb{R}$ ($f^{(k)}$ is piecewise continuous on each bounded interval), then the Fourier coefficients c_n, (a_n, b_n) of f satisfy the following:*

$$\sum_{n=1}^{\infty} |n^k c_n|^2 < \infty,$$

$$\sum_{n=1}^{\infty} |n^k a_n|^2 < \infty, \quad \sum_{n=1}^{\infty} |n^k b_n|^2 < \infty.$$

In particular,

$$\lim_{n \to \infty} n^k a_n = 0,$$

$$\lim_{n \to \infty} n^k b_n = 0, \quad \lim_{n \to \infty} n^k c_n = 0.$$

On the other hand, suppose that the Fourier coefficients c_n ($n \neq 0$), (a_n, b_n) satisfy

$$|c_n| \leq \frac{M}{n^{k+\alpha}}$$

or equivalently

$$|a_n| \leq \frac{C}{n^{k+\alpha}} \quad and \quad |a_n| \leq \frac{C}{n^{k+\alpha}}$$

for some constants $M > 0$ and $\alpha > 1$. Then f and its first k derivatives $f', \ldots, f^{(k)}$ are all continuous on $\mathbb{R}$.

Proof. We will consider only the complex coefficients c_n. The real Fourier coefficients a_n and b_n are treated similarly.

For the first part, applying Theorem 1.3.3 k times we have that the Fourier coefficients $c_n^{(k)}$ of the kth derivative $f^{(k)}$ are given by $c_n^{(k)} = (in)^k c_n$. From Bessel's inequality applied to the function $f^{(k)}$ it follows that

$$\sum_{n=1}^{\infty} |n^k c_n|^2 = \sum_{n=-\infty}^{\infty} |c_n^{(k)}|^2 \leq \frac{1}{2\pi} \int_{-\pi}^{\pi} |f^{(k)}(x)|^2 \, dx < \infty,$$

which establishes the first part.

For the second part, let $j \leq k$. Since $\alpha > 1$ we have

$$\sum_{\substack{n=-\infty \\ n\neq 0}}^{\infty} |n^j c_n| \leq M \sum_{\substack{n=-\infty \\ n\neq 0}}^{\infty} n^j \frac{1}{n^{k+\alpha}} \leq 2M \sum_{n=1}^{\infty} \frac{1}{n^\alpha} < \infty.$$

By the Weierstrass M-test it follows that the series

$$\sum_{n=-\infty}^{\infty} (in)^j c_n e^{inx}$$

are absolutely and uniformly convergent on $\mathbb{R}$ for every $j \leq k$. Therefore these series define continuous functions, which are the jth derivatives $f^{(j)}$ of the function

$$f(x) = \sum_{n=-\infty}^{\infty} c_n e^{inx}. \quad \blacksquare$$

The next example illustrates the above theorem.

Example 1.3.4. Discuss the rate of convergence of the Fourier series for the 2-periodic function f which on the interval $[-1,1]$ is defined by

$$f(x) = \begin{cases} (1+x)x, & -1 \leq x \leq 0 \\ (1-x)x, & 0 \leq x \leq 1. \end{cases}$$

Solution. Notice that this function has a first derivative everywhere, but it does not have a second derivative whenever x is an integer.

Since f is an odd function, $a_n = 0$ for every $n \geq 0$. By computation we have

$$b_n = \begin{cases} \frac{8}{\pi^3 n^3}, & \text{if } n \text{ is odd} \\ 0, & \text{if } n \text{ is even.} \end{cases}$$

Therefore the Fourier series of f is

$$f(x) = \frac{8}{\pi^3} \sum_{n=1}^{\infty} \frac{1}{(2n-1)^3} \sin(2n-1)\pi x.$$

This series converges very fast. If we plot the partial sum up to the third harmonic, that is, the function

$$S_2 f(x) = \frac{8}{\pi^3} \sin(\pi x) + \frac{8}{27\pi^3} \sin(3\pi x),$$

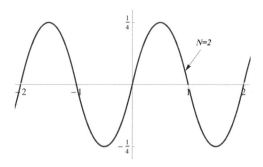

Figure 1.3.1

from Figure 1.3.1 we see that the graphs of f and $S_2\, f(x)$ are almost indistinguishable.

In fact, the coefficient $\frac{8}{27\pi^3}$ is already just 0.0096 (approximately). The reason for this fast convergence is the n^3 term in the denominator of the nth coefficient b_n, so the coefficients b_n tend to zero as fast as n^{-3} tends to zero.

Example 1.3.5. Discuss the convergence of the Fourier series of the 2π-periodic function f defined by the Fourier series

$$f(x) = \sum_{n=1}^{\infty} \frac{1}{n^3} \sin\, nx.$$

Solution. By the Weierstrass M-test it follows that the above series is a uniformly convergent series of continuous functions on $\mathbb{R}$. Therefore f is a continuous function on $\mathbb{R}$. The convergence rate of this series is like n^{-3}. Now, since the series obtained by term-by-term differentiation of the above series is uniformly convergent we have

$$f'(x) = \sum_{n=1}^{\infty} \frac{1}{n^2} \cos(nx),$$

and f' is a continuous function on $\mathbb{R}$. The convergence rate of this series is like n^{-2}. If we differentiate again (wherever we can) we obtain the following series:

$$\sum_{n=1}^{\infty} -\frac{1}{n} \sin(nx).$$

Even though at most points the derivative of f' is defined, at the points $x = (4n - 1)\frac{\pi}{2}$, $n \in \mathbb{N}$, the above series is divergent. At these points the function f' is not differentiable, so the function f'' fails to be continuous (has jumps). Finally if we try to differentiate term by term the last series we would obtain the series

$$\sum_{n=0}^{\infty} \left(-\cos nx \right),$$

which does not converge anywhere except at the points $x = (2n - 1)\frac{\pi}{2}$.

Now let us revisit the Bessel inequality and discuss the question of equality in it. We will need a few definitions first.

Definition 1.3.3. A function f is called *square integrable* on an interval $[a, b]$ if

$$\int_a^b |f(x)|^2 \, dx < \infty.$$

Continuous and piecewise continuous functions are examples of integrable and square integrable functions on any finite interval.

Remark. From the obvious inequality $(x - y)^2 \geq 0$ which is valid for all $x, y \in \mathbb{R}$ follows the inequality

$$2xy \leq x^2 + y^2, \quad x, y \in \mathbb{R}.$$

Therefore, every square integrable on any interval $[a, b]$ (finite or infinite) is also a Riemann integrable on that interval.

Example 1.3.6. The function $f(x) = x^{-\frac{1}{2}}$ is a Riemann integrable function on $[0, 1]$, but it is not a square integrable function on $[0, 1]$.

Solution. From

$$\int_0^1 f(x) \, dx = \int_0^1 x^{-\frac{1}{2}} \, dx = 2\sqrt{x} \Big|_0^1 = 2$$

it follows that $f(x)$ is Riemann integrable on $[0, 1]$.

On the other hand, from

$$\int_0^1 f^2(x) \, dx = \int_0^1 \frac{1}{x} \, dx = \infty$$

it follows that $f(x)$ is not square integrable on $[0, 1]$.

In engineering and some physical sciences one of the fundamental quantities is *power (energy)*.

Definition 1.3.4. The *energy* or *energy content* of a square integrable function f on $[a, b]$ is defined by

$$\int\limits_a^b |f(x)|^2 \, dx.$$

Before addressing the question of equality in the Bessel inequality, we will consider the following problem:

Problem. *Let f be a 2π-periodic function which is continuous and piecewise smooth on $[-\pi, \pi]$ and let N be a given natural number. Among all trigonometric polynomials t_N of the form*

$$t_N(x) = A_0 + \sum_{n=1}^N \left[A_n \cos nx + B_n \sin nx \right],$$

find that trigonometric polynomial t_N, i.e., find those coefficients A_n and B_n for which the energy content

$$E_N := \int\limits_{-\pi}^{\pi} \left(f(x) - t_N(x) \right)^2 dx$$

is minimal.

Solution. To compute E_N, we first expand the integral:

$$(1.3.1) \qquad E_N = \int\limits_{-\pi}^{\pi} f^2(x) \, dx - 2 \int\limits_{-\pi}^{\pi} f(x) t_N(x) \, dx + \int\limits_{-\pi}^{\pi} t_N^2(x) \, dx.$$

Since f is continuous on $[-\pi, \pi]$ we have that the Fourier series $S f(x)$ of f is equal to $f(x)$ at every point $x \in [-\pi, \pi]$:

$$f(x) = \frac{a_0}{2} + \sum_{n=1}^{\infty} \left[a_n \cos nx + b_n \sin nx \right],$$

where a_n and b_n are the Fourier coefficients of the function f. By the orthogonality of the sine and cosine functions and by direct integration we have

$$\int\limits_{-\pi}^{\pi} t_N^2(x) \, dx = \pi \left[2A_0^2 + \sum_{n=1}^N (A_n^2 + B_n^2) \right],$$

and

$$\int_{-\pi}^{\pi} f(x)t_N(x)\, dx = \pi\left[2A_0 a_0 + \sum_{n=1}^{N}(A_n a_n + B_n b_n)\right].$$

Therefore

$$E_N = \int_{-\pi}^{\pi} f^2(x)\, dx - 2\pi\left[2A_0 a_0 + \sum_{n=1}^{N}(A_n a_n + B_n b_n)\right] + \pi\left[2A_0^2 + \sum_{n=1}^{N}(A_n^2 + B_n^2)\right],$$

that is,

$$E_N = \int_{-\pi}^{\pi} f^2(x)\, dx - \pi\left[\frac{a_0^2}{2} + \sum_{n=1}^{N}(a_n^2 + b_n^2)\right]$$

$$+ \pi\left\{\frac{1}{2}(A_0 - a_0)^2 + \sum_{n=1}^{N}\left[(A_n - a_n)^2 + (B_n - b_n)^2\right]\right\}.$$

Since

$$\frac{1}{2}(A_0 - a_0)^2 + \sum_{n=1}^{N}\left[(A_n - a_n)^2 + (B_n - b_n)^2\right] \geq 0,$$

the energy E_N takes its minimum value if we choose $A_n = a_n$ for $n = 0, 1, \ldots, N$, and $B_n = b_n$ for $n = 1, \ldots, N$.

Therefore, the energy E_N becomes minimal if the coefficients A_n and B_n in t_N are chosen to be the corresponding Fourier coefficients a_n and b_n of the function f. Thus the trigonometric polynomial t_N should be chosen to be the Nth partial sums of the Fourier series of f:

$$t_N(x) = \frac{a_0}{2} + \sum_{n=1}^{N}\left[a_n \cos nx + b_n \sin nx\right]$$

in order to minimize E_N. Now when we know which choice of A_n and B_n minimizes E_N, we can find its minimum value. After a little algebra we find

$$(1.3.2) \qquad \min E_N = \int_{-\pi}^{\pi} f^2(x)\, dx - \pi\left[\frac{a_0^2}{2} + \sum_{n=1}^{N}(a_n^2 + b_n^2)\right]. \qquad \blacksquare$$

Even though the next theorem is valid for any square integrable function, we will prove it only for continuous functions which are piecewise smooth. The proof for the more general class of square integrable functions involves several important results about approximation of square integrable functions by trigonometric polynomials and for details the interested reader is referred to the book by T. M. Apostol, *Mathematical Analysis* [12].

Theorem 1.3.10 (Parseval's Identity). *Let f be a 2π-periodic, continuous and piecewise smooth function on $[-\pi, \pi]$. If c_n are the complex Fourier coefficients of f (a_n, b_n are the real Fourier coefficients), then*

$$\frac{1}{\pi} \int_{-\pi}^{\pi} |f(x)|^2 \, dx = \sum_{n=-\infty}^{\infty} |c_n|^2 = \frac{a_0^2}{2} + \sum_{n=1}^{\infty} (a_n^2 + b_n^2).$$

Proof. We prove the result only for the complex Fourier coefficients c_n. The case for the real Fourier coefficients a_n and b_n is very similar.

Since f is a continuous and piecewise smooth function on $[-\pi, \pi]$, by the Fourier convergence theorem we have

$$f(x) = \sum_{n=-\infty}^{\infty} c_n e^{inx}$$

for every $x \in [-\pi, \pi]$, and thus

$$|f(x)|^2 = \sum_{n=-\infty}^{\infty} \sum_{m=-\infty}^{\infty} c_n \overline{c_m} e^{i(n-m)x}.$$

By Theorem 1.3.6 the above series can be integrated term by term and using the orthogonality property of the system $\{e^{inx} : n \in \mathbb{Z}\}$ we obtain the required formula

$$\frac{1}{\pi} \int_{-\pi}^{\pi} |f(x)|^2 \, dx = \sum_{n=-\infty}^{\infty} |c_n|^2. \qquad \blacksquare$$

Using the above result now we have a complete answer to our original question about minimizing the mean error E_N. From (1.3.2) and Parseval's identity we have

$$\min E_N = \pi \sum_{n=N+1}^{\infty} (a_n^2 + b_n^2)$$

for all 2π-periodic, continuous and piecewise smooth functions f on $[-\pi, \pi]$. Now, by Bessel's inequality, both series $\sum_{n=1}^{\infty} a_n^2$ and $\sum_{n=1}^{\infty} b_n^2$ are convergent and therefore we have

$$\lim_{N \to \infty} \min E_N = 0.$$

The last equation can be restated as

$$\lim_{N \to \infty} \int_{-\pi}^{\pi} |f(x) - S_N f(x)|^2 \, dx = 0,$$

and usually we say that *the Fourier series of f converges to f in the mean* or *in L^2*.

Example 1.3.7. Using the Parseval identity for the function $f(x) = x^2$ on $[-\pi, \pi]$ find the sum of the series

$$\sum_{n=1}^{\infty} \frac{1}{n^4}.$$

Solution. The Fourier series of the function x^2 is given by

$$x^2 = \frac{\pi^2}{3} + \sum_{n=1}^{\infty} \frac{(-1)^n}{n^2} \cos nx.$$

By the Parseval identity we have

$$\frac{1}{\pi} \int_{-\pi}^{\pi} x^2 \, dx = \frac{4\pi^4}{18} + 16 \sum_{n=1}^{\infty} \frac{1}{n^4}.$$

Therefore

$$\frac{8}{45} \pi^4 = \frac{2\pi^4}{9} + 16 \sum_{n=1}^{\infty} \frac{1}{n^4},$$

and so

$$\frac{\pi^4}{90} = \sum_{n=1}^{\infty} \frac{1}{n^4}.$$

We close this section with an application of the Fourier series in determining a particular solution of a differential equation describing a physical system in which the input driving force is a periodic function.

Periodically Forced Oscillation. An undamped spring/mass system with mass m, a spring constant k and a damping constant c is driven by a $2L$-periodic external force $f(t)$ (think, for example, when we are pushing a child on a swing). The differential equation which models this oscillation is given by

$$(1.3.3) \qquad\qquad mx''(t) + cx'(t) + kx(t) = f(t).$$

We know that the general solution $x(t)$ of (1.3.3) is of the form

$$x(t) = x_h(t) + x_p(t),$$

where $x_h(t)$ is the general solution of the corresponding homogeneous equation

$$mx''(t) + cx'(t) + kx(t) = 0,$$

and $x_p(t)$ is a particular solution of (1.3.3) (see Appendix D). For $c > 0$, the solution $x_h(t)$ will decay as time goes on. Therefore, we are mostly interested in finding a particular solution $x_p(t)$ of which does not decay and is periodic with the same period as f.

For simplicity, let us suppose that $c = 0$. The problem with $c > 0$ is very similar. The general solution of the equation

$$mx''(t) + kx(t) = 0$$

is given by

$$x(t) = A\cos(\omega_0 t) + B\sin(\omega_0 t)$$

where $\omega_0 = \sqrt{k/m}$. Any solution of the non-homogeneous equation

$$mx''(t) + kx(t) = f(t)$$

will be of the form $A\cos(\omega_0 t) + B\sin(\omega_0 t) + x_p$. To find x_p we expand $f(t)$ in Fourier series

$$f(t) = \frac{a_0}{2} + \sum_{n=1}^{\infty}\left[a_n\cos\frac{n\pi}{L} + b_n\sin\frac{n\pi}{L}\right].$$

We look for a particular solution $x_p(t)$ of the form

$$x_p(t) = \frac{a_0}{2} + \sum_{n=1}^{\infty}\left[a_n\cos\frac{n\pi}{L}t + b_n\sin\frac{n\pi}{L}t\right],$$

where the coefficients a_n and b_n are unknown. We substitute $x_p(t)$ into the differential equation and solve for the coefficients c_n and d_n. This process is perhaps best understood by an example.

Example 1.3.8. Suppose that $k = 2$ and $m = 1$. There is a jetpack strapped to the mass which fires with a force of 1 Newton for the first time period of 1 second and is off for the next time period of 1 second, then it fires with a force of 1 Newton for 1 second and is off for 1 second, and so on. We need to find that particular solution which is periodic and which does not decay with time.

Solution. The differential equation describing this oscillation is given by

$$x''(t) + 2x = f(t),$$

where $f(t)$ is the step function

$$f(t) = \begin{cases} 1, & 0 < t < 1 \\ 0, & 1 < t < 2, \end{cases}$$

extended periodically on the whole real line $\mathbb{R}$. The Fourier series of $f(t)$ is given by

$$f(t) = \frac{1}{2} + \sum_{n=1}^{\infty} \frac{2}{(2n-1)\pi} \sin{(2n-1)\pi t}.$$

Now we look for a particular solution $x_p(t)$ of the given differential equation of the form

$$x_p(t) = \frac{a_0}{2} + \sum_{n=1}^{\infty} \left[a_n \cos{(n\pi t)} + b_n \sin{(n\pi t)} \right].$$

If we substitute $x_p(t)$ and the Fourier expansion of $f(t)$ into the differential equation

$$x''(t) + 2x = f(t)$$

by comparison, first we find $a_n = 0$ for $n \geq 1$ since there are no corresponding terms in the series for $f(t)$. Similarly we find $b_{2n} = 0$ for $n \geq 1$. Therefore

$$x_p(t) = \frac{a_0}{2} + \sum_{n=1}^{\infty} b_{2n-1} \sin{(2n-1)\pi t}.$$

Differentiating $x_p(t)$ twice, after rearrangement we find

$$x_p''(t) + 2x_p(t) = a_0 + \sum_{n=1}^{\infty} b_{2n-1}[2 - (2n-1)^2\pi^2] \sin{(2n-1)\pi t}.$$

If we compare the above series with the Fourier series obtained for $f(t)$ (the Uniqueness Theorem for Fourier series) we have that $a_0 = \frac{1}{2}$ and for $n \geq 1$

$$b_n = \frac{2}{(2n-1)\pi \left[2 - (2n-1)^2\pi^2 \right]}.$$

So the required particular solution $x_p(t)$ has the Fourier series

$$x_p(t) = \frac{1}{4} + \sum_{n=1}^{\infty} \frac{2}{(2n-1)\pi[2 - (2n-1)^2\pi^2]} \sin{(2n-1)\pi t}.$$

Resonance of Periodically Forced Oscillation. Again let us consider the equation

$$mx''(t) + kx(t) = f(t).$$

If it happens that the general solution of the corresponding homogeneous equation is of the form

$$x_h(t) = A \cos{\omega_N t} + B \sin{\omega_N t}$$

where $\omega_N = N\pi/L$ for some natural number N, then some of the terms in the Fourier expansion of $f(t)$ will coincide with the solution $x_h(t)$. In this case we have to modify the form of the particular solution $x_p(t)$ in the following way:

$$x_p(t) = \frac{a_0}{2} + t \cdot \left[a_N \cos \omega_0 t + b_N \sin \omega_0 t \right] + \sum_{\substack{n=1 \\ n \neq N}}^{\infty} \left[a_n \cos \frac{n\pi}{L} t + b_n \sin \frac{n\pi}{L} t \right].$$

In other words, we multiply the duplicating term by t. Notice that the expansion of $x_p(t)$ is no longer a Fourier series. After that we proceed as before.

Let us take an example.

Example 1.3.9. Find that particular solution which is periodic and does not decay in time of the equation

$$2x''(t) + 18\pi^2 x = f(t),$$

where $f(t)$ is the step function

$$f(t) = \begin{cases} 1, & 9 < t < 1 \\ 0, & 1 < t < 1, \end{cases}$$

extended periodically with period 2 on the whole real line $\mathbb{R}$.

Solution. The Fourier series of $f(t)$ is

$$f(t) = \sum_{n=1}^{\infty} \frac{4}{(2n-1)\pi} \sin(2n-1)\pi t.$$

The general solution of the given nonhomogeneous differential equation is of the form

$$x(t) = A \cos 3\pi t + B \sin 3\pi t + x_p(t),$$

where x_p is a particular solution of the corresponding homogeneous differential equation.

If we try as before that x_p has a Fourier series, then it will not work since we have duplication when $n = 3$. Therefore we look for a particular solution x_p of the form

$$x_p(t) = t \cdot \left(a_3 \cos 3\pi t + b_3 \sin 3\pi t \right) + \sum_{n=2}^{\infty} b_{2n+1} \sin(2n+1)\pi t.$$

Differentiating x_p twice and substituting in the nonhomogeneous equations along with the Fourier expansion of $f(t)$ we get

$$\sum_{n=2}^{\infty} \left[-2(2n+1)^2 \pi^2 + 18\pi^2 \right] b_{2n+1} \sin(2n+1)\pi t$$

$$- 12 a_3 \pi \sin 3\pi t + 12 b_3 \pi \cos 3\pi t = \sum_{n=1}^{\infty} \frac{4}{(2n-1)\pi} \sin(2n-1)\pi t.$$

Comparing the corresponding coefficients we find

$$a_3 = -\frac{1}{9\pi^2}, \quad b_3 = 0,$$

$$b_{2n+1} = \frac{2}{\pi^3 (2n+1)\left[9 - (2n+1)^2\right]}, \quad n = 2, 3, \ldots.$$

Therefore,

$$x_p(t) = -\frac{1}{9\pi^2} t \cos{(3\pi t)} + \sum_{n=2}^{\infty} \frac{2}{\pi^3 (2n+1)\left[9 - (2n+1)^2\right]} \sin{(2n+1)\pi t}.$$

Exercises for Section 1.3.

1. Expand the given function in Fourier series and, using the Parseval identity, find the sum of the given series

(a) $f(x) = \dfrac{x}{2}, \ |x| < \pi, \ \displaystyle\sum_{n=1}^{\infty} \frac{1}{n^2}.$

(b) $f(x) = x^3 - \pi^2 x, \ |x| < \pi, \ \displaystyle\sum_{n=1}^{\infty} \frac{1}{n^6}.$

2. Evaluate the following series by applying the Parseval equation to an appropriate Fourier expansion:

(a) $\displaystyle\sum_{n=1}^{\infty} \frac{1}{n^4}.$

(b) $\displaystyle\sum_{n=1}^{\infty} \frac{1}{(2n-1)^2}.$

(c) $\displaystyle\sum_{n=1}^{\infty} \frac{1}{(2n-1)^4}.$

(d) $\displaystyle\sum_{n=1}^{\infty} \frac{1}{(4n^2-1)^2}.$

3. Let f be the 4-periodic function defined on $(-2, 2)$ by

$$f(x) = \begin{cases} 2x + x^2, & -2 < x < 0 \\ 2x - x^2, & 0 < x < 2. \end{cases}$$

(a) Show that f is an odd function.
(b) Show that the Fourier series of f is

$$f(x) = \frac{32}{\pi^3} \sum_{n=1}^{\infty} \frac{\sin{\frac{(2n-1)\pi x}{2}}}{(2n-1)^3}, \quad x \in \mathbb{R} \backslash \{ \pm 2, \pm 4, \pm 6, \ldots \}.$$

(c) Use the Parseval identity to show that

$$\frac{\pi^6}{960} = \sum_{n=1}^{\infty} \frac{1}{(2n-1)^6}.$$

4. Evaluate the following series by applying the Parseval identity to certain Fourier expansions

(a) $\displaystyle\sum_{n=1}^{\infty} \frac{n^2}{(n^2+1)^2}.$

(b) $\displaystyle\sum_{n=1}^{\infty} \frac{\sin^2 na}{n^2},\quad 0 < |a| < \pi.$

5. Show that each of the following Fourier series expansions is valid in the range indicated and, for each expansion, apply the Parseval identity.

(a) $\displaystyle x\cos x = -\frac{1}{2}\sin x + 2\sum_{n=2}^{\infty}(-1)^n \frac{n}{n^2-1}\sin nx,\quad -\pi < x < \pi.$

(b) $\displaystyle x\sin x = 1 - \frac{1}{2}\cos x - 2\sum_{n=2}^{\infty}(-1)^n \frac{n}{n^2-1}\cos nx,\quad -\pi \le x \le \pi.$

6. Let

$$f(x) = \sum_{n=1}^{\infty} \frac{1}{n^3}\cos(nx).$$

(a) Is the function f continuous and differentiable everywhere?

(b) Find the derivative f' (wherever it exists) and justify your answer.
(c) Answer the same questions for the second derivative f''.

7. Let

$$f(x) = \sum \frac{1}{n}\cos(nx).$$

(a) Is the function f continuous and differentiable everywhere? (b) Find the derivative f' (wherever it exists) and justify your answer.

8. Suppose that (a_n) and (b_n) are sequences that tend to 0 as $n \to \infty$. If c is any positive number, show that the series

$$a_0 + \sum_{n=1}^{\infty} e^{-nc}\big[a_n\cos nx + b_n\sin nx\big]$$

converges uniformly and absolutely on the whole real line $\mathbb{R}$.

9. Expand the function

$$f(x) = \begin{cases} a^{-2}(a - |x|), & |x| < a \\ 0, & a < |x| < \pi \end{cases}$$

in Fourier series and, using the Parseval identity, find the sum of the series

$$\sum_{n=1}^{\infty} \frac{(1 - \cos na)^2}{n^4}.$$

10. For $x \in \mathbb{R}$, let $f(x) = |\sin x|$.

   (a) Plot the graph of f.

   (b) Find the Fourier series of f.

   (c) At each point x find the sum of the Fourier series. Where does the Fourier series converge uniformly?

   (d) Compute f' and find the Fourier series of f'.

   (e) Show that, for $0 \le x \le \pi$,

   $$\frac{\pi}{4}(\cos x - 1) + \frac{x}{2} = \sum_{n-1}^{\infty} \frac{\sin 2nx}{(2n - 1)2n(2n + 1)}.$$

   (f) Show that

   $$\sum_{n-1}^{\infty} \frac{(-1)^{n+1}}{(4n - 3)(4n - 2)n(4n - 1)} = \frac{\pi}{8}(\sqrt{2} - 1).$$

11. Define the function f by

   $$f(x) = \begin{cases} 0, & -\pi \le x \le 0 \\ \sin x, & 0 \le x \le \pi. \end{cases}$$

   (a) Plot the function f.

   (b) Find the Fourier series of f.

   (c) At what points does the Fourier series converge? Where is the convergence uniform?

(d) Integrate the Fourier series for f term by term and thus find the Fourier series of

$$F(x) = \int\limits_0^x f(t)\, dt.$$

12. Let α be any positive number which is not an integer. Define f by $f(x) = \sin(\alpha x)$ for $x \in (-\pi, \pi)$. Let $f(-\pi) = f(\pi) = 0$ and extend f periodically.

(a) Plot the function f when $0 < \alpha < 1$ and when $1 < \alpha < 2$.

(b) Show that the Fourier series of f is

$$f(x) = \frac{\sin(\alpha\pi)}{\pi} \sum_{n=1}^{\infty} (-1)^n \frac{n \sin(nx)}{\alpha^2 - n^2}.$$

(c) Prove that the Fourier series converges uniformly to f on any interval $[a, b]$ that does not contain an odd multiple of π.

(d) Let $x = \pi/2$, $t = \alpha/2$. Use the formula $\sin 2\pi t = 2 \sin \pi t \cos \pi t$ in order to show that

$$\pi \sec t\, \pi t = -4 \sum_{n=1}^{\infty} \frac{(-1)^{n+1}}{4t^2 - (2n-1)^2}.$$

13. For which real numbers α is the series

$$\sum \frac{1}{n^\alpha} \cos nx$$

uniformly convergent on each interval?

14. For which positive numbers a does the function $f(x) = |x|^{-a}$, $0 < |x| < \pi$, have a Fourier series?

15. Does there exist an integrable function on the interval $(-\pi, \pi)$ that has the series

$$\sum_{n=1}^{\infty} \sin nx$$

as its Fourier series?

16. Solve the following differential equations by Fourier series. The forcing function f is the periodic function given by

$$f(t) = \begin{cases} 1, & 0 < t < \pi \\ 0, & \pi < t < 2\pi \end{cases}$$

and $f(t) = f(t + 2\pi)$ for every other t.

(a) $y'' - y = f(t)$.

(b) $y'' - 3y' + 2y = f(t)$.

17. Solve the following differential equations by complex Fourier series. The forcing function f is the periodic function given by

$$f(t) = |t|, \quad -\pi < t < \pi$$

and $f(t) = f(t + 2\pi)$ for any other t.

(a) $y'' + 9y = f(t)$.

(b) $y'' + 2y = f(t)$.

18. An object radiating into its surroundings has a temperature $y(t)$ governed by the equation

$$y'(t) + ky(t) = a_0 + \sum_{n=1}^{\infty} \left[a_n \cos(n\omega t) + b_n \sin)n\omega t) \right],$$

where k is the heat loss coefficient and the Fourier series describes the temporal variation of the atmospheric air temperature and the effective sky temperature. Find $y(t)$ if $y(0) = T_0$.

1.4 Fourier Sine and Cosine Series.

Very often, a function f is given only on an interval $[0, L]$ but still we want the function f to be represented in the form of a Fourier series. There are many ways of doing this. One way, for example, is to define a new function which is equal to 0 on $[-L, 0]$ and coincides with the function f on $[0, L]$. But two other ways are especially simple and useful: we extend the given function to a function which is defined on the interval $[-L, L]$ by making the extended function either odd or even.

Definition 1.4.1. Let f be a given function on $(0, L)$. The *odd extension* f_o of f is defined by

$$f_o(x) = \begin{cases} -f(-x), & -L \le x < 0 \\ f(x), & 0 \le x \le L. \end{cases}$$

The *even extension* f_e of f is defined by

$$f_e(x) = \begin{cases} f(-x), & -L \le x < 0 \\ f(x), & 0 \le x \le L. \end{cases}$$

Graphically, the even extension is made by reflecting the graph of the original function around the vertical axis, and the odd extension is made by reflecting the graph of the original function around the coordinate origin. See Figure 1.4.1.

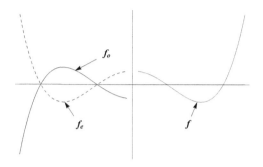

Figure 1.4.1

After we extend the original function f to an even or odd function, we extend these new functions f_o and f_e on the whole real line to $2L$-periodic functions by

$$f_e(x + 2nL) = f_e(x), \quad x \in [-L, L], \ n = 0, \pm 1, \pm 2, \ldots,$$
$$f_o(x + 2nL) = f_o(x), \quad x \in [-L, L], \ n = 0, \pm 1, \pm 2, \ldots.$$

Notice that if the original function f is piecewise continuous or piecewise smooth on the interval $[0, L]$, then the $2L$-periodic extensions f_e and f_o are piecewise continuous or piecewise smooth (possibly with extra points of discontinuity at nL) on the whole real line $\mathbb{R}$.

Now we expand the functions f_o and f_e in Fourier series. The advantage of using these functions is that the Fourier coefficients are simple. Indeed, it follows from Lemma 1.1.3 that

$$\int_{-L}^{L} f_e(x) \cos \frac{n\pi}{L} x \, dx = 2 \int_{0}^{L} f(x) \cos \frac{n\pi}{L} x \, dx,$$

$$\int_{-L}^{L} f_e(x) \sin \frac{n\pi}{L} x \, dx = 0,$$

$$\int_{-L}^{L} f_o(x) \sin \frac{n\pi}{L} x \, dx = 2 \int_{0}^{L} f(x) \sin \frac{n\pi}{L} x \, dx,$$

$$\int_{-L}^{L} f_o(x) \cos \frac{n\pi}{L} x \, dx = 0.$$

Therefore the Fourier series of f_e involves only cosines and the Fourier series of f_o involves only sines; moreover, the Fourier coefficients of these functions can be computed in terms of the value of the original function f over the interval $[0, L]$.

The above discussion naturally leads to the following definition.

Definition 1.4.2. Suppose that f is an integrable function on $[0, L]$. The series

$$\frac{1}{2}a_0 + \sum_{n=1}^{\infty} a_n \cos \frac{n\pi}{L} x, \quad a_n = \frac{2}{L} \int_0^L f(x) \cos \frac{n\pi}{L} x \, dx$$

is called the *half-range Fourier cosine series (Fourier cosine series)* of f on $[0, L]$.

The series

$$\sum_{n=1}^{\infty} b_n \sin \frac{n\pi x}{L}, \quad b_n = \frac{2}{L} \int_0^L f(x) \sin \frac{n\pi}{L} x \, dx$$

is called the *half-range Fourier sine (Fourier sine series)* of f on $[0, L]$.

Example 1.4.1. Find the Fourier sine and cosine series of the function

$$f(x) = \begin{cases} 1, & 0 < x < 1 \\ 2, & 1 < x < 2 \end{cases}$$

on the interval $[0, 2]$.

Solution. In this example $L = 2$, and using very simple integration in evaluating a_n and b_n we obtain the Fourier sine series

$$\frac{2}{\pi} \sum_{n=1}^{\infty} \frac{1 + \cos \frac{n\pi}{2} - 2(-1)^n}{n} \sin \frac{n\pi x}{2},$$

and the Fourier cosine series

$$\frac{3}{2} + \frac{2}{\pi} \sum_{n-1}^{\infty} \frac{(-1)^n}{2n-1} \cos \frac{(2n-1)\pi x}{2}.$$

Example 1.4.2. Find the Fourier sine and Fourier cosine series of the function $f(x) = \sin x$ on the interval $[0, \pi]$.

Solution. In this example $L = \pi$. It is obvious that the Fourier sine series of this function is simply $\sin x$. After simple calculations we obtain that the Fourier cosine series is given by

$$\frac{2}{\pi} - \frac{4}{\pi} \sum_{n=1}^{\infty} \frac{1}{4n^2 - 1} \cos 2nx.$$

Since the Fourier sine and Fourier cosine series are only particular Fourier series, all the theorems for convergence of Fourier series are also true for Fourier sine and Fourier cosine series. Therefore we have the following results.

Theorem 1.4.1. *If f is a piecewise smooth function on the interval $[0, L]$, then the Fourier sine and Fourier cosine series of f converge to*

$$\frac{f(x^-) + f(x^+)}{2}$$

at every $x \in (0, L)$. In particular, both series converge to $f(x)$ at every point $x \in (0, L)$ where f is continuous.

The Fourier cosine series of f converges to $f(0^+)$ at $x = 0$ and to $f(L^-)$ at $x = L$.

The Fourier sine series of f converges to 0 at both of these points.

Example 1.4.3. Apply Theorem 1.4.1 to the function in Example 1.4.1.

Solution. Since the function $f(x)$ in this example is continuous on the set $(0, 1) \cup (1, 2)$, from Theorem 1.4.1, applied to the Fourier Cosine series, we obtain the following:

$$f(x) = 1 = \frac{3}{2} + \frac{2}{\pi} \sum_{n=1}^{\infty} \frac{(-1)^n}{2n-1} \cos \frac{(2n-1)\pi x}{2}, \text{ for every } 0 < x < 1,$$

$$f(x) = 1 = \frac{3}{2} + \frac{2}{\pi} \sum_{n=1}^{\infty} \frac{(-1)^n}{2n-1} \cos \frac{(2n-1)\pi x}{2}, \text{ for every } 1 < x < 2.$$

Theorem 1.4.2 (Parseval's Formula). *If f is a square integrable function on $[0, \pi]$, and a_n and b_n are the Fourier coefficients of the Fourier cosine and Fourier sine expansion of f, respectively, then*

$$\int_0^{\pi} f^2(x)\, dx = \frac{\pi}{4} a_0^2 + \frac{\pi}{2} \sum_{n=1}^{\infty} a_n^2 = \frac{\pi}{2} \sum_{n=1}^{\infty} b_n^2.$$

Example 1.4.4. Find the Fourier sine and Fourier cosine series of the function f defined on $[0, \pi]$ by

$$f(x) = \begin{cases} x, & 0 \le x \le \frac{\pi}{2}, \\ \pi - x, & \frac{\pi}{2} \le x \le \pi. \end{cases}$$

Solution. For a_0 we have

$$a_0 = \frac{2}{\pi} \int_0^{\pi} f(x)\, dx = \frac{2}{\pi} \int_0^{\frac{\pi}{2}} x\, dx + \int_{\frac{\pi}{2}}^{\pi} (\pi - x)\, dx = \frac{\pi}{2}.$$

For $n = 1, 2, \ldots$ by the integration by parts formula we have

$$a_n = \frac{2}{\pi} \int_0^{\pi} f(x) \cos nx \, dx = \frac{2}{\pi} \int_0^{\frac{\pi}{2}} x \cos nx \, dx + \frac{2}{\pi} \int_{\frac{\pi}{2}}^{\pi} (\pi - x) \cos nx \, dx$$

$$= \frac{4}{n^2 \pi} \cos \frac{n\pi}{2} - \frac{2}{n^2 \pi} (1 + \cos n\pi).$$

Therefore the Fourier cosine series of f on $[0, \pi]$ is given by

$$\frac{\pi}{4} + \sum_{n=1}^{\infty} \left[\frac{4}{n^2 \pi} \cos \left(\frac{n\pi}{2} \right) - \frac{2}{n^2 \pi} (1 + \cos n\pi) \cos nx \right].$$

The plot of the first two partial sums of the Fourier cosine series of f, along with the plot of the function f, is given in Figure 1.4.2.

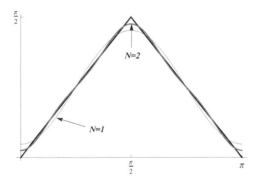

Figure 1.4.2

The plot of the 10th partial sum of the Fourier cosine series of f, along with the plot of the function f, is given in Figure 1.4.3.

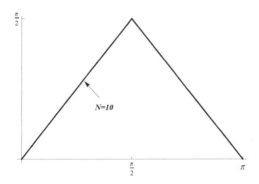

Figure 1.4.3

The computation of the coefficients b_n in the Fourier sine series is like that of the coefficients a_n, and we find that the Fourier sine series of $f(x)$ is

$$\sum_{n=1}^{\infty} \frac{2}{n^2 \pi} \sin \frac{n\pi}{2} \sin nx.$$

The plot of the first two partial sums of the Fourier sine series of f, along with the plot of the function f, is given in Figure 1.4.4

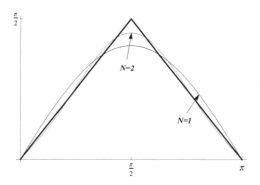

Figure 1.4.4

The plot of the 10th partial sum of the Fourier sine series of f, along with the plot of the function f , is given in Figure 1.4.5.

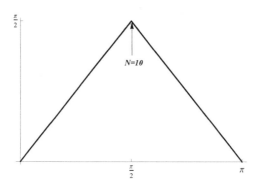

Figure 1.4.5

The following question is quite natural: How do we know whether and when to use the Fourier cosine or the Fourier sine series?

The answer to this question is related to Theorem 1.3.8. In order to have better (faster) approximation of a given function, the periodic extension of the function needs to be as smooth as possible. The smoother the extension, the faster the Fourier series will converge.

In Example 1.4.4, the odd periodic extension did not produce a sharp turn at $x = 0$ (the extension is smooth at the origin), while the even periodic extension is not smooth at the origin.

Example 1.4.5. Find the half-range Fourier cosine and sine expansions of the function $f(x) = 1$ on the interval $(0, \pi)$.

Solution. It is obvious the Fourier cosine of the given function is 1.
For any $n \in \mathbb{N}$ we find

$$b_n = \frac{2}{\pi} \int_0^\pi 1 \sin nx\, dx = -\frac{2}{n\pi} \left[(-1)^n - 1 \right],$$

and so the Fourier sine expansion of the function is

$$\frac{4}{\pi} \sum_{n=1}^\infty \frac{1}{2n-1} \sin(2n-1)\pi x.$$

Example 1.4.6. Find the half-range Fourier cosine and sine expansions of the function $f(x) = x$ on the interval $(0, 2)$.

Solution. For the Fourier cosine expansion of f we use the even extension $f_e(x) = |x|$ of the function f. By computation we find the Fourier cosine coefficients.

$$a_0 = \frac{2}{2} \int_0^2 x\, dx = 2.$$

For $n \in \mathbb{N}$, by the integration by parts formula, we have

$$a_n = \frac{2}{2} \int_0^2 x \cos \frac{n\pi x}{2}\, dx = \frac{4}{n^2\pi^2} \left[(-1)^n - 1 \right],$$

and so the Fourier cosine expansion of the function is

$$1 - \frac{8}{\pi^2} \sum_{n=1}^\infty \frac{1}{(2n-1)^2} \cos \frac{2n-1}{2}\pi x.$$

The plot of the first two partial sums of the Fourier cosine series of f, along with the plot of the the function f, is given in Figure 1.4.6.

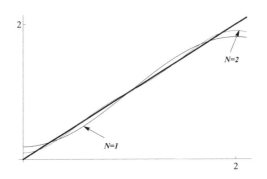

Figure 1.4.6

The plot of the 10th and the 20th partial sums of the Fourier cosine series of f, along with the plot of the function f, is given in Figure 1.4.7.

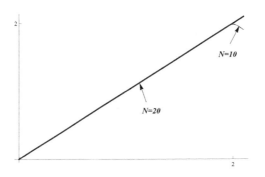

Figure 1.4.7

For the Fourier sine expansion of f we use the odd extension

$$f_o(x) = x$$

of the function f.

Using the integration by parts formula we find

$$b_n = \frac{2}{2} \int_0^2 x \sin \frac{n\pi x}{2} \, dx = \frac{4(-1)^{n+1}}{n\pi}.$$

The Fourier sine series of f is then

$$\sum_{n=1}^{\infty} \frac{4(-1)^{n+1}}{n\pi} \sin \frac{n\pi x}{2}.$$

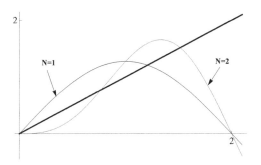

Figure 1.4.8

The plot of the first two partial sums of the Fourier cosine series of f, along with the plot of the function f, is given in Figure 1.4.8

The plot of the 10th and the 20th partial sums of the Fourier cosine series of f, along with the plot of the function f, is given in Figure 1.4.9.

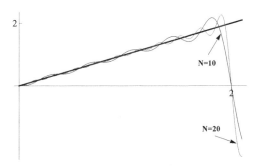

Figure 1.4.9

Example 1.4.7. Suppose f is a piecewise continuous function on $[0, \pi]$, such that $f(x) = f(\pi - x)$ for every $x \in [0, \pi]$. In other words, the graph of f is symmetric with respect to the vertical line $x = \frac{\pi}{2}$. Let a_n and b_n be the Fourier cosine and Fourier sine coefficients of f, respectively. Show that $a_n = 0$ if n is odd, and $b_n = 0$ if n is even.

Solution. We show only that the coefficients a_n have the required property, and leave the other part as an exercise. Let n be a natural number. Then

$$a_{2n-1} = \frac{2}{\pi} \int_0^\pi f(x) \cos{(2n-1)}x \, dx$$

$$= \frac{2}{\pi} \int_0^{\frac{\pi}{2}} f(x) \cos{(2n-1)}x \, dx + \frac{2}{\pi} \int_{\frac{\pi}{2}}^\pi f(x) \cos{(2n-1)}x \, dx$$

$$= \frac{2}{\pi} \int_0^{\frac{\pi}{2}} f(x) \cos{(2n-1)}x \, dx + \frac{2}{\pi} \int_{\frac{\pi}{2}}^\pi f(\pi-x) \cos{(2n-1)}x \, dx.$$

If in the last integral we introduce a new variable t by $t = \pi - x$, then we have

$$a_{2n-1} = \frac{2}{\pi} \int_0^{\frac{\pi}{2}} f(x) \cos{(2n-1)}x \, dx + \frac{2}{\pi} \int_{\frac{\pi}{2}}^0 f(t) \cos{\big((2n-1)(\pi-t)\big)} \, dt$$

$$= \frac{2}{\pi} \int_0^{\frac{\pi}{2}} f(x) \cos{(2n-1x}\, dx - \frac{2}{\pi} \int_0^{\frac{\pi}{2}} f(t) \cos{\big(2n-1)t\big)} \, dt = 0.$$

Example 1.4.8. Let f be a function which has a continuous first derivative on $[0, \pi]$. Further, let $f(0) = f(\pi) = 0$. Show that

$$\int_0^\pi f^2(x) \, dx \le \int_0^\pi (f'(x))^2 \, dx.$$

For what functions f does equality hold?

Solution. Since f is continuous on $[0, \pi]$ and $f(0) = 0$, we have that the Fourier sine expansion of f on $[0, \pi]$ is

$$f(x) = \sum_{n=1}^\infty a_n \sin{nx}.$$

By the Parseval identity we have

$$\int_0^\pi f^2(x) \, dx = \frac{\pi}{2} \sum_{n=1}^\infty a_n^2.$$

Since the Fourier series of the first derivative f' is

$$\sum_{n=1}^{\infty} nb_n \cos nx,$$

again from the Parseval identity we have

$$\int_0^{\pi} (f'(x))^2 \, dx = \frac{\pi}{2} \sum_{n=1}^{\infty} n^2 a_n^2.$$

Now from the obvious inequality

$$\sum_{n=1}^{\infty} a_n^2 \leq \sum_{n=1}^{\infty} n^2 a_n^2$$

it follows that

$$\int_0^{\pi} f^2(x) \, dx \leq \int_0^{\pi} (f'(x))^2 \, dx.$$

Equality in the above inequality holds if and only if

$$\sum_{n=1}^{\infty} a_n^2 = \sum_{n=1}^{\infty} n^2 a_n^2.$$

From the last equation it follows $a_n = 0$ for $n \geq 2$. Therefore

$$f(x) = a_1 \sin x.$$

Example 1.4.9. Let $f : [0, \pi] \to \mathbb{R}$ be the function given by

$$f(x) = x^2 - 2x.$$

Find the Fourier cosine and Fourier sine series of the function f.

Solution. For the Fourier cosine series we have

$$a_0 = \frac{2}{\pi} \int_0^{\pi} f(x) \, dx = \frac{1}{2\pi} \int_0^{\pi} (x^2 - 2x) \, dx = \frac{2}{3}\pi^2 - 2\pi,$$

and for $n \geq 1$, by the integration by parts formula

$$a_n = \frac{2}{\pi} \int_0^{\pi} (x^2 - 2x) \cos nx \, dx$$

$$= \frac{2}{n\pi} \left[(x^2 - 2x) \sin nx \Big|_{x=0}^{x=\pi} - 2 \int_0^{\pi} (x - 1) \sin nx \, dx \right]$$

$$= \frac{4}{n\pi^2} \left[(x - 1) \cos nx \right]_{x=0}^{x=\pi} - \frac{4}{n\pi^2} \int_0^{\pi} \cos nx \, dx$$

$$= \frac{4}{n\pi^2} \left[(-1)^n (\pi - 1) + 1 \right].$$

Therefore the Fourier cosine series of f on $[0, \pi]$ is given by

$$\frac{\pi^2}{3} - \pi + \frac{4}{\pi} \sum_{n=1}^{\infty} \frac{(-1)^n (\pi - 1) + 1}{n^2} \cos nx.$$

Since f is continuous on $[0, \pi]$, by the Fourier cosine convergence theorem, it follows that

$$x^2 - 2x = \frac{\pi^2}{3} - \pi + \frac{4}{\pi} \sum_{n=1}^{\infty} \frac{(-1)^n (\pi - 1) + 1}{n^2} \cos nx$$

for every $x \in [0, \pi]$.

The plot of the first two partial sums of the Fourier cosine series of f, along with the plot of the function f, is given in Figure 1.4.10.

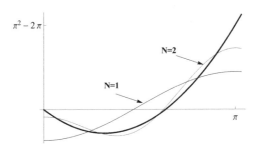

Figure 1.4.10

The plot of the 10th and the 20th partial sums of the Fourier cosine series of f, along with the plot of the function f, is given in Figure 1.4.11.

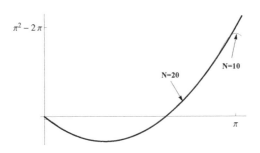

Figure 1.4.11

For the Fourier sine series of f we use the odd extension f_o of f. Working similarly as for the Fourier cosine series we obtain

$$x^2 - 2x = \sum_{n=1}^{\infty} \left\{ 2(-1)^{n-1} \frac{\pi - 2}{n} - \frac{4}{\pi n^3} \left[1 - (-1)^n \right] \right\} \sin nx$$

for every $x \in [0, \pi]$.

The plot of the first two partial sums of the Fourier sine series of f, along with the plot of the function f, is given in Figure 1.4.12.

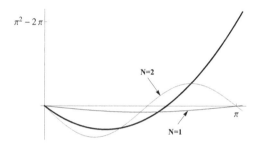

Figure 1.4.12

The plot of the 10th and the 50th partial sums of the Fourier sine series of f, along with the plot of the function f, is given in Figure 1.4.13.

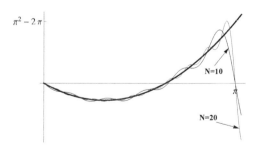

Figure 1.4.13

Exercises for Section 1.4.

In Exercises 1–5, find both the Fourier cosine series and the Fourier sine series of the given function on the interval $[0, \pi]$. Find the values of these series when $x = 0$ and $x = \pi$.

1. $f(x) = \pi - x$.

2. $f(x) = \sin x$.

3. $f(x) = \cos x$.

4. $f(x) = x^2$.

5. $f(x) = \begin{cases} x, & 0 \le x \le \frac{\pi}{2} \\ \pi - x, & \frac{\pi}{2} \le x \le \pi. \end{cases}$

In Exercise 6–11, find both the Fourier cosine series and the Fourier sine series of the given function on the indicated interval.

6. $f(x) = x$; $[0, 1]$.

7. $f(x) = 3$; $[0, \pi]$.

8. $f(x) = \begin{cases} 1, & 0 < x < 2 \\ 0, & 2 < x < 4 \end{cases}$; $[0, 4]$.

9. $f(x) = \begin{cases} x, & 0 \le x \le 1 \\ 1, & 1 \le x \le 2 \end{cases}$; $[0, 2]$.

10. $f(x) = \begin{cases} 0, & 0 < x < 1 \\ 1, & 1 < x < 2 \end{cases}$; $[0, 2]$.

11. $f(x) = e^x$; $[0, \pi]$.

1.5 Projects Using Mathematica.

In this section we will see how Mathematica can be used to solve many problems involving Fourier series. In particular, we will develop several *Mathematica notebooks* which automate the computation of partial sums of Fourier series. For a brief overview of the computer software Mathematica consult Appendix H.

Project 1. Let f be the 2π-periodic function which on the interval $[-\pi, \pi]$ is given by $f(x) = |x|$. Using Mathematica solve the following problems.

(a) Plot the function f on the interval $[-3\pi, 3\pi]$.

(b) Find the Fourier coefficients of f.

(c) Plot several Fourier partial sums $S_N f$.

Solution. First define an expression for the function we want to expand in a Fourier series.

$In[1] := ab[x\_] := Piecewise[\{\{Abs[x], Abs[x] \le Pi\}\}]$

We also want to plot the periodic extension. The following function will replicate a function over several periods.

$In[2] := periodicExtension[func\_, nPeriods\_] = Sum[func[x + 2\, kPi], \{k, -n\, Periods, n\, Periods\}]$

$In[3] := Plot[periodicExtension[ab, 4], \{x, -4\, Pi, 4\, Pi\}, PlotRange \to All]$

In Figure 1.5.1 we plot the function on the interval $[-4\pi, 4\pi]$.

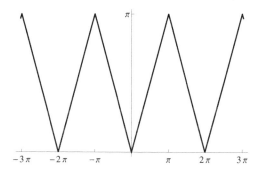

Figure 1.5.1

Further, we define the Fourier basis consisting of the cosine and sine functions:

$In[4] := s[n\_, x\_] = Sin[n\,x]$

$In[5] := c[n\_, x\_] = Cos[n\,x]$

Next we define the inner product:

$In[6] := IP[f\_, g\_] := 1/2 Integrate[f\, g, \{x, -2, 2\}]$

Next we define the Fourier coefficients a_n and b_n:

$In[7] := a0[func\_] := IP[func[x], c[0, x]]$

$In[8] := aFC[func\_, n\_] := IP[func[x], c[n, x]]$

$In[9] : bFC[func\_, n\_] := IP[func[x], s[n, x]]$

We define the Nth Fourier partial sum.

$In[10] := fourierSeries[func\_, N\_, x\_] := a0[func]/2$
$+Sum[aFC[func, n]c[n, x] + bFC[func, n]s[n, x], \{n, 1, N\}]$

If in an expression appears $\sin(n\pi)$, $\cos n\pi$, $\sin(2n-1)\pi/2$, and n is an integer, then we must tell Mathematica that n is an integer. This is done as follows:

$In[11] := Simplify[Sin[n\,Pi], Element[n, Integers]]$

$In[12] := Simplify[Cos[n\,Pi], Element[n, Integers]]$

$In[13] := Simplify[Sin[(2n-1)\,Pi/2], Element[n, Integers]]$

Next, compute the 1st, 5th, and the 50th partial sums:

$In[14] := fs1[x\_] = fourierSeries[1, ab, x]$

$In[15] := fs5[x\_] = fourierSeries[5, ab, x]$

$In[16] := fs50[x\_] = fourierSeries[50, ab, x]$

The plot of the first two partial sums of the Fourier series of f, along with the plot of the function f, is given in Figure 1.5.2.

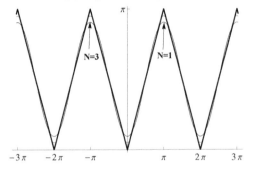

Figure 1.5.2

The plot of the partial sum of the Fourier series of f for $N = 4$, along with the plot of the function f, is given in Figure 1.5.3.

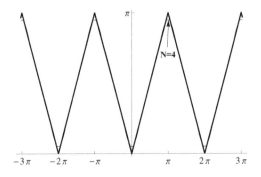

Figure 1.5.3

Project 2. Let f be the 4-periodic function which on the interval $[-2, 2]$ is given by

$$f(x) = \begin{cases} 2, & \text{for } x = -2, -1, 0, 1, 2 \\ 3, & \text{for } x \in (-2, -1) \cup (0, 1) \\ 1, & \text{for } x \in (-1, 0) \cup (1, 2). \end{cases}$$

Using Mathematica solve the following problems.

(a) Plot the function f on the interval $[-6, 6]$.

(b) Find the Fourier coefficients of f.

(c) Plot the Fourier partial sums $S_N f$ for several values of N.

(d) Investigate the Gibbs phenomenon at the point $x = 1$.

Solution. First we define our function f:

$In[1] := f[x\_] := Piecewise[\{\{3, -2 < x < 0\}, \{1, 0 < x < 2\}, \{2, x == -2\}, \{2, x == 0\}, \{2, x == 2\}\}]$

We also want to plot the periodic extension (displayed in Figure 1.5.4).

$In[2] := e[x\_] = periodicExtension[func\_, nPeriods\_] = Sum[func[x + 4k], \{k, -nPeriods, nPeriods\}]$

$In[3] := Plot[periodicExtension[f, 4], \{x, -6, 6\}, PlotRange- > \{\{-6, 6\}, \{0, 3.5\}\}]$

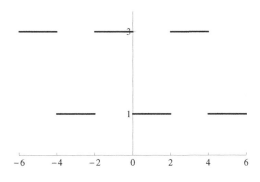

Figure 1.5.4

We define the Fourier basis consisting of the cosine and sine functions:

$In[4] := s[n\_, x\_] = Sin[n\, Pi\, x/2]$

$In[5] := c[n\_, x\_] = Cos[n\, Pi\, x/2]$

Next we define the inner product:

$In[6] := IP[f\_, g\_] := Integrate[f\,g, \{x, -2, 2\}]$

Next we define the Fourier coefficients a_n and b_n:

$In[7] := a0[func\_] := IP[func[x], c[0, x]]$

$In[8] := aFC[func\_, n\_] := IP[func[x], c[n, x]]$

$In[9] := bFC[func\_, n\_] := IP[func[x], s[n, x]]$

We define the Nth Fourier partial sum:

$In[10] := FSeries[func\_, N\_, x\_] := a0[func]/2$
$+Sum[aFC[func, n]c[n, x] + bFC[func, n]s[n, x], \{n, 1, N\}]$

$In[11] := Simplify[Sin[nPi], Element[n, Integers]]$

$In[12] := Simplify[Cos[nPi], Element[n, Integers]]$

$In[13] := Simplify[Sin[(2n - 1)Pi/2], Element[n, Integers]]$

$In[14] := Simplify[Sin[(2n - 1)Pi/2], Element[n, Integers]]$

Next, compute the 1st, 3rd, 5th, 10th and the 50th partial sums:

$In[15] := fs1[x\_] = fourierSeries[1, f, x]$

$In[16] := fs3[x\_] = fourierSeries[3, f, x]$

$In[17] := fs5[x\_] = fourierSeries[5, f, x]$

$In[18] := fs10[x\_] = fourierSeries[10, f, x]$

$In[19] := fs50[x\_] = fourierSeries[50, f, x]$

The plot of the partial sums of the Fourier series of f for $N = 1$ and $N = 3$, along with the plot of the function f, is given in Figure 1.5.5.

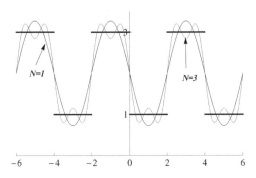

Figure 1.5.5

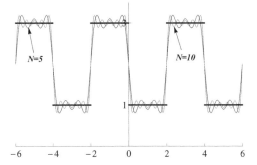

Figure 1.5.6

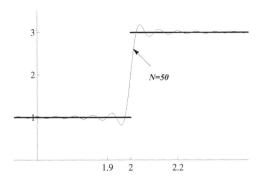

Figure 1.5.7

The plot of the 5th and the 10th partial sums of the Fourier series of f, along with the plot of the function f, is given in Figure 1.5.6.

The plot of the 50th partial sum of the Fourier series of f, along with the plot of the function f, is given in Figure 1.5.7.

Project 3. Let f be the 2π-periodic function which on the interval $[-\pi, \pi]$ is defined by

$$f(x) = \begin{cases} 0 & \text{for } x = 0, \frac{\pi}{2}, \pi \\ 0 & \text{for } x \in (-\pi, 0) \cup (\frac{\pi}{2}, \pi) \\ 1 & \text{for } x \in (0, \frac{\pi}{2}). \end{cases}$$

Using Mathematica solve the following problems:

(a) Plot the function f on the interval $[-3\pi, 3\pi]$.

(b) Find the Fourier coefficients of f.

(c) Plot the Fourier partial sums $S_N f$ for several values of N.

(d) Investigate the Gibbs phenomenon at the point $x = 0$.

Solution. First we define the function by

$In[1] := f[x\_] := Piecewise[\{\{0, -Pi < x < 0\}, \{1, 0 < x < Pi/2\},$
$\{0, x == 0\}, \{0, x == Pi/2\}, \{0, x == Pi\}\}]$

Next we define the periodic extension of the function f.

$In[2] := e[x\_] = periodicExtension[func\_, nPeriods\_]$
$= Sum[func[x + 4k], \{k, -nPeriods, nPeriods\}]$

We plot the function f on the segment $[-3\pi, 3\pi]$.
See Figure 1.5.8.
$In[3] := Plot[periodicExtension[f, 4], \{x, -3\pi, 3\pi\},$
$PlotRange \rightarrow \{\{-3\pi, 3\pi\}, \{0, 1\}\}]$

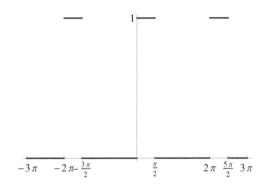

Figure 1.5.8

We compute the Fourier coefficients of the function f:

$In[4] := IP[f\_, g\_] = 1/Pi\, Integrate[f\, g, \{x, -Pi, Pi\}];$

$In[5] := a0 = IP[f[x], 1]$

$Out[6] := 1/2$

$In[7] := a[n\_] := IP[f[x], Cos[n\, x]]$

$Out[8] := \dfrac{Sin[\frac{n\pi}{2}]}{n\pi}$

$In[9] := b[n\_] := IP[f[x], Sin[n\, x]]$

$Out[10] = 2\dfrac{Sin^2[\frac{n\pi}{4}]}{n\pi}$

We form the Nth partial sum:

$In[11] := SN = Table[a0/2 + Sum[a[n]\, Cos[n\, x] + b[n]\, Sin[n\, x],$
$\{n, 1, N\}], \{N, 1, 200\}]$

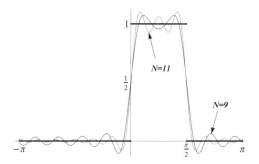

Figure 1.5.9

The plot of the partial sums S_N of the Fourier series of f for $N = 9$ and $N = 11$, along with the plot of the function f, is given in Figure 1.5.9.

The plot of the partial sums of the Fourier series of f for $N = 20$ and $N = 30$, along with the plot of the function f, is given in Figure 1.5.10.

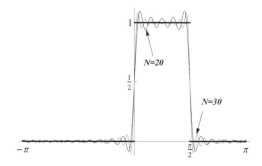

Figure 1.5.10

The plot of the partial sum of the Fourier series of f for $N = 150$, along with the plot of the function f, is given in Figure 1.5.11.

The plot of the partial sum of the Fourier series of f for $N = 500$ and a the function f on the interval $\left(-\frac{\pi}{100}, \frac{\pi}{100}\right)$ is given in Figure 1.5.12.

From these figures, in particular from Figure 1.5.12, we can clearly see the Gibbs phenomenon in the neighborhood of the point $x = 0$.

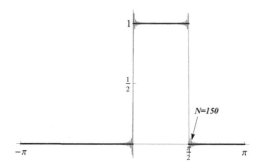

Figure 1.5.11

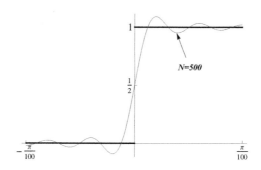

Figure 1.5.12

CHAPTER 2

INTEGRAL TRANSFORMS

Many functions in analysis are defined and expressed as improper integrals of the form

$$F(y) = \int_{-\infty}^{\infty} K(x,y)f(y)\,dx.$$

A function F defined in this way (y may be either a real or a complex variable) is called an *integral transform* of f. The function K which appears in the integrand is referred to as the *kernel* of the transform.

Integral transforms are used very extensively in both pure and applied mathematics, as well as in science and engineering. They are especially useful in solving certain boundary value problems, partial differential equations and some types of integral equations. An integral transform is a linear and invertible transformation, and a partial differential equation can be reduced to a system of algebraic equations by application of an integral transform. The algebraic problem is easy to solve for the transformed function F, and the function f can be recovered from F by some inversion formula.

Some of the more commonly used integral transforms are the following:

1. The *Fourier Transform* $\mathcal{F}\{f\} \equiv \widehat{f}$ of a function f is defined by

$$\left(\mathcal{F}\{f\}\right)(\omega) \equiv \widehat{f}(\omega) = \int_{-\infty}^{\infty} e^{-ix\omega} f(x)\,dx.$$

2. The *Fourier Cosine Transform* $\mathcal{F}_c\{f\} \equiv \widehat{f}_c$ of a function f is defined by

$$\left(\mathcal{F}_c\{f\}\right)(\omega) \equiv \widehat{f}_c(\omega) = \int_{0}^{\infty} \cos(\omega x) f(x)\,dx.$$

3. The *Fourier Sine Transform* $\mathcal{F}_s\{f\} \equiv \widehat{f}_s$ of a function f is defined by

$$\left(\mathcal{F}_s\{f\}\right)(\omega) \equiv \widehat{f}_s(\omega) = \int_{0}^{\infty} \sin(\omega x) f(x)\,dx.$$

4. The *Laplace Transform* $\mathcal{L}\{f\} \equiv \widehat{f}$ of a function f is defined by

$$(\mathcal{L}\{f\})(s) \equiv \widehat{f}(s) = \int_0^\infty e^{-xs} f(x)\, dx.$$

5. The *Hankel Transform* $\mathcal{H}_n\{f\}$ of order n of a function f is defined by

$$(\mathcal{H}_n\{f\})(s) \equiv \widehat{f_{\mathcal{H}_n}}(s) = \int_0^\infty x f(x) J_n(sx)\, dx,$$

where J_n is the Bessel function of order n of the first kind.

Sometimes, the integral transforms are defined with different constants (for normalization purposes), particulary in the case of the Fourier transforms.
Since

$$e^{-ixy} = \cos(xy) - i\sin(xy),$$

the Fourier sine and Fourier cosine transforms are special cases of the Fourier transform in which the function f vanishes on the negative real axis.

The Laplace transform is also related to the Fourier transform. Indeed, if we consider the complex number $s = u + iv$, where u and v are real, then we have

$$\int_0^\infty e^{-xs} f(x)\, dx = \int_0^\infty e^{-ixv} e^{-xu} f(x)\, dx = \int_0^\infty e^{-ixv} \phi_u(x)\, dx,$$

where $\phi_u(x) = e^{-xu} f(x)$. Therefore the Laplace transform can also be regarded as a special case of the Fourier transform.

2.1 The Laplace Transform.

In this part we will study the Laplace transform and some of its applications. Many more applications of the Laplace transform will be discussed later in the chapters dealing with partial differential equations.

2.1.1 Definition and Properties of the Laplace Transform.

Definition 2.1.1. Given a function $f(t)$ defined for all $t \geq 0$, the *Laplace transform* of f is the function $F(s)$ defined by

$$(2.1.1) \qquad (\mathcal{L}\{f\})(s) = F(s) = \int_0^\infty e^{-st} f(t)\, dt$$

for all values of s for which the improper integral converges.

Usually we use the capital letters for the Laplace transform of a given function $f(t)$. For example, we write

$$\mathcal{L}\, f(t) = F(s), \quad \mathcal{L}\, g(t) = G(s), \quad \mathcal{L}\, y(t) = Y(s).$$

Example 2.1.1. Find the Laplace transform of the function
$$f(t) = 1, \quad t \geq 0.$$

Solution. By the definition of the Laplace transform we have

$$\left(\mathcal{L}\{1\}\right)(s) = \int_0^\infty e^{-st} \cdot 1 \, dt = -\frac{1}{s} e^{-st} \Big|_{t=0}^{t=\infty}$$

$$= \lim_{t \to \infty} \left(\frac{1}{s} e^{-st} + \frac{1}{s}\right) = \frac{1}{s} \quad \text{for } s > 0$$

and therefore,

$$\left(\mathcal{L}\{1\}\right)(s) = \frac{1}{s}, \quad \text{for } s > 0.$$

Example 2.1.2. Find the Laplace transform of the function
$$f(t) = t^a, \quad t \geq 0, \text{ and } a > -1.$$

Solution. By the definition of the Laplace transform we have

$$\left(\mathcal{L}\{t^a\}\right)(s) = \int_0^\infty e^{-st} t^a \, dt.$$

If we use the substitution $u = st$, then from the above equation we obtain

$$\left[\left(\mathcal{L}\{t^a\}\right)(s) = \frac{1}{s^{a+1}} \int_0^\infty e^{-u} u^a \, du = \frac{\Gamma(a+1)}{s^{a+1}}, \quad s > 0,\right.$$

where Γ is the Euler Gama function. For discussion of the Gama function Γ see Appendix G.

Since $\Gamma(n+1) = n!$ if $n \in \mathbb{N} \cup \{0\}$, from Example 2.1.2. we obtain

$$\left(\mathcal{L}\{t^n\}\right)(s) = \frac{n!}{s^{n+1}}, \quad \text{for } s > 0.$$

Below are some basic Laplace transforms.

$$\left(\mathcal{L}\{\sin(kt)\}\right)(s) = \frac{k}{s^2 + k^2}, \quad \left(\mathcal{L}\{\cos(kt)\}\right)(s) = \frac{s}{s^2 + k^2},$$

$$\left(\mathcal{L}\{\cosh(kt)\}\right)(s) = \frac{s}{s^2 - k^2}, \quad \left(\mathcal{L}\{\sinh(kt)\}\right)(s) = \frac{k}{s^2 - k^2},$$

$$\left(\mathcal{L}\{e^{at}\}\right)(s) = \frac{1}{s - a}.$$

A more comprehensive table of Laplace transforms is given in Table A of Appendix A.

There are a number of useful operational properties of the Laplace transform.

Theorem 2.1.1 Linear Property. *If* a_1 *and* a_2 *are any constants, then*

$$\mathcal{L}\{a\,f_1 + b\,f_2\} = a\mathcal{L}\{f_1\} + b\mathcal{L}\{f_2\}$$

for all functions f_1 *and* f_2 *whose Laplace transforms exist.*

Proof. The proof of this theorem follows immediately from the linearity property of the operations integration and taking the limit. ■

Example 2.1.3. Find the Laplace transform of the function

$$f(t) = 3t^4 + 4t^{\frac{3}{2}}, \quad t \geq 0.$$

Solution. Using the linearity property of the Laplace transform and the result of Example 2.1.2 we have

$$\left(\mathcal{L}\{3t^2 + 4t^{\frac{3}{2}}\}\right)(s) = 3\frac{2!}{s^3} + 4\frac{\Gamma(\frac{5}{2})}{s^{\frac{5}{2}}}$$

$$= \frac{6}{s^3} + 3\sqrt{\frac{\pi}{s^5}}$$

since

$$\Gamma(\frac{5}{2}) = \frac{3}{2} \cdot \frac{1}{2}\Gamma(\frac{1}{2}) = \frac{3}{4}\sqrt{\pi}.$$

(See Appendix G for the Gamma function).

Theorem 2.1.2 The First Shift Theorem. *If* $F = \mathcal{L}\{f\}$ *and* c *is a constant, then*

$$\left(\mathcal{L}\{e^{-ct}f(t)\}\right)(s) = F(s + c).$$

Proof. The proof of this theorem follows immediately from the definition of the Laplace transform. ■

The proofs of the next two theorems follow directly from the definition of the Laplace transform.

Theorem 2.1.3 The Second Shift Theorem. *If* $F = \mathcal{L}\{f\}$ *and* c *is a positive constant, then*

$$\left(\mathcal{L}\{f(t - c)\}\right)(s) = e^{-cs}F(s).$$

Theorem 2.1.4. *If the Laplace transform of a function* f *is* F, *and* c *is a positive constant, then*

$$\left(\mathcal{L}\{f(ct)\}\right)(s) = \frac{1}{s}F\left(\frac{s}{c}\right).$$

In addition to the shifting theorems, there are two other useful theorems that involve the derivative and integral of the Laplace transform $F(s)$.

Theorem 2.1.5. *If* $F = \mathcal{L}\{f\}$, *then*

$$F'(s) = -\big(\mathcal{L}\{tf(t)\}\big)(s),$$

and in general,

$$F^{(n)}(s) = (-1)^n \big(\mathcal{L}\{t^n f(t)\}\big)(s),$$

for any natural number n.

Proof. If we use the definition of the Laplace transform, then

$$F'(s) = \frac{d}{ds}\left(\int_0^\infty e^{-st} f(t)\,dt\right) = \int_0^\infty \frac{\partial}{\partial s}\big(e^{-st} f(t)\big)\,dt$$

$$= -\int_0^\infty e^{-st} t f(t)\,dt = -\big(\mathcal{L}\{tf(t)\}\big)(s).$$

For $n \in \mathbb{N}$ use the principle of mathematical induction.   ∎

Theorem 2.1.6. *If* $F = \mathcal{L}\{f\}$, *then*

$$\int_s^\infty F(z)\,dz = \left(\mathcal{L}\left\{\frac{f(t)}{t}\right\}\right)(s).$$

Using these theorems, along with the Laplace transforms of some functions listed in Table A of Appendix A, we can compute the Laplace transforms of many other functions.

Example 2.1.4. Find the Laplace transform of

$$e^{at} t^n, \quad n \in \mathbb{N}.$$

Solution. Since

$$\big(L\{t^n\}\big)(s) = \frac{n!}{s^{n+1}},$$

from Theorem 2.1.2 it follows that

$$\big(L\{e^{at} t^n\}\big)(s) = \frac{n!}{(s-a)^{n+1}}, \quad s > a.$$

Example 2.1.5. Find the Laplace transform of

$$e^{at} \cos kt.$$

Solution. Since

$$(\mathcal{L}\{\cos kt\})(s) = \frac{s}{s^2 + k^2},$$

from Theorem 2.1.2 it follows that

$$(\mathcal{L}\{e^{at} \cos kt\})(s) = \frac{s - a}{(s - a)^2 + k^2}, \quad s > a.$$

Example 2.1.6. Find the Laplace transform of

$$t \sin at.$$

Solution. Let $f(t) = \sin at$ and $F = \mathcal{L}\{f\}$. From Theorem 2.1.5 we have

$$(\mathcal{L}\{t \sin at\})(s) = -F'(s) = -\frac{d}{ds}\left(\frac{a}{s^2 + a^2}\right) = \frac{2as}{(s^2 + a^2)^2}.$$

Example 2.1.7. Find the Laplace transform of the following function.

$$f(t) = \frac{1 - \cos at}{t}.$$

Solution. If $F = \mathcal{L}\{f\}$, then from Theorem 2.1.6 we have

$$F(s) = \int_s^\infty (\mathcal{L}\{1 - \cos at\})(z)\, dz = \int_s^\infty \left(\frac{1}{z} - \frac{z}{z^2 + a^2}\right) dz$$

$$= \ln \frac{z}{\sqrt{z^2 + a^2}}\Bigg|_{z=s}^{z=\infty} = -\ln \frac{s}{\sqrt{s^2 + a^2}}.$$

Remark. The improper integral which defines the Laplace transform does not have to converge.

For example, neither

$$\mathcal{L}\{\frac{1}{t}\} \quad \text{nor} \quad \mathcal{L}\{e^{t^2}\}$$

exists.

Sufficient conditions which guarantee the existence of $\mathcal{L}\{f\}$ are that f be piecewise continuous on $[0, \infty)$ and that f be of exponential order. Recall that a function f is piecewise continuous on $[0, \infty)$ if f is continuous on any closed bounded interval $[a, b] \subset [0, \infty)$ except at finitely many points. The concept of *exponential order* is defined as follows:

Definition 2.1.2. A function f is said to be of *exponential order* c if there exist constants $c > 0$, $M > 0$ and $T > 0$ such that

$$|f(t)| \leq Me^{ct} \quad \text{for all} \quad t > T.$$

Remark. If f is an increasing function, then the statement that f is of exponential order means that the graph of the function f on the interval $[T, \infty)$ does not grow faster than the graph of the exponential function Me^{ct}.

Every polynomial function, e^{-t}, $2\cos t$ are a few examples of functions which are of exponential order. The function e^{t^2} is an example of a function which is not of exponential order.

Theorem 2.1.7 Existence of Laplace Transform. *If f is a piecewise continuous function on $[0, \infty)$ and of exponential order c, then the Laplace transform*

$$F(s) = (\mathcal{L}\{f\})(s)$$

exists for $s > c$.

Proof. By the definition of the Laplace transform and the additive property of definite integrals we have

$$\left(\mathcal{L}\{f(t)\}\right)(s) = \int_0^T e^{-st}f(t)\,dt + \int_T^\infty e^{-st}f(t)\,dt = I_1 + I_2.$$

Since f is of exponential order c, there exist constants $c > 0$, $M > 0$ and $T > 0$ such that $|f(t)| \leq Me^{ct}$ for all $t > T$. Therefore

$$|I_2| \leq \int_T^\infty |e^{-st}f(t)|\,dt \leq M\int_T^\infty e^{-st}e^{ct}\,dt = M\frac{e^{-(s-c)T}}{s-c}, \quad s > c.$$

By the comparison test for improper integrals, the integral I_2 converges for $s > c$.

The integral I_1 exists because it can be written as a finite sum of integrals over intervals on which the function $e^{st}f(t)$ is continuous. This proves the theorem. ∎

In the proof of the above theorem we have also shown that

$$|F(s)| \leq \int_0^\infty |e^{-st}f(t)|\,dt \leq \frac{M}{s-c}, \quad s > c.$$

If we take limits from both sides of the above inequality as $s \to \infty$, then we obtain the following result.

Corollary 2.1.1. *If* $f(t)$ *satisfies the hypotheses of Theorem 2.1.7, then*

$$(2.1.2) \qquad\qquad \lim_{s\to\infty} F(s) = 0.$$

The condition (2.1.2) restricts the functions that can be Laplace transforms. For example, the functions s^2, e^s cannot be Laplace transforms of any functions because their limits as $s\to\infty$ are ∞, not 0.

On the other hand, the hypotheses of Theorem 2.1.7 are sufficient but not necessary conditions for the existence of the Laplace transform. For example, the function

$$f(t) = \frac{1}{\sqrt{t}}, \quad t > 0$$

is not piecewise continuous on the interval $[0, \infty)$ but nevertheless, from Example 2.1.2 it follows that its Laplace transform

$$\left(\mathcal{L}\left\{\frac{1}{\sqrt{t}}\right\}\right)(s) = \frac{\Gamma\left(\frac{1}{2}\right)}{s^{\frac{1}{2}}} = \sqrt{\frac{\pi}{s}}$$

exists.

Up to this point, we were dealing with finding the Laplace transform of a given function. Now we want to reverse the operation: For a given function $F(s)$ we want to find (if possible) a function $f(t)$ such that

$$\big(\mathcal{L}\{f(t)\}\big)(s) = F(s).$$

Definition 2.1.3. The *Inverse Laplace Transform* of a given function $F(s)$ is a function $f(t)$, if such a function exists, such that

$$\big(\mathcal{L}\{f(t)\}\big)(s) = F(s).$$

We denote the inverse Laplace transform of $F(s)$ by

$$\mathcal{L}^{-1}\left\{F(s)\}\right\}(t).$$

Remark. For the inverse Laplace transform the linearity property holds.

The most common method of inverting the Laplace transform is by decomposition into partial fractions, along with Table A.

Example 2.1.8. Find the inverse transform of

$$F(s) = \frac{4}{(s-1)^2(s^2+1)^2}.$$

Solution. We decompose $F(s)$ into partial fractions as follows:

$$\frac{4}{(s-1)^2(s^2+1)^2} = \frac{A}{s-1} + \frac{B}{(s-1)^2} + \frac{Cs+D}{s^2+1} + \frac{Es+F}{(s^2+1)^2}.$$

By comparison, we find that the coefficients are $A = -2$, $B = 1$, $C = 2$, $D = 1$, $E = 2$ and $F = 0$. Therefore

$$F(s) = -\frac{2}{s-1} + \frac{1}{(s-1)^2} + \frac{2s+1}{s^2+1} + \frac{2s}{(s^2+1)^2}.$$

If we recall the first shifting property (Theorem 2.1.2), and the familiar Laplace transforms of the functions $\sin t$ and $\cos t$, then we find that the required inverse transform of the given function $F(s)$ is

$$-2e^t + te^t + 2\cos t + \sin t + t\sin t.$$

Another method to recover a function f from its Laplace transform F is to use the inversion formula with contour integration. The inversion formula is given by the following theorem.

Theorem 2.1.8 The Laplace Inversion Formula. *Suppose that $f(t)$ is a piecewise smooth function of exponential order c which vanishes for $t < 0$. If $F = \mathcal{L}\{f\}$ and $b > c$, then*

$$(2.1.3) \qquad f(t) = \frac{1}{2\pi i} \int_{b-i\infty}^{b+i\infty} F(z)\,e^{zt}\,dz := \frac{1}{2\pi i} \lim_{y\to\infty} \int_{b-iy}^{b+iy} F(z)\,e^{zt}\,dz,$$

at every point t of continuity of the function f.

Later in this chapter we will prove this theorem by an easy application of the Fourier transform. This theorem can also be proved by Cauchy's theorem for analytic functions. (See Appendix F).

The line integral in the inversion formula (2.1.3) is called the *Bromwich integral.* We choose b large enough so that the straight line $Re(z) = b$ lies in the half-plane where $F(z)$ is an analytic function. Usually, the line integral in the inversion formula is transformed into an integral along a closed contour (*Bromwich contour*) and after that the Cauchy residue theorem is applied. (See Appendix F for the Cauchy residue theorem.)

Let us take an example to illustrate the inversion formula.

Example 2.1.9. Find the inverse Laplace transform of

$$F(s) = \frac{1}{s \sinh \pi s}.$$

Solution. Consider the function

$$F(z) = \frac{1}{z \sinh \pi z}.$$

This function is analytic everywhere except at the point $z = 0$ and at all points z for which

$$\sinh \pi z \equiv \frac{e^{\pi z} - e^{-\pi z}}{2} = 0.$$

There are infinitely many solutions z_n of the above equation, $z_n = ni$, $n = 0, \pm 1, \pm 2, \ldots$. The singularity at $z = 0$ is a double pole of $F(z)$, and the other singularities are simple poles.

Now, consider the contour integral

(2.1.4) $$\frac{1}{2\pi i} \int_C F(z) \, e^{tz} \, dz,$$

where C is the closed contour shown in Figure 2.1.1.

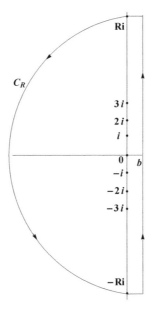

Figure 2.1.1

On the semi-circle $C_R = \left\{ z = R \, e^{i\theta} : \dfrac{\pi}{2} \le \theta \le \dfrac{3\pi}{2} \right\}$ we have

$$\left| \int_{C_R} F(z) \, e^{tz} \, dz \right| \le \int_{\frac{\pi}{2}}^{\frac{3\pi}{2}} 2 \, \frac{e^{Rt \cos \theta}}{\left| e^{R \cos \theta} - e^{-R \cos \theta} \right|} \, d\theta.$$

Since $\cos\theta \le 0$ for $\dfrac{\pi}{2} \le \theta \le \dfrac{3\pi}{2}$ and since $t > 0$, from the above inequality it follows that

$$(2.1.5) \qquad \lim_{R\to\infty} \int_{C_R} F(z)\, e^{tz}\, dz = 0.$$

Now, taking $R \to \infty$ in (2.1.4), from (2.1.5) and the residue theorem (see Appendix F) we obtain

$$(2.1.6) \qquad f(t) = \frac{1}{2\pi i} \int_{b-i\infty}^{b+i\infty} F(z)\, e^{zt}\, dz = \sum_{n=-\infty}^{\infty} \operatorname{Res}\left(F(z)\, e^{zt},\ z = z_n \right).$$

Since $z = 0$ is a double pole of $F(z)$, using the formula for residues (given in Appendix F) and l'Hôpital's rule we have

$$(2.1.7) \qquad \operatorname{Res}\left(F(z)\, e^{zt},\ z = 0 \right) = \lim_{z\to 0} \frac{d}{dz}\left(\frac{(z-0)^2 e^{zt}}{z^2 \sinh z} \right) = \frac{t}{\pi}.$$

For the other simple poles $z = z_n = \pm ni$, $n = 1, 2, \ldots$, using l'Hôpital's rule we find

$$(2.1.8) \qquad \operatorname{Res}\left(F(z)\, e^{zt},\ z = z_n \right) = \lim_{z\to n} \frac{(z-n)e^{tz}}{z \sinh z} = (-1)^n \frac{e^{nit}}{ni}.$$

If we substitute (2.1.7) and (2.1.8) into (2.1.6) we obtain (after small rearrangements)

$$f(t) = \frac{t}{\pi} + \frac{2}{\pi} \sum_{n=1}^{\infty} \frac{(-1)^n}{n} \sin nt.$$

The next theorem, which is a consequence of Theorem 2.1.3, has theoretical importance, particularly when solving initial boundary value problems using a Laplace transform.

Theorem 2.1.9 (Uniqueness of the Inverse Laplace Transform). *Let the functions $f(t)$ and $g(t)$ satisfy the hypotheses of Theorem 2.1.7, so that their Laplace transforms $F(s)$ and $G(s)$ exist for $s > c$. If $F(s) = G(s)$, then $f(t) = g(t)$ at all points t where f and g are both continuous.*

As a consequence of the previous theorem it follows that there is a one-to-one correspondence between the continuous functions of exponential order and their Laplace transforms.

Exercises for Section 2.1.1.

1. In Problems (a)–(g) apply the definition 2.1.1 to find directly the Laplace transforms of the given functions.

   (a) $f(t) = t$.

   (b) $f(t) = e^{3t+1}$.

   (c) $f(t) = \cos 2t$.

   (d) $f(t) = \sin^2 t$.

   (e) $f(t) = \begin{cases} 1, & 0 \le t \le 2 \\ 0, & t > 2. \end{cases}$

   (f) $f(t) = \begin{cases} t, & 0 \le t < 2 \\ 3, & t > 2. \end{cases}$

   (g) $f(t) = \begin{cases} 0, & 0 \le t \le 1 \\ 1, & t > 1. \end{cases}$

2. Use the shifting property of the Laplace transform to compute the following:

   (a) $\mathcal{L}\{e^t \sin 2t\}$.

   (b) $\mathcal{L}\{e^{-t} \cos 2t\}$.

   (c) $\mathcal{L}\{e^t \cos 3t\}$.

   (d) $\mathcal{L}\{e^{-2t} \cos 4t\}$.

   (e) $\mathcal{L}\{e^{-t} \sin 5t\}$.

   (f) $\mathcal{L}\{e^t t\}$.

3. Using properties of the Laplace transform compute the Laplace transforms of the following functions:

   (a) $f(t) = -18e^{3t}$.

   (b) $f(t) = t^2 e^{-3t}$.

   (c) $f(t) = 3t \sin 2t$.

(d) $f(t) = e^{-2t} \cos 7t$.

(e) $f(t) = 1 + \cos 2t$.

(f) $f(t) = t^3 \sin 2t$.

4. Use properties of the Gamma function (see Appendix G) to find the Laplace transform of:

(a) $f(t) = t^{-\frac{1}{2}}$.

(b) $f(t) = t^{\frac{1}{2}}$.

5. In the following exercises, compute the inverse Laplace transform of the given function.

(a) $F(s) = \dfrac{1}{s^6}$.

(b) $F(s) = \dfrac{1}{(s^2 - 4s + 4)}$.

(c) $F(s) = \dfrac{1}{(s^2 + 9)^2}$.

(d) $F(s) = \dfrac{4}{(s - 6s + 9)}$.

(e) $F(s) = \dfrac{1}{s^2 + 15s + 56}$.

(f) $F(s) = \dfrac{1}{s^2 + 16s + 36}$.

(g) $F(s) = \dfrac{2s - 7}{2s^2 - 14s + 55}$.

(h) $F(s) = \dfrac{s + 2}{s^2 + 4s + 12}$.

6. Find the following inverse Laplace transforms.

(a) $\mathcal{L}^{-1}\left\{ \dfrac{2s - 5}{(s - 1)^3(s^2 + 4)} \right\}$.

(b) $\mathcal{L}^{-1}\left\{ \dfrac{3s - 5}{s^2(s^2 + 9)(s^2 + 1)} \right\}$.

7. Determine if $F(s)$ is the Laplace transform of a piecewise continuous function of exponential order:

(a) $F(s) = \dfrac{s}{4 - s}$.

(b) $f(s) = \dfrac{s^2}{(s-2)^2}$.

(c) $f(s) = \dfrac{s^2}{s^2+9}$.

8. Show that the following functions are of exponential order:

   (a) $f(t) = t^2$.

   (b) $f(t) = t^n$, where n is any positive integer.

9. Show that the following functions are not of exponential order:

   (a) $f(t) = e^{t^2}$.

   (b) $f(t) = e^{t^n}$, where n is any positive integer.

10. Show that
$$\mathcal{L}^{-1}\left\{e^{-a\sqrt{s}}\right\} = \frac{a}{2\sqrt{\pi t^3}}e^{-\frac{a^2}{4t}}.$$

2.1.2 Step and Impulse Functions.

In many applications in electrical and mechanical engineering functions which are discontinuous or quite large in very small intervals are frequently present. An important and very useful function in such physical situations is the *Heaviside function* or *unit step function*. This function, denoted by H, is defined by
$$H(t) = \begin{cases} 0, & t < 0 \\ 1, & t \geq 0. \end{cases}$$

The graph of the Heaviside function $y = H(t-2)$ is shown in Figure 2.1.2

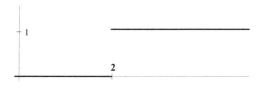

Figure 2.1.2

Remark. $u_a(t)$ is another notation for the Heaviside function $H(t-a)$.

It is very easy to find the Laplace transform of $H(t-a)$.

$$\mathcal{L}(\{H(t-a)\})(s) = \int_0^\infty H(t-a)e^{-st}\,ds$$

$$= \int_a^\infty e^{-st}\,ds = \frac{e^{-as}}{s}.$$

A useful observation for the Heaviside function is that the function

$$H(t-a)f(t-a) = \begin{cases} 0, & t < a \\ f(t-a), & t \geq a \end{cases}$$

is simply a translation of the function $f(t)$.

The Heaviside step function can be used to express the Laplace transform of a translation of a function $f(t)$ and the Laplace transform of the function f.

Theorem 2.1.10. *Let* $F(s) = \mathcal{L}\{f(t)\}$ *exist for* $s > c \geq 0$, *and let* $a > 0$ *be a constant. Then*

$$\mathcal{L}\left(\{H(t-a)f(t-a)\}\right)(s) = e^{-as}F(s), \quad s > c.$$

Proof. We apply the definition of the Laplace transform.

$$(\mathcal{L}\{H(t-a)f(t-a)\})(s) = \int_0^\infty e^{-st}H(t-a)f(t-a)\,dt = \int_a^\infty e^{-st}f(t-a)\,dt.$$

If we make the substitution $v = t - a$ in the last integral, then we have

$$\mathcal{L}\{H(t-a)f(t-a)\}(s) = \int_0^\infty e^{-s(v+a)}f(v)\,dv$$

$$= e^{-as}\int_0^\infty e^{-sv}f(v)\,dv = e^{-cas}F(s). \quad \blacksquare$$

Example 2.1.10. Find the Laplace transform of the function f defined by

$$f(t) = \begin{cases} t, & 0 \leq t < 2, \\ t + (t-2)^2, & 2 \leq t < \infty. \end{cases}$$

Solution. Consider the following translation of the function t^2:

$$H(t-2)(t-2)^2 = \begin{cases} 0, & 0 \leq t < 2, \\ (t-2)^2, & t \geq 2. \end{cases}$$

Using this translation we have

$$f(t) = t + H(t-2)(t-2)^2.$$

Therefore,

$$\begin{aligned} \left(\mathcal{L}\{f(t)\}\right)(s) &= \left(\mathcal{L}\{t\}\right)(s) + \left(\mathcal{L}\{H(t-2)(t-2)^2\}\right)(s) \\ &= \left(\mathcal{L}t\}\right)(s) + e^{-2s}\left(\mathcal{L}\{t^2\}\right)(s). \end{aligned}$$

Using the Laplace transforms of t and t^2 from Table A in Appendix A, we obtain

$$\begin{aligned} \left(\mathcal{L}\{f(t)\}\right)(s) &= \frac{1}{s^2} + e^{-2s}\frac{2}{s^3} \\ &= \frac{s + e^{-2s}}{s^3}. \end{aligned}$$

Example 2.1.11. Find the inverse Laplace transform of the function

$$F(s) = \frac{1 + e^{-5s}}{s^4}.$$

Solution. Since the inverse Laplace transform is linear we have

$$\begin{aligned} f(t) = \mathcal{L}^{-1}\{F(s)\} &= \mathcal{L}^{-1}\{\frac{1}{s^4}\} + \mathcal{L}^{-1}\left\{\frac{e^{-5s}}{s^4}\right\} \\ &= \frac{t^3}{3!} + H(t-5)\frac{(t-5)^3}{3!}. \end{aligned}$$

Now, we will introduce (only intuitively) a "function" which is of great importance in many disciplines, such as quantum physics, electrostatics and mathematics.

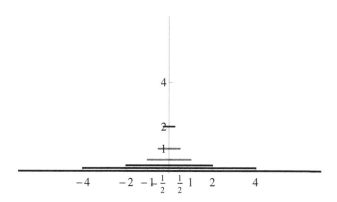

Figure 2.1.3

Let us consider a function $f(t)$ which has relatively big values in a relatively short interval around the origin $t = 0$ and it is zero outside that interval. For natural numbers n, define the functions $f_n(t)$ by

$$(2.1.9) \qquad f_n(t) = \begin{cases} n, & |t| < \frac{1}{2n}, \\ 0, & |t| \geq \frac{1}{2n}. \end{cases}$$

Figure 2.1.3 shows the graphs of f_n for several values of n.

Each of the functions $f_n(t)$ has the following properties:

$$(2.1.10) \qquad \begin{cases} \lim_{n \to \infty} f_n(t) = 0, & t \neq 0 \\ \\ I_n \equiv \int_{-\infty}^{\infty} f_n(t)\, dt = \int_{-\frac{1}{2n}}^{\frac{1}{2n}} f_n(t)\, dt = 1, & n = 1, 2, \ldots \end{cases}$$

Now, since $I_n = 1$ for every $n \in \mathbb{N}$, it follows that

$$(2.1.11) \qquad \lim_{n \to \infty} I_n = 1.$$

Using equations (2.1.10) and (2.1.11) we can "define" the *Dirac delta function* $\delta(t)$, concentrated at $t = 0$, by

$$\begin{cases} \delta(t) = 0, & t \neq 0; \\ \\ \int_{-\infty}^{\infty} \delta(t)\, dt = 1. \end{cases}$$

The Dirac delta function, concentrated at any other point $t = a$ is defined by $\delta(t - a)$.

It is obvious that there is no function that satisfies both of the above conditions.

Even though the Dirac function δ is not at all an ordinary function, its Laplace transform can be defined. Namely, one of the ways to define the Laplace transform of the Dirac delta function $\delta(t - a)$ is to use the limiting process:

$$(2.1.12) \qquad \mathcal{L}\{\delta(t - a)\} = \lim_{n\to\infty} \mathcal{L}\{f_n(t - a)\},$$

where f_n are the functions defined by (2.1.9).

If we suppose that $a > 0$, then $a - \frac{1}{2n} > 0$ for some $n \in \mathbb{N}$ and so from (2.1.9) and the definition of the Laplace transform, for these values of n we have

$$\mathcal{L}\{f_n(t - a)\} = \int_0^\infty e^{-st} f_n(t - a)\, dt$$

$$= \int_{a-\frac{1}{2n}}^{a+\frac{1}{2n}} e^{-st} f_n(t - a)\, dt = n \int_{a-\frac{1}{2n}}^{a+\frac{1}{2n}} e^{-st}\, dt$$

$$= n \left. \frac{e^{-st}}{-s} \right|_{t=a-\frac{1}{2n}}^{t=a+\frac{1}{2n}} = ne^{-sa} \frac{e^{\frac{s}{2n}} - e^{-\frac{s}{2n}}}{s}.$$

Therefore,

$$(2.1.13) \qquad \mathcal{L}\{f_n(t - a)\} = ne^{-sa} \frac{e^{\frac{s}{2n}} - e^{-\frac{s}{2n}}}{s}.$$

If we take the limit in (2.1.13) as $n \to \infty$, then from (2.1.12) we obtain that

$$(2.1.14) \qquad \mathcal{L}\{\delta(t - a)\} = e^{-sa}.$$

If we take the limit as $a \to 0$ in (2.1.14) we obtain

$$\mathcal{L}\{\delta(t)\} = 1.$$

If f is any continuous function, then it can be shown that

$$(2.1.15) \qquad \int_{-\infty}^\infty \delta(t - a) f(t)\, dt = f(a),$$

where the improper integral is defined by

$$\int_{-\infty}^{\infty} \delta(t-a)f(t)\,dt = \lim_{n\to\infty} \int_{-\infty}^{\infty} f_n(t-a)f(t)\,dt.$$

For the proof of (2.1.15) the interested reader is referred to the book by W. E. Boyce and R. C. DiPrima [1].

Exercises for Section 2.1.2.

1. Sketch the graphs of the following functions on the interval $t \geq 0$.

   (a) $H(t-1) + 2H(t-3) - 6H(t-4)$.

   (b) $(t-3)H(t-2) - (t-2)H(t-3)$.

   (c) $(t-\pi)^2 H(t-\pi)$.

   (d) $\sin(t-3)\,H(t-3))$.

   (e) $2(t-1)H(t-2)$.

2. Find the Laplace transform of the following functions:

   (a) $f(t) = \begin{cases} 0, & t < 2 \\ (t-2)^2, & t \geq 2. \end{cases}$

   (b) $f(t) = \begin{cases} t, & t < 1 \\ t^2 - 2t + 2, & t \geq 1. \end{cases}$

   (c) $f(t) = \begin{cases} 0, & 0 \leq t \leq 1 \\ 1, & t > 1. \end{cases}$

   (d) $f(t) = \begin{cases} 0, & t < \pi \\ t - \pi, & \pi \leq t < 2\pi \\ 0, & t \geq 2\pi. \end{cases}$

   (e) $f(t) = H(t-1) + 2H(t-3) - H(t-4)$.

   (f) $f(t) = (t-3)H(t-3) - (t-2)H(t-2)$.

   (g) $f(t) = t - (t-1)H(t-1)$.

3. In the following exercises compute the inverse Laplace transform of the given function.

(a) $F(s) = \dfrac{3!}{(s-2)^4}$.

(b) $F(s) = \dfrac{e^{-2s}}{s^2 + s - 2}$.

(c) $F(s) = \dfrac{2(s-1)e^{-2s}}{s^2 - 2s + 2}$.

(d) $F(s) = \dfrac{2e^{-2s}}{s^2 - 4}$.

4. In the following exercises compute the Laplace transform of the given function.

   (a) $\delta(t - \pi)$.

   (b) $\delta(t - 3)$.

   (c) $2\delta(t-3) - \delta(t-1)$.

   (d) $\sin t \cdot u_{2\pi}(t) + \delta(t - \frac{\pi}{2})$.

   (e) $\cos 2t \cdot u_\pi(t) + \delta(t - \pi)$.

2.1.3 Initial-Value Problems and the Laplace Transform.

In this section we will apply the Laplace transform to solve initial-value problems. For this purpose we first need the following theorem.

Theorem 2.1.11. *Suppose that f is continuous and is of exponential order c for $t > T$. Also, suppose that f' is piecewise continuous on any closed subinterval of $[0, \infty)$. Then, for $s > c$*

$$\left(\mathcal{L}\{f'(t)\}\right)(s) = s\left(\mathcal{L}\{f(t)\}\right)(s) - f(0).$$

Proof. Let M be any positive number. Consider the integral

$$\int_0^M e^{-st} f'(t)\, dt.$$

Let $0 < t_1 < t_2 < \ldots < t_n \le M$ be the points where the function f' is possibly discontinuous. Using the continuity of the function f and the integration by parts formula on each of the intervals $[t_{j-1}, t_j]$, $t_0 = 0$, $t_n = M$ we have

$$\int_0^M e^{-st} f'(t)\, dt = e^{-sM} f(M) - f(0) + s \int_0^M e^{-st} f(t)\, dt.$$

Since f is of exponential order c we have $e^{-sM}f(M) \to 0$ as $M \to \infty$, whenever $s > c$. Therefore, for $s > c$,

$$\int_0^\infty e^{-st}f'(t)\,dt = s\big(\mathcal{L}\{f(t)\}\big)(s) - f(0),$$

which establishes the theorem. ∎

Continuing this process, we can find similar expressions for the Laplace transform of higher-order derivatives. This leads to the following corollary to Theorem 2.1.11.

Corollary 2.1.2 Laplace Transform of Higher Derivatives. *Suppose that the function f and its derivatives $f', f'', \ldots, f^{(n-1)}$ are of exponential order b on $[0, \infty)$ and the function $f^{(n)}$ is piecewise continuous on any closed subinterval of $[0, \infty)$. Then for $s > b$ we have*

$$\int_0^\infty e^{-st}f^{(n)}(t)\,dt = s^n \mathcal{L}\{f(t)\}(s) - s^{n-1}f(0) - \ldots - f^{(n-1)}(0).$$

Now we show how the Laplace transform can be used to solve initial-value problems. Typically, when we solve an initial-value problem that involves $y(t)$, we use the following steps:

1. Compute the Laplace transform of each term in the differential equation.
2. Solve the resulting equation for $\big(\mathcal{L}\{y(t)\}\big)(s) = Y(s)$.
3. Find $y(t)$ by computing the inverse Laplace transform of $Y(s)$.

Example 2.1.12. Solve the initial-value problem

$$y''' + 4y' = -10e^{-t}, \quad y(0) = 2,\ y'(0) = 2,\ y''(0) = -10.$$

Solution. Let $Y = \mathcal{L}\{y\}$. Taking the Laplace transform of both sides of the differential equation, and using the formula in Corollary 2.1.2, we obtain

$$s^3 Y(s) - s^2 y(0) - sy'(0) - y''(0) + 4\big[sY(s) - y(0)\big] = \frac{10}{s-1}.$$

Using the given initial conditions and solving the above equation for $Y(s)$ we have

$$Y(s) = \frac{2s^3 - 4s - 8}{s(s-1)(s^2+4)}.$$

The partial fraction decomposition of the above fraction is

$$\frac{2s^3 - 4s - 8}{s(s-1(s^2+4)} = \frac{A}{s} + \frac{B}{s-1} + \frac{Cs+D}{s^2+4}.$$

We find the coefficients: $A = 2$, $B = -2$, $C = 2$ and $D = 4$ and so

$$\frac{2s^3 - 4s - 8}{s(s-1)(s^2+4)} = \frac{2}{s} + \frac{-2}{s-1} + \frac{2s+4}{s^2+4}.$$

Finding the inverse Laplace transform of both sides of the above equation and using the linearity of the inverse Laplace transform we obtain

$$y(t) = 2 - 2e^t + 2\cos 2t + 2\sin 2t.$$

Let us take another example.

Example 2.1.13. Solve the initial-value problem

$$y'' + 2y' + y = 6, \quad y(0) = 5, \ y'(0) = 10.$$

Solution. Let $Y = \mathcal{L}\{y\}$. Taking the Laplace transform of both sides of the differential equation, and using the formula in Corollary 2.1.2, we obtain

$$s^2 Y(s) - sy(0) - y'(0) + 2\big[sY(s) - y(0)\big] + Y(s) = \frac{6}{s}.$$

Using the given initial conditions and solving the above equation for $Y(s)$ we have

$$Y(s) = \frac{5s^2 + 20s + 6}{s(s+1)^2}.$$

The partial fraction decomposition of the above fraction is

$$\frac{s^2 + 20s + 6}{s(s+1)^2} = \frac{6}{s} + \frac{-1}{s+1} + \frac{9}{(s+1)^2}.$$

Finding the inverse Laplace transform of both sides of the above equation and using the linearity of the inverse Laplace transform we obtain

$$y(t) = 6 - e^{-t} + 9te^{-t}.$$

In the next few examples we consider initial-value problems in which the nonhomogeneous term, or forcing function, is discontinuous.

Example 2.1.14. Find the solution of

$$y''(t) + 9y = h(t),$$

subject to the conditions $y(0) = y'(0) = 0$, where

$$h(t) = \begin{cases} 1, & 0 \le t < \pi, \\ 0, & t \ge \pi. \end{cases}$$

Solution. Because $h(t)$ is a piecewise continuous function, it is more convenient to write it in terms of the Heaviside functions as

$$h(t) = H(t) - H(t - \pi) = 1 - H(t - \pi).$$

Then

$$\left(\mathcal{L}\{h(t)\}\right)(s) = \left(\mathcal{L}\{1\}\right)(s) - \left(\mathcal{L}\{H(t - \pi)\}\right)(s)$$

$$= \frac{1}{s} - \frac{e^{-\pi s}}{s}.$$

Let $\mathcal{L}\{y\} = Y$ and let us take the Laplace transform of both sides of the given differential equation. Using the formula in Corollary 2.1.2 we have

$$s^2 Y(s) - s y(0) - y'(0) + 9Y(s) = \frac{1}{s} - \frac{e^{-\pi s}}{s}.$$

Using the prescribed initial conditions and solving for $Y(s)$ we obtain

$$Y(s) = \frac{1}{s(s^2 + 9)} - \frac{e^{-\pi s}}{s(s^2 + 9)} = F(s) - e^{-\pi s} F(s),$$

where

$$F(s) = \frac{1}{s(s^2 + 9)} = \frac{1}{9s} - \frac{ss}{s^2 + 9}.$$

Therefore,

$$y(t) = \mathcal{L}^{-1}\{F(s)\} - \mathcal{L}^{-1}\{e^{-\pi s} F(s)\}.$$

First,

$$f(t) = \mathcal{L}^{-1}\{F(s)\} = \frac{1}{9} - \frac{1}{9}\cos 3t,$$

and by Theorem 2.1.10 we have

$$\mathcal{L}^{-1}\left\{e^{-\pi s} F(s)\right\} = f(t - \pi) u_\pi(t) = \left[\frac{1}{9} - \frac{1}{9}\cos 3(t - \pi)\right] u_\pi(t)$$

$$= \left[\frac{1}{9} + \frac{1}{9}\cos 3t\right] u_\pi(t).$$

Combining these results we find that the solution of the original problem is given by

$$y(t) = \frac{1}{9} - \frac{1}{9}\cos 3t - \left[\frac{1}{9} + \frac{1}{9}\cos 3t\right] u_\pi(t).$$

Example 2.1.15. Find the solution of

$$y''(t) + y'(t) + y = h(t),$$

subject to the conditions $y(0) = y'(0) = 0$, where

$$h(t) = \begin{cases} 1, & 0 \le t < 1, \\ 0, & t \ge 1. \end{cases}$$

Solution. Again as in the previous example we write the function in the form $h(t) = 1 - H(t - 1)$. Let $\mathcal{L}\{y\} = Y$ and let us take the Laplace transform of both sides of the given differential equation. Using the formula in Corollary 2.1.2 we have

$$s^2 Y(s) - sy(0) - y'(0) + sY(s) - y(0) + Y(s) = \frac{1}{s} - \frac{e^{-s}}{s}.$$

Using the prescribed initial conditions and solving for $Y(s)$ we obtain

$$Y(s) = \frac{1}{s(s^2 + s + 1)} - \frac{e^{-s}}{s(s^2 + s + 1)} = F(s) - e^{-s}F(s),$$

where

$$F(s) = \frac{1}{s(s^2 + s + 1)}.$$

Then, if $f = \mathcal{L}^{-1}\{F\}$, from Theorem 2.1.10, we find the solution of the given initial-value problem is

$$y(t) = f(t) - f(t - 1)H(t - 1).$$

To determine $f(t)$ we use the partial fraction decomposition of $F(s)$:

$$F(s) = \frac{1}{s} - \frac{s+1}{s^2 + s + 1} = \frac{1}{s} - \frac{s+1}{(s+\frac{1}{2})^2 + \frac{3}{4}})$$

$$= \frac{1}{s} - \left[\frac{s + \frac{1}{2}}{(s+\frac{1}{2})^2 + \frac{3}{4}} + \frac{1}{2} \frac{1}{(s+\frac{1}{2})^2 + \frac{3}{4}} \right].$$

Now, using the first shifting property of the Laplace transform and referring to Table A in Appendix A we obtain

$$h(t) = 1 - e^{-\frac{t}{2}} \cos \frac{\sqrt{3}}{2} t - \frac{1}{\sqrt{3}} e^{-\frac{t}{2}} \sin \frac{\sqrt{3}}{2} t.$$

Example 2.1.16. Find a nontrivial solution of the following linear differential equation of second order, with non-constant coefficients.

$$ty''(t) + (t - 2)y'(t) + y(t) = 0, \quad y(0) = 0.$$

Solution. If $Y = \mathcal{L}\{y\}$, then

$$\mathcal{L}\{y'(t)\} = sY(s), \quad \mathcal{L}\{y''(t)\} = s^2Y(s) - y'(0),$$

and so by Theorem 2.1.5 we have that

$$\mathcal{L}\{ty'(t)\} = -\frac{d}{ds}\big(sY(s)\big), \quad \mathcal{L}\{ty''(t)\} = -\frac{d}{ds}\big(s^2Y(s) - y'(0)\big).$$

The result is the following differential equation.

$$-\frac{d}{ds}\big(s^2Y(s) - y'(0)\big) - \frac{d}{ds}\big(sY(s)\big) - 2Y(s) + Y(s) = 0,$$

or after simplification

$$(s + 1)Y'(s) + 4Y(s) = 0.$$

The general solution of the above differential equation is

$$Y(s) = \frac{C}{(s + 1)^4},$$

where C is any numerical constant. Taking the inverse Laplace transform of the above $Y(s)$ we obtain
$$y(t) = Ct^3e^{-t}.$$

In the next example we apply the Laplace transform to find the solution of a harmonic oscillator equation with an impulse forcing.

Example 2.1.17. Find the solution of the initial-value problem

$$y''(t) + y'(t) + 3y = \delta_4(t), \quad y(0) = 1, \quad y'(0) = 0,$$

where $\delta_4(t) = \delta(t - 4)$ is the Dirac delta function.

Solution. As usual, let $\mathcal{L}\{y\} = Y$, and take the Laplace transform of both sides of the equation. Using the formula for the Laplace transform of a derivative, taking into account the given initial conditions and using the fact that
$$\big(\mathcal{L}\{\delta_4(t)\}\big)(s) = e^{-4s},$$

we have

$$s^2 Y(s) - s + sY(s) - 1 + 3Y(s) = e^{-4t}.$$

Solving the above equation for $Y(s)$ we obtain

$$Y(s) = \frac{s+1}{s^2 + s + 3} + \frac{e^{-4s}}{s^2 + s + 3}.$$

Therefore,

$$y(t) = \mathcal{L}^{-1}\left\{\frac{s+1}{s^2 + s + 3}\right\} + \mathcal{L}^{-1}\left\{\frac{e^{-4s}}{s^2 + s + 3}\right\}$$

$$= \mathcal{L}^{-1}\left\{\frac{s+\frac{1}{2}}{(s+\frac{1}{2})^2 + \frac{11}{4}}\right\} + \mathcal{L}^{-1}\left\{\frac{\frac{1}{2}}{(s+\frac{1}{2})^2 + \frac{11}{4}}\right\}$$

$$+ \mathcal{L}^{-1}\left\{\frac{e^{-4s}}{(s+\frac{1}{2})^2 + \frac{11}{4}}\right\}.$$

Using the shifting theorems 2.1.2 and 2.1.3 and Table A in Appendix A we obtain that the solution of the original initial-value problem is given

$$y(t) = e^{-\frac{1}{2}t} \cos\left(\frac{\sqrt{11}}{2}t\right) + \frac{1}{\sqrt{11}} e^{-\frac{1}{2}t} \sin\left(\frac{\sqrt{11}}{2}t\right)$$

$$+ H(t-4)\frac{2}{\sqrt{11}} e^{-\frac{1}{2}(t-4)} \sin\left(\frac{\sqrt{11}}{2}(t-4)\right).$$

Exercises for Section 2.1.3.

1. Using the Laplace transform solve the initial-value problems:

   (a) $y'(t) + y = e^{-2t}$, $y(0) = 2$.

   (b) $y'(t) + 7y = H(t-2)$, $y(0) = 3$.

   (c) $y'(t) + 7y = H(t-2)e^{-2(t-2)}$, $y(0) = 1$.

2. Using the Laplace transform solve the initial-value problems:

   (a) $y''(t) + y'(t) + 7y = \sin 3t$, $y(0) = 2$, $y'(0) = 0$.

   (b) $y''(t) + 3y = H(t-4)\cos 5(t-4)$, $y(0) = 0$, $y'(0) = -2$.

   (c) $y''(t) + 9y = H(t-5)\sin 3(t-5)$, $y(0) = 2$, $y'(0) = 0$.

(d) $y^{iv}(t) - y = H(t-1) - H(t-2)$, $y(0) = y'(0) = y''(0) = y'''(0) = 0$.

(e) $y''(t) + 3y = w(t)$, $y(0) = 0$, $y'(0) = -2$, $w(t) = \begin{cases} t, & 0 \le t < 1 \\ 1, & t \ge 1. \end{cases}$

3. Suppose that f is a periodic function with period T. If the Laplace transform of f exists show that

$$(\mathcal{L}\{f\})(s) = \frac{1}{1 - e^{-Ts}} \int_0^T f(t) e^{-st}\, dt.$$

4. Find the Laplace transform of the 2-periodic function w defined by

$$w(t) = \begin{cases} 1, & 2n \le t < 2n+1 \text{ for some integer } n \\ -1, & 2n+1 \le t < 2n \text{ for some integer } n. \end{cases}$$

5 Find the solution for each of the following initial-value problems.

(a) $y''(t) - 2y'(t) + y = 3\delta_2(t)$, $y(0) = 0$, $y'(0) = 1$.

(b) $y''(t) + 2y'(t) + 6y = 3\delta_2(t) - 4\delta_5(5)$, $y(0) = 0$, $y'(0) = 1$.

(c) $y''(t) + 2y'(t) + y = \delta_\pi(t)$, $y(0) = 1$, $y'(0) = 0$.

(d) $y''(t) + 2y'(t) + 3y = \sin t + \delta_\pi(t)$, $y(0) = 0$, $y'(0) = 1$.

(e) $y''(t) + 4y = \delta_\pi(t) - \delta_{2\pi}(t)$, $y(0) = 0$, $y'(0) = 0$.

(f) $y''(t) + y = \delta_\pi(t) \cos t$, $y(0) = 0$, $y'(0) = 1$.

2.1.4 The Convolution Theorem for the Laplace Transform.

In this section we prove the *Convolution Theorem*, which is a fundamental result for the Laplace transform.

First we introduce the concept of *convolution*. This concept is inherent in fields of the physical sciences and engineering. For example, in mechanics, it is known as the super position or Duhamel integral. In system theory, it plays a crucial role as the impulse response and in optics as the point spread or smearing function.

Definition 2.1.4. If f and g are functions on $\mathbb{R}$, their *convolution* is the function, denoted by $f * g$, defined by

$$(f * g)(x) = \int\limits_0^x f(x - y)g(y)\, dy,$$

provided that the integral exists.

Remark. The convolution of f and g exists for every $x \in \mathbb{R}$ if the functions satisfy certain conditions. For example, it exists in the following situations:

(1) If f is an integrable function on $\mathbb{R}$ and g is a bounded function.

(2) If g is an integrable function on $\mathbb{R}$ and f is a bounded function.

(3) If f and g are both square integrable functions on $\mathbb{R}$.

(4) If f is a piecewise continuous function on $\mathbb{R}$ and g is a bounded function which vanishes outside a closed bounded interval.

Theorem 2.1.12. The Convolution Theorem. *Suppose that $f(t)$ and $g(t)$ are piecewise continuous functions on $[0, \infty)$ and both are of exponential order. If $\mathcal{L}\{f\} = F$ and $\mathcal{L}\{g\} = G$, then*

$$F(s)G(s) = \big(\mathcal{L}\{f * g\}\big)(s).$$

Proof. If we compute the product $F(s)G(s)$ by the definition of the Laplace transform; then we have that

$$F(s)G(s) = \left(\int\limits_0^\infty e^{-sx} f(x)\, dx\right) \cdot \left(\int\limits_0^\infty e^{-sy} g(y)\, dy\right),$$

which can be written as the following iterated integral:

$$\int\limits_0^\infty \int\limits_0^\infty e^{-s(x+y)} f(x)g(y)\, dx\, dy.$$

If in the above double integral we change the variable x with a new variable t by the transformation $x = t - y$ and keep the variable y, then we obtain

$$F(s)G(s) = \iint\limits_R e^{-st} f(t - y)g(y)\, dt\, dy = \int\limits_0^\infty \int\limits_y^\infty e^{-st} f(t - y)g(y)\, dt\, dy,$$

Figure 2.1.4

where the region of integration R is the unbounded region shown in Figure 2.1.4.

If we interchange the order of integration in the last iterated integral, then we obtain

$$F(s)G(s) = \int\limits_0^\infty \int\limits_0^t e^{-st} f(t-y)g(y)\,dy\,dt,$$

which can be written as

$$F(s)G(s) = \int\limits_0^\infty \left(e^{-st} \int\limits_0^t f(t-y)g(y)\,dy \right) dt$$

$$= \mathcal{L}\left\{ \int\limits_0^t f(t-y)g(y)\,dy \right\} = \mathcal{L}\{f*g\}. \quad \blacksquare$$

Remark. The convolution $f*g$ has many of the properties of ordinary multiplication. For example, it is easy to show that

$$f*g = g*f \qquad \text{(commutative law)}.$$
$$f*(g+h) = f*g + f*h \qquad \text{(distributive law)}.$$
$$(f*g)*h = f*(g*h) \qquad \text{(associative law)}.$$
$$f*0 = 0.$$

The proofs of these properties are left as an exercise.

But, there are other properties of ordinary multiplication of functions which the convolution does not have. For example, it is not true that

$$1*f = f$$

for every function f.

Indeed, if $f(t) = t$, then

$$(1 * f)(t) = \int_0^t 1 \cdot f(x)\, dx = \int_0^t x\, dx = \frac{t^2}{2},$$

and so, for this function we have $1 * f \neq f$.

Example 2.1.18. Compute $f * g$ and $g * f$ if $f(t) = e^{-t}$ and $g(t) = \sin t$. Verify the Convolution Theorem for these functions.

Solution. By using the definition of convolution and the integration by parts formula, we obtain

$$(f * g)(t) = \int_0^t f(t - x)g(x)\, dx = \int_0^t e^{-(t-x)} \sin x\, dx = e^{-t} \int_0^t e^x \sin x\, dx$$

$$= e^{-t} \left[\frac{e^x}{2} (\sin x - \cos x) \right]_{x=0}^{x=t} = \frac{1}{2}(\sin t - \cos t) + \frac{1}{2} e^{-t}.$$

Similarly, we find

$$(g * f)(t) = \int_0^t g(t - x)f(x)\, dx = \int_0^t \sin(t - x)e^x\, dx = \frac{1}{2}(\sin t - \cos t) + \frac{1}{2} e^{-t},$$

which shows that $f * g = g * f$.

To check the result in the Convolution Theorem we let $F = \mathcal{L}\{f\}$ and $G = \mathcal{L}\{g\}$. Then

$$F(s) = \mathcal{L}\{f\} = \mathcal{L}\{e^{-t}\} = \frac{1}{s+1}, \quad G(s) = \mathcal{L}\{g\} = \mathcal{L}\{\sin t\} = \frac{1}{s^2+1}$$

and so

$$\mathcal{L}^{-1}\{F(s)G(s)\} = \mathcal{L}^{-1}\left\{ \frac{1}{s+1} \frac{1}{s^2+1} \right\}.$$

We compute

$$\mathcal{L}^{-1}\left\{ \frac{1}{s+1} \frac{1}{s^2+1} \right\}$$

by the partial fraction decomposition

$$\frac{1}{s+1} \frac{1}{s^2+1} = \frac{1}{2} \frac{1}{s+1} + \frac{1}{2} \frac{-s+1}{s^2+1}.$$

Therefore,

$$\mathcal{L}^{-1}\{\frac{1}{s+1}\frac{1}{s^2+1}\} = \frac{1}{2}\mathcal{L}^{-1}\{\frac{1}{s+1} - \frac{-s+1}{s^2+1}\} = \frac{1}{2}\mathcal{L}^{-1}\{\frac{1}{s+1}\}$$

$$-\frac{1}{2}\mathcal{L}^{-1}\{\frac{s}{s^2+1}\} + \frac{1}{2}\mathcal{L}^{-1}\{\frac{1}{s^2+1}\}$$

$$= \frac{1}{2}e^{-t} - \frac{1}{2}\cos t + \frac{1}{2}\sin t,$$

which is the same result as that obtained for $(f * g)(t)$.

Example 2.1.19. Using the Convolution Theorem find the Laplace transform of the function h defined by

$$h(t) = \int_0^t \sin x \cos(t - x)\, dx.$$

Solution. Notice that $h(t) = (f * g)(t)$, where $f(t) = \cos t$ and $g(t) = \sin t$. Therefore, by the Convolution Theorem,

$$\mathcal{L}\{h(t)\} = \mathcal{L}\{(f * g)(t)\} = (\mathcal{L}\{f(t)\})(\mathcal{L}\{g(t)\})$$

$$= (\mathcal{L}\{\cos t\})(\mathcal{L}\{\sin t\}) = \frac{s}{(s^2+1)^2}.$$

Example 2.1.20. Find the solution of the initial-value problem

$$y''(t) + 16y = f(t), \quad y(0) = 5,\ y'(0) = -5,$$

where f is a given function.

Solution. Let $\mathcal{L}\{y\} = Y$ and $\mathcal{L}\{f\} = F$. By taking the Laplace transform of both sides of the differential equation and using the initial conditions, we obtain

$$s^2 Y(s) - 5s + 5 + 16Y(s) = F(s).$$

Solving for $Y(s)$ we obtain

$$Y(s) = \frac{5s - 5}{s^2 + 16} + \frac{F(s)}{s^2 + 16}.$$

Therefore, by linearity and the Convolution Theorem of the Laplace transform it follows that

$$y(t) = 5\mathcal{L}^{-1}\{\frac{s}{s^2+16}\} - 5\mathcal{L}^{-1}\{\frac{1}{s^2+16}\} + \mathcal{L}^{-1}\{\frac{F(s)}{s^2+16}\}$$

$$= 5\cos 4t - \frac{5}{4}\sin 4t + \left(\mathcal{L}^{-1}\{\frac{1}{s^2+16}\} * f\right)(t)$$

$$= 5\cos 4t - \frac{5}{4}\sin 4t + \frac{1}{4}\left(\sin 4t * f(t)\right)$$

$$= 5\cos 4t - \frac{5}{4}\sin 4t + \frac{1}{4}\int_0^t f(t - x)\sin 4x\, dx.$$

The Convolution Theorem is very helpful also in solving *integral equations.*

Example 2.1.21. Use the Convolution Theorem to solve the following integral equation.

$$y(t) = 4t + \int_0^t y(t - x) \sin x \, dx.$$

Solution. If $f(t) = \sin t$, then the integral equation can be written as

$$y(t) = 4t + (y * f)(t).$$

Therefore, if $Y = \mathcal{L}\{y\}$ and we apply the Laplace transform to both sides of the equation, and using the Convolution Theorem, we obtain

$$Y(s) = \frac{4}{s^2} + \frac{Y(s)}{s^2 + 1}.$$

Solving for $Y(s)$, we have

$$Y(s) = 4 \frac{s^2 + 1}{s^4} = \frac{4}{s^2} + \frac{4}{s^4}.$$

By computing the inverse Laplace transform, we find

$$y(t) = 4t + \frac{2}{3} t^3.$$

Exercises for Section 2.1.4.

1. Establish the commutative, distributive and associative properties of the convolution.

2. Compute the convolution $f * g$ of the following functions.

   (a) $f(t) = 1$, $g(t) = t^2$.

   (b) $f(t) = e^{-3t}$, $g(t) = 2$.

   (c) $f(t) = t$, $g(t) = e^{-t}$.

   (d) $f(t) = \cos t$, $g(t) = \sin t$.

   (e) $f(t) = t$, $g(t) = H(t - 1) - H(t - 2)$.

3. Find the Laplace transform of each of the following functions.

(a) $f(t) = \int\limits_{0}^{t} (t - x)^2 \cos 2x\, dx$

(b) $f(t) = \int\limits_{0}^{t} e^{-(t-x)} \sin x\, dx$

(c) $f(t) = \int\limits_{0}^{t} (t - x)e^x\, dx$

(d) $f(t) = \int\limits_{0}^{t} \sin (t - x) \cos x\, dx$

(e) $f(t) = \int\limits_{0}^{t} e^x (t - x)^2, dx$

(f) $f(t) = \int\limits_{0}^{t} e^{-(t-x)} \sin^2 x\, dx$

4. Find the inverse Laplace transform of each of the following functions by using the Convolution Theorem.

(a) $F(s) = \dfrac{1}{s^2(s + 1)}$

(b) $F(s) = \dfrac{s^2}{(s^2 + 1)^2}$

(c) $F(s) = \dfrac{1}{(s^2 + 4)(s + 1)}$

(d) $F(s) = \dfrac{1}{s^4(s + ^2 + 16)}$

(e) $F(s) = \dfrac{s}{(s^2 + 4)(s + 1)}$

5 Solve the given integral equations using Laplace transforms.

(a) $y(t) - 4t = -3 \int\limits_{0}^{t} y(x) \sin (t - x)\, dx$

(b) $y(t) = \dfrac{t^2}{2} - \int\limits_{0}^{t} (t - x)y(x)\, dx$

(c) $y(t) - e^{-t} = - \int\limits_{0}^{t} y(x) \cos (t - x)\, dx$

(d) $y(t) = t^3 + \int\limits_{0}^{t} y(t - x)x \sin x\, dx$

(e) $y(t) = 1 + 2 \int\limits_{0}^{t} e^{-2(t-x)}y(x)\, dx$

6 Solve the given integro-differential equations using Laplace transforms.

(a) $y'(t) - t = \int\limits_0^t y(x) \cos{(t-x)}dx, \; y(0) = 4.$

(b) $\int\limits_0^t \frac{y'(\tau)}{\sqrt{t-\tau}} d\tau = 1 - 2\sqrt{t}, \; y(0) = 0.$

2.2 Fourier Transforms.

In this section we study Fourier transforms, which provide integral representations of functions defined on either the whole real line $(-\infty, \infty)$ or on the half-line $(0, \infty)$. In Chapter 1, Fourier series were used to represent a function f defined on a bounded interval $(-L, L)$ or $(0, L)$. When f and f' are piecewise continuous on an interval, the Fourier series represents the function on that interval and converges to the periodic extension of f outside that interval. We will now derive, in a non-rigorous manner, a method of representing some non-periodic functions defined either on the whole line or on the half-line by some integrals.

Fourier transforms are of fundamental importance in a broad range of applications, including both ordinary and partial differential equations, quantum physics, signal and image processing and control theory.

We begin this section by investigating how the Fourier series of a given function behaves as the length $2L$ of the interval $(-L, L)$ goes to infinity.

2.2.1 Definition of Fourier Transforms.

Suppose that a "reasonable" function f, defined on $\mathbb{R}$, is $2L$-periodic. From Chapter 1 we know that the complex Fourier series expansion of f on the interval $(-L, L)$ is

$$S(f, x) = \sum_{n=-\infty}^{\infty} c_n e^{i\frac{n\pi}{L}x},$$

where

$$c_n = \frac{1}{2L} \int\limits_{-L}^{L} f(x)e^{-i\frac{n\pi}{L}x} \, dx.$$

If we introduce the quantity

$$\omega_n = \frac{n\pi}{L}, \quad \Delta\omega_n = \omega_{n+1} - \omega_n = \frac{\pi}{L},$$

then

(2.2.1) $$S(f, x) = \frac{1}{2\pi} \sum_{n=-\infty}^{\infty} F(\omega_n)e^{i\omega_n x} \Delta\omega_n,$$

where

$$F(\omega_n) = \int_{-L}^{L} f(x)e^{-i\omega_n x}\,dx.$$

Now, we expand the interval $(-L, L)$ by letting $L \to \infty$ in such a way so that $\Delta\omega_n \to 0$. Notice that the sum in (2.2.1) is very similar to the Riemann sum of the improper integral

$$\frac{1}{2\pi} \int_{-\infty}^{\infty} F(\omega)e^{i\omega x}\,d\omega,$$

where

$$(2.2.2) \qquad F(\omega) = \int_{-\infty}^{\infty} f(x)e^{-i\omega x}\,dx.$$

Therefore, we have

$$(2.2.3) \qquad S(f, x) = \frac{1}{2\pi} \int_{-\infty}^{\infty} F(\omega)\,d\omega.$$

The function F in (2.2.2) is called the *Fourier transform* of the function f on $(-\infty, \infty)$. The above discussion naturally leads to the following definition.

Definition 2.2.1. Let $f : \mathbb{R} \to \mathbb{R}$ be an integrable function. The *Fourier transform* of f, denoted by $F = \mathcal{F}\{f\}$, is given by the integral

$$(2.2.4) \qquad F(\omega) = (\mathcal{F}\{f\})(\omega) = \int_{-\infty}^{\infty} f(x)e^{-i\omega x}\,dx,$$

for all $x \in \mathbb{R}$ for which the improper integral exists.

Remark. Notice that the placement of the factor 1 in (2.2.4) is arbitrary. In some literature, instead, it is chosen to be $\frac{1}{2\pi}$:

$$F(\omega) = (\mathcal{F}\{f\})(\omega) = \frac{1}{2\pi} \int_{-\infty}^{\infty} f(x)e^{-i\omega x}\,dx, \quad S(f, x) = \int_{-\infty}^{\infty} F(\omega)\,d\omega.$$

In this book we will use the notation given in (2.2.4).

Remark. Other common notations for $\mathcal{F}\{f\}$ are $\tilde{f}$ and $\hat{f}$.

For a given Fourier transform F, the function (if it exists) given by

$$\frac{1}{2\pi} \int_{-\infty}^{\infty} F(\omega)e^{i\omega x}\,d\omega$$

is called the *inverse Fourier transform*, and it is denoted by $\mathcal{F}^{-1}$.

Similarly as in a Fourier series we have the following Dirichlet condition for existence of the inverse Fourier transform.

Theorem 2.2.1. *Suppose that* $f : \mathbb{R} \to \mathbb{R}$ *is an integrable function and satisfies the Dirichlet condition: on any finite interval, the functions* f *and* f' *are piecewise continuous with finitely many maxima and minima. If* $F = \mathcal{F}\{f\}$, *then*

$$\frac{f(x^+) + f(x^-)}{2} = \frac{1}{2\pi} \int_{-\infty}^{\infty} F(\omega)e^{i\omega x} \, d\omega.$$

Remark. Notice that, if the integrable function f is even or odd, then the Fourier transform $\mathcal{F}\{f\}$ of f is even or odd, respectively, and

$$\mathcal{F}\{f\}(\omega) = 2 \int_0^{\infty} f(x) \cos \omega x \, dx, \quad \text{if } f \text{ is even}$$

and

$$\mathcal{F}\{f\}(\omega) = 2i \int_0^{\infty} f(x) \sin \omega x \, dx, \quad \text{if } f \text{ is odd}.$$

This suggests the following definition.

Definition 2.2.2. Suppose that f is an integrable function on $(0, \infty)$. Define *Fourier cosine transform* $\mathcal{F}_c\{f\}$ and *Fourier sine transform* $\mathcal{F}_s\{f\}$ by

$$\mathcal{F}_c\{f\}(\omega) = \int_0^{\infty} f(x) \cos \omega x \, dx, \quad \mathcal{F}_s\{f\}(\omega) = \int_0^{\infty} f(x) \sin \omega x \, dx.$$

The inversion formulas for the Fourier cosine and Fourier sine transforms are

$$f(x) = \frac{1}{2\pi} \int_0^{\infty} \mathcal{F}_c\{f\}(\omega) \cos \omega x \, dx$$

$$= \frac{1}{2\pi} \int_0^{\infty} \mathcal{F}_s\{f\}(\omega) \sin \omega x \, dx.$$

Let us take several examples.

Example 2.2.1. Find the Fourier transform of $f(x) = e^{-|x|}$ and hence using Theorem 2.2.1, deduce that

$$\int_0^{\infty} \frac{\omega \sin(\omega x)}{1 + x^2} \, d\omega = \frac{\pi}{2} e^{-a}, \quad x > 0.$$

Solution. By the definition of the Fourier transform we have

$$F(\omega) = \int_{-\infty}^{\infty} f(x)e^{-i\omega x}\, dx = \int_{-\infty}^{0} e^{-x}e^{-i\omega x}\, dx + \int_{0}^{\infty} e^{x}e^{-i\omega x}\, dx$$

$$= \int_{-\infty}^{0} e^{x(1-i\omega)}\, dx + \int_{0}^{\infty} e^{-x(1+i\omega)}\, dx$$

$$= \left[\frac{e^{x(1-i\omega)}}{1-i\omega}\right]_{x=-\infty}^{x=0} + \left[\frac{e^{-x(1+i\omega)}}{-(1+i\omega)}\right]_{x=0}^{x=\infty} = \frac{2}{1+\omega^2}.$$

Since the function $f(x) = e^{-|x|}$ is continuous on the whole real line, by the inversion formula in Theorem 2.2.1 we have

$$e^{-|x|} = \frac{1}{2\pi} \int_{-\infty}^{\infty} F(\omega)e^{i\omega x}\, d\omega = \frac{1}{2\pi} \int_{-\infty}^{\infty} 2\frac{e^{i\omega x}}{1+\omega^2}\, d\omega$$

$$= \frac{1}{2\pi} \int_{-\infty}^{0} 2\frac{e^{i\omega x}}{1+\omega^2}\, d\omega + \frac{1}{2\pi} \int_{0}^{\infty} 2\frac{e^{i\omega x}}{1+\omega^2}\, d\omega$$

$$= \frac{1}{\pi} \int_{0}^{\infty} \frac{e^{-i\omega x}}{1+\omega^2}\, d\omega + \frac{1}{\pi} \int_{0}^{\infty} \frac{e^{i\omega x}}{1+\omega^2}\, d\omega$$

$$= \frac{1}{\pi} \int_{0}^{\infty} \frac{e^{i\omega x} + e^{-i\omega x}}{1+\omega^2}\, d\omega = \frac{2}{\pi} \int_{0}^{\infty} \frac{\cos(\omega x)}{1+\omega^2}.$$

Example 2.2.2. Find the Fourier transform of $f(x) = e^{-x^2}$.

Solution. By the definition of the Fourier transform we have

$$F(\omega) = \int_{-\infty}^{\infty} f(x)e^{-i\omega x}\, dx = \int_{-\infty}^{\infty} e^{-x^2}e^{-i\omega x}\, dx = \int_{-\infty}^{\infty} e^{-x^2-i\omega x}\, dx$$

$$= \int_{-\infty}^{\infty} e^{-(x+\frac{i\omega}{2})^2 - \frac{\omega^2}{4}}\, dx = e^{-\frac{\omega^2}{4}} \int_{-\infty}^{\infty} e^{-(x+\frac{i\omega}{2})^2}\, dx$$

$$= e^{-\frac{\omega^2}{4}} \int_{-\infty}^{\infty} e^{-u^2}\, du = \sqrt{\pi}e^{-\frac{\omega^2}{4}}.$$

Example 2.2.3. Find the Fourier transform of the following step function.

$$f(x) = \begin{cases} 1, & |x| < a \\ 0, & |x| > a. \end{cases}$$

Solution. Let $F = \mathcal{F}\{f\}$. From the definition of the Fourier transform we have

$$F(\omega) = \int_{-\infty}^{\infty} f(x)e^{-i\omega x}\, dx = \int_{-a}^{a} e^{-i\omega x}\, dx$$

$$= \frac{e^{\omega a i} - e^{-\omega a i}}{\omega i} = 2\frac{\sin \omega a}{\omega}.$$

Example 2.2.4. Find the Fourier transform of

$$\frac{1}{a^2 + x^2},$$

where a is a positive constant.

Solution. Let F be the Fourier transform of f. From the definition of the Fourier transform,

$$F(\omega) = \int_{-\infty}^{\infty} \frac{e^{-i\omega x}}{a^2 + x^2}\, dx.$$

To evaluate the above improper integral, we evaluate the following contour integral.

$$I(R) = \int_{\gamma} \frac{e^{-i\omega z}}{a^2 + z^2}\, dz,$$

where $\gamma = C_R \cup [-R, R]$ is the closed contour formed by the upper semicircle $C_R = \{z \in \mathbb{C} : z = Re^{i\theta},\ 0 \le \theta \le \pi\}$ and the interval $[-R, R]$. See Figure 2.2.1. Let us consider first the case when $\omega < 0$.

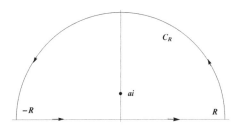

Figure 2.2.1

If R is large enough, the only singularity (simple pole) of the function

$$f(z) = \frac{e^{-i\omega z}}{a^2 + z^2}$$

inside the contour $C_R \cup [-R, R]$ is the point $z_0 = ai$. Therefore, by the Cauchy Residue Theorem (see Appendix F) we have

$$I(R) = 2\pi i \, Res\big(f(z), z_0 = ai\big).$$

From

$$Res\left(\frac{e^{-i\omega z}}{a^2 + z^2}, \, z_0 = ai\right) = \frac{e^{-i\omega ai}}{2ai} = -\frac{i}{2a}e^{\omega a}$$

we have

(2.2.5)
$$I(R) = \frac{\pi}{a}e^{\omega a}.$$

We decompose the integral $I(R)$ along the semicircle C_R and the interval $[-R, R]$:

(2.2.6)
$$I(R) = \int_{C_R} \frac{e^{-i\omega z}}{a^2 + z^2}\, dz + \int_{-R}^{R} \frac{e^{-i\omega x}}{a^2 + x^2}\, dx.$$

Denote the first integral in (2.2.6) by I_{C_R}. For the integral I_{C_R} along the semicircle C_R we have

$$I_{C_R} = \left| \int_0^\pi \frac{e^{-i\omega Re^{i\theta}}}{a^2 + R^2 e^{2i\theta}} Rie^{i\theta}\, d\theta \right| \leq \int_0^\pi \left| \frac{e^{-i\omega Re^{i\theta}}}{a^2 + R^2 e^{2i\theta}} \right| R\, d\theta \leq \int_0^\pi R\frac{e^{R\omega \sin\theta}}{R^2 - a^2}\, d\theta.$$

Since $\omega < 0$ and $\sin\theta > 0$ for $0 < \theta < \pi$ we have

$$\lim_{R\to\infty} R\frac{e^{R\omega \sin\theta}}{R^2 - a^2} = 0$$

and so from the above inequality it follows that

$$\lim_{R\to\infty} I_{C_R} = 0.$$

Therefore from (2.2.6) we have

$$\lim_{R\to\infty} I(R) = \int_{-\infty}^{\infty} \frac{e^{-i\omega x}}{a^2 + x^2}\, dx,$$

so (2.2.5) implies

$$(2.2.7) \qquad \int\limits_{-\infty}^{\infty} \frac{e^{-i\omega x}}{a^2 + x^2}\, dx = \frac{\pi}{a} e^{\omega a}.$$

For the case when $\omega > 0$, working similarly as above, but integrating along the lower semicircle $z = Re^{i\theta}$, $\pi \leq \theta \leq 2\pi$, we obtain

$$(2.2.8) \qquad \int\limits_{-\infty}^{\infty} \frac{e^{-i\omega x}}{a^2 + x^2}\, dx = \frac{\pi}{a} e^{-\omega a}.$$

Therefore, from (2.2.8) and (2.2.7) it follows that

$$\mathcal{F}\{f\}(\omega) = \frac{\pi}{a} e^{-|\omega| a}.$$

In the treatment of Fourier transforms until now we considered only functions that are integrable on the real line $\mathbb{R}$. It is an obvious fact that such simple functions as cosine, sine and the Heaviside or the constant functions, even though bounded, are not integrable. Does this mean that these functions do not possess a Fourier transform? In the rest of this section we will try to overcome this difficulty and make some sense of Fourier transforms of such functions. We start with the Dirac delta "function" $\delta(x)$:

$$\delta(x) = \begin{cases} 0, & x \neq 0 \\ \infty, & x = 0. \end{cases}$$

This function, introduced in Section 2.1.2 was defined as the "limit" of the following sequence of functions.

$$f_n(x) = \begin{cases} \frac{n}{2}, & |x| \leq \frac{1}{n} \\ 0, & |x| > \frac{1}{n}. \end{cases}$$

For these integrable functions f_n we have

$$\mathcal{F}\{f_n\}(\omega) = \int\limits_{-\infty}^{\infty} f_n(x) e^{-i\omega x}\, dx = \frac{n}{2} \int\limits_{-\frac{1}{n}}^{\frac{1}{n}} e^{-i\omega x}\, dx = \frac{\sin\left(\frac{\omega}{n}\right)}{\omega} n.$$

Notice that

$$\lim_{n \to \infty} \frac{n \sin\left(\frac{\omega}{n}\right)}{\omega} = 1.$$

Therefore, it is natural to define

(2.2.9) $$\mathcal{F}\{\delta\}(\omega) = \lim_{n\to\infty} \mathcal{F}\{f_n\}(\omega) = 1.$$

In Section 2.1.2, when we discussed the Dirac delta function we "stated" that

$$\int_{-\infty}^{\infty} \delta(x-a)f(x)\,dx = \int_{-\infty}^{\infty} \delta_a f(x)\,dx = f(a),$$

for every continuous function f. In particular, we have

(2.2.10) $$\int_{-\infty}^{\infty} \delta_a(x)e^{-i\omega x}\,dx = e^{-i\omega a},$$

and therefore

(2.2.11) $$\mathcal{F}\{\delta_a\}(\omega) = e^{-i\omega a}.$$

For the inverse Fourier transform, from (2.2.10) it follows that

$$\int_{-\infty}^{\infty} \delta_{\omega_0}(\omega)e^{ix\omega}\,d\omega = e^{i\omega_0 x}.$$

Hence

$$\mathcal{F}^{-1}\{\delta_{\omega_0}\}(x) = \frac{1}{2\pi}e^{i\omega_0 x},$$

that is,

(2.2.12) $$\mathcal{F}\{e^{i\omega_0 x}\}(\omega) = 2\pi\delta(\omega - \omega_0).$$

If we set $\omega_0 = 0$ we have the following result.

$$\mathcal{F}\{1\}(\omega) = 2\pi\delta(\omega).$$

Example 2.2.5. Find the Fourier transforms of the functions

$$\sin\omega_0 x, \quad \cos\omega_0 x, \ \omega_0 \neq 0, \ x \in \mathbb{R}.$$

Solution. We will compute the Fourier transform of $\sin\omega_0 x$, leaving the other function as an exercise. The function $\sin\omega_0 x$ is not integrable, and so we need to use the Dirac delta function. From Euler's formula, the linearity property of the Fourier transform, and (2.2.12) we have

$$\mathcal{F}\{\sin\omega_0 x\}(\omega) = \frac{1}{2i}\left[\mathcal{F}\{e^{i\omega_0 x}\}(\omega) - \mathcal{F}\{e^{-i\omega_0 x}\}(\omega)\right]$$
$$= -\pi i\left[\delta(\omega - \omega_0) - \delta(\omega + \omega_0)\right].$$

Example 2.2.6. By approximation, find the Fourier transforms of the sign function $sgn(x)$, where

$$sgn(x) = \begin{cases} 1, & x > 0 \\ 0, & x = 0 \\ -1, & x < 0. \end{cases}$$

Solution. The function $sgn(x)$ is not integrable on $\mathbb{R}$, but the function

$$f_\epsilon(x) = e^{-\epsilon|x|} sgn(x),$$

where $\epsilon > 0$, is integrable. Indeed

$$\int_{-\infty}^{\infty} \left| e^{-\epsilon|x|} sgn(x) \right| dx = 2 \int_0^{\infty} e^{-\epsilon x} dx = \frac{2}{\epsilon}.$$

Also, for the function f_ϵ we have

$$\lim_{\epsilon \to 0} f_\epsilon(x) = sgn(x), \qquad x \in \mathbb{R}.$$

First, let us find the Fourier transform F_ϵ of the function f_ϵ.

$$F_\epsilon = \mathcal{F}\{f_\epsilon\}(\omega) = \int_{-\infty}^{0} e^{-\epsilon(-x)} \cdot (-1) \cdot e^{-i\omega x} dx + \int_0^{\infty} e^{-\epsilon x} \cdot 1 \cdot e^{-i\omega x} dx$$

$$= -\left[\frac{1}{\epsilon - i\omega} e^{(\epsilon - i\omega)x} \right]_{x=-\infty}^{x=0} + \left[\frac{1}{-\epsilon - i\omega} e^{(-\epsilon - i\omega x)} \right]_{x=0}^{x=\infty}$$

$$= -\frac{1}{\epsilon - i\omega} - \frac{1}{-\epsilon - i\omega} = -\frac{2i\omega}{\epsilon^2 + \omega^2}.$$

Therefore,

$$\mathcal{F}\{sgn\}(\omega) = \lim_{\epsilon \to 0} F_\epsilon(\omega) = -\lim_{\epsilon \to 0} \frac{2i\omega}{\epsilon^2 + \omega^2}.$$

If $\omega = 0$, the above lim equals 0, and if $\omega \neq 0$, the above lim equals $\dfrac{2}{i\omega}$. Therefore we conclude that

$$\mathcal{F}\{sgn\}(\omega) = \begin{cases} \frac{2}{i\omega}, & \omega \neq 0 \\ 0, & \omega = 0. \end{cases}$$

Exercises for Section 2.2.1.

1. Find the Fourier transform of the following functions and for each apply Theorem 2.2.1.

   (a) $f(x) = \begin{cases} 0, & x < -1 \\ -1, & -1 < x < 1 \\ 2, & x > 1. \end{cases}$

   (b) $f(x) = \begin{cases} 0, & x < 0 \\ x, & 0 < x < 3 \\ 0, & x > 3. \end{cases}$

   (c) $f(x) = \begin{cases} 0, & x < 0 \\ e^{-x}, & x > 0. \end{cases}$

   (d) $f(x) = e^{-\frac{ax^2}{2}}$, $a > 0$.

2. For given $a > 0$, let

$$f(x) = \begin{cases} e^{-x}x^{a-1}, & x > 0 \\ 0, & x \le 0. \end{cases}$$

   Show that
$$\mathcal{F}\{f\}(\omega) = \Gamma(a)(1 + i\omega)^{-a}.$$

3. Show that the Fourier transform of

$$f(x) = \begin{cases} e^{2x}, & x < 0 \\ e^{-2x}, & x > 0 \end{cases}$$

   is

$$\mathcal{F}\{f\}(\omega) = \frac{3}{(2 - i\omega)(1 + i\omega)}.$$

4. Show that the Fourier transform of

$$f(x) = \begin{cases} \cos(ax), & |x| < 1 \\ 0, & |x| > 1 \end{cases}$$

   is

$$\mathcal{F}\{f\}(\omega) = \frac{\sin(\omega - a)}{\omega - a} + \frac{\sin(\omega + a)}{\omega + a}.$$

5. Let g be a function defined on $[0, \infty)$, and let its Laplace transform $\mathcal{L}\{g\}$ exist. On $(-\infty, \infty)$ define the function f by

$$f(t) = \begin{cases} 0, & t > 0 \\ g(t), & t \geq 0. \end{cases}$$

Show formally that

$$\mathcal{L}\{g\}(\omega) = F(-i\omega),$$

where $F(\omega) = \mathcal{F}\{f\}(\omega)$.

6. Find the Fourier transform of

$$f(x) = \begin{cases} \cos x, & |x| < \pi \\ 0, & |x| > \pi, \end{cases}$$

and, using, the result in Theorem 2.2.1, evaluate the integral

$$\int\limits_0^\infty \frac{x \sin(\pi x) \cos(xy)}{1 - x^2} \, dx.$$

7. If $a > 0$ is a constant, then compute the following Fourier transforms.

(a) $\mathcal{F}_c\{e^{-ax}\}$.

(b) $\mathcal{F}_s\{e^{-ax}\}$.

(c) $\mathcal{F}_s\{xe^{-ax}\}$.

(d) $\mathcal{F}_c\{(1+x)e^{-ax}\}$.

8. Find the Fourier transform of the Heaviside unit step function u_0:

$$u_0(x) = \begin{cases} 1, & x > 0 \\ 0, & x < 0. \end{cases}$$

9. Verify that

(a) $\mathcal{F}\{\sin(\omega_0 x)u_0(x)\}(\omega) = \dfrac{\omega_0}{\omega_0^2 - \omega^2} \dfrac{\pi i}{2}[\delta(\omega + \omega_0) - \delta(\omega - \omega_0)],$

and

(b) $\mathcal{F}\{\cos(\omega_0 x)u_0(x)\}(\omega) = \dfrac{i\omega_0}{\omega_0^2 - \omega^2} \dfrac{\pi i}{2}[\delta(\omega + \omega_0) - \delta(\omega - \omega_0)].$

2.2.2. Properties of Fourier Transforms.

In principle, we can compute the Fourier transform of any function from the definition. However, it is much more efficient to derive some properties of the Fourier transform. This is the purpose of this section.

The following theorem establishes the linearity property of the Fourier transform.

Theorem 2.2.2 Linear Property. *If f_1 and f_2 are integrable functions on $\mathbb{R}$, then*

$$\mathcal{F}\{c_1 f_1 + c_2 f_2\} = c_1 \mathcal{F}\{f_1\} + c_2 \mathcal{F}\{f_2\}$$

for any constants c_1 and c_2.

Proof. This result follows directly from the definition of the Fourier transform and the linearity property of integrals.   ∎.

The next theorem summarizes some of the more important and useful properties of the Fourier transform.

Theorem 2.2.3. *Suppose that f and g are integrable functions on $\mathbb{R}$. Then*

(a) *For any $c \in \mathbb{R}$,* $\mathcal{F}\{f(x-c)\}(\omega) = e^{-ic\omega}\mathcal{F}\{f\}(\omega)$ *(Translation).*

(b) *For any $c \in \mathbb{R}$,* $\mathcal{F}\{e^{icx}f(x)\}(\omega) = \mathcal{F}\{f\}(\omega - c)$ *(Modulation).*

(c) *If $c > 0$ and $f_c(x) = \frac{1}{c}f\left(\frac{x}{c}\right)$, then $\mathcal{F}\{f_c\}(\omega) = \mathcal{F}\{f\}(c\omega)$ (Dilation).*

(d) $\mathcal{F}\{f * g\} = \left(\mathcal{F}\{f\}\right)\left(\mathcal{F}\{g\}\right)$ *(Convolution).*

(e) *If f is continuous and piecewise smooth and f' is integrable, then*
$\mathcal{F}\{f'\}(\omega) = i\omega\mathcal{F}\{f\}(\omega)$ *(Differentiation)*

(f) *If also $xf(x)$ is integrable, then $\mathcal{F}\{xf(x)\}(\omega) = i\left(\mathcal{F}\{f\}\right)'(\omega)$.*

Proof. The proof of part (a) we leave as an exercise.
For part (b), we have

$$\mathcal{F}\{e^{icx}f(x)\}(\omega) = \int_{-\infty}^{\infty} e^{icx}f(x)e^{-i\omega x}\,dx = \int_{-\infty}^{\infty} e^{-ix(\omega - c)}f(x)\,dx$$

$$= \mathcal{F}\{f\}(\omega - c).$$

For part (c), by direct computation we find

$$\mathcal{F}\{f_c\}(\omega) = \int_{-\infty}^{\infty} \frac{1}{c}f\left(\frac{x}{c}\right)e^{-i\omega x}\,dx \qquad \text{(substitution } x = cy\text{)}$$

$$= \int_{-\infty}^{\infty} f(y)e^{-ic\omega y}\,dy = \mathcal{F}\{f\}(c\omega).$$

For part (d) we have

$$\mathcal{F}\{f * g\}(\omega) = \int_{-\infty}^{\infty} \left(\int_{-\infty}^{\infty} f(x-y)g(y)\,dy \right) e^{-ix\omega}\,dx$$

$$= \int_{-\infty}^{\infty} \left(\int_{-\infty}^{\infty} f(x-y)g(y)e^{-ix\omega}\,dx \right) dy$$

$$= \int_{-\infty}^{\infty} \left(\int_{-\infty}^{\infty} f(z)e^{-i(z+y)\omega}\,dz \right) g(y)\,dy \quad (\ z = x - y)$$

$$= \int_{-\infty}^{\infty} \left(\int_{-\infty}^{\infty} f(z)e^{-iz\omega}\,dz \right) g(y)e^{-iy\omega}\,dy$$

$$= \int_{-\infty}^{\infty} \mathcal{F}\{f\}(\omega)g(y)e^{-iy\omega}\,dy$$

$$= \mathcal{F}\{f\}(\omega) \int_{-\infty}^{\infty} g(y)e^{-iy\omega}\,dy$$

$$= \left(\mathcal{F}\{f\}(\omega)\right) \left(\mathcal{F}\{g\}(\omega)\right).$$

In the proof (e) we will use the fact that

$$\lim_{x \to \pm\infty} f(x) = 0$$

for every integrable function f on $\mathbb{R}$. Now, based on this fact, by direct computation and the integration by parts formula we have

$$\mathcal{F}\{f'\}(\omega) = \int_{-\infty}^{\infty} f'(x)e^{-i\omega x}\,dx = \left[f(x)e^{-i\omega x} \right]_{x=-\infty}^{x=\infty}$$

$$- \int_{-\infty}^{\infty} (-i\omega)f(x)e^{-i\omega x}\,dx$$

$$= 0 + i\omega \int_{-\infty}^{\infty} f(x)e^{-i\omega x}\,dx = i\omega\mathcal{F}\{f\}(\omega).$$

For the last part (f),

$$\mathcal{F}\{xf(x)\}(\omega) = \int_{-\infty}^{\infty} xf(x)e^{-i\omega x}\,dx = \int_{-\infty}^{\infty} if(x)\frac{d}{d\omega}\left(e^{-i\omega x}\right)dx$$

$$= i\frac{d}{d\omega}\left(\int_{-\infty}^{\infty} f(x)e^{-i\omega x}\,dx \right) = i\left(\mathcal{F}\{f\}\right)'(\omega). \quad \blacksquare$$

Remark. Parts (e) and (f) in the above theorem can be generalized as follows:

(e') If f and the derivatives $f', \cdots, f^{(n-1)}$ are continuous on $\mathbb{R}$ and $f^{(n)}$ is piecewise continuous and integrable on $\mathbb{R}$, then

$$\mathcal{F}\{f^{(n)}\}(\omega) = (i\omega)^n \mathcal{F}\{f\}.$$

(f') If f is an integrable function on $\mathbb{R}$, such that the function $x^n f(x)$ is also integrable, then

$$\mathcal{F}\{x^n f(x)\}(\omega) = i^n \left(\mathcal{F}\{f\}\right)^{(n)}(\omega).$$

For the Fourier sine and Fourier cosine transform we have the following theorem.

Theorem 2.2.4. *Suppose that f is continuous and piecewise smooth and that f and f' are integrable functions on $(0, \infty)$. Then*

(a) $\mathcal{F}_c\{f'\}(\omega) = \omega \mathcal{F}_s\{f\}(\omega) - f(0).$

(b) $\mathcal{F}_s\{f'\}(\omega) = -\omega \mathcal{F}_c\{f\}(\omega).$

Proof. We prove only (a), leaving (b) as an exercise.

Since f is an integrable function on $(0, \infty)$ we have

$$\lim_{x \to \infty} f(x) = 0.$$

Using this condition and the integration by parts formula we have

$$\mathcal{F}_c\{f'\}(\omega) = \int_0^\infty f'(x) \cos \omega x \, dx = \left[f(x) \cos \omega x \right]_{x=0}^{x=\infty}$$

$$+ \omega \int_0^\infty f(x) \sin \omega x \, dx$$

$$= -f(0) + \omega \mathcal{F}_s\{f\}(\omega). \qquad \blacksquare$$

As a consequence of this theorem we have the following.

Corollary 2.2.1. *Suppose that the functions f and f' are continuous and piecewise smooth and that f and f' are integrable on $(0, \infty)$. Then*

(a) $\mathcal{F}_s\{f''\}(\omega) = -\omega^2 \mathcal{F}_s\{f\}(\omega) + \omega f(0).$

(b) $\mathcal{F}_c\{f''\}(\omega) = -\omega^2 \mathcal{F}_c\{f\}(\omega) - f'(0).$

Let us take now several examples.

Example 2.2.7. Let

$$f(x) = \begin{cases} 1, & |x| \le 1 \\ 0, & |x| > 1 \end{cases}$$

and

$$g(x) = \begin{cases} |x|, & |x| \le 2 \\ 0, & |x| > 2. \end{cases}$$

Compute the Fourier transform $\mathcal{F}\{f * g\}$.

Solution. First we find $\mathcal{F}\{f\}$ and $\mathcal{F}\{g\}$. In Example 2.2.3 we have already found that

$$\mathcal{F}\{f\}(\omega) = 2\frac{\sin \omega}{\omega}.$$

For $\mathcal{F}\{g\}$ we have

$$\mathcal{F}\{g\}(\omega) = \int_{-\infty}^{\infty} g(x)e^{-i\omega x}\, dx = \int_{-2}^{0} (-x)e^{-i\omega x}\, dx + \int_{0}^{2} xe^{-i\omega x}\, dx$$

$$= \left[\left(\frac{1}{\omega^2} + \frac{ix}{\omega} \right) e^{-i\omega x} \right]_{x=-2}^{x=0} + \left[\left(\frac{1}{\omega^2} + \frac{ix}{\omega} \right) e^{-i\omega x} \right]_{x=0}^{x=2}$$

$$= \frac{2\cos 2\omega - 2}{\omega^2}.$$

Thus, by the convolution property of the Fourier transform in Theorem 2.2.3,

$$\mathcal{F}\{f * g\}(\omega) = \left(\mathcal{F}\{f\}(\omega) \right) \left(\mathcal{F}\{g\}(\omega) \right)$$

$$= 2\frac{\sin \omega}{\omega} \left[\frac{2\cos 2\omega - 2}{\omega^2} \right].$$

Example 2.2.8. Find

$$\mathcal{F}^{-1} \left\{ \frac{e^{-\frac{\omega^2}{4}}}{1 + \omega^2} \right\}.$$

Solution. From Examples 2.2.1 and 2.2.2 we have

$$\mathcal{F}\{e^{-|x|}\}(\omega) = \frac{2}{1 + \omega^2} \quad \text{and} \quad \mathcal{F}\{e^{-x^2}\}(\omega) = \sqrt{\pi}e^{-\frac{\omega^2}{4}}.$$

Therefore, from the convolution property of the Fourier transform

$$\mathcal{F}^{-1} \left\{ \frac{e^{-\frac{\omega^2}{4}}}{1 + \omega^2} \right\}(x) = \mathcal{F}^{-1} \left\{ e^{-\frac{\omega^2}{4}} \cdot \frac{1}{1 + \omega^2} \right\}(x)$$

$$= \left(\mathcal{F}^{-1} \left\{ e^{-\frac{\omega^2}{4}} \right\}(x) \right) * \left(\mathcal{F}^{-1} \left\{ \frac{1}{1 + \omega^2} \right\}(x) \right)$$

$$= \frac{1}{\sqrt{\pi}} \left(e^{-x^2} * \frac{1}{2}e^{-|x|} \right)$$

$$= \frac{1}{2\sqrt{\pi}} \int_{-\infty}^{\infty} e^{-(x-y)^2} e^{-|y|}\, dy.$$

Example 2.2.9. Using properties (e) and (f) in Theorem 2.2.3 find the Fourier transform of the function

$$y(x) = e^{-x^2}.$$

Solution. The function $y(x) = e^{-x^2}$ satisfies the following differential equation.

$$y'(x) + 2xy(x) = 0.$$

Applying the Fourier transform to both sides of this equation, with $\mathcal{F}\{y\} = Y$, from (e) and (f) of Theorem 2.2.3 it follows that

$$i\omega Y(\omega) + 2iY'(\omega) = 0,$$

i.e.,

$$Y'(\omega) + 2\omega Y(\omega) = 0.$$

The general solution of the above equation is

$$Y(\omega) = Ce^{-\frac{\omega^2}{4}}.$$

For the constant C we have

$$C = Y(0) = \int_{-\infty}^{\infty} y(x)e^{-ix\cdot 0}\, dx$$

$$= \int_{-\infty}^{\infty} e^{-x^2}\, dx = \sqrt{\pi}. \quad (\text{See Appendix G})$$

Therefore,

$$Y(\omega) = \sqrt{\pi}e^{-\frac{\omega^2}{4}}.$$

In the definition of the Fourier transform of a function f we required that the function was integrable on the real line. We mention only (without any further discussion) that theory has been developed that allows us to apply the Fourier transform also to functions that are square integrable. The next theorem, even though valid for square integrable functions, will be stated only for functions that are both integrable and square integrable.

Theorem 2.2.5 (Parseval's Formula). *Suppose that f is a function which is both integrable and square integrable on $\mathbb{R}$. If $F = \mathcal{F}\{f\}$ is the Fourier transform of f, then*

$$\int_{-\infty}^{\infty} |f(x)|^2\, dx = \frac{1}{2\pi} \int_{-\infty}^{\infty} |F(\omega)|^2\, d\omega.$$

Proof. From the definition of the inverse Fourier transform

$$f(x) = \frac{1}{2\pi} \int\limits_{-\infty}^{\infty} F(\omega)e^{i\omega x} \, d\omega$$

we have

$$\int\limits_{-\infty}^{\infty} |f(x)|^2 \, dx = \frac{1}{2\pi} \int\limits_{-\infty}^{\infty} f(x) \left(\int\limits_{-\infty}^{\infty} F(\omega)e^{i\omega x} \, d\omega \right) dx$$

$$= \frac{1}{2\pi} \int\limits_{-\infty}^{\infty} F(\omega) \left(\int\limits_{-\infty}^{\infty} f(x)e^{i\omega x} \, dx \right) d\omega$$

$$= \frac{1}{2\pi} \int\limits_{-\infty}^{\infty} F(\omega) \, \overline{F(\omega)} \, d\omega$$

$$= \frac{1}{2\pi} \int\limits_{-\infty}^{\infty} |F(\omega)|^2 \, d\omega. \quad \blacksquare$$

Example 2.2.10. Using the Fourier transform of the unit step function on the interval $(-a, a)$ show that

$$\int\limits_{-\infty}^{\infty} \frac{\sin^2(a\omega)}{\omega^2} \, d\omega = a\pi.$$

Solution. Let $\mathbf{1}_a$ be the unit step function on $(-a, a)$:

$$\mathbf{1}_a(x) = \begin{cases} 1, & |x| < a| \\ 0, & |x| > a. \end{cases}$$

In Example 2.2.3 we showed that

$$\mathcal{F}\{\mathbf{1}_a\}(\omega) = \frac{2\sin(\omega a)}{\omega}.$$

Therefore, by Parseval's formula we have

$$\frac{2}{\pi} \int\limits_{-\infty}^{\infty} \frac{\sin^2(a\omega)}{\omega^2} \, d\omega = \frac{1}{2\pi} \int\limits_{-\infty}^{\infty} |F(\omega)|^2 \, d\omega = \int\limits_{-\infty}^{\infty} |\mathbf{1}_a(x)|^2 \, dx$$

$$= \int\limits_{-a}^{a} 1 \, dx = 2a.$$

Now we examine Poisson's summation theorem, which provides a beautiful and important connection between the Fourier series and the Fourier transform.

Theorem 2.2.6. *Let f be absolutely integrable and continuous on the real line $\mathbb{R}$ with Fourier transform $F = \mathcal{F}\{f\}$. Assume that there exist constants $p > 1$, $M > 0$ and $A > 0$ such that*

$$|f(x)| \le M|x|^{-p} \quad for \ |x| > A.$$

Also assume that the series

$$(2.2.13) \qquad\qquad \sum_{n=-\infty}^{\infty} F(n)$$

converges. Then

$$(2.2.14) \qquad \sum_{n=-\infty}^{\infty} f(2\pi n) = \frac{1}{2\pi} \sum_{n=-\infty}^{\infty} F(n),$$

and both series are absolutely convergent.

Proof. First we define the function g by

$$(2.2.15) \qquad\qquad g(x) = \sum_{n=-\infty}^{\infty} f(x + 2n\pi).$$

If we replace x by $x + 2\pi$ in (2.6.15) we see that we will obtain the same series; therefore g is 2π-periodic. Without proof we state that, based on the conditions of the function f, the function $g(x)$ exists for every x, and it is continuous on $\mathbb{R}$. Now, since g is a continuous and periodic function with a period 2π, it can be represented by the complex Fourier series:

$$(2.2.16) \qquad\qquad g(x) = \sum_{n=-\infty}^{\infty} c_n e^{inx}.$$

If we evaluate $g(0)$ from (2.2.15) and (2.6.16), then we obtain

$$(2.2.17) \qquad \sum_{n=-\infty}^{\infty} f(2n\pi) = \sum_{n=-\infty}^{\infty} c_n.$$

Next, using the definition of the function $g(x)$ we compute the coefficient c_n:
(2.2.18)

$$c_n = \frac{1}{2\pi} \int_{-\pi}^{\pi} g(x) e^{-inx}\, dx = \frac{1}{2\pi} \int_{-\pi}^{\pi} \left(\sum_{n=-\infty}^{\infty} f(x + 2n\pi) \right) e^{-inx}\, dx$$

$$= \frac{1}{2\pi} \sum_{n=-\infty}^{\infty} \int_{-\pi}^{\pi} f(x + 2n\pi) e^{-inx}\, dx \quad \text{(integration term by term)}$$

$$= \frac{1}{2\pi} \sum_{n=-\infty}^{\infty} \int_{2n\pi-\pi}^{2n\pi+\pi} f(t) e^{-in(t-2n\pi)}\, dt \quad \text{(substitution } x = t - 2n\pi)$$

$$= \frac{1}{2\pi} \int_{-\infty}^{\infty} f(t) e^{-int}\, dt = \frac{1}{2\pi} F(n).$$

If we substitute (2.2.18) into (2.2.17), then we obtain

$$\sum_{n=-\infty}^{\infty} f(2\pi n) = \frac{1}{2\pi} \sum_{n=-\infty}^{\infty} F(n). \quad \blacksquare$$

Remark. Using the dilation property (c) of Theorem 2.2.3, Poisson's summation formula (2.2.14) in Theorem 2.2.6 can be written in the following form.

(2.2.19) $$\sum_{n=-\infty}^{\infty} f(cn) = \frac{1}{c} \sum_{n=-\infty}^{\infty} F\left(\frac{2n\pi}{c}\right),$$

where c is any nonzero number.

Among other applications, Poisson's summation formula can be used to evaluate some infinite series.

Example 2.2.11. Using Poisson's summation formula for the function

$$f(x) = \frac{a}{a^2 + x^2}, \quad a > 0$$

evaluate the following infinite series.

$$\sum_{n=-\infty}^{\infty} \frac{1}{a^2 + \omega^2}.$$

Solution. From the result of Example 2.2.4 it follows that

$$\mathcal{F}\{f\}(\omega) = \frac{\pi}{a} e^{-|\omega| a}.$$

First, we check the assumptions in Theorem 2.2.6.

$$\sum_{n=-\infty}^{\infty} F(n) = \sum_{n=-\infty}^{\infty} \pi e^{-|2n| a} = 1 + 2\pi \sum_{n=1}^{\infty} e^{-2na}$$

$$= 1 + 2\pi \frac{e^{-2a}}{1 - e^{-2a}} < \infty$$

Therefore, condition (2.2.13) of Theorem 2.2.6 is satisfied. The condition for the growth of the function f is easily verified. Indeed, for $x \neq 0$, we have

$$f(x) = \frac{1}{a^2 + x^2} \leq \frac{1}{x^2},$$

and so this condition is satisfied taking $p = 2$, $M = 1$, and $A > 0$ to be arbitrary.

Now, if we substitute the following expressions

$$f(n) = \frac{a}{a^2 + n^2}$$

and

$$F(2n\pi) = \pi e^{-|2n\pi|a}$$

into the Poisson summation formula (2.2.14) we obtain

$$\sum_{n=-\infty}^{\infty} \frac{1}{a^2 + n^2} = \frac{\pi}{a} \sum_{n=-\infty}^{\infty} e^{-|2n|a\pi} = \frac{\pi}{a} \left(1 + 2 \sum_{n=1}^{\infty} e^{-2na\pi} \right)$$

$$= \frac{\pi}{a} \left(1 + 2 \frac{e^{-2a\pi}}{1 - e^{-2a\pi}} \right) = \frac{\pi}{a} \frac{1 + e^{-2a\pi}}{1 - e^{-2a\pi}}.$$

Therefore,

$$\sum_{n=-\infty}^{\infty} \frac{1}{a^2 + n^2} = \frac{\pi}{a} \frac{1 + e^{-2a\pi}}{1 - e^{-2a\pi}}.$$

We rewrite the last formula as follows.

$$\frac{1}{a^2} + 2 \sum_{n=1}^{\infty} \frac{1}{a^2 + n^2} = \frac{\pi}{a} \frac{1 + e^{-2a\pi}}{1 - e^{-2a\pi}}$$

and so

$$\sum_{n=1}^{\infty} \frac{1}{a^2 + n^2} = \frac{\pi}{2a} \frac{1 + e^{-2a\pi}}{1 - e^{-2a\pi}} - \frac{1}{2a^2}.$$

Although in the next several chapters we will apply the Fourier transform in solving partial differential equations, let us take a few examples of the application of the Fourier transform in solving ordinary differential equations.

Example 2.2.12. Solve the following boundary value problem

$$\begin{cases} y''(x) + a^2 y = f(x), & -\infty < x < \infty, \\ \lim_{x \to \pm\infty} y(x) = 0, \end{cases}$$

where $a > 0$ is a positive constant and f is a given function.

Solution. Let $\mathcal{F}\{y\} = Y$ and $\mathcal{F}\{f\} = F$. Applying the Fourier transform to both sides of the differential equation we obtain

$$\omega^2 Y(\omega) + a^2 Y(\omega) = F(\omega).$$

Solving for $Y(\omega)$, it follows that

(2.2.20) $$Y(\omega) = \frac{F(\omega)}{a^2 + \omega^2}.$$

Therefore, we can reconstruct the solution by applying the inverse Fourier transform to (2.2.20):

$$(2.2.21) \qquad y(x) = \frac{1}{2\pi} \int_{-\infty}^{\infty} \frac{F(\omega)}{a^2 + \omega^2} e^{i\omega x} \, d\omega.$$

Alternatively, in order to find the solution $y(x)$ from (2.2.21), we can use the convolution property for the Fourier transform:

$$y(x) = \mathcal{F}^{-1} \left\{ \frac{1}{a^2 + \omega^2} F(\omega) \right\} = \mathcal{F}^{-1} \left\{ \frac{1}{a^2 + \omega^2} \right\}(x) * f(x)$$

$$= \frac{1}{2a} e^{-ax} * f(x) = \frac{1}{2a} \int_{\infty}^{\infty} e^{-a(x-y)} f(y) \, dy.$$

The improper integral (2.2.21) can also be evaluated using the Cauchy Residue Theorem.

Example 2.2.13. Find the solution $y(x)$ of the boundary value problem in Example 2.2.12 if the forcing function f is given by

$$f(x) = e^{-|x|},$$

and $a \neq \pm 1$.

Solution. From Example 2.2.1 we have

$$\mathcal{F}\left\{ e^{-|x|} \right\}(\omega) = \frac{2}{1 + \omega^2},$$

and so from (2.2.21) it follows that

$$y(x) = \frac{1}{2\pi} \int_{-\infty}^{\infty} \frac{2}{(1 + \omega^2)(a^2 + \omega^2)} e^{i\omega x} \, d\omega.$$

Using partial fraction decomposition in the above integral, we obtain

$$\int_{-\infty}^{\infty} \frac{2}{(1 + \omega^2)(a^2 + \omega^2)} e^{i\omega x} \, d\omega = \frac{2}{a^2 - 1} \left(\int_{-\infty}^{\infty} \frac{e^{i\omega x}}{1 + \omega^2} \, d\omega - \int_{-\infty}^{\infty} \frac{e^{i\omega x}}{a^2 + \omega^2} \, d\omega \right)$$

$$= \frac{2}{a^2 - 1} \left(\frac{1}{2\pi} \mathcal{F}^{-1} \left\{ \frac{1}{1 + \omega^2} \right\}(x) - \frac{1}{2\pi} \mathcal{F}^{-1} \left\{ \frac{1}{a^2 + \omega^2} \right\}(x) \right)$$

$$= \frac{2}{a^2 - 1} \left(\frac{1}{2\pi} \pi e^{-|x|} - \frac{1}{2\pi} \pi \frac{e^{-|x|a}}{a} \right) \qquad \text{(from Example 2.2.1).}$$

Therefore, the solution of the given boundary value problem is

$$y(x) = \frac{1}{2\pi(a^2 - 1)}\left(e^{-|x|} - \frac{e^{-|x|a}}{a}\right).$$

Two more basic properties of Fourier transforms of integrable functions should be mentioned. First, it is relatively easy to show that if f is integrable, then $\mathcal{F}\{f\}$ is a bounded and continuous function on $\mathbb{R}$ (see Exercise 1 of this section). But something more is true, which we state without a proof.

Theorem 2.2.7 The Riemann–Lebesgue Lemma. *If f is an integrable function, then*

$$\lim_{\omega \to \infty} \mathcal{F}\{f\}(\omega) = 0.$$

Theorem 2.2.8. The Fourier Transform Inversion Formula. *Suppose that f is an integrable function on $\mathbb{R}$. If the Fourier transform $F = \mathcal{F}\{f\}$ is also integrable on $\mathbb{R}$, then f is continuous on $\mathbb{R}$ and*

$$f(x) = \frac{1}{2\pi} \int_{-\infty}^{\infty} F(\omega)\, e^{i\omega x}\, d\omega, \quad x \in \mathbb{R}.$$

Remark. A consequence of the Fourier Inversion Formula is the Laplace Inversion Formula which was given by Theorem 2.1.7.

Exercises for Section 2.2.2.

1. Prove that the Fourier transform of an integrable function is a bounded and continuous function.

2. Given that

$$\mathcal{F}\{\frac{1}{1 + x^2}\}(\omega) = \pi e^{-|\omega|},$$

find the Fourier transform of

(a) $\dfrac{1}{1 + a^2 x^2}$, a is a real constant.

(b) $\dfrac{\cos(ax)}{1 + x^2}$, a is a real constant.

3. Let $H(x)$ be the Heaviside unit step function and let $a > 0$. Use the modulation property of the Fourier transform and the fact that

$$\mathcal{F}\{e^{-ax}H(x)\}(\omega) = \frac{1}{1+i\omega}$$

to show that

$$\mathcal{F}\{e^{-ax}\sin(bx)H(x)\}(\omega) = \frac{b}{(a+i\omega)^2 + b^2}.$$

4. Let $a > 0$. Use the function

$$f(x) = e^{-ax}\sin(bx)H(x)$$

and Parseval's equality to show that

$$\int_{-\infty}^{\infty} \frac{dx}{(x^2 + a^2 - b^2)^2 + 4a^2b^2} = \frac{\pi}{2a(a^2 + b^2)}.$$

5 Use the fact that

$$\mathcal{F}\{\frac{1}{1+x^2}\}(\omega) = \pi e^{-|\omega|}$$

and Parseval's inequality to show that

$$\int_{-\infty}^{\infty} \frac{dx}{(x^2 + 1)^2} = \frac{\pi}{2}.$$

6. Find the inverse of the following Fourier transforms:

(a) $\dfrac{1}{\omega^2 - 2ib\omega - a^2 - b^2}$.

(b) $\dfrac{i\omega}{(1+i\omega)(1-i\omega)}$.

(c) $\dfrac{1}{(1+i\omega)(1+2i\omega)^2}$.

7. By taking the appropriate closed contour, find the inverse of the following Fourier transforms by the Cauchy Residue Theorem. The parameter a is positive.

(a) $\dfrac{\omega}{\omega^2 + a^2}$.

(b) $\dfrac{3}{(2 - i\omega)(1 + i\omega)}$.

(c) $\dfrac{\omega^2}{(\omega^2 + a^2)^2}$.

8. Find the inverse Fourier transform of

$$F(\omega) = \frac{\cos(i\omega)}{\omega^2 + a^2}, \quad a > 0.$$

Hint: Use the Cauchy Residue Theorem.

9. Using the Fourier transform, find particular solutions of the following differential equations:

(a) $y'(x) + y = \frac{1}{2}e^{-|x|}$.

(b) $y'(x) + y = \frac{1}{2}e^{-|x|}$, where we assume that the given function f has the Fourier transform F.

(c) $y''(x) + 4y'(x) + 4y = \frac{1}{2}e^{-|x|}$.

(d) $y''(x) + 3y'(x) + 2y = e^{-x}H(x)$, H is the Heaviside step function.

10. Suppose that f is continuous and piecewise smooth and that f and f' are integrable on $(0, \infty)$. Show that

$$\mathcal{F}_s\{f'\}(\omega) = -\omega\{F_c\}(\omega)$$

and

$$\mathcal{F}_c\{f'\}(\omega) = \omega\{F_s\}(\omega) - f(0).$$

11. State and prove Parseval's formulas for the Fourier cosine and Fourier sine transforms.

2.3. Projects Using Mathematica.

In this section we will see how Mathematica can be used to evaluate the Laplace and Fourier transforms, as well as their inverse transforms. For a brief overview of the computer software Mathematica consult Appendix H.

Let us start with the Laplace transform.

Mathematica's commands for the Laplace transform and the inverse Laplace transform

$$(2.3.1) \qquad F(s) = \int_0^\infty f(t)e^{-st}\,dt, \quad f(t) = \frac{1}{2\pi}\int_{-\infty}^\infty F(s)e^{st}\,ds$$

are

$\quad In[] := \text{LaplaceTransform } [f[t], t, s];$

$\quad In[] := \text{InverseLaplaceTransform } [F[s], s, t];$

Project 2.3.1. Using Mathematica find the Laplace transform of the function

$$f(t) = \begin{cases} t, & 0 \le t < 1 \\ 0, & 1 < t < \infty \end{cases}$$

in two ways:

(a) By (2.3.1).

(b) By *Mathematica's* command for the Laplace transform.

Solution. (a). First define the function $f(t)$:

$\quad In[1] := f[t\_] := \text{Piecewise } [\{\{t, 0 \le t < 1\}, \{0, 1 < t\}\}];$

Now define the Laplace transform $F(s)$ of $f(t)$:

$\quad In[2] := F[s\_] := \text{Integrate } f[t]\,e^{-st}, \{t, 0, \infty\},$
$\quad \text{Assumptions} \to \{\text{ Im } [s] == 0,\ s > 0\}]$

$\quad Out[2] = \frac{e^{-s}(-1+e^s-s)}{s^2}$

Part (b):

$\quad In[3]:=\text{LaplaceTransform } [f[t],\ t,\ s]$

$\quad Out[3] = \frac{e^{-s}(-1+e^s-s)}{s^2}$

Project 2.3.2. Using Mathematica find the inverse Laplace transform of the function

$$F(s) = \frac{2s^2 - 3s + 1}{s^2(s^2 + 9)}$$

in two ways:

(a) By *Mathematica's* command for the inverse Laplace transform.

(b) By the following formula for the inverse Laplace transform.

$$(2.3.2) \qquad f(t) = \frac{1}{2\pi} \int\limits_{c-\infty\, i}^{c+\infty\, i} F(s)e^{st}\, ds = \sum_{n=0}^{\infty} Res\left(F(z)e^{zt},\, z = z_n\right),$$

where z_n are all the singularities of $F(z)$ in the half-plane $\{z \in \mathbb{C} : Re\,(z) < c\}$ and $c > 0$ is any fixed number.

Solution. (a). First clear all previous f and F:

$In[1] := \mathrm{Clear}\,[f, F];$

Next define the function $F(s)$:

$In[2] := F[s\_] := \frac{2s^2 - 3s + 1}{s^3(s^2 + 9)}];$

$In[3] := \mathrm{InverseLaplaceTransform}\,\left[F[s], s, t\right]$

$Out[3] = \frac{1}{162}\left(34 - 54\,t + 9\,t^2 - 34\,Cos[3\,t] + 18\,Sin[3t]\right)$

$In[4] := \mathrm{Expand}\,[\%]$

$Out[4] = \frac{17}{81} - \frac{t}{3} + \frac{t^2}{18} - \frac{17}{81}\,Cos[3\,t] + \frac{1}{9}\,Sin[3\,t]$

Part (b):

Find the singularities of the function $F(z)$:

$In[5] := \mathrm{Solve}\,[z^3\,(z^2 + 9) == 0, z]$

$Out[5] = \{\{z \to 0\}, \{z \to 0\}, \{z \to 0\}, \{z \to -3\,i\}, \{z \to 3\,i\}\}$

Find the residues at these singularities

$In[6] := r0 = \mathrm{Residue}\,[F[z]\,e^{z\,t}, \{z, 0\}]$

$Out[6] := \frac{1}{162}\left(34 - 54\,t + 9\,t^2\right)$

$In[7] := r1 = \mathrm{Residue}\,[F[z]\,e^{z\,t}, \{z, -3\,i\}]$

$Out[7] = \left(-\frac{17}{162} + \frac{i}{18}\right)e^{-3\,i\,t}$

$In[8] := r2 = \mathrm{Residue}\,[F[z]\,e^{z\,t}, \{z, 3\,i\}]$

$Out[8] = \left(-\frac{17}{162} - \frac{i}{18}\right)e^{3\,i\,t}$

Now add the above residues to get the inverse Laplace transform.

$In[9] := r0 + r1 + r$

$Out[9] = \left(-\frac{17}{162} + \frac{i}{18}\right)e^{-3\,i\,t} + \left(-\frac{17}{162} - \frac{i}{18}\right)e^{3\,i\,t} + \frac{1}{162}\left(34 - 54\,t + 9\,t^2\right)$

We can simplify the above expression:

$In[10] := \text{FullSimplify} [\%]$

$Out[10] = \frac{1}{162}\left(34 + 9\left(-6 + t\right)t - 34\, Cos[3\,t] + 18\, Sin[3\,t]\right).$

$In[11] := \text{Expand} [\%]$

$Out[11] = \frac{17}{81} - \frac{t}{3} + \frac{t^2}{18} - \frac{17}{81}\, Cos[3\,t] + \frac{1}{9}\, Sin[3\,t]$

Using the Laplace transform we can solve differential equations.

Project 2.3.3. Using Mathematica solve the differential equation

$$y''(t) + 9y(t) = \cos t.$$

Solution. Take the Laplace transform of both sides of the equation.

$In[1] := \text{LaplaceTransform} [y''[t] + 9\, y[t] == Cos[t], t, s]]$

$Out[1] = -s\, y[0] + 9\, \text{LaplaceTransform} [y[t], t, s]]$
$+s^2\ \text{LaplaceTransform} [y[t], t, s] - y'[0] == \frac{s}{s^2+9}$

Solve for the Laplace transform.

$In[2] := \text{Solve} [\%, \text{LaplaceTransform} [y[t], t, s]]$

$Out[2] = \{\{ \text{LaplaceTransform} [y[t], t, s] \rightarrow \frac{s+s\, y[0]+s^2\, y'[0]}{(1+s^2)\,(9+s^2)} \}\}$

Find the inverse Laplace transform.

$In[3] := \text{InverseLaplaceTransform} [\%, s, t]$

$Out[3] = \{\{y[t] \rightarrow \frac{1}{24}\left(3Cos[t] - 3Cos[3\,t] + 24Cos[3\,t]\, y[0] + 8Sin[3\,t]y'[0]\right)\}\}$

$In[4] := \text{Expand} [\%]$

$Out[4] = \{\{y[t] \rightarrow \frac{Cos[t]}{8} - \frac{1}{8}\, Cos[3\,t] + Cos[3\,t]\, y[0] + \frac{1}{3}\, Sin[3\,t]\, y'[0]\right)\}\}$

Now we discuss the Fourier transform in Mathematica.

In Mathematica the Fourier transform of a function $f(t)$ and its inverse are by default defined to be

$$\frac{1}{\sqrt{2\pi}} \int\limits_{-\infty}^{\infty} f(t)e^{i\omega t}\, dt, \qquad \frac{1}{\sqrt{2\pi}} \int\limits_{-\infty}^{\infty} F(\omega)^{-i\omega t}\, d\omega,$$

respectively.

The command to find the Fourier transform is

$In[] := \text{FourierTransform} [f[t], t, \omega]$

The default Mathematica command

$In[] := $ InverseFourierTransform $[f[t], t, \omega]$

gives the inverse Fourier transform of $f(t)$.

The Fourier transform and its inverse, defined by

$$\int\limits_{-\infty}^{\infty} f(t)e^{-i\omega t}\, dt, \quad \frac{1}{2\pi} \int\limits_{-\infty}^{\infty} F(\omega)^{i\omega t}\, d\omega,$$

in Mathematica are evaluated by

$In[] := $ FourierTransform $[f[t], t\omega, \text{FourierParameters} \rightarrow \{1, -1\}]$

$In[] := $ InverseFourierTransform $[F[\omega], \omega, t, \text{FourierParameters} \rightarrow \{1, -1\}]$, respectively.

The next project shows how Mathematica can evaluate Fourier transforms and inverse Fourier transforms of a big range of functions: algebraic, exponential and trigonometric functions, step and impulse functions.

Project 2.3.4. Find the Fourier transform of each of the following functions. Take the constant in the Fourier transform to be 1. From the obtained Fourier transforms find their inverses.

(a) e^{-t^2}

(b) $\frac{1}{\sqrt{t}}$.

(c) $sinc(t) = \frac{\sin t}{t}$.

(d) $sign(t) = \begin{cases} 1, & t > 0 \\ -1, & t < 0. \end{cases}$

(e) $\delta(t)$—the Dirac Delta function, concentrated at $t = 0$.

(f) e^{-t}.

Solution. First we clear any previous variables and functions.

$In[1] := $ Clear $[t, f, g, h, s]$;

Part (a).

$In[2] := $ FourierTransform $[e^{-t^2}, t, \omega, \text{FourierParameters} \rightarrow \{1, -1\}]$

$Out[2] = e^{-\frac{\omega^2}{4}} \sqrt{\pi}$

$In[3] := $ InverseFourierTransform $[\%], \omega, t, \text{FourierParameters} \rightarrow \{1, -1\}]$

$Out[3] = e^{-t^2}$

Part (b).

$In[4] := \text{FourierTransform } [\frac{1}{\sqrt{Abs[t]}}, t, \omega, \text{FourierParameters} \to \{1, -1\}]$

$Out[4] = \frac{\sqrt{2\pi}}{\sqrt{Abs[\omega]}}$

$In[5] := \text{FourierTransform } [\text{HeavisideTheta } [t], t, \omega,$
$\text{FourierParameters} \to \{1, -1\}]$

$Out[5] = -\frac{i}{\omega} + \pi \text{ DiracDelta } [\omega]$

$In[6] := \text{InverseFourierTransform } [\%], \omega, t, \text{FourierParameters} \to \{1, -1\}]$

$Out[6] = \frac{1}{2}(1 + Sign[t])$

Part (c).

$In[7] := \text{FourierTransform } [Sinc[t], t, \omega, \text{FourierParameters} \to \{1, -1\}]$

$Out[7] = \frac{1}{2}\pi Sign[1 - \omega] + \frac{1}{2}\pi Sign[1 + \omega]$

$In[8] := \text{InverseFourierTransform } [\%, \omega, t, \to \{1, -1\}]$

$Out[8] = \frac{Sin[t]}{t}$

Part (d).

$In[9] := \text{FourierTransform } [Sign[t], t, \omega, \text{FourierParameters} \to \{1, -1\}]$

$Out[9] = -\frac{2i}{\omega}$

$In[10] := \text{InverseFourierTransform } [\%, \omega, t, \to \{1, -1\}]$

$Out[10] = \text{Sign } [t]$

Part (e).

$In[11] := \text{FourierTransform } [DiracDelta[t], t, \omega,$
$\text{FourierParameters} \to \{1, -1\}]$

$Out[11] = 1$

$In[12] := \text{InverseFourierTransform } [\%, \omega, t, \to \{1, -1\}]$

$Out[12] = \text{DiracDelta } [t]$

Part (f).

$In[13] := \text{FourierTransform } [Exp[-Abs[t]], t, \omega,$
$\text{FourierParameters} \to \{1, -1\}]$

$Out[13] = \frac{2}{1+\omega^2}$

$In[14] := \text{InverseFourierTransform } [\%, \omega, t, \to \{1, -1\}]$

$Out[14] = e^{-Abs[t]}.$

CHAPTER 3

STURM–LIOUVILLE PROBLEMS

In the first chapter we saw that the trigonometric functions sine and cosine can be used to represent functions in the form of Fourier series expansions. Now we will generalize these ideas.

The methods developed here will generally produce solutions of various boundary value problems in the form of infinite function series. Technical questions and issues, such as convergence, termwise differentiation and integration and uniqueness, will not be discussed in detail in this chapter. Interested readers may acquire these detail from advanced literature on these topics, such as the book by G. B. Folland [6].

3.1 Regular Sturm–Liouville Problems.

In mathematical physics and other disciplines fairly large numbers of problems are defined in the form of boundary value problems involving second order ordinary differential equations. Therefore, let us consider the differential equation

$$(3.1.1) \qquad a(x)y''(x) + b(x)y'(x) + \big[c(x) + \lambda r(x)\big]y = 0, \quad a < x \le b$$

subject to some boundary conditions on a bounded interval $[a, b]$. We suppose that the real functions $a(x)$, $r(x)$ are continuous on the interval $[a, b]$, λ a parameter, and we suppose that $a(x)$ is not zero for every $x \in [a, b]$. It turns out that it is much more convenient to rewrite the differential equation (3.1.1) in its equivalent form, the so-called *Sturm–Liouville form*

$$\Big(p(x)y'(x)\Big)' + [q(x) + \lambda r(x)]y = 0,$$

where the real functions $p(x)$, $p'(x)$, $r(x)$ are continuous on $[a, b]$, and $p(x)$ and $r(x)$ are positive on $[a, b]$.

Remark. Any differential equation of the form (3.1.1) can be written in Sturm–Liouville form.

Indeed, first divide (3.1.1) by $a(x)$ to obtain

$$y''(x) + \frac{b(x)}{a(x)}y'(x) + \left[\frac{c(x)}{a(x)} + \lambda\frac{r(x)}{a(x)}\right]y = 0.$$

Multiplying the last equation by

$$\mu(x) = e^{\int \frac{b(x)}{a(x)} \, dx} \quad (\text{ ignore the integration constant })$$

we obtain

$$\mu(x)y''(x) + \mu(x)\frac{b(x)}{a(x)}y'(x) + \left[\frac{\mu(x)c(x)}{a(x)} + \lambda\frac{\mu(x)r(x)}{a(x)}\right]y = 0.$$

Now, using the fact that

$$\mu'(x) = \mu(x)\frac{b(x)}{a(x)},$$

it follows that

$$\mu(x)y''(x) + \mu'(x)y'(x) + \left[\frac{\mu(x)c(x)}{a(x)} + \lambda\frac{\mu(x)r(x)}{a(x)}\right]y = 0,$$

and thus, by the product rule for differentiation we have

$$\left(\mu(x)y'(x)\right)' + \left[\frac{\mu(x)c(x)}{a(x)} + \lambda\frac{\mu(x)r(x)}{a(x)}\right]y = 0.$$

But, the last equation is indeed in Sturm–Liouville form.

Definition 3.1.1. A *regular Sturm–Liouville* problem is a second order homogeneous linear differential equation of the form

$$(3.1.2) \qquad \left(p(x)y'(x)\right)' + \left[q(x) + \lambda r(x)\right]y(x) = 0, \quad a < x < b$$

where $p(x)$, $p'(x)$ and $r(x)$ are real continuous functions on a finite interval $[a,b]$, and $p(x)$, $r(x) > 0$ on $[a,b]$, together with the set of homogeneous boundary conditions of the form

$$(3.1.3) \qquad \begin{cases} \alpha_1 y(a) + \beta_1 y'(a) = 0, \\ \alpha_2 y(b) + \beta_2 y'(b) = 0, \end{cases}$$

where α_i and β_i are constants. We regard λ as an undetermined constant parameter.

Notation. Sometimes, we will denote by L the following *linear differential operator.*

$$(3.1.4) \qquad L[y] = \left(p(x)y'(x)\right)' + q(x)y(x).$$

Using this operator the differential equation (3.1.2) can be expressed in the following form.

$$(3.1.5) \qquad\qquad L[y] = -\lambda\, r\, y.$$

We use the term linear differential operator because of the following important linear property of the operator L.

$$L[c_1 y_1 + c_2 y_2] = c_1 L[y_1] + c_2 L[y_2]$$

for any constants c_1 and c_2 and any differentiable functions y_1 and y_2.

Remark. For different values of the constants $\alpha's$ and $\beta's$ in the boundary conditions (3.1.3) we have special types of boundary conditions.
For $\beta_1 = \beta_2 = 0$, (3.1.3) these are called *Dirichlet boundary conditions*.
For $\alpha_1 = \alpha_2 = 0$, (3.1.3) these are called *Neumann boundary conditions*.
Other types of boundary conditions that are often encountered are *periodic boundary conditions* when

$$y(a) = y(b), \quad y'(a) = y'(b).$$

Our goal is to find all solutions of the Sturm–Liouville problem. It is clear that $y \equiv 0$ is one, the trivial solution of (3.1.2) for every λ, and satisfies the boundary conditions (3.1.3). However, we are interested here to find those parameters λ for which the Sturm–Liouville problem has non-trivial solutions y. Those values of λ and the corresponding solutions $y(x)$ have special names:

Definition 3.1.2. If $y(x) \not\equiv 0$ is a solution of a regular Sturm–Liouville problem (3.1.2), (3.1.3) corresponding to some constant λ, then this solution is called an *eigenfunction* corresponding (associated) to the *eigenvalue* λ.

Let us take now an example.

Example 3.1.1. Find all eigenvalues and the corresponding eigenfunctions of the following problem.

$$y''(x) + \lambda y(x) = 0, \quad 0 < x < l,$$

subject to the boundary conditions $y(0) = 0$ and $y'(l) = 0$.

Solution. First we observe that this is a regular Sturm–Liouville problem. Indeed, the differential equation can be written in the form

$$\big(1 \cdot y'(x)\big)' + \big[0 + 1 \cdot \lambda\big] y = 0,$$

so $p(x) = 1$, $q(x) = 0$ and $r(x) = 1$.

The given differential equation is a simple homogeneous linear differential equation of second order with constant coefficients. Its characteristic equation is

$$m^2 + \lambda = 0,$$

whose solutions are

$$m = \pm\sqrt{-\lambda}.$$

The general solution $y(x)$ of the differential equation depends on whether $\lambda = 0$, $\lambda < 0$ or $\lambda > 0$.

Case 1^0. $\lambda = 0$. In this case the differential equation is simply $y'' = 0$, which has a general solution $y = A + Bx$. The two boundary conditions $y(0) = y'(l) = 0$ imply $A = 0$ and $B = 0$, which yields $y \equiv 0$ on the interval $[0, l]$, and so $\lambda = 0$ is not an eigenvalue of the given problem.

Case 2^0. $\lambda < 0$. In this case we can write $\lambda = -\mu^2$ where $\mu > 0$. The solutions of the characteristic equation in this case are $m = \pm\mu$, and so the general solution of the differential equation is

$$y(x) = A_1 e^{\mu x} + B_1 e^{-\mu x} = A \cosh(\mu x) + B \sinh(\mu x), \quad A,\ B \text{ are constants.}$$

The boundary condition $y(0) = 0$ implies $A = 0$, so $y(x) = B \sinh(\mu x)$. The other boundary condition $y'(l) = 0$ implies that $B\mu \cosh(\mu l) = 0$, and since $\cosh(\mu l) > 0$ we have $B = 0$. Therefore, $y(x) \equiv 0$, and so any $\lambda < 0$ cannot be an eigenvalue of the problem.

Case 3^0. $\lambda > 0$. In this case we have $\lambda = \mu^2$, with $\mu > 0$. The differential equation in this case has a general solution

$$y(x) = A \sin(\mu x) + B \cos(\mu x), \quad A \text{ and } B \text{ are constants.}$$

The boundary condition $y(0) = 0$ implies $B = 0$, so $y(x) = A \sin(\mu x)$. The other boundary condition $y'(l) = 0$ implies that $A\mu \cos(\mu l) = 0$. To avoid triviality, we want A to be nonzero, and so we must have

$$\cos(\mu l) = 0.$$

If the solutions μ of the last equation are denoted by μ_n, then we have

$$\mu_n = \frac{(2n - 1)\pi}{2l}, \quad n \in \mathbb{N}.$$

Therefore, the eigenvalues for this problem are

$$\lambda_n = \mu_n^2 = \frac{(2n - 1)^2 \pi^2}{4l^2}, \quad n = 1, 2, \ldots.$$

The corresponding eigenfunctions are functions of the form $A_n \sin(\mu_n x)$, and ignoring the factors A_n, the eigenfunctions of the given problem are

$$y_n(x) = \sin(\mu_n x), \quad \mu_n = \frac{(2n - 1)\pi}{2l}, \quad n = 1, 2, \ldots.$$

Notice that all the eigenvalues in Example 3.1.1 are real numbers. Actually, this is not an accident, and this fact is true for any regular Sturm–Liouville problem.

We introduce a very useful shorthand notation:

For two square integrable complex functions f and g on an interval $[a, b]$, the expression

$$(f, g) = \int_a^b f(x)\overline{g(x)}\, dx$$

$(\overline{g(x)}$ is the complex conjugate of $g(x))$ is called the *inner product* of f and g.

The inner product satisfies the following useful properties:

(a) $(f, f) \geq 0$ for any square integrable f and $(f, f) = 0$ only if $f \equiv 0$.

(b) $(f, g) = \overline{(g, f)}$.

(c) $(\alpha f + \beta g, h) = \alpha(f, h) + \beta(g, h)$, for any scalars α and β.

The next theorem shows that the eigenvalues of a regular Sturm–Liouville problem are real numbers.

Theorem 3.1.1. *Every eigenvalue of the regular Sturm–Liouville problem* (3.1.2), (3.1.3) *is a real number.*

Proof. Let λ be an eigenvalue of the boundary value problem (3.1.2), (3.1.3), with a corresponding eigenfunction $y = y(x)$. If we multiply both sides of the equation

$$L[y] = -\lambda r y$$

by $\overline{y(x)}$ and integrate from a to b, then we obtain

$$\int_a^b \left[(p(x)y'(x))' + q(x)y \right]\overline{y(x)}\, dx = -\lambda(y, r\overline{y}).$$

If we integrate by parts, then from the last equation we have

$$\left[p(x)\overline{y(x)}y'(x) \right]_{x=a}^{x=b} - \int_a^b p(x)y'(x)\overline{y'(x)}\, dx$$

$$+ \int_a^b q(x)y(x)\overline{y(x)}\, dx = -\lambda(y, r\overline{y}),$$

i.e.,

(3.1.6) $p(b)\overline{y(b)}y'(b) - p(a)\overline{y(a)}y'(a) - (y', py') + (y, q\overline{y}) = -\lambda(y, r\overline{y}).$

Now, taking the complex conjugate of the above equation, and using the fact that p, q and r are real functions, along with properties (b) and (c) of the inner product, we have

$$(3.1.7) \qquad p(b)y(b)\overline{y'(b)} - p(a)y(a)\overline{y'(a)} - (y', py') + (y, qy) = -\overline{\lambda}(y, ry).$$

If we subtract (3.1.7) from (3.1.6) we obtain
(3.1.8)
$$p(b)\left[\overline{y(b)}y'(b) - y(b)\overline{y'(b)}\right] - p(a)\left[y(a)\overline{y'(a)} - \overline{y(a)}y'(a)\right] = (\overline{\lambda} - \lambda)(y, r\overline{y}).$$

The boundary condition at the point b together with its complex conjugate gives

$$\begin{cases} \alpha_2 y(b) + \beta_2 y'(b) = 0, \\ \alpha_2 \overline{y(b)} + \beta_2 \overline{y'(b)} = 0. \end{cases}$$

Both α_2, β_2 cannot be zero, otherwise there would be no boundary condition at $x = b$. Therefore,

$$(3.1.9) \qquad\qquad\qquad y(b)\overline{y'(b)} - y'(b)\overline{y(b)} = 0.$$

Similar considerations (working with the boundary condition at $x = a$) give

$$(3.1.10) \qquad\qquad\qquad y(a)\overline{y'(a)} - y'(a)\overline{y(a)} = 0.$$

Inserting conditions (3.1.9) and (3.1.10) into (3.1.8) we have

$$(3.1.11) \qquad\qquad\qquad (\overline{\lambda} - \lambda)(y, r\overline{y}) = 0.$$

Now, since $y(x) \not\equiv 0$ on $[a, b]$, the continuity of $y(x)$ at some point of $[a, b]$ implies that $y(x)\overline{y(x)} > 0$ for every x in some interval $(c, d) \subseteq [a, b]$. Therefore, from $r(x) > 0$ on $[a, b]$ and the continuity of $r(x)$ it follows that $y(x)r(x)\overline{y(x)} > 0$ for every $x \in (c, d)$. Hence $(y, r\overline{y}) > 0$, and so (3.1.11) forces $\lambda - \overline{\lambda} = 0$, i.e., $\lambda = \overline{\lambda}$. Therefore λ is a real number. ∎

Regular Sturm–Liouville problems have several important properties. However, we will not prove them all here. An obvious question regarding a regular Sturm–Liouville problem is that about the existence of eigenvalues and eigenfunctions. One property, related to this question and whose proof is beyond the scope of this book, but can be found in more advanced books, is the following theorem.

Theorem 3.1.2. *A regular Sturm–Liouville problem has infinitely many real and simple eigenvalues* λ_n, $n = 0, 1, 2, \ldots$, *which can be arranged as a monotone increasing sequence*

$$\lambda_0 < \lambda_1 < \lambda_2 < \ldots < \lambda_n < \ldots,$$

such that

$$\lim_{n \to \infty} \lambda_n = \infty.$$

For each eigenvalue λ_n there exists only one eigenfunction $y_n(x)$ (up to a multiplicative constant).

Several other properties will be presented in the next section.

Prior to that we take up a few more illustrative examples.

Example 3.1.2. Find the eigenvalues and the corresponding eigenfunctions of the following problem.

$$x^2 y''(x) + xy'(x) + (\lambda + 2)y = 0, \quad 1 < x < 2,$$

subject to the boundary conditions $y'(1) = 0$ and $y'(2) = 0$.

Solution. First we observe that this is a regular Sturm–Liouville problem. Indeed, the differential equation can be written in the form

$$\left(x \cdot y'\right)' + \left[\frac{2}{x} + \frac{1}{x} \cdot \lambda\right] y = 0,$$

so

$$p(x) = x, \quad q(x) = \frac{2}{x}, \quad r(x) = \frac{1}{x}.$$

The given differential equation is a Euler-Cauchy equation. Its characteristic equation is

$$m^2 + \lambda + 2 = 0,$$

whose solutions are

$$m = \pm\sqrt{-\lambda - 2}.$$

(See Appendix D.) The general solution $y(x)$ of the differential equation depends on whether $\lambda = -2$, $\lambda < -2$ or $\lambda > -2$.

Case 1^0. $\lambda = -2$. In this case the differential equation is simply $xy'' + y'(x) = 0$ whose general solution is given by $y = A + B\ln(x)$. The two boundary conditions give $B = 0$, which yields $y(x) \equiv A$, and so $\lambda = -2$ is an eigenvalue and the corresponding eigenfunction is $y(x) = 1$.

Case 2^0. $\lambda < -2$. In this case we have $\lambda + 2 = -\mu^2$ with $\mu > 0$. The solutions of the characteristic equation in this case are $m = \pm\mu$, and so a general solution of the differential equation is

$$y(x) = Ax^\mu + Bx^{-\mu}, \quad A \text{ and } B \text{ are constants.}$$

The boundary condition $y'(1) = 0$ implies that $A\mu - B\mu = 0$. The other boundary condition $y'(2) = 0$ implies $A\mu 2^{\mu-1} - B\mu 2^{-\mu-1} = 0$. Solving the linear system for A and B we obtain that $A = B = 0$. Therefore $y(x) = 0$, and so any $\lambda < -2$ cannot be an eigenvalue of the problem.

Case 3^0. $\lambda > -2$. In this case we have $\lambda + 2 = \mu^2$, with $\mu > 0$. The differential equation in this case has a general solution

$$y(x) = A \sin(\mu \ln x) + B \cos(\mu \ln x), \quad A \text{ and } B \text{ are constants.}$$

From the boundary condition $y'(1) = 0$ we have $A = 0$, and therefore, $y(x) = B \cos(\mu \ln x)$. The other boundary condition $y'(2) = 0$ implies

$$\sin(\mu \ln 2) = 0.$$

From the last equation it follows that

$$\mu = \frac{n\pi}{\ln(2)},$$

and so, denoting μ by μ_n, we have

$$\mu_n = \frac{n\pi}{\ln 2}, \quad n \in \mathbb{N}.$$

Therefore, the eigenvalues of this problem are

$$\lambda_n = \mu_n^2 = \frac{n^2 \pi^2}{\ln^2 2}, \quad n = 1, 2, \ldots,.$$

The corresponding eigenfunctions are functions of the form $A_n \cos(\mu_n \ln x)$, and ignoring the coefficients A_n, the eigenfunctions of the given problem are

$$y_n(x) = \cos(\mu_n \ln x), \quad \mu_n = \frac{n\pi}{\ln 2}, \quad n = 1, 2, \ldots,.$$

Example 3.1.3. Find the eigenvalues and the corresponding eigenfunctions of the following boundary value problem.

$$y''(x) - 2y'(x) + \lambda y = 0, \quad 0 < x < 1, \quad y(0) = y'(1) = 0.$$

Solution. The given differential equation is a second order homogeneous linear differential equation with constant coefficients. Its characteristic equation is $m^2 - 2m + \lambda = 0$, whose solutions are

$$m = 1 \pm \sqrt{1 - \lambda}.$$

The general solution $y(x)$ depends on whether $\lambda = 1$, $\lambda < 1$ or $\lambda > 1$.

Case 1^0. $\lambda = 1$. The differential equation in this case has a general solution

$$y = Ae^x + Bxe^x.$$

The two boundary conditions imply that

$$\begin{cases} 0 = y(0) = A \\ 0 = y'(1) = A + 2eB. \end{cases}$$

Thus $A = B = 0$ and so $\lambda = 1$ is not an eigenvalue.

Case 2^0. $\lambda < 1$. In this case we have $1 - \lambda = \mu^2$ with $\mu > 0$. The solutions of the characteristic equation in this case are $m = 1 \pm \mu$, and so a general solution of the differential equation is

$$y(x) = Ae^{(1+\mu)x} + Be^{(1-\mu)x}, \quad A \text{ and } B \text{ are constants.}$$

The boundary conditions imply

$$\begin{cases} 0 = y(0) = A + B \\ 0 = y'(1) = A(1 + \mu)e^{1+\mu} + B(1 - \mu)e^{1-\mu}. \end{cases}$$

Thus $B = -A$ and so

$$A\left[(1 + \mu)e^{1+\mu} - (1 - \mu)e^{1-\mu}\right] = 0.$$

Now, since

$$(1 + \mu)e^{1+\mu} > (1 - \mu)e^{1-\mu}$$

for all $\mu > 0$, it follows that $A = 0$, and hence the given boundary value problem has only a trivial solution in this case also.

Case 3^0. $\lambda > 1$. In this case we have $1 - \lambda = -\mu^2$, with $\mu > 0$. The differential equation in this case has a general solution

$$y(x) = e^x\left[A\cos(\mu x) + B\sin(\mu x)\right].$$

The boundary condition $y(0) = 0$ implies $A = 0$, and since

$$y'(x) = Be^x\sin(\mu x) + B\mu e^x\cos(\mu x),$$

the other boundary condition $y'(1) = 0$ implies

$$B\left(\sin\mu + \mu\cos\mu\right) = 0.$$

To avoid again the trivial solution, we want B to be nonzero, and so we must have

$$\sin\mu + \mu\cos\mu = 0,$$

i.e.,

(3.1.12) $$\tan\mu = -\mu.$$

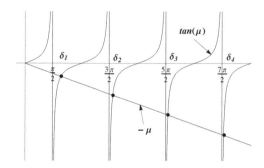

Figure 3.1.1

Therefore, the eigenvalues for this problem are $\lambda = 1 + \mu^2$, where the positive number μ satisfies the transcendental equation (3.1.12). Although the solutions of (3.1.12) cannot be found explicitly, from the graphical sketch of $\tan \mu$ and $-\mu$ in Figure 3.1.1 we see that there are infinitely many solutions $\mu_1 < \mu_2 < \dots$. Also, the following estimates are valid.

$$\frac{\pi}{2} < \mu_1 < \pi, \quad \frac{3\pi}{2} < \mu_2 < 2\pi, \quad \dots, \quad \frac{2n-1}{2}\pi < \mu_n < n\pi, \dots$$

If $\mu = \mu_n$ is the solution of (3.1.12) such that

$$\frac{2n-1}{2}\pi < \mu_n < n\pi,$$

then

$$\lambda_n = 1 + \mu_n^2, \quad y_n(x) = e^x \sin\left(\sqrt{1 + \lambda_n}\, x\right)$$

are the eigenvalues and the associated eigenfunctions.

From the estimates

$$\frac{2n-1}{2}\pi < \mu_n < n\pi$$

it is clear that

$$\lambda_1 < \lambda_2 < \dots < \lambda_n < \dots$$

and

$$\lim_{n \to \infty} \lambda_n = \infty.$$

Example 3.1.4. Find the eigenvalues and the corresponding eigenfunctions of the following boundary value problem.

$$y''(x) + \lambda y = 0, \quad 0 < x < 1,$$

subject to the boundary conditions

$$y'(0) = 0, \ y(1) - y'(1) = 0.$$

Solution. The general solution $y(x)$ of the given differential equation depends on the parameter λ:

For the case when $\lambda = 0$ the solution of the equation is $y(x) = A + Bx$. When we impose the two boundary conditions we find $B = 0$ and $A+B-B = 0$. But this means that both A and B must be zero, and therefore $\lambda = 0$ is not an eigenvalue of the problem.

If $\lambda = -\mu^2 < 0$, with $\mu > 0$, then the solution of the differential equation is

$$y(x) = A \sinh(\mu x) + B \cosh(\mu x).$$

Since

$$y'(x) = A\mu \cosh(\mu x) + B\mu \sinh(\mu x)$$

the boundary condition at $x = 0$ implies that $A = 0$, and so

$$y(x) = B \cosh(\mu x).$$

The other boundary condition at $x = 1$ yields

$$\mu \sinh(\mu) = \cosh(\mu),$$

i.e.,

$$\tanh(\mu) = \frac{1}{\mu}, \ \mu > 0.$$

As we see from the Figure 3.1.2, there is a single solution of the above equation, which we will denote by μ_0. Thus, in this case there is a single eigenvalue $\lambda_0 = -\mu_0^2$ and a corresponding eigenfunction $y_0(x) = \cosh(\mu_0 x)$.

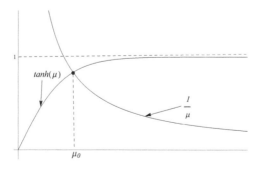

Figure 3.1.2

Finally, if $\lambda = \mu^2 > 0$, with $\mu > 0$, then the solution of the differential equation is

$$y(x) = A \sin (\mu x) + B \cos (\mu x).$$

Since

$$y'(x) = A\mu \cos (\mu x) - B\mu \sin (\mu x)$$

the boundary condition at $x = 0$ implies $A = 0$, and so

$$y(x) = B \cos (\mu x).$$

The other boundary condition at $x = 1$ yields

$$-\mu B \sin (\mu) = B \cos (\mu),$$

i.e.,

(3.1.13) $$- \tan (\mu) = \frac{1}{\mu}, \quad \mu > 0.$$

As before, the positive solutions of this equation correspond to the intersections of the curves

$$y = \tan \mu \quad y = -\frac{1}{\mu}.$$

As we see from Figure 3.1.3, there are infinitely many solutions, which we will denote by $\mu_1 < \mu_2 < \dots$. So, the eigenvalues in this case are $\lambda_n = \mu_n^2$ and the associated eigenfunctions are given by

$$y_n(x) = \cos (\mu_n x), \quad n = 1, 2 \dots,$$

where μ_n are the positive solutions of Equation (3.1.13).

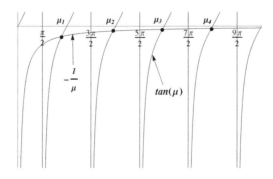

Figure 3.1.3

Next, we examine an example of a boundary value problem which is not a regular Sturm–Liouville problem.

Example 3.1.5. Find all eigenvalues and the corresponding eigenfunctions of the following boundary value problem.

$$y'' + \lambda y = 0, \quad 0 < x < \pi,$$

subject to the periodic boundary conditions

$$y(-\pi) = y(\pi), \quad y'(-\pi) = y'(\pi).$$

Solution. First we observe that this is not a regular Sturm–Liouville problem because the boundary conditions are periodic and they are not of the canonical boundary conditions (3.1.3). But yet this boundary value problem has infinitely many eigenvalues and eigenfunctions.

The general solution $y(x)$ of the given differential equation depends on the parameter λ:

If $\lambda = 0$, then the solution of the equation is

$$y(x) = A + Bx.$$

The boundary condition $y(-\pi) = y(\pi)$ implies $A - B\pi = A + B\pi$, and so $B = 0$.

The other boundary condition is satisfied, and therefore $\lambda = 0$ is an eigenvalue of the problem, and $y(x) = 1$ is the corresponding eigenfunction.

If $\lambda = -\mu^2 < 0$, then the solution of the differential equation is

$$y(x) = A \sinh \mu x + B \cosh \mu x.$$

The boundary condition $y(-\pi) = y(\pi)$ implies that

$$-A \sinh \mu\pi + B \cosh \mu\pi = A \sinh \mu\pi + B \cosh \mu\pi,$$

from which it follows that $A = 0$. Now, the other boundary condition $y'(-\pi) = y'(\pi)$ implies

$$-B\mu \sinh \mu\pi = B\mu \sinh \mu\pi.$$

From the last equation we obtain $B = 0$, and therefore any $\lambda < 0$ cannot be an eigenvalue.

Now, let $\lambda = \mu^2 > 0$. In this case the general solution of the differential equation is given by

$$y = A \sin \mu x + B \cos \mu x.$$

From the boundary condition, $y(-\pi) = y(\pi)$ it follows that $A\sin \mu\pi = 0$. From the second boundary condition, $y'(-\pi) = y'(\pi)$, we obtain

$$A\mu\cos \mu\pi + B\mu\sin \mu\pi = A\mu\cos \mu\pi - B\mu\sin \mu\pi.$$

Therefore $\sin \mu\pi = 0$. From the last equation, in view the fact that $\mu \neq 0$, it follows that $\mu = n, \ n = \pm 1, \pm 2, \ldots$. Now since A and B are arbitrary constants, to each eigenvalue $\lambda_n = n^2$ there are two linearly independent eigenfunctions

$$y_n(x) = \{\sin nx, \ \cos nx, \ n = 1, 2, 3, \ldots\}.$$

The negative integers n give the same eigenvalues and the same associated eigenfunctions.

Exercises for Section 3.1.

1. Show that the differential equation (3.1.1) is equivalent to the Sturm-Liouville form.

2. Find the eigenvalues and eigenfunctions for each of the following.

   (a) $y'' + \lambda y = 0, \ y'(0) = 0, \ y(1) = 0$.

   (b) $y'' + \lambda y = 0, \ y(0) + y'(0) = 0, \ y(\pi) + y'(\pi) = 0$.

   (c) $y'' + \lambda y = 0, \ y'(0) = 0, \ y(\pi) - y'(\pi) = 0$.

   (d) $y^{(iv)} + \lambda y = 0, \ y(0) = y''(0) = 0, \ y(\pi) = y''(\pi) = 0$.

3. Consider the following boundary value problem.

$$y'' + \lambda y = 0, \quad y(0) + 2y'(0) = 0, \ 3y(2) + 2y'(2) = 0.$$

Show that the eigenvalues of this problem are all negative $\lambda = -\mu^2$, where μ satisfies the transcendental equation

$$\tanh(2\mu) = \frac{4\mu}{3 - 4\mu^2}.$$

4. Find an equation that the eigenvalues for each of the following boundary value problems satisfy.

   (a) $y'' + \lambda y = 0, \ y(0) = 0, \ y(\pi) + y'(\pi) = 0$.

(b) $y'' + \lambda y = 0$, $y(0) + y'(0) = 0$, $y'(\pi) = 0$.

(c) $y'' + \lambda y = 0$, $y(0) + y'(0) = 0$, $y(\pi) - y'(\pi) = 0$.

5. Find the eigenvalues and corresponding eigenfunctions for each of the following Sturm–Liouville problems.

(a) $\left(x^3 y'\right)' + \lambda x y = 0$, $y(1) = y(e^\pi) = 0$, $1 \le x \le e^\pi$.

(b) $\left(\frac{1}{x} y'\right)' + \frac{\lambda}{x} y = 0$, $y(1) = y(e) = 0$, $1 \le x \le e$.

(c) $\left(x y'\right)' + \frac{\lambda}{x} y = 0$, $y(1) = y'(e) = 0$, $1 \le x \le e$.

6. Show that the eigenvalues λ_n and the associated eigenfunctions of the boundary value problem

$$\left((1 + x^2) y'(x)\right)' + \lambda y = 0, \quad 0 < x < 1, \quad y(0) = y(1) = 0$$

are given by

$$\lambda_n = \left(\frac{n\pi}{\ln 2}\right) + \frac{1}{4}, \quad y_n(x) = \frac{\sin\left(n\pi \ln(1 + x)\right)}{(\ln 2)\sqrt{1 + x}}, \quad n = 1, 2, \ldots$$

7. Find the eigenvalues of the following boundary value problem problem.

$$x^2 y'' + 2\lambda x y' + \lambda y = 0 \quad 1 < x < 2, \quad y(1) = y(2) = 0.$$

8. Find all eigenvalues and the corresponding eigenfunctions of the following periodic boundary value problem.

$$y'' + (\lambda - 2) y = 0, \quad y(-\pi) = y(\pi), \ y'(-\pi) = y'(\pi).$$

3.2 Eigenfunction Expansions.

In this section we will investigate additional properties of Regular Sturm–Liouville Problems. We begin with the following result, already mentioned in the previous section.

Theorem 3.2.1. *Every eigenvalue of the regular Sturm–Liouville problem (3.1.2) and (3.1.3) is simple, i.e., if λ is an eigenvalue and y_1 and y_2 are* *two corresponding eigenfunctions to* λ, *then* y_1 *and* y_2 *are linearly dependent (one function is a scalar multiple of the other function).*

Proof. Since y_1 and y_2 both are solutions of (3.1.2), we have

$$\begin{cases} \big(p(x)y_1'(x)\big)' + q(x)y_1(x) + \lambda r(x)y_1(x) = 0 \\ \big(p(x)y_2'(x)\big)' + q(x)y_2(x) + \lambda r(x)y_2(x) = 0. \end{cases}$$

Multiplying the first equation by $y_2(x)$ and the second by $y_1(x)$ and subtracting, we get

$$(3.2.1) \qquad \big(p(x)y_1'(x)\big)' y_2(x) - \big(p(x)y_2'(x)\big)' y_1(x) = 0.$$

However, since

$$\big[p(x)y_1'(x)\, y_2(x) - p(x)y_2'(x)\, y_1(x)\big]' = \big(p(x)y_1'(x)\big)' y_2(x) - \big(p(x)y_2'(x)\big)' y_1(x)$$

from (3.2.1) it follows that

$$\big[p(x)y_1'(x)y_2(x) - p(x)y_2'(x)\, y_1(x)\big]' = 0, \quad a \le x \le b$$

and hence

$$(3.2.2) \qquad p(x)\big[y_1'(x)\, y_2(x) - y_2'(x)\, y_1(x)\big] = C = constant, \quad a \le x \le b.$$

To find C, we use the fact that y_1 and y_2 satisfy the boundary condition at the point $x = a$:

$$\begin{cases} \alpha_1 y_1(a) + \beta_1 y_1'(a) = 0 \\ \alpha_1 y_2(a) + \beta_1 y_2'(a) = 0. \end{cases}$$

Since at least one of the coefficients α_1 and β_1 is not zero from the above two equations it follows that

$$y_1(a)y_2'(a) - y_2(a)y_1'(a) = 0.$$

Thus, from (3.2.2) we have $C = 0$ and hence

$$p(x)\big[y_1'(x)\, y_2(x) - y_2'(x)\, y_1(x)\big] = 0, \quad a \le x \le b.$$

Since $p(x) > 0$, for $x \in [a, b]$ we must have

$$y_1'(x)\, y_2(x) - y_2'(x)y_1(x) = 0, \quad a \le x \le b,$$

from which we conclude that $y_2(x) = Ay_1(x)$ for some constant A. ■

The next result is about the "orthogonality" property of the eigenfunctions.

Theorem 3.2.2. *If λ_1 and λ_2 are two distinct eigenvalues of the regular Sturm–Liouville problem* (3.1.2) *and* (3.1.3), *with corresponding eigenfunctions y_1 and y_2, then*

$$\int_a^b y_1(x)y_2(x)r(x)\,dx = 0.$$

Proof. Since y_1 and y_2 both are solutions of (3.1.2), we have

$$\begin{cases} (p(x)y_1'(x))' + q(x)y_1(x) + \lambda_1 r(x)y_1(x) = 0 \\ (p(x)y_2'(x))' + q(x)y_2(x) + \lambda_2 r(x)y_2(x) = 0. \end{cases}$$

Multiplying the first equation by $y_2(x)$ and the second by $y_1(x)$ and subtracting, we get

$$(3.2.3) \quad (p(x)y_1'(x))'y_2(x) - (p(x)y_2'(x))'y_1(x) + (\lambda_1 - \lambda_2)r(x)y_1(x)y_2(x) = 0.$$

Using the identity

$$[p(x)y_1'(x)y_2(x) - p(x)y_2'(x)y_1(x)]' = (p(x)y_1'(x))'y_2(x) - (p(x)y_2'(x))'y_1(x)$$

of the previous theorem, Equation (3.2.3) becomes

$$[p(x)y_1'(x)y_2(x) - p(x)y_2'(x)y_1(x)]' + (\lambda_1 - \lambda_2)r(x)y_1(x)y_2(x) = 0.$$

Integration of the last equation implies

$$(3.2.4) \qquad p(x)\left[y_1'y_2 - y_2'y_1 \right]_{x=a}^{x=b} + (\lambda_1 - \lambda_2)\int_a^b r(x)y_1(x)y_2(x)\,dx = 0.$$

Now, following the argument in Theorem 3.2.1, from the boundary conditions for the functions $y_1(x)$ and $y_2(x)$ at $x = a$ and $x = b$ we get

$$\begin{cases} y_1(a)y_2'(a) - y_2(a)y_1'(a) = 0 \\ y_1(b)y_2'(b) - y_2(b)y_1'(b) = 0. \end{cases}$$

Inserting the last two equations in (3.2.4) we have

$$(\lambda_1 - \lambda_2)\int_a^b r(x)y_1(x)y_2(x)\,dx = 0,$$

and since $\lambda_1 \neq \lambda_2$, it follows that

$$\int_a^b r(x)y_1(x)y_2(x)\,dx = 0. \quad \blacksquare$$

Example 3.2.1. Show that the eigenfunctions of the Sturm–Liouville problem in Example 3.1.1 are orthogonal.

Solution. In Example 3.1.1 we found that the eigenfunctions of the boundary value problem

$$y''(x) + \lambda = 0, \quad 0 < x < l,$$

subject to the boundary conditions $y(0) = 0$ and $y'(l) = 0$, are given by

$$y_n(x) = \sin \mu_n x, \quad \mu_n = \frac{(2n-1)\pi}{2l}, \quad n = 1, 2, \dots,.$$

The weight function r in this example is $r(x) \equiv 1$. Observe, that, if m and n are two distinct natural numbers, then using the trigonometric identity

$$\frac{1}{2}(\sin x)(\sin y) = \cos(x - y) - \cos(x + y)$$

we have

$$\int_0^l y_m(x) y_n(x) w(x)\, dx = \int_0^l \sin \frac{(2m-1)\pi x}{2l} \sin \frac{(2n-1)\pi x}{2l}\, dx$$

$$= 2 \int_0^l \left[\cos \frac{(m-n)\pi x}{l} - \cos \frac{(m+n-1)\pi x}{l} \right] dx$$

$$= -\frac{2l}{(m-n)\pi} \left[\sin \frac{(m-n)\pi x}{l} \right]_0^l + \frac{2l}{(m+n-1)\pi} \left[\sin \frac{(m+n-1)\pi x}{l} \right]_0^l$$

$$= 0.$$

Now, with the help of this theorem we can expand a given function f in a series of eigenfunctions of a regular Sturm–Liouville problem. We have had examples of such expansions in Chapter 1. Namely, if f is a continuous and piecewise smooth function on the interval $[0, 1]$, and satisfies the boundary conditions $f(0) = f(1) = 0$, then f can be expanded in the Fourier sine series

$$f(x) = \sum_{n=1}^{\infty} b_n \sin n\pi x,$$

where the coefficients b_n are given by

$$b_n = 2 \int_0^1 f(x) \sin n\pi x,\, dx.$$

Since $f(x)$ is continuous we have that the Fourier sine series converges to $f(x)$ for every $x \in [0, 1]$. Notice that the functions $\sin n\pi x$, $n = 1, 2 \ldots$ are the eigenfunctions of the boundary value problem

$$y'' + \lambda y = 0, \quad y(0) = y(1) = 0.$$

We have had similar examples for expanding functions in Fourier cosine series.

Let $f(x)$ be a function defined on the interval $[a, b]$. For a sequence

$$\{y_n(x) : n \in \mathbb{N}\}$$

of eigenfunctions of a given Sturm–Liouville problem (3.1.2) and (3.1.3), we wish to express f as an infinite linear combination of the eigenfunctions $y_n(x)$.

Let the function f be such that

$$\int_a^b f(x) y_n(x) r(x) \, dx < \infty \quad \text{for each } n \in \mathbb{N}$$

and let

(3.2.5)
$$f(x) = \sum_{k=1}^\infty c_k y_k(x), \quad x \in (a, b).$$

But the question is how to compute each of the coefficients c_k in the series (3.2.5). To answer this question we work very similarly to the case of Fourier series. Let n be any natural number. If we multiply Equation (3.2.5) by $r(x) y_n(x)$, and if we assume that the series can be integrated term by term, we obtain that

(3.2.6)
$$\int_a^b f(x) y_n(x) r(x) \, dx = \sum_{k=1}^\infty c_k \int_a^b y_k(x) y_n(x) r(x) \, dx.$$

From the orthogonality property of the eigenfunctions $\{y_k(x) : k = 1, 2, \ldots\}$ given in Theorem 3.2.2, it follows that

$$\int_a^b y_k(x) y_n(x) r(x) \, dx = \begin{cases} \int_a^b (y_n(x))^2 r(x) \, dx & \text{if } k = n \\ 0 & \text{if } k \neq n, \end{cases}$$

and therefore, from (3.2.6) we have

(3.2.7)
$$c_n = \frac{\int_a^b f(x) y_n(x) r(x) \, dx}{\int_a^b (y_n(x))^2 r(x) \, dx}.$$

Remark. The series in (3.2.5) is called the *generalized Fourier series* (also called *the eigenfunction expansions*) of the function f with respect to the eigenfunctions $y_k(x)$ and c_k, given by (3.2.7), are called *generalized Fourier coefficients* of f.

The study of pointwise and uniform convergence of this kind of generalized Fourier series is a challenging problem. We present here the following theorem without proof, dealing with the pointwise and uniform convergence of such series.

Theorem 3.2.3. *Let λ_n and $y_n(x)$, $n = 1, 2, \ldots$, be the eigenvalues and associated eigenfunctions, respectively, of the regular Sturm–Liouville problem* (3.1.2) *and* (3.1.3)*. Then*

(i) *If both $f(x)$ and $f'(x)$ are piecewise continuous on $[a, b]$, then f can be expanded in a convergent generalized Fourier series* (3.2.5)*, whose generalized Fourier coefficients c_n are given by* (3.2.7)*, and moreover*

$$\sum_{n=1}^{\infty} c_n y_n(x) = \frac{f(x^-) + f(x^+)}{2}$$

at any point x in the open interval (a, b).

(ii) *If f is continuous and $f'(x)$ is piecewise continuous on the interval $[a, b]$, and if f satisfies both boundary conditions* (3.1.3) *of the Sturm–Liouville problem, then the series converges uniformly on $[a, b]$.*

(iii) *If both $f(x)$ and $f'(x)$ are piecewise continuous on $[a, b]$, and f is continuous on a subinterval $[\alpha, \beta] \subset [a, b]$, then the generalized Fourier series of f converges uniformly to f on the subinterval $[\alpha, \beta]$.*

Let us illustrate this theorem with a few examples.

Example 3.2.2. Expand the function $f(x) = x$, $0 \leq x \leq \pi$, in terms of the eigenfunctions of the regular Sturm–Liouville problem

$$y'' + \lambda y = 0, \ 0 < x < \pi, \ y(0) = y'(\pi) = 0,$$

and discuss the convergence of the associated generalized Fourier series.

Solution. In Example 3.1.1 we showed that the eigenfunctions of this boundary value problem are

$$y_n(x) = \sin nx, \quad n = 1, 2, \ldots.$$

To compute the generalized Fourier coefficients c_n in (3.2.7) of the given function, first we compute

$$\int_0^\pi \left(y_n(x) \right)^2 r(x)\, dx = \int_0^\pi \sin^2 nx\, dx$$

$$= \left[\frac{x}{2} - \frac{\sin 2nx}{4n} \right]_{x=0}^{x=\pi} = \frac{\pi}{2}.$$

Therefore,

$$c_n = \frac{\int\limits_0^\pi f(x) y_n(x) r(x)\,dx}{\int\limits_0^\pi \left(y_n(x)\right)^2 r(x)\,dx} = \frac{2}{\pi} \int\limits_0^\pi x \sin nx\,dx$$

$$= \frac{2}{\pi}\left[-x\frac{\cos nx}{n} + \frac{\sin nx}{n^2}\right]_{x=0}^{x=\pi}$$

$$= \frac{2}{\pi}\left[-\pi \cos n\pi\right] = \frac{2}{n}(-1)^{n-1}.$$

Now, since $f(x) = x$ is continuous on $[0, \pi]$ by the above theorem we have

$$x = 2\sum_{n=1}^\infty \frac{(-1)^{n-1}}{n}\sin nx, \quad \text{for every } x \in (0, \pi).$$

Example 3.2.3. Expand the constant function $f(x) = 1$, $1 \le x \le e$, in terms of the eigenfunctions of the regular Sturm–Liouville problem

$$\left(xy'(x)\right)' + \frac{\lambda}{x}y = 0, \quad 1 < x < e, \ y(1) = y'(e) = 0,$$

and discuss the convergence of the associated generalized Fourier series.

Solution. The eigenvalues and corresponding eigenfunctions, respectively, of the given eigenvalue problem are given by

$$\lambda_n = \frac{(2n-1)^2\pi^2}{4}, \quad y_n(x) = \sin\left(\frac{(2n-1)\pi}{2}\ln x\right), \ n = 1, 2, \ldots$$

(See Exercise 5 (c) of Section 3.1.)

To compute the generalized Fourier coefficients c_n in (3.2.7) of the given function, first we compute

$$\int\limits_1^e y_n^2(x)\, r(x)\, dx = \int\limits_1^e \sin^2\left(\frac{(2n-1)\pi}{2}\ln x\right)\frac{1}{x}\,dx \quad \left(t = \frac{(2n-1)\pi}{2}\ln x\right)$$

$$= \frac{2}{(2n-1)\pi}\int\limits_0^{(2n-1)\pi/2} \sin^2 t\,dt = \frac{1}{(2n-1)\pi}\left[t - \frac{1}{2}\sin(2t)\right]_{t=0}^{t=(2n-1)\pi/2} = \frac{1}{2}.$$

Therefore,

$$c_n = \frac{\int\limits_1^e f(x) y_n(x) r(x)\, dx}{\int\limits_1^e y_n^2(x)\, r(x)\, dx} = 2\int\limits_1^e \sin\left(\frac{(2n-1)\pi}{2}\ln x\right)\frac{1}{x}\, dx$$

$$= \frac{4}{(2n-1)\pi}\int\limits_0^{\frac{(2n-1)\pi}{2}} \sin t\, dt = -\frac{4}{(2n-1)\pi}\Big[\cos t\Big]_{t=0}^{t=(2n-1)\pi/2}$$

$$= \frac{4}{(2n-1)\pi}.$$

Since the given function $f = 1$ does not satisfy the boundary conditions at $x = 1$ and $x = e$, we do not have uniform convergence of the Fourier series on the interval $[1, e]$. However, we have pointwise convergence to 1 for all x in the interval $(1, e)$:

$$1 = \sum_{n=1}^{\infty} \frac{4}{(2n-1)\pi}\sin\left(\frac{(2n-1)\pi}{2}\ln x\right),\quad 1 < x < e.$$

Now, if, for example, we take $x = \sqrt{e} \in (1, e)$ in the above series, we obtain

$$1 = \sum_{n=1}^{\infty} \frac{4}{(2n-1)\pi}\sin\left(\frac{(2n-1)\pi}{4}\right)$$

and since

$$\sin\left(\frac{(2n-1)\pi}{4}\right) = \begin{cases} (-1)^k \frac{\sqrt{2}}{2} & \text{if } n = 2k \\ (-1)^{k+1}\frac{\sqrt{2}}{2} & \text{if } n = 2k-1 \end{cases}$$

it follows that

$$\frac{\pi}{4} = \frac{\sqrt{2}}{2}\left[1 + \frac{1}{3} - \frac{1}{5} - \frac{1}{7} + \frac{1}{9} + \frac{1}{11} - \cdots\right].$$

Therefore, we have the following result.

$$1 + \frac{1}{3} - \frac{1}{5} - \frac{1}{7} + \frac{1}{9} + \frac{1}{11} - \cdots = \frac{\sqrt{2}}{4}\pi.$$

Exercises for Section 3.2.

1. The eigenvalues and eigenfunctions of the problem

$$\left(xy'(x)\right)' + \frac{\lambda}{x}y = 0, \quad 1 < x < 2, \ y(1) = y(2) = 0$$

are given by

$$\lambda_n = \left(\frac{n\pi}{\ln 2}\right)^2, \quad y_n(x) = \sin\left(\sqrt{\lambda_n}\,\ln x\right).$$

Find the expansion of the function $f(x) = 1$ in terms of these eigen-functions. What values does the series converge to at the points $x = 1$ and $x = b$?

2. Using the eigenfunctions of the boundary value problem

$$y'' + \lambda y = 0, \quad y(0) = y'(1) = 0,$$

find the eigenfunction expansion of each of the following functions.

(a) $f(x) = 1, \ 0 \le x \le 1$.

(b) $f(x) = x, \ 0 \le x \le 1$.

(c) $f(x) = \begin{cases} 1, & 0 \le x < \frac{1}{2} \\ 0, & \frac{1}{2} \le x \le 1. \end{cases}$

(d) $f(x) = \begin{cases} 2x, & 0 \le x < \frac{1}{2} \\ 1, & \frac{1}{2} \le x \le 1. \end{cases}$

3. The Sturm–Liouville problem

$$y'' + \lambda y = 0, \quad y(0) = y(l) = 0$$

has eigenfunctions

$$y_n(x) = \sin\left(\frac{n\pi x}{l}\right) \quad n = 1, 2, \ldots.$$

Find the eigenfunction expansion of the function $f(x) = x$ using these eigenfunctions.

4. The Sturm–Liouville problem

$$y'' + \lambda y = 0, \quad y(0) = y'(l) = 0$$

has eigenfunctions

$$y_n(x) = \sin\left(\frac{(2n-1)\pi x}{2l}\right) \quad n = 1, 2, \ldots .$$

Find the eigenfunction expansion of the function $f(x) = x$ using these eigenfunctions.

5. The eigenfunctions of the Sturm–Liouville problem

$$y'' + \lambda y = 0, \quad 0 < x < 1, \ y'(0) = 0, \ y(1) + y'(1) = 0$$

are

$$y_n(x) = \cos \sqrt{\lambda_n},$$

where λ_n are the positive solutions of $\cos \sqrt{\lambda} - \sqrt{\lambda} \sin \sqrt{\lambda} = 0$.
   Find the eigenfunction expansion of the following functions in terms of these eigenfunctions.

(a) $f(x) = 1$.

(b) $f(x) = x^c$, $c \in \mathbb{R}$.

3.3 Singular Sturm–Liouville Problems.

In this section we study Singular Sturm–Liouville Problems.

3.3.1 Definition of Singular Sturm–Liouville Problems.

In the previous sections we have discussed regular Sturm–Liouville problems and we have mentioned an example of a boundary value problem which is not regular. Because of their importance in applications, it is worthwhile to discuss Sturm–Liouville problems which are not regular. A problem which is not a regular Sturm–Liouville problem is called a *singular Sturm–Liouville problem*. A systematic study of such Sturm–Liouville problems is quite lengthy and technical and so we restrict ourselves to some special cases such as Legendre's and Bessel's equations. This class of boundary value problems has found significant physical applications. First we define precisely the notion of a singular Sturm–Liouville problem.

Definition 3.3.1. A Sturm–Liouville differential equation

$$\Big(p(x)y'(x)\Big)' + \big[q(x) + \lambda r(x)\big]y = 0$$

is called a *singular Sturm–Liouville problem* on an interval $[a, b]$ if at least one of the following conditions is satisfied.

(i)   The interval $[a, b]$ is unbounded, i.e., either $a = -\infty$ and/or $b = \infty$.

(ii) $p(x) = 0$ for some $x \in [a, b]$ or $r(x) = 0$ for some $x \in [a, b]$.

(iii) $|p(x)| \to \infty$ and/or $r(x) \to \infty$ as $x \to a$ and/or $x \to b$.

For singular Sturm-Liouville problems, appropriate boundary conditions need to be specified. In particular, if $p(x)$ vanishes at $x = a$ or $x = b$, we should require that $y(x)$ and $y'(x)$ remain bounded as $x \to a$, or $x \to b$, respectively.

As for regular Sturm–Liouville problems, the orthogonality property of the eigenfunctions with respect to the weight function $r(x)$ holds:

If $y_1(x)$ and $y_2(x)$ are any two linearly independent eigenfunctions of a singular Sturm-Liouville problem (with appropriate boundary conditions) then

$$\int_a^b y_1(x)y_2(x)r(x)\,dx = 0.$$

There are many differences between regular and singular Sturm-Liouville problems. But one, the most profound difference, is that the *spectrum* (the set of all eigenvalues) of a singular Sturm-Liouville problem can happen not to be a sequence of numbers. In other words, every number λ in some interval can be an eigenvalue of a singular Sturm-Liouville problem.

Example 3.3.1. Solve the following singular Sturm–Liouville problem.

$$\begin{cases} y''(x) + \lambda y = 0, \quad 0 < x < \infty, \\ y'(0) = 0, \quad |y'(x)| \text{ is bounded on } (0, \infty). \end{cases}$$

Solution. For $\lambda = 0$, the general solution of the given differential equation is $y(x) = A + Bx$. The boundary condition $y'(0) = 0$ implies that $B = 0$. Therefore $\lambda = 0$ is an eigenvalue and $y(x) \equiv 1$ on $[0, \infty)$ is the associated eigenfunction of the problem.

For $\lambda = -\mu^2 < 0$, the general solution of the given differential equation is given by

$$y(x) = Ae^{\mu x} + Be^{-\mu x}.$$

From the boundary condition $y'(0) = 0$ we obtain $A - B = 0$, and so $y(x) = A(e^{\mu x} - e^{-\mu x})$. If $A \neq 0$, then from $e^{\mu x} - e^{-\mu x} \to \infty$ as $x \to \infty$ it follows that $|y(x)|$ is not bounded on $(0, \infty)$. Therefore, any $\lambda < 0$ is not an eigenvalue of the problem.

If $\lambda > 0$, then the general solution of the given differential equation is

$$y(x) = A \sin\left(\sqrt{\lambda}x\right) + B \cos\left(\sqrt{\lambda}x\right).$$

The boundary condition $y'(0) = 0$ implies $A = 0$, and therefore $y(x) = B \cos\left(\sqrt{\lambda}x\right)$. Since $\left|\cos(\sqrt{\lambda}x)\right| \leq 1$ for every x it follows that any $\lambda > 0$ is an eigenvalue of the given boundary value problem, with corresponding eigenfunction $y(x) = \cos(\sqrt{\lambda}x)$. Therefore, the set of all eigenvalues of the given problem is the half-line $[0, \infty)$.

Since the most frequently encountered singular Sturm–Liouville problems in mathematical physics are those involving Legendre and Bessel's differential equations, the next two parts of this section are devoted to these differential equations. The functions which are the solutions of these two equations are only two examples of so called *special functions of mathematical physics*. Besides the Legendre and Bessel functions, there are other important special functions (which will be not discussed in this book), such as the Tchebychev, Hermite, Laguerre and Jacobi functions.

3.3.2 Legendre's Differential Equation.

In this section we will discuss in some detail Legendre's differential equation and its solutions.

Definition 3.3.2. For a number λ with $\lambda > -1/2$, the differential equation

$$(3.3.1) \qquad (1 - x^2)y''(x) - 2xy'(x) + \lambda(\lambda + 1)y = 0, \quad -1 \leq x \leq 1.$$

or in the Sturm–Liouville form

$$(3.3.2) \qquad \left((1 - x^2)y'(x)\right)' + \lambda(\lambda + 1)y = 0$$

is called *Legendre's differential equation of order* λ.

Since $p(x) = 1 - x^2$ and $p(-1) = p(1) = 0$, Legendre's differential equation is an equation of a singular Sturm-Liouville problem. There is not a simple solution of (3.3.1). Therefore we try for a solution $y(x)$ of the form of power series

$$(3.3.3) \qquad y(x) = \sum_{k=0}^{\infty} a_k x^k.$$

If we substitute the derivatives

$$y'(x) = \sum_{k=0}^{\infty} k a_k x^{k-1} \text{ and } y''(x) = \sum_{k=0}^{\infty} k(k-1) a_k x^{k-2}$$

in (3.3.1) we obtain

$$\sum_{k=0}^{\infty} (1-x^2) k(k-1) a_k x^{k-2} - 2x \sum_{k=0}^{\infty} k a_k x^{k-1} + \sum_{k=0}^{\infty} \lambda(\lambda+1) a_k x^k = 0,$$

or after a rearrangement

$$\sum_{k=0}^{\infty} k(k-1) a_k x^{k-2} + \sum_{k=0}^{\infty} \left[\lambda(\lambda+1) - k(k+1) \right] a_k x^k = 0.$$

If we re-index the first sum, after a rearrangement we obtain

$$\sum_{k=0}^{\infty} \left\{ (k+2)(k+1) a_{k+2} + \left[\lambda(\lambda+1) - k(k+1) \right] a_k \right\} x^k = 0.$$

Since the above equation holds for every $x \in (-1,1)$, each coefficient in the above power series must be zero, leading to the following recursive relation between the coefficients a_k:

$$(3.3.4) \qquad a_{k+2} = \frac{k(k+1) - \lambda(\lambda+1)}{(k+2)(k+1)} a_k, \quad k = 0,1,2,\ldots.$$

If k is an even number, then from the recursive formula (3.3.4) by induction, we have that

$$(3.3.5) \qquad a_{2m} = (-1)^m \frac{\prod_{k=1}^{m} (\lambda - 2k + 2)}{(2m)!} a_0.$$

If k is an odd number, then from the recursive formula (3.3.4), again by induction, we have that

$$(3.3.6) \qquad a_{2m+1} = (-1)^m \frac{\prod_{k=1}^{m} (\lambda - 2k + 1)}{(2m+1)!} a_1.$$

Inserting (3.3.5) and (3.3.6) into (3.3.3) we obtain that the general solution of Legendre's equation is

$$y(x) = a_0 y_0(x) + a_1 y_1(x)$$

where two particular solutions $y_0(x)$ and $y_1(x)$ are given by

$$(3.3.7) \quad y_0(x) = \sum_{m=0}^{\infty} (-1)^m \frac{\left(\prod_{k=1}^{m} (\lambda - 2k + 2)\right)\left(\prod_{k=1}^{m} (\lambda + 2k - 1)\right)}{(2m)!} x^{2m},$$

and

$$(3.3.8) \quad y_1(x) = \sum_{m=0}^{\infty} (-1)^m \frac{\left(\prod_{k=1}^{m} (\lambda - 2k + 1)\right)\left(\prod_{k=1}^{m} (\lambda + 2k)\right)}{(2m+1)!} x^{2m+1}.$$

Remark. The symbol $\prod$ (read "product") in the above formulas has the following meaning:

For a sequence $\{c_k : k = 1, 2, \ldots\}$ of numbers c_k we define

$$\prod_{k=1}^{m} c_k = c_1 \cdot c_2 \cdot \ldots \cdot c_m.$$

For example,

$$\prod_{k=1}^{m} (\lambda - 2k + 2) = \lambda(\lambda - 2) \cdot \ldots \cdot (\lambda - 2m + 2).$$

Remark. It is important to observe that if n is a nonnegative even integer and $\lambda = n$, then the power series for $y_0(x)$ in (3.3.7) terminates with the term involving x^n, and so $y_0(x)$ is a polynomial of degree n. Similarly, if n is an odd integer and $\lambda = n$, then the power series for $y_1(x)$ in (3.3.8) terminates with the term involving x^n.

The solutions of the differential equation (3.3.2) when $\lambda = n$ is a natural number are called *Legendre functions*.

Therefore, if n is a nonnegative integer, the polynomial solution $P_n(x)$ of the equation

$$(3.3.9) \qquad (1 - x^2)y''(x) - 2xy'(x) + n(n+1)y = 0, \quad -1 \le x \le 1$$

such that $P_n(1) = 1$ is called *Legendre's polynomial of degree n*, or *Legendre's function of the first kind* and is given by

$$(3.3.10) \qquad P_n(x) = \frac{1}{2^n} \sum_{k=0}^{m} (-1)^k \frac{(2n - 2k)!}{k!(n-k)!(n-2k)!} x^{n-2k},$$

where $m = n/2$ if n is even, and $m = (n-1)/2$ if n is odd.

The other solution of (3.3.9) which is linearly independent of $P_n(x)$ is called *Legendre's function of the second kind.*

The first six Legendre polynomials are listed below.

$$P_0(x) = 1; \ P_1(x) = x; \ P_2(x) = \frac{1}{2}(3x^2 - 1); \ P_3(x) = \frac{1}{2}(5x^3 - 3x);$$

$$P_4(x) = \frac{1}{8}(35x^4 - 30x^2 + 3); \ P_5(x) = \frac{1}{8}(63x^5 - 70x^3 + 15x).$$

The graphs of the first six polynomials are shown in Figure 3.3.1.

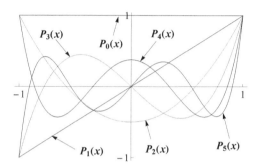

Figure 3.3.1

Now we state and prove several important properties of Legendre's polynomials.

Theorem 3.3.1. *Let n be a nonnegative even integer. Then any polynomial solution $y(x)$ which has only even powers of x of Legendre's differential equation (3.3.1) is a constant multiple of $P_n(x)$. Similarly, if n is a nonnegative odd integer, then any polynomial solution $y(x)$ of Legendre's differential equation (3.3.1) which has only odd powers of x is a constant multiple of $P_n(x)$.*

Proof. Suppose that n is a nonnegative even integer. Let $y(x)$ be a polynomial solution of (3.3.1) with only even powers of x. Then for some constants c_0 and $c-1$ we have

$$y(x) = c_0 y_0(x) + c_1 y_1(x),$$

where $y_0(x) = P_n(x)$ is a polynomial of degree n with even powers of x and $y_1(x)$ is a power series solution of (3.3.1) with odd powers of x. Now, since $y(x)$ is a polynomial we must have $c_1 = 0$, and so $y(x) = c_0 P_n(x)$.

The case when n is an odd integer is treated similarly. ∎

Theorem 3.3.2. Rodrigues' Formula. *Legendre's polynomials* $P_n(x)$ *for* $n = 0, 1, 2, \ldots$ *are given by*

(3.3.11) $$P_n(x) = \frac{1}{2^n n!} \frac{d^n}{dx^n} \left((x^2 - 1)^n \right).$$

Proof. If $y(x) = (x^2 - 1)^n$, then $y'(x) = 2nx(x^2 - 1)^{n-1}$ and therefore

$$(x^2 - 1)y'(x) = 2nxy(x).$$

If we differentiate the last equation $(n + 1)$ times, using the Leibnitz rule for differentiation, we obtain

$$(x^2 - 1)y^{(n+2)}(x) + \binom{n+1}{1} 2xy^{(n+1)}(x) + 2\binom{n+1}{2} y^{(n)}(x)$$

$$= 2nxy^{(n+1)}(x) + 2n\binom{n+1}{1} y^{(n)}(x).$$

If we introduce a new function $w(x)$ by $w(x) = y^{(n)}(x)$, then from the last equation it follows that

$$(x^2 - 1)w''(x) + 2(n+1)xw'(x) + (n+1)nw(x) = 2nxw'(x) + 2n(n+1)w(x),$$

or

$$(1 - x^2)w''(x) - 2xw'(x) + n(n+1)w(x) = 0.$$

So $w(x)$ is a solution of Legendre's equation, and since $w(x)$ is a polynomial, by Theorem 3.3.1 it follows that

$$\frac{d^n}{dx^n} \left((x^2 - 1)^n \right) = w(x) = c_n P_n(x)$$

for some constant c_n. To find the constant c_n, again we apply the Leibnitz rule

$$\frac{d^n}{dx^n} \left((x^2 - 1)^n \right) = \frac{d^n}{dx^n} \left(\left((x - 1)(x + 1) \right)^n \right)$$

$$= n!(x + 1)^n + R_n,$$

where R_n denotes the sum of the n remaining terms each having the factor $(x - 1)$. Therefore,

$$\frac{d^n}{dx^n} \left((x^2 - 1)^n \right) \Big|_{x=1} = 2^n n!,$$

and so from

$$P_n(1) = c_n 2^n n!$$

and $P_n(1) = 1$ we obtain

$$c_n = \frac{1}{2^n n!}. \qquad \blacksquare$$

Example 3.3.2. Compute $P_2(x)$ using Rodrigues' formula (3.3.11),

Solution. From (3.3.11) with $n = 2$,

$$P_2(x) = \frac{1}{2^2 2!} \frac{d^2}{dx^2} \left((x^2 - 1)^2 \right) = \frac{1}{8} \frac{d^2}{dx^2} \left(x^4 - 2x^2 + 1 \right) = \frac{1}{2} (3x^2 - 1).$$

Legendre's polynomials can be defined through their *generating function*

$$(3.3.12) \qquad\qquad G(t, x) \equiv \frac{1}{\sqrt{1 - 2xt + t^2}}.$$

Theorem 3.3.3. *Legendre's polynomials* $P_n(x)$, $n = 0, 1, 2, \cdots$ *are exactly the coefficients in the expansion*

$$(3.3.13) \qquad\qquad \frac{1}{\sqrt{1 - 2xt + t^2}} = \sum_{n=0}^{\infty} P_n(x) t^n.$$

Proof. If we use the binomial expansion for the function $G(t, x)$, then we have

$$
\begin{aligned}
G(t, x) &= (1 - 2xt + t^2)^{-\frac{1}{2}} = \sum_{n=0}^{\infty} \frac{1 \cdot 3 \cdots \cdots (2n - 1)}{2^n n!} (2xt - t^2)^n \\
&= \sum_{n=0}^{\infty} \frac{1 \cdot 3 \cdots \cdots (2n - 1)}{2^n n!} \sum_{k=0}^{n} \frac{n!}{k!(n - k)!} (2x)^{n-k} (-t^2)^k \\
&= \sum_{n=0}^{\infty} \left[\sum_{k=0}^{m} (-1)^k \frac{(2n - 2k)!}{2^n k!(n - 2k)!(n - k)!} x^{n-2k} \right] t^n,
\end{aligned}
$$

where $m = \dfrac{n}{2}$ if n is even and $m = \dfrac{n - 1}{2}$ if n is odd. But the coefficient of t^n is exactly the Legendre polynomial $P_n(x)$. ∎

Theorem 3.3.4. *Legendre polynomials satisfy the recursive relations*

$$(3.3.14) \qquad (n+1)P_{n+1}(x) = (2n + 1)x P_n(x) - n P_{n-1}(x).$$

$$(3.3.15) \qquad P'_{n+1}(x) + P'_{n-1}(x) = 2x P'_n(x) + P_n(x).$$

$$(3.3.16) \qquad P'_{n+1}(x) - P'_{n-1}(x) = (2n+1) P_n(x).$$

$$(3.3.17) \qquad \int_{-1}^{1} P_n^2(x) \, dx = \frac{2}{2n + 1}.$$

Proof. We leave the proof of identity (3.3.16) as an exercise (Exercise 4 of this section).

First we prove (3.1.14). If we differentiate Equation (3.3.13) with respect to t we have that

$$\frac{\partial G(t,x)}{\partial t} = \frac{x - t}{(1 - 2xt + t^2)^{\frac{3}{2}}} = \sum_{n=0}^{\infty} nP_n(x)t^{n-1},$$

which can be rewritten as

$$(1 - 2xt + t^2) \sum_{n=0}^{\infty} nP_n(x)t^{n-1} - \frac{x - t}{\sqrt{1 - 2xt + t^2}} = 0,$$

or, using again the expansion (3.3.13) for the generating function $G(t,x)$,

$$(1 - 2xt + t^2) \sum_{n=0}^{\infty} nP_n(x)t^{n-1} - (x - t) \sum_{n=0}^{\infty} P_n(x)t^n = 0.$$

After multiplication, the last equation becomes

$$\sum_{n=0}^{\infty} nP_n(x)t^{n-1} - 2x \sum_{n=0}^{\infty} nP_n(x)t^n + \sum_{n=0}^{\infty} nP_n(x)t^{n+1}$$

$$+ \sum_{n=0}^{\infty} P_n(x)t^{n+1} - \sum_{n=0}^{\infty} xP_n(x)t^n = 0,$$

or after re-indexing, we obtain

$$\sum_{n=0}^{\infty} \left[\sum_{n=0}^{\infty} nP_n(x)t^{n-1}(n + 1)P_{n+1}(x) - (2n + 1)xP_n(x) + nP_{n-1}(x) \right] t^n = 0.$$

Since the left hand side vanishes for all t, the coefficients of each power of t must be zero, giving (3.3.14).

To prove (3.3.15) we differentiate Equation (3.3.13) with respect to x:

$$\frac{\partial G(t,x)}{\partial x} = \frac{t}{(1 - 2xt + t^2)^{\frac{3}{2}}} = \sum_{n=0}^{\infty} P_n'(x)t^n.$$

The last equation can be written as

$$(1 - 2xt + t^2) \sum_{n=0}^{\infty} P_n'(x)t^n - \frac{t}{\sqrt{1 - 2xt + t^2}} = 0,$$

or, in view of the expansion (3.3.13) for the generating function $g(t,x)$,

$$(1 - 2xt + t^2) \sum_{n=0}^{\infty} P_n'(x)t^n - t \sum_{n=0}^{\infty} P_n(x)t^n = 0.$$

Setting to zero each coefficient of t gives (3.3.15).

Formula (3.3.17) is easily verified for $n = 0$ and $n = 1$. Suppose $n \geq 2$. From (3.3.14) we have

$$nP_n(x) - (2n-1)xP_n(x) + (n-1)P_{n-2}(x) = 0,$$
$$(n+1)P_{n+1}(x) - (2n+1)xP_n(x) + nP_{n-1}(x) = 0,$$

for all $x \in (-1, 1)$. Now, if we multiply the first equation by $P_n(x)$ and the second by $P_{n-1}(x)$, integrate over the interval $[-1, 1]$ and use the orthogonality property of Legendre's polynomials we obtain

$$n \int_{-1}^{1} P_n^2(x)\, dx = (2n-1) \int_{-1}^{1} x P_{n-1}(x) P_n(x)\, dx,$$

$$n \int_{-1}^{1} P_{n-1}^2(x)\, dx = (2n+1) \int_{-1}^{1} x P_{n-1}(x) P_n(x)\, dx.$$

From the last two equations it follows that

$$(2n+1) \int_{-1}^{1} P_n^2(x)\, dx = (2n-1) \int_{-1}^{1} P_{n-1}^2(x)\, dx$$

from which, recursively, we obtain

$$(2n-1) \int_{-1}^{1} P_{n-1}^2(x)\, dx = (2n-3) \int_{-1}^{1} P_{n-2}^2(x)\, dx = \ldots = 3 \int_{-1}^{1} P_1^2(x)\, dx = 2,$$

which is the required assertion (3.3.17).   ■.

Example 3.3.3. Compute $P_3(x)$ using some recursive formula.

Solution. From (3.3.14) with $n = 2$, we have

$$3P_3(x) = 5xP_2(x) - 2P_1(x).$$

But, from $P_1(x) = x$ and $P_2(x) = \frac{1}{2}(3x^2 - 1)$, computed in Example 3.3.2, we have

$$3P_3(x) = 5x\frac{1}{2}(3x^2 - 1) - 2x = \frac{1}{2}(15x^3 - 9x).$$

Using the orthogonality property of Legendre's polynomials we can represent a piecewise continuous function $f(x)$ on the interval $(-1, 1)$ by the series

$$f(x) = \sum_{n=0}^{\infty} a_n P_n(x), \quad -1 \leq x \leq 1.$$

Indeed, using the orthogonality property and (3.3.16) we have that

$$a_n = \frac{2n+1}{2} \int_{-1}^{1} f(x) P_n(x) \, dx, \quad n = 0, 1, \ldots.$$

Remark. If the function f and its first n derivatives are continuous on the interval $(-1, 1)$, then using Rodrigues' formula (3.3.11) and integration by parts, it follows that

$$(3.3.18) \qquad\qquad a_n = \frac{(-1)^n}{2^n n!} \int_{-1}^{1} (x^2 - 1)^n f^{(n)}(x) \, dx.$$

Example 3.3.4. Represent $f(x) = x^2$ as a series of Legendre's polynomials.

Solution. Because of (3.3.15), we need to calculate only a_0, a_1 and a_2. From (3.3.18) we can easily calculate these coefficients and obtain that

$$x^2 = \frac{1}{3} + 0 \cdot x + \frac{1}{3}(3x^2 - 1).$$

Example 3.3.5. Expand the Heaviside function in a Fourier-Legendre series.

Solution. Recall that the Heaviside function is defined as

$$H(x) = \begin{cases} 1, & \text{if } x > 0, \\ 0, & \text{if } x < 0. \end{cases}$$

We need to compute

$$a_n = \frac{2n+1}{2} \int_{-1}^{1} H(x) P_n(x) \, dx = \frac{2n+1}{2} \int_{0}^{1} P_n(x) \, dx, \quad n = 0, 1, 2, \ldots.$$

For $n = 0$, we have

$$a_0 = \frac{1}{2} \int_{0}^{1} P_0(x) \, dx = \frac{1}{2} \int_{0}^{1} 1 \, dx = \frac{1}{2}.$$

For $n \geq 1$, from the identity (3.3.15) we have that

$$\frac{2n+1}{2} \int_{0}^{1} P_n(x) \, dx = \frac{1}{2} \int_{0}^{1} \left[P'_{n+1}(x) - P'_n(x) \right] \, dx$$

$$= \frac{1}{2} \left[P_{n+1}(1) - P_n(1) \right] - \frac{1}{2} \left[P_{n+1}(0) - P_n(0) \right]$$

$$= \frac{1}{2} \left[P_n(0) - P_{n+1}(0) \right].$$

Now, since

$$P_n(0) = \begin{cases} 0, & \text{if } n = 2k+1 \\ (-1)^k \frac{(2k-1)!!}{(2k)!!}, & \text{if } n = 2k, \end{cases}$$

(see Exercise 7 of this section), we have $a_{2k} = 0$, for $k = 0, 1, 2, \ldots$, and

$$\begin{aligned} a_{2k-1} &= (-1)^{k-1} \frac{(2k-3)!!}{(2k+2)!!} - (-1)^k \frac{(2k-1)!!}{(2k)!!} \\ &= (-1)^{k-1} \frac{(2k-3)!!}{(2k+2)!!} \frac{4k-1}{2k}. \end{aligned}$$

Remark. The symbol !! is defined as follows.

$$(2k-1)!! = 1 \cdot 3 \cdot \ldots \cdot (2k-1) \quad \text{and} \quad (2k)!! = 2 \cdot 4 \cdot \ldots \cdot (2k).$$

Therefore, the Fourier-Legendre expansion of the Heaviside function is

$$H(x) \sim \frac{1}{2} + \frac{1}{2} \sum_{k=1}^{\infty} (-1)^{k-1} \frac{(2k-3)!!}{(2k+2)!!} \frac{4k-1}{2k} P_{2k-1}(x).$$

The sum of the first 23 terms is shown in Figure 3.3.2. We note the slow convergence of the series to the Heaviside function. Also, we see that the Gibbs phenomenon is present due to the jump discontinuity at $x = 0$.

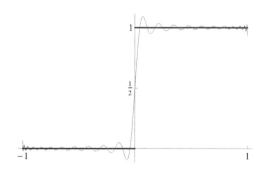

Figure 3.3.2

We conclude this section with a description of the *associated Legendre poly-nomials* which are related to the Legendre polynomials and which have important applications in solving some partial differential equations.

Consider the differential equation

$$(3.3.19) \qquad \frac{d}{dx}\left((1-x^2)\frac{dy}{dx}\right) + \left(\lambda - \frac{m^2}{1-x^2}\right)y = 0, \quad -1 < x < 1,$$

where $m = 0, 1, 2, \ldots$ and λ is a real number. The nontrivial and bounded functions on the interval $(-1, 1)$ which are solutions of Equation (3.3.19) are called *associated Legendre functions of order* m. Notice that if $m = 0$ and $\lambda = n(n+1)$ for $n \in \mathbb{N}$, then equation (3.3.19) is the Legendre equation of order n and so the associated Legendre functions in this case are the Legendre polynomials $P_n(x)$. To find the associated Legendre functions for any other m we use the following substitution.

$$(3.3.20) \qquad y(x) = (1-x^2)^{\frac{m}{2}} v(x).$$

If we substitute this $y(x)$ in Equation (3.3.19) we obtain the following equation.

$$(3.3.21) \quad (1-x^2)v''(x) - 2(m+1)xv'(x) + \left[\lambda - m(m+1)\right]v = 0, \quad -1 < x < 1.$$

To find a particular solution $v(x)$ of Equation (3.3.21) we consider the following Legendre equation.

$$(3.3.22) \qquad (1-x^2)\frac{d^2w}{dx^2} - 2x\frac{dw}{dx} + \lambda w = 0.$$

If we differentiate Equation (3.3.22) m times (using the Leibnitz rule) we obtain

$$(1-x^2)\frac{d^{m+2}w}{dx^{m+2}} - 2(m+1)x\frac{d^{m+1}w}{dx^{m+1}} + \left[\lambda - m(m+1)\right]\frac{d^mw}{dx^m} = 0,$$

which can be written in the form

$$(3.3.23) \quad (1-x^2)\left(\frac{d^mw}{dx^m}\right)'' - 2(m+1)x\left(\frac{d^mw}{dx^m}\right)' + \left[\lambda - m(m+1)\right]\frac{d^mw}{dx^m} = 0.$$

From (3.3.21) and (3.3.22) we have that $v(x)$ is given by

$$(3.3.24) \qquad v(x) = \frac{d^mw}{dx^m},$$

where $w(x)$ is a particular solution of the Legendre equation (3.3.23). We have already seen that the nontrivial bounded solutions $w(x)$ of the Legendre equation (3.3.22) are only if $\lambda = n(n+1)$ for some $n = 0, 1, 2, \ldots$ and in this case we have that $w(x) = P_n(x)$ is the Legendre polynomial of order n. Therefore from (3.3.20) and (3.3.23) it follows that

$$(3.3.25) \qquad y(x) = (1-x^2)^{\frac{m}{2}} \frac{d^m P_n(x)}{dx^m}$$

are the bounded solutions of (3.3.19). Since $P_n(x)$ are polynomials of degree n, from (3.3.25) using the Leibnitz rule it follows that $y(x)$ are nontrivial only if $m \leq n$, and so the associated Legendre functions, usually denoted by $P_n^{(m)}(x)$, are given by

$$(3.3.26) \qquad P_n^{(m)}(x) = (1 - x^2)^{\frac{m}{2}} \frac{d^m P_n(x)}{dx^m}.$$

For the associated Legendre functions we have the following orthogonality property.

Theorem 3.3.5. *For any m, n and k we have the following integral relation.*

$$(3.3.27) \qquad \int_{-1}^{1} P_n^{(m)}(x) \, P_k^{(m)}(x) \, dx = \begin{cases} 0, & k \neq n \\ \dfrac{2(n+m)!}{(2n+1)(n-m)!}, & k = n. \end{cases}$$

Proof. From the definition of the associated Legendre polynomials $P_n^{(m)}$ we have

$$(1-x^2)\frac{d^{m+2} P_n(x)}{dx^{m+2}} - 2(m+1)x\frac{d^{m+1} P_n(x)}{dx^{m+1}} + \left[n(n+1) - m(m+1)\right]\frac{d^m P_n(x)}{dx^m} = 0.$$

If we multiply the last equation by $(1 - x^2)^m$, then we obtain

$$(1 - x^2)^{m+1} \frac{d^{m+2} P_n(x)}{dx^{m+2}} - 2(m+1)x(1 - x^2)^m \frac{d^{m+1} P_n(x)}{dx^{m+1}}$$
$$+ \left[n(n+1) - m(m+1)\right] (1 - x^2)^m \frac{d^m P_n(x)}{dx^m} = 0,$$

which can be written as

$$\frac{d}{dx}\left((1 - x^2)^{m+1} \frac{d^{m+1} P_n(x)}{dx^m}\right)$$
$$(3.3.28)$$
$$= -\left[n(n+1) - m(m+1)\right] (1 - x^2)^m \frac{d^m P_n(x)}{dx^m}.$$

Now let

$$A_{n,k}^{(m)} = \int_{-1}^{1} P_n^{(m)} \, P_k^{(m)} \, dx.$$

If we use relation (3.3.26) for the associated Legendre polynomials $P_n^{(m)}$ and $P_k^{(m)}$, using the integration by parts formula we obtain

$$A_{n,k}^{(m)} = \int_{-1}^{1} (1 - x^2)^m \frac{d^m P_n(x)}{dx^m} \frac{d^m P_k(x)}{dx^m} dx$$

$$= \left[(1 - x^2)^m \frac{d^{m-1} P_n(x)}{dx^{m-1}} \frac{d^m P_k(x)}{dx^m} \right]_{x=-1}^{x=1}$$

$$- \int_{-1}^{1} \frac{d^{m-1} P_k(x)}{dx^{m-1}} \frac{d}{dx} \left((1 - x^2)^m \frac{d^m P_n(x)}{dx^m} \right) dx$$

$$= - \int_{-1}^{1} \frac{d^{m-1} P_k(x)}{dx^{m-1}} \frac{d}{dx} \left((1 - x^2)^m \frac{d^m P_n(x)}{dx^m} \right) dx.$$

The last equation, in view of (3.3.28), can be written in the form

$$A_{n,k}^{(m)} = [n(n + 1) - m(m - 1)] \int_{-1}^{1} (1 - x^2)^{m-1} \frac{d^{m-1} P_n(x)}{dx^{m-1}} \frac{d^{m-1} P_k(x)}{dx^{m-1}} dx$$

$$= (n + m)(n - m - 1) A_{n,k}^{(m-1)}.$$

By recursion, from the above formula we obtain

$$A_{n,k}^{(m)} = (n+m)(n-m-1)(n+m-1)(n-m-2) A_{n,k}^{(m-2)} = \ldots = \frac{(n + m)!}{(n - m)!} A_{n,k}^{(0)}.$$

Therefore, using the already known orthogonality property (3.3.17) for the Legendre polynomials

$$A_{n,k}^{(0)} = \int_{-1}^{1} P_n(x) P_k(x) \, dx = \begin{cases} 0, & n \neq k \\ \frac{2}{n+1}, & n = k \end{cases}$$

we obtain that

$$A_{n,k}^{(m)} = \frac{2(n + m)!}{(n + 1)(n - m)!}. \qquad \blacksquare$$

3.3.3 Bessel's Differential Equation.

Bessel's differential equation is one of the most important equations in mathematical physics. It has applications in electricity, hydrodynamics, heat propagation, wave phenomena and many other disciplines.

Definition 3.3.3. Suppose that μ is a nonnegative real number. Then the differential equation

$$(3.3.29) \qquad x^2 y''(x) + x y'(x) + (x^2 - \mu^2) y = 0, \quad x > 0$$

or in the Sturm–Liouville form

$$(3.3.30) \qquad \left(x y'(x) \right)' + \left(x - \frac{\mu^2}{x} \right) y = 0$$

is called *Bessel's differential equation of order* μ.

We will find the solution of Bessel's equation in the form of a Frobenius series (see Appendix D).

$$y(x) = x^r \sum_{k=0}^{\infty} a_k x^k = \sum_{k=0}^{\infty} a_k x^{k+r}$$

with $a_0 \neq 0$ and a constant r to be determined. If we substitute $y(x)$ and its derivatives into Bessel equation (3.3.29) we obtain

$$\sum_{k=0}^{\infty} (k+r)(k+r-1) a_k x^{k+r} + \sum_{k=0}^{\infty} (k+r) a_k x^{k+r}$$

$$+ \sum_{k=0}^{\infty} a_k x^{k+r+2} - \mu^2 \sum_{k=0}^{\infty} a_k x^{k+r} = 0.$$

If we write out explicitly the terms for $k = 0$ and $k = 1$ in the first sum, and re-index the other sums, then we have

$$(r^2 - \mu^2) a_0 + \left[(r+1)^2 - \mu^2 \right] a_1 x + \sum_{k=2}^{\infty} \left\{ \left[(k+r)^2 - \mu^2 \right] a_k + a_{k-2} \right\} x^k = 0.$$

In order for this equation to be satisfied identically for every $x > 0$, the coefficients of all powers of x must vanish. Therefore,

$$(3.3.31) \qquad (r^2 - \mu^2) a_0 = 0,$$

$$(3.3.32) \qquad \left[(r+1)^2 - \mu^2 \right] a_1 = 0,$$

$$(3.3.33) \qquad \left[(k+r)^2 - \mu^2 \right] a_k + a_{k-2} = 0, \quad k \geq 2.$$

Since $a_0 \neq 0$, from (3.3.31) we have $r^2 - \mu^2 = 0$, i.e., $r = \pm \mu$.

Case 1^0. $r = \mu$. Since $(r+1)^2 - \mu^2 \neq 0$ in this case, from (3.3.32) we have $a_1 = 0$, and therefore from (3.3.33), iteratively, it follows that

$$a_{2k+1} = 0, \quad k = 2, 3, \ldots$$

and

$$a_{2k} = -\frac{1}{2^2 k(k+\mu)} a_{2k-2} = \frac{1}{2^4 k(k-1)(k+\mu)(k-1+\mu)} a_{2k-4}$$

$$= \frac{(-1)^k}{2^{2k} k!(k+\mu)(k-1+\mu)\cdots(1+\mu)} a_0 = (-1)^k \frac{\Gamma(1+\mu)}{2^{2k} k!\Gamma(k+1+\mu)} a_0.$$

(See Appendix G for the Gamma function Γ.)

Therefore, one solution of the Bessel equation, denoted by J_μ, is given by

$$J_\mu(x)(x) = a_0 x^\mu \sum_{k=0}^{\infty} (-1)^k \frac{\Gamma(1+\mu)}{2^{2k} k!\Gamma(k+1+\mu)} x^{2k}.$$

For normalization reasons we choose

$$a_0 = \frac{1}{2^\mu \Gamma(\mu+1)}$$

and so we have that the solution $J_\mu(x)$ is given by

(3.3.34) $$J_\mu(x) = \sum_{k=0}^{\infty} \frac{(-1)^k}{k!\Gamma(k+\mu+1)} \left(\frac{x}{2}\right)^{2k+\mu}.$$

The function $J_\mu(x)$, $\mu \geq 0$, is called the *Bessel function of the first kind of order* μ.

If $\mu = n$ is a nonnegative integer, then using the properties of the Gamma function Γ, from (3.3.34) it follows that

$$J_n(x) = \sum_{k=0}^{\infty} \frac{(-1)^k}{k!(k+n)!} \left(\frac{x}{2}\right)^{2k+n}.$$

The plots of the Bessel functions $J_0(x)$, $J_1(x)$, $J_2(x)$ and $J_3(x)$ are given in Figure 3.3.3.

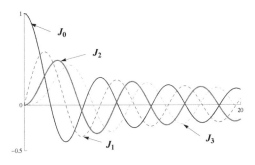

Figure 3.3.3

Case 2^0. $r = -\mu$. Consider the coefficient $(k + r)^2 - \mu^2 = k(k - 2\mu)$ of a_k in the iterative formula (3.3.33). So we have two cases to consider in this situation: either $k - 2\mu \neq 0$ for every $k \geq 2$, or $k - 2\mu = 0$ for some $k \geq 2$.

Case 2_a^0. $k - 2\mu \neq 0$ *for every* $k \geq 2$. In this case, $(k + r)^2 - \mu^2 \neq 0$ for every $k = 2, 3, \ldots$, and so from $a_1 = 0$ and (3.3.33), similarly as in Case 1^0 it follows that

$$a_{2k+1} = 0, \quad k = 2, 3, \ldots$$

and

$$a_{2k} = (-1)^k \frac{\Gamma(1 - \mu)}{2^{2k} k! \Gamma(k - \mu + 1)} a_0, \quad k = 1, 2, \ldots$$

Again, if we choose

$$a_0 = \frac{1}{2^\mu \Gamma(1 - \mu)},$$

then

$$J_{-\mu}(x) = \sum_{k=0}^{\infty} \frac{(-1)^k}{k! \Gamma(k - \mu + 1)} \left(\frac{x}{2}\right)^{2k - \mu}$$

is a solution of the Bessel equation in this case. Since

$$\lim_{x \to 0+} J_\mu(x) = 0, \quad \lim_{x \to 0+} J_{-\mu}(x) = \infty,$$

the Bessel functions J_μ and $J_{-\mu}$ are linearly independent, and so

$$y(x) = c_1 J_\mu(x) + c_2 J_{-\mu}(x)$$

is the general solution to the Bessel equation (3.3.29) in this case.

Case 2_b^0. $k - 2\mu = 0$ *for some* $k \geq 2$. Working similarly as in Case 2_a^0 it can be shown that the general solution to the Bessel equation (3.3.29) is a linear combination of the linearly independent Bessel functions J_μ and $J_{-\mu}$ if μ is not an integer. But if μ is an integer, then from

$$(3.3.35) \qquad J_{-\mu}(x) = (-1)^\mu J_\mu(x)$$

it follows that we have only one solution $J_\mu(x)$ of the Bessel equation and we don't have yet another, linearly independent solution. One of the several methods to find another, linearly independent solution of the Bessel equation in the case when 2μ is an integer, and μ is also an integer, is the following. If ν is not an integer number, then we define the function

$$(3.3.36) \qquad Y_\nu(x) = \frac{J_\nu(x) \cos(\mu\pi) - J_{-\nu}(x)}{\sin(\nu\pi)}.$$

This function, called the *Bessel function of the second kind of order* ν, is a solution of the Bessel equation of order ν, and it is linearly independent of the other solution $J_\nu(x)$. Next, for an integer number n, we define $Y_n(x)$ by the limiting process:

$$(3.3.37) \qquad\qquad Y_n(x) = \lim_{\substack{\nu \to n \\ \nu \text{ not integer}}} Y_\nu(x).$$

It can be shown that $Y_n(x)$ is a solution of the Bessel equation of order n, linearly independent of $J_n(x)$. Using l'Hôpital's rule in (3.3.36) with respect to ν it follows that

$$Y_n(x) = \frac{1}{\pi} \lim_{\nu \to n} \left[\frac{\partial}{\partial \mu} J_\nu(x) - \tan(\nu\pi) - \frac{\frac{\partial}{\partial \nu} J_{-\nu}(x)}{\cos(\nu\pi)} \right]$$

$$= \frac{1}{\pi} \left[\frac{\partial J_\nu(x)}{\partial \nu} - (-1)^n \frac{\partial J_{-\nu}(x)}{\partial \nu} \right]_{\nu=n}.$$

The plots of the Bessel functions $Y_0(x)$, $Y_1(x)$, $Y_2(x)$ and $Y_3(x)$ are given in Figure 3.3.4.

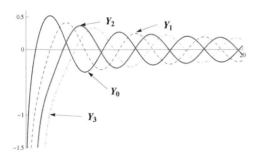

Figure 3.3.4

The following theorem summarizes the above discussion.

Theorem 3.3.6. *Suppose that μ is a nonnegative real number. If μ is not an integer, then $J_\mu(x)$, defined by*

$$J_\mu(x) = \sum_{k=0}^{\infty} \frac{(-1)^k}{k!\,\Gamma(k+\mu+1)} \left(\frac{x}{2}\right)^{2k+\mu}$$

is one solution of the Bessel differential equation (3.3.29) of order μ. Another solution of the Bessel equation, independent of $J_\mu(x)$, is $J_{-\mu}(x)$, obtained when μ in the above series is replaced by $-\mu$.

If $\mu = n$ is an integer, then J_n is one solution of the Bessel differential equation (3.3.20) of order n. Another solution of the Bessel equation, independent of $J_n(x)$, is $Y_n(x)$, which is defined by

$$Y_n(x) = \lim_{\substack{\mu \to n \\ \mu \text{ is not integer}}} Y_\mu(x),$$

where for a noninteger number μ, the function $Y_\mu(x)$ is defined by

$$Y_\mu(x) = \frac{J_\mu(x)\cos(\mu\pi) - J_{-\mu}(x)}{\sin(\mu\pi)}.$$

Let us take a few examples.

Example 3.3.6. Find explicitly the Bessel functions $J_{\frac{1}{2}}(x)$ and $J_{-\frac{1}{2}}(x)$.

Solution. From equation (3.3.34) and the results in Appendix G for the Gamma function Γ it follows that

$$J_{\frac{1}{2}}(x) = \sum_{j=0}^{\infty} \frac{(-1)^j}{j!\,\Gamma(j+\frac{3}{2})} \left(\frac{x}{2}\right)^{2j+\frac{1}{2}} = \sqrt{\frac{2}{\pi x}} \sum_{j=0}^{\infty} (-1)^j \frac{2^{j+1}}{j! \cdot (2j+1)!!} \left(\frac{x}{2}\right)^{2j+1}$$

$$= \sqrt{\frac{2}{\pi x}} \sum_{j=0}^{\infty} \frac{(-1)^j}{(2j+1)!} x^{2j+1} = \sqrt{\frac{2}{\pi x}} \sin x.$$

Similarly

$$J_{-\frac{1}{2}}(x) = \sum_{j=0}^{\infty} \frac{(-1)^j}{j!\,\Gamma(j+\frac{1}{2})} \left(\frac{x}{2}\right)^{2j-\frac{1}{2}} = \sqrt{\frac{2}{\pi x}} \sum_{j=0}^{\infty} (-1)^j \frac{2^j}{j!\cdots(2j-1)!!} \left(\frac{x}{2}\right)^{2j}$$

$$= \sqrt{\frac{2}{\pi x}} \sum_{j=0}^{\infty} \frac{(-1)^j}{(2j)!} x^{2j} = \sqrt{\frac{2}{\pi x}} \cos x.$$

Example 3.3.7. *Parametric Bessel Equation of Order μ.* Express the general solution of the *Parametric Bessel equation of order* μ

(3.3.38) $$x^2 y''(x) + x y'(x) + (\lambda^2 x^2 - \mu^2)y = 0$$

in terms of the Bessel functions.

Solution. If we use the substitution $t = \lambda x$ in (3.3.38), then by the chain rule for differentiation

$$y'(x) = \lambda y'(t), \quad y''(x) = \lambda^2 y''(t),$$

it follows that (3.3.38) becomes

$$t^2 y''(t) + t y'(t) + (t^2 - \mu^2)y = 0.$$

The last equation is the Bessel equation of order μ whose solution is

$$y(t) = c_1 J_\mu(t) + c_2 Y_\mu(t).$$

Therefore, the general solution of Equation (3.3.38) is

$$y(x) = c_1 J_\mu(\lambda x) + c_2 Y_\mu(\lambda x).$$

Example 3.3.8. *The Aging Spring.* Consider unit mass attached to a spring with a constant c. If the spring oscillates for a long period of time, then clearly the spring would weaken. The linear differential equation that describes the displacement $y = y(t)$ of the attached mass at moment t is given by

$$y''(t) + ce^{-at}y(t) = 0.$$

Find the general solution of the aging spring equation in terms of the Bessel functions.

Solution. If we introduce a new variable τ by

$$\tau = \frac{2}{a}\sqrt{c}\,e^{-\frac{a}{2}t},$$

then the differential equation of the aging spring is transformed into the differential equation

$$\tau^2 y''(\tau) + \tau y'(\tau) + \tau^2 y = 0.$$

We see that the last equation is the Bessel equation of order $\mu = 0$. The general solution of this equation is

$$y(\tau) = AJ_0(\tau) + BY_0(\tau).$$

If we go back to the original time variable t, then the general solution of the aging spring equation is given by

$$y(t) = AJ_0\left(\frac{2}{a}\sqrt{c}\,e^{-\frac{a}{2}t}\right) + BY_0\left(\frac{2}{a}\sqrt{c}\,e^{-\frac{a}{2}t}\right).$$

We state and prove some most elementary and useful recursive formulas for the Bessel functions.

Theorem 3.3.7. *Suppose that μ is any real number. Then for every $x > 0$ the following recursive formulas for the Bessel functions hold.*

$$(3.3.39) \qquad \frac{d}{dx}\left(x^\mu J_\mu(x)\right) = x^\mu J_{\mu-1}(x),$$

$$(3.3.40) \qquad \frac{d}{dx}\left(x^{-\mu} J_\mu(x)\right) = -x^{-\mu} J_{\mu+1}(x),$$

$$(3.3.41) \qquad x J_{\mu+1}(x) = \mu J_\mu(x) - x J_\mu'(x),$$

$$(3.3.42) \qquad x J_{\mu-1}(x) = \mu J_\mu(x) + x J_\mu'(x),$$

$$(3.3.43) \qquad 2\mu J_\mu(x) = x J_{\mu-1}(x) + x J_{\mu+1}(x),$$

$$(3.3.44) \qquad 2 J_\mu'(x) = J_{\mu-1}(x) - J_{\mu+1}(x).$$

Proof. To prove (3.3.39) we use the power series (3.3.34) for $J_\mu(x)$:

$$\frac{d}{dx}\left(x^\mu J_\mu(x)\right) = \frac{d}{dx}\sum_{k=0}^\infty \frac{(-1)^k}{2^{2k+2\mu}k!\Gamma(k+\mu+1)}x^{2k+\mu}$$

$$= \sum_{k=0}^\infty \frac{(-1)^k(2k+2\mu)}{2^{2k+2\mu}k!\Gamma(k+\mu+1)}x^{2k+2\mu-1}$$

$$= x^\mu \sum_{k=0}^\infty \frac{(-1)^k}{2^{2k+\mu-1}k!\Gamma(k+\mu)}x^{2k+\mu-1} = x^\mu J_{\mu-1}.$$

The proof of (3.3.40) is similar and it is left as an exercise.

From (3.3.39) and (3.3.40), after differentiation we obtain

$$\begin{cases} \mu x^{\mu-1}J_\mu(x) + x^\mu J_\mu'(x) = x^\mu J_{\mu-1}(x), \\ -\mu x^{-\mu-1}J_\mu(x) + x^{-\mu}J_\mu'(x) = -x^{-\mu}J_{\mu+1}(x). \end{cases}$$

If we multiply the first equation by $x^{1-\mu}$ we obtain (3.3.42). To obtain (3.3.41), multiply the second equation by $x^{1+\mu}$. By adding and subtracting (3.3.41) and (3.3.42) we obtain (3.3.43) and (3.3.43), respectively. ∎

Example 3.3.9. Find $J_{3/2}$ in terms of elementary functions.

Solution. From (3.3.44) with $\mu = \dfrac{1}{2}$ we have

$$J_{\frac{3}{2}} = \frac{1}{x}J_{\frac{1}{2}} - J_{-\frac{1}{2}}.$$

In Example 3.3.6 we have found

$$J_{\frac{1}{2}}(x) = \sqrt{\frac{2}{\pi x}}\sin x \qquad J_{-\frac{1}{2}}(x) = \sqrt{\frac{2}{\pi x}}\cos x$$

and therefore

$$J_{\frac{3}{2}} = \sqrt{\frac{2}{\pi x}}\left(\frac{\sin x}{x} - \cos x\right).$$

Theorem 3.3.8. *If x is a positive number and t is a nonzero complex number, then*

(3.3.45) $$e^{\frac{x}{2}\left(t-\frac{1}{t}\right)} = \sum_{n=-\infty}^\infty J_n(x)t^n.$$

Proof. From the Taylor expansion of the exponential function we have that

$$e^{\frac{xt}{2}} = \sum_{j=0}^\infty \frac{t^j}{j!}\left(\frac{x}{2}\right)^j, \qquad e^{-\frac{x}{2t}} = \sum_{k=0}^\infty \frac{(-1)^k}{k!\,t^k}\left(\frac{x}{2}\right)^k.$$

Since both power series are absolutely convergent, they can be multiplied and the terms in the resulting double series can be added in any order:

$$
e^{\frac{x}{2}\left(t-\frac{1}{t}\right)} = e^{\frac{xt}{2}} e^{-\frac{x}{2t}} = \left(\sum_{j=0}^{\infty} \frac{t^j}{j!}\left(\frac{x}{2}\right)^j\right)\left(\sum_{k=0}^{\infty} \frac{(-1)^k}{k!\,t^k}\left(\frac{x}{2}\right)^k\right)
$$

$$
= \sum_{j=0}^{\infty}\sum_{k=0}^{\infty} \frac{(-1)^k}{j!\,k!}\left(\frac{x}{2}\right)^{j+k} t^{j-k}.
$$

If we introduce in the last double series a new index n by $j - k = n$, then $j = k + n$ and $j + k = 2k + n$, and therefore we have

$$
e^{\frac{x}{2}\left(t-\frac{1}{t}\right)} = \sum_{n=-\infty}^{\infty}\left\{\sum_{k=0}^{\infty}(-1)^k \frac{1}{(n+k)!\,k!}\left(\frac{x}{2}\right)^{2k+n}\right\} t^n.
$$

Since the Bessel function $J_n(x)$ is given by

$$
J_n(x) = \sum_{k=0}^{\infty}(-1)^k \frac{1}{(n+k)!\,k!}\left(\frac{x}{2}\right)^{2k+n}
$$

we obtain (3.3.45).   ∎

A consequence of the previous theorem is the following formula, which gives integral representation of the Bessel functions.

Theorem 3.3.9. *If n is any integer and x is any positive number, then*

$$
J_n(x) = \frac{1}{\pi}\int_0^{\pi} \cos(x\sin\varphi - n\varphi)\,d\varphi.
$$

Proof. If we introduce a new variable t by $t = e^{i\varphi}$, then

$$
t - \frac{1}{t} = e^{i\varphi} - e^{-i\varphi} = 2i\sin\varphi,
$$

formula (3.3.45) implies

$$
\cos\left(x\sin\varphi\right) + i\sin\left(x\sin\varphi\right) = e^{ix\sin\varphi} = e^{\frac{x}{2}\left(t-\frac{1}{t}\right)}
$$

$$
= \sum_{n=-\infty}^{\infty} J_n(x)e^{in\varphi} = J_0(x) + \sum_{n=1}^{\infty}\left[J_n(x)e^{in\varphi} + J_{-n}(x)e^{-in\varphi}\right]
$$

$$
= J_0(x) + \sum_{n=1}^{\infty}\left[J_n(x)e^{in\varphi} + (-1)^n J_n(x)e^{-in\varphi}\right]
$$

$$
= J_0(x) + \sum_{n=1}^{\infty} J_n(x)\left[e^{in\varphi} + (-1)^n e^{-in\varphi}\right]
$$

$$
= J_0(x) + 2\sum_{k=1}^{\infty} J_{2k}(x)\cos\left(2k\varphi\right) + 2i\sum_{k=1}^{\infty} J_{2k-1}(x)\sin\left(2k-1\right)\varphi.
$$

Therefore

$$(3.3.46) \qquad \cos(x \sin \varphi) = J_0(x) + 2 \sum_{k=1}^{\infty} J_{2k}(x) \cos(2k\varphi),$$

and

$$(3.3.47) \qquad \sin(x \sin \varphi) = 2 \sum_{k=1}^{\infty} J_{2k-1}(x) \sin(2k-1)\varphi,$$

Since the systems $\{1, \cos 2k\varphi, \ k = 1, 2, \ldots\}$ and $\{\sin(2k-1)\varphi, \ k = 1, 2, \ldots\}$ are orthogonal on $[0, \pi]$, from (3.3.46) and (3.3.47) it follows that

$$\frac{1}{\pi} \int_0^{\pi} \cos(x \sin \varphi) \cos n\varphi \, d\varphi = \begin{cases} J_n(x), & n \text{ is even} \\ 0, & n \text{ is odd} \end{cases}$$

and

$$\frac{1}{\pi} \int_0^{\pi} \sin(x \sin \varphi) \cos n\varphi \, d\varphi = \begin{cases} 0, & n \text{ is even} \\ J_n(x), & n \text{ is odd}. \end{cases}$$

If we add the last two equations we obtain

$$J_n(x) = \frac{1}{\pi} \int_0^{\pi} \left[\cos(x \sin \varphi) \cos n\varphi + \sin(x \sin \varphi) \sin n\varphi \right] d\varphi$$

$$= \frac{1}{\pi} \int_0^{\pi} \cos(x \sin \varphi - n\varphi) \, d\varphi. \quad \blacksquare$$

Now we will examine an eigenvalue problem of the Bessel equation. Consider again the parametric Bessel differential equation

$$(3.3.48) \qquad x^2 y'' + xy' + (\lambda^2 x^2 - \mu^2)y = 0$$

on the interval $[0, 1]$. This is a singular Sturm–Liouville problem because the leading coefficient x^2 vanishes at $x = 0$. Among several ways to impose the boundary values at the boundary points $x = 0$ and $x = 1$ is to consider the problem of finding the solutions $y(x)$ of (3.3.48) which satisfy the boundary conditions

$$(3.3.49) \qquad | \lim_{x \to 0+} y(x) | < \infty, \quad y(1) = 0.$$

The following result, whose proof can be found in the book by G. B. Folland [6], will be used to show that there are infinitely many values of λ for which there exist pairwise orthogonal solutions of (3.3.38) which satisfy the boundary conditions (3.3.49).

Theorem 3.3.10. *If* μ *is a real number, then the Bessel function* $J_\mu(x)$ *of order* μ *has infinitely many positive zeros* λ_n. *These zeros are simple and can be arranged in an increasing sequence*

$$\lambda_1 < \lambda_2 \cdots < \lambda_n < \cdots,$$

such that

$$\lim_{n\to\infty} \lambda_n = \infty.$$

As discussed in Example 3.3.7, the general solution of the parametric Bessel differential equation (3.3.48) is

$$y(x) = c_1 J_\mu(\lambda x) + c_2 Y_\mu(\lambda x).$$

We have to have $c_2 = 0$, otherwise the solutions $y(x)$ will become infinite at $x = 0$.

For these functions we have the following result.

Theorem 3.3.11. *Suppose that* μ *is a nonnegative real number, and that* $J_\mu(x)$ *is the Bessel function of order* μ. *If* $\lambda_1, \lambda_2, \ldots, \lambda_n, \ldots$ *are the positive zeros of* $J_\mu(x)$, *then each function* $J_\mu(\lambda_n x)$ *of the system*

$$J_\mu(\lambda_1\, x),\ J_\mu(\lambda_2\, x), \ldots, J_\mu(\lambda_n\, x), \ldots \quad x \in (0,1)$$

satisfies the parametric Bessel differential equation (3.3.48) *with* $\lambda = \lambda_n$, *and the boundary conditions* (3.3.49). *Furthermore, this system is orthogonal on* $(0,1)$ *with respect to the weight function* $r(x) = x$.

Proof. That each function $J_\mu(\lambda_n\, x)$ satisfies Equation (3.3.48) with $\lambda = \lambda_n$ was shown in Example 3.3.7. It remains to show the orthogonality property. Suppose that λ_1 and λ_2 are any two distinct and positive zeroes of $J_\mu(x)$. If $y_1(x) = J_\mu(\lambda_1\, x)$ and $y_2(x) = J_\mu(\lambda_2\, x)$, then

$$x^2 y_1''(x) + x y_1'(x) + (\lambda_1^2 x^2 - \mu^2)y_1(x) = 0, \quad 0 < x < 1,$$
$$x^2 y_2''(x) + x y_2'(x) + (\lambda_2^2 x^2 - \mu^2)y_2(x) = 0, \quad 0 < x < 1.$$

If we multiply the first equation by y_2 and the second by y_1, and subtract, then

$$x\big(y_2\, y_1'' - y_1\, y_2''\big) + y_2\, y_1' - y_1\, y_2' = (\lambda_2^2 - \lambda_1^2)x\, y_1\, y_2.$$

Notice that the last equation can be written in the form

$$\frac{d}{dx}\big(x\,(y_2\, y_1' - y_1\, y_2')\big) = (\lambda_2^2 - \lambda_1^2)\, x y_1\, y_2.$$

If we integrate the last equation and substitute the functions y_1 and y_2 we obtain that

$$x\Big[J_\mu(\lambda_2\, x)J_\mu'(\lambda_1\, x) - J_\mu(\lambda_1\, x)J_\mu'(\lambda_2\, x)\Big]_0^1 = (\lambda_2^2 - \lambda_1^2)\int_0^1 x J_\mu(\lambda_1\, x)\, J_\mu(\lambda_2\, x)\, dx.$$

Since λ_1 and λ_2 were such that $J_\mu(\lambda_1) = J_\mu(\lambda_2) = 0$, it follows that

$$(\lambda_2^2 - \lambda_1^2) \int_0^1 x\, J_\mu(\lambda_1\, x)\, J_\mu(\lambda_2\, x)\, dx = 0$$

and therefore

$$\int_0^1 x\, J_\mu(\lambda_1\, x)\, J_\mu(\lambda_2\, x)\, dx = 0. \qquad \blacksquare$$

This theorem can be used to construct a Fourier–Bessel expansion of a function $f(x)$ on the interval $[0, 1]$.

Theorem 3.3.12. *Suppose that f is a piecewise smooth function on the interval $[0, 1]$. Then f has a Fourier–Bessel expansion on the interval $[0, 1]$ of the form*

$$(3.3.50) \qquad\qquad f(x) \sim \sum_{k=1}^\infty c_k J_\mu(\lambda_k x),$$

where λ_k are the positive zeros of $J_\mu(x)$.
    The coefficients c_k in (3.3.50) can be computed by the formula

$$(3.3.51) \qquad\qquad c_k = \frac{2}{J_{\mu+1}^2(\lambda_k)} \int_0^1 x f(x) J_\mu(\lambda_k x)\, dx.$$

The series in (3.3.50) converges to

$$\frac{f(x^+) + f(x^-)}{2} \qquad \text{for any point } x \in (0, 1).$$

Proof. From the orthogonality property of $J_\mu(\lambda_k x)$ it follows that

$$(3.3.52) \qquad\qquad c_k = \frac{1}{\int_0^1 x J_\mu^2(\lambda_k\, x)\, dx} \int_0^1 x f(x) J_\mu(\lambda_k\, x)\, dx.$$

To evaluate the integral

$$\int_0^1 x J_\mu^2(\lambda_k\, x)\, dx$$

we proceed as in Theorem 3.3.11. If $u(x) = J_\mu(\lambda_k x)$, then

$$x^2 u'' + x u' + (\lambda_k^2 x^2 - \mu^2)u = 0.$$

If we multiply the above equation by $2u'$, then we obtain

$$2x^2 u'' u' + 2x u'^2 + 2(\lambda_k^2 x^2 - \mu^2)uu' = 0,$$

which can be written in the form

$$\frac{d}{dx}\left[x^2 u'^2 + (\lambda_k^2 x^2 - \mu^2)u^2\right] - 2\lambda_k^2 x u^2 = 0.$$

Integrating the above equation on the interval $[0, 1]$ we obtain

$$\left[x^2 u'^2 + (\lambda_k^2 x^2 - \mu^2)u^2\right]_{x=0}^{x=1} - 2\lambda_k^2 \int_0^1 x u^2(x)\, dx = 0,$$

or

$$u'^2(1) + (\lambda_k^2 - \mu^2)u^2(1) = 2\lambda_k^2 \int_0^1 x u^2(x)\, dx.$$

If we substitute the function $u(x)$ and use the fact that λ_k are the zeros of $J_\mu(x)$, then it follows that

$$\lambda_k^2 J_\mu'^2(\lambda_k) = 2\lambda_k^2 \int_0^1 x J_\mu^2(\lambda_k x)\, dx,$$

i.e.,

$$\int_0^1 x J_\mu^2(\lambda_k x)\, dx = \frac{1}{2}J_\mu'^2(\lambda_k).$$

From Equation (3.3.41) in Theorem 3.3.7, with $x = \lambda_k$ we have

$$J_\mu'^2(\lambda_k) = J_{\mu+1}^2(\lambda_k).$$

Thus,

$$\int_0^1 x J_\mu^2(\lambda_k x)\, dx = \frac{1}{2}J_{\mu+1}^2(\lambda_k).$$

Therefore, from (3.3.52) it follows that

$$c_k = \frac{2}{J_{\mu+1}^2(\lambda_k)} \int_0^1 x f(x) J_\mu(\lambda_k x)\, dx. \qquad \blacksquare$$

Let us illustrate this theorem with an example.

Example 3.3.10. Expand the function $f(x) = x$, $0 < x < 1$ in the series

$$x \sim \sum_{k=1}^{\infty} c_k J_1(\lambda_k x),$$

where λ_k, $k = 1, 2, \cdots$ are the zeros of $J_1(x)$.

Solution. From (3.3.51) we have

$$c_k = \frac{2}{J_2^2(\lambda_k)} \int_0^1 x^2 J_1(\lambda_k x)\, dx = \frac{2}{\lambda_k^3 J_2^2(\lambda_k)} \int_0^{\lambda_k} t^2 J_1(t)\, dt.$$

However, from (3.3.39),

$$\frac{d}{dx}\left[x^2 J_2(x)\right] = x^2 J_1(x)$$

with $\mu = 2$, it follows that

$$c_k = \frac{2}{\lambda_k^3 J_2^2(\lambda_k)} \int_0^{\lambda_k} \frac{d}{dt}\left[t^2 J_2(t)\right] dt = \frac{2}{\lambda_k^3 J_2^2(\lambda_k)} t^2 J_2(t)\Big|_{t=0}^{t=\lambda_k} = \frac{2}{\lambda_k J_2(\lambda_k)}.$$

Therefore, the resulting series is

$$x = 2\sum_{k=1}^{\infty} \frac{J_1(\lambda_k x)}{\lambda_k J_2(\lambda_k)}.$$

Example 3.3.11. Show that for any $p \geq 0$, $a > 0$ and $\lambda > 0$, we have

$$\int_0^a (a^2 - t^2) t^{p+1} J_p\left(\frac{\lambda}{a} t\right) dt = 2\frac{a^{p+4}}{\lambda^2} J_{p+2}(\lambda).$$

Solution. If we introduce a new variable x by $x = \frac{\lambda}{a} t$, then

$$\int_0^a (a^2 - t^2) t^{p+1} J_p\left(\frac{\lambda}{a} t\right) dt = \frac{a^{p+4}}{\lambda^{p+2}} \int_0^{\lambda} x^{p+1} J_p(x)\, dx$$

(3.3.53)

$$- \frac{a^{p+4}}{\lambda^{p+2}} \int_0^{\lambda} x^2 x^{p+1} J_p(x)\, dx.$$

For the first integral in the right hand side of (3.3.53), from identity (3.3.39) in Theorem 3.3.7 we have

$$(3.3.54) \qquad \int_0^\lambda x^{p+1} J_p(x)\, dx = \lambda^{p+1} J_{p+1}(\lambda).$$

For the second integral in the right hand side of (3.3.53) we apply the integration by parts formula with

$$u = x^2, \qquad dv = x^{p+1} J_p(x)\, dx.$$

Then, by identity (3.3.39), we have $v = x^{p+1} J_{p+1}(x)$ and so,

$$\int_0^\lambda x^2\, x^{p+1} J_p(x)\, dx = x^{p+3} J_{p+1}(x)\Big|_{x=0}^{x=\lambda} - 2\int_0^\lambda x^{p+2} J_{p+1}(x)\, dx$$

$$= \lambda^{p+3} J_{p+1}(\lambda) - 2\int_0^\lambda x^{p+2} J_{p+1}(x)\, dx \quad \text{(use of (3.3.39))}$$

$$= \lambda^{p+3} J_{p+1}(\lambda) - 2\lambda^{p+2} J_{p+2}(\lambda).$$

If we substitute (3.3.54) and the last expression in (3.3.53) we obtain the required formula. ∎

Exercises for Section 3.3.

1. Verify the expressions for the first six Legendre polynomials.

2. Show by means of Rodrigues' formula that

$$P_n'(1) = \frac{n(n+1)}{2}, \quad \text{and} \quad P_n'(-1) = (-1)^{n-1}\frac{n(n+1)}{2}.$$

3. If $n \geq 1$ show that $\int_{-1}^{1} P_n(x)\, dx = 0$.

4. Prove the recursive formula (3.3.16) in Theorem 3.3.4:

$$P_{n+1}'(x) - P_{n-1}'(x) = (2n+1)P_n(x).$$

5. Find the sum of the series

$$\sum_{n=0}^{\infty} \frac{x^{n+1}}{n+1} P_n(x).$$

6. Find the first three nonvanishing coefficients in the Legendre polynomial expansion for the following functions.

(a) $f(x) = \begin{cases} 0, & -1 < x < 0 \\ x, & 0 < x < 1. \end{cases}$

(b) $f(x) = |x|, \; -1 < x < 1.$

(c) $f(x) = \begin{cases} -1, & -1 < x < 0 \\ 0, & 0 < x < 1. \end{cases}$

7. Show that

(a) $P_n(1) = 1.$

(b) $P_n(-1) = (1)^n.$

(c) $P_{2n-1}(0) = 0.$

(d) $P_{2n}(0) = (-1)^n \dfrac{(2n)!}{2^{2n} n! n!}.$

8. Prove that

$$\int_x^1 P_n(t)\, dt = \frac{1}{2n+1}\big[P_{n-1}(x) - P_{n+1}(x)\big].$$

9. Establish the following properties of the Bessel functions.

(a) $J_0(0) = 1.$

(b) $J_\mu(0) = 0$ for $\mu > 0.$

(c) $J_0'(x) = -J_1(x).$

(d) $\dfrac{d}{dx}\big(x J_1(x)\big) = x J_0(x).$

10. Show that the differential equation

$$y'' + xy = 0, \quad x > 0$$

has the general solution

$$y = c_1 x^{\frac{1}{2}} J_{\frac{1}{3}}\Big(\frac{2}{3}x^{\frac{3}{2}}\Big) + c_2 x^{\frac{1}{2}} J_{-\frac{1}{3}}\Big(\frac{2}{3}x^{\frac{3}{2}}\Big).$$

11. Show that the differential equation

$$xy'' + (x^2 - 2)y = 0, \quad x > 0$$

has the general solution

$$y = c_1 x^{\frac{1}{2}} J_{\frac{3}{2}}(x) + c_2 x^{\frac{1}{2}} J_{-\frac{3}{2}}(x).$$

12. Show that the function $cJ_1(x)$ is a solution of the differential equation

$$xy'' - y' + xy = 0, \ x > 0.$$

13. Solve the following differential equations.

(a) $x^2 y'' + 3xy' + (\mu^2 x^2 + 1)y = 0, \quad \mu > 0, \ x > 0.$

(b) $xy'' + 5y' + xy = 0, \quad x > 0.$

(c) $xy'' + y' + \mu^2 x^{2\mu-1} y = 0, \quad \mu \neq 0, \ x > 0.$

(d) $x^2 y'' + xy' + (x^2 - 9)y = 0, \quad x > 0.$

14. Show that

$$\int\limits_0^{\frac{\pi}{2}} J_0(x \cos \theta) \, d\theta = \frac{\sin x}{x}, \quad x > 0.$$

15. Find $J_{-\frac{3}{2}}(x)$ in terms of elementary functions.

16. Suppose that x and y are positive numbers and n an integer. Show that

$$J_n(x + y) = \sum_{k=-\infty}^{\infty} J_k(x) J_{n-k}(y).$$

17. Expand the following functions

(a) $f(x) = 1, \quad 0 < x < 1$

(b) $f(x) = x^2, \quad 0 < x < 1$

(c) $f(x) = \dfrac{1 - x^2}{8}, \quad 0 < x < 1$

in the Fourier–Bessel series

$$\sum_{k=1}^{\infty} c_k J_0(\lambda_k x),$$

where λ_k denote the positive zeros of $J_0(x)$.

18. Show that

$$x^3 = 2 \sum_{k=1}^{\infty} \frac{\lambda_k^2 - 8}{\lambda_k^3 J_2(\lambda_k)} J_1(\lambda_k x) \quad x > 0,$$

where λ_k denote the positive zeros of $J_1(x)$.

19. From the recursive formulas, show the following relations.

(a) $2J_0''(x) = J_2(x) - J_0(x)$.

(b) $2J_0''(x) = J_2(x) - J_0(x)$.

(c) $J_2(x) = J_2''(x) - \frac{J_0'(x)}{x}$.

(d) $4J_n''(x) = J_{n-2}(x) - 2J_n(x) + J_{n+2}(x)$.

(e) $2J_0''(x) = J_2(x) - J_0(x)$.

20. Show that

(a) $\int\limits_{0}^{1} x J_0(\lambda x)\, dx = \frac{1}{\lambda} J_1(\lambda)$.

(b) $\int\limits_{0}^{1} x^3 J_0(\lambda x)\, dx = \frac{\lambda^2 - 4}{\lambda^3} J_1(\lambda) + \frac{2}{\lambda^2} J_0(\lambda)$.

21. If λ is any solution of the equation $J_0(x) = 0$, show that

(a) $\int\limits_{0}^{1} J_1(\lambda x)\, dx = \frac{1}{\lambda}$.

(b) $\int\limits_{0}^{\lambda} J_0(\lambda x)\, dx = 1$.

3.4 Projects Using Mathematica.

In this section we will see how Mathematica can be used to solve some eigenvalue problems. Legendre polynomials and Bessel functions also are investigated in this section using Mathematica. For a brief overview of the computer software Mathematica consult Appendix H.

Project 3.4.1. Using Mathematica solve the eigenvalue problem

$$y''(x) + xy'(x) + \lambda y = 0, \quad 1 < x < e, \ y(1) = y(e) = 0.$$

Solution. We find the general solution of the Cauchy–Euler differential equation

$In[1] := GS = First[DSolve[x^2\, y''[x] + x\, y'[x] + \lambda\, y[x] == 0, y, x]]$

$Out[1] = \{y-> \text{Function } [\{x\}, C[1]\, Cos[\sqrt{\lambda}]\, Log[x]]$
$+C[2]\, Sin[\sqrt{\lambda}]\, Log[x]]$

Next define the boundary conditions at $x = 1$ and $x = e$:

$In[2] := bc1 = (y[1] == 0)/.\ GS$

$Out[2] = C[1] = 0$

$In[3] := bce = (y[Exp[1]] == 0)/.\ GS$

$Out[3] = C[1]Cos[\sqrt{\lambda}] + C[2]Sin[\sqrt{\lambda}] == 0$

Now substitute $C[1] = 0$ and find the eigenvalues:

$In[4] := \text{Reduce } [Sin[\sqrt{\lambda}] == 0, \lambda, \text{Reals }]$

$Out[4] = \lambda = 0 \,||\, (C[1] \in \text{Integers} \&\& ((C[1] \geq 1 \&\& \lambda == 4\pi^2 C[1]^2) \,||\, (C[1] \geq 0 \&\& \lambda == (\pi + 2\pi C[1])^2)))$

Notice that the eigenvalues are given by $\lambda_n = n^2\, \pi^2$ $n = 1, 2, \ldots$.

The eigenfunctions are therefore

$$y_n(x) = \sin\left(n\pi \ln x\right).$$

In the next project we use Mathematica to solve a problem involving the Legendre polynomials.

Project 3.4.2. Let $f(x)$ be the function defined on $[-1, 1]$ by

$$f(x) = \begin{cases} -2x - 2, & -1 \leq x < -\frac{1}{2} \\ 2x, & -\frac{1}{2} \leq x \leq \frac{1}{2} \\ -2x + 2, & \frac{1}{2} \leq x \leq 1. \end{cases}$$

Expand the function $f(x)$ into the Legendre series

$$f(x) = \sum_{n=0}^{\infty} c_n P_n(x),$$

where $P_n(x)$ is the Legendre polynomial of degree n. Using Mathematica plot the function $f(x)$ and several partial sums of the Legendre series.

Solution. First define the function $f(x)$:

$In[1] := f[x\_] := \text{Piecewise } [\{\{-1, -1 \leq x < 0\}, \{x, 0 < x \leq 1\}\}];$

Next define the Legendre polynomial $P_n(x)$:

$In[2] := L[x\_, n\_] := \text{Legendre } [n, x];$

Evaluate the integral

$$\int_{-1}^{1} P_n^2(x)\, dx :$$

$In[3] := a[n\_] := \text{Integrate } [(L[x,n])^2, \{x, -1, 1\}];$

Now we define the nth coefficient in the Legendre expansion:

$In[4] := c[n\_] := (1/a[n]) \text{ Integrate}[f[x]\, L[x,n], \{x, -1, 1\}];$

Now we define the Nth Legendre partial sum:

$In[5] := S[x\_, N\_] := Sum[c[n]\, L[x,n], \{n, 0, N\}];$

Plots of the function $f(x)$ and its several partial Legendre sums are displayed in Figure 3.4.1.

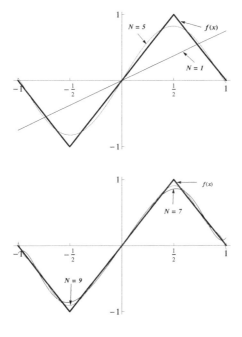

Figure 3.4.1

Mathematica has many options to work with Bessel functions. It can plot the Bessel functions of the first and second kind. It can evaluate numerically the zeros of the Bessel functions. It can solve some differential equations whose solutions are expressed in terms of the Bessel functions. It can evaluate some integrals in terms of the Bessel functions.

In the next project we explore some of these options.

Project 3.4.3. Explore some of the capabilities of Mathematica in solving problems related to Bessel functions.

Solution. The Mathematica commands

$In[] :=$ BesselJ $[n, x]$;

$In[] :=$ BesselY $[n, x]$;

give the Bessel function $J_n(x)$ of the first kind of order n and the Bessel function $Y_n(x)$ of the second kind of order n.

Example 1. We can plot several Bessel functions.

$In[1] :=$ Plot $[\{$ BesselJ$[0, x]$, BesselJ$[1, x]\}, \{x, 0, 50\}$,

Ticks $\rightarrow \{\{0, 10, 20, 40, 50\}, \{-0.4, -0.2, 0, 0.2, 0.4\}\}$, AxesLabel $\rightarrow \{"x"\}]$

$In[2] :=$ Plot $[\{$ BesselY$[0, x]$, BesselY$[1, x]\}, \{x, 0, 50\}$,

The plots are displayed in Figure 3.4.2.

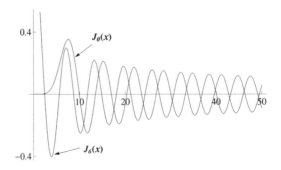

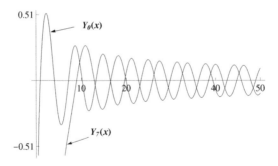

Figure 3.4.2

The command

$In[] :=$ N[BesselJZero$[n, k, x_0]$]

gives the kth zero of the Bessel function $J_n(x)$ which is greater than x_0.

Example 2. Write out the first 5 positive zeros of $J_1(x)$.

Solution. Although we can specify a desired precision, let us work with the Mathematica default.

$In[1] := \text{Table}[N[\text{BesselJZero}[1, k, 0]], k, 1, 10]$

$Out[1] = \{3.83171, 7.01559, 10.1735, 16.4706, 13.3237,$
$19.6159, 22.7601, 25.9037, 29.0468, 32.1897\}$

Example 3. Solve the differential equation

$$s^2 y''(x) + xy'(x) + (x^2 - 1)y(x) = 0.$$

Solution. This is the Bessel equation of order 1 and the solution of this equation is expressed in terms of the Bessel function which can been seen from the following.

$In[2] := \text{DSolve}[x^2 y''[x] + x y'[x] + (x^2 - 1) y[x] == 0, y[x], x]$

$Out[2] = \{\{y[x] - > \text{BesselJ}[1, x]C[1] + \text{BesselY}[1, x]C[2]\}\}$

CHAPTER 4

PARTIAL DIFFERENTIAL EQUATIONS

Many problems in physics and engineering are described by partial differential equations (PDEs) with appropriate boundary and initial value conditions. Partial differential equations have a major role in electromagnetic theory, fluid dynamics, traffic flow and many other disciplines. Therefore there is a need to study the theory of partial differential equations and the rest of the book is devoted entirely to this important topic.

A major part of this chapter is devoted to the study of a particular class of first order partial differential equation, namely, linear partial differential equations. We will start with basic concepts and terminology and finally we will give some generalities of linear partial differential equations of the second order and their classification.

In this chapter, as well as in the remaining chapters, all functions will be real-valued functions of one or more real variables unless we indicate otherwise.

4.1 Basic Concepts and Terminology.

A *domain* Ω in the plane $\mathbb{R}^2$ or in the space $\mathbb{R}^3$ is a nonempty, open and connected set in $\mathbb{R}^2$ or in $\mathbb{R}^3$. A set Ω is called *open* if for every point $P \in \Omega$, there is an open disc/ball (without the circle/sphere) with center at P which is a subset of Ω. A set Ω is called *connected* if any two points in Ω can be joined by a polygonal line which lies entirely in Ω.

If $u = u(x, y, \ldots)$, $(x, y, \ldots) \in \Omega$ is a function of two or more variables, then the partial derivatives of u of the first order will be denoted by

$$u_x, \ u_y,$$

instead of

$$\frac{\partial u}{\partial x}, \frac{\partial u}{\partial y}.$$

Similarly, for the partial derivatives of the second order we will use the notation

$$u_{xx}, \ u_{yx}, \ u_{yy},$$

instead of

$$\frac{\partial^2 u}{\partial x^2}, \frac{\partial^2 u}{\partial y \partial x}, \frac{\partial^2 u}{\partial y^2}.$$

We say that a function $u = u(x, y)$, $(x, y) \in \Omega \subseteq \mathbb{R}^2$ belongs to the class $C^k(\Omega)$, $k = 1, 2, \ldots$, if all partial derivatives of u up to the order of k are continuous functions in Ω.

These concepts can be generalized for functions $u = u(x, y, z, \ldots)$ of three or more variables $x, y, z, \ldots$

A *partial differential equation* for a function $u = u(x, y, z, \ldots)$ is an equation which contains the function u and some of its partial derivatives:

$$(4.1.1) \qquad F(x, y, u, u_x, u_y, u_{xx}, u_{xy}, u_{yy}, \ldots) = 0, \quad (x, y, \ldots) \in \Omega.$$

The *order* of a partial differential equation is the order of the highest partial derivative that appears in the equation.

As in ordinary differential equations, a *solution* of the partial differential Equation (4.1.1) of order k is a function $u = u(x, y, z, \ldots) \in C^k(\Omega)$ which together with its partial derivatives satisfies Equation (4.1.1) for all $(x, y, \ldots) \in \Omega$. The *general solution* is the set of all solutions.

Example 4.1.1. Find the general solution of the differential equation

$$u_{yy}(x, y) = 0, \quad (x, y) \in \mathbb{R}^2.$$

Solution. The given equation simply means that $u_y(x, y)$ does not depend on y. Therefore, $u_y(x, y) = A(x)$, where $A(x)$ is an arbitrary function. Integrating the last equation with respect to y (keeping x constant), it follows that the general solution of the equation is given by

$$u(x, y) = A(x)y + B(x),$$

where $A(x)$ and $B(x)$ are arbitrary functions.

Example 4.1.2. Show that the function

$$u(x, y, z) = \frac{1}{\sqrt{x^2 + y^2 + z^2}}, \quad (x, y, z) \neq 0$$

is a solution of the differential equation

$$u_{xx}(x, y, z) + u_{yy}(x, y, z) + u_{zz}(x, y, z) = 0, \quad (x, y, z) \in \mathbb{R}^3 \setminus \{(0, 0, 0)\}.$$

Solution. If we let $r = \sqrt{x^2 + y^2 + z^2}$, then $u = \dfrac{1}{r}$. Using the chain rule we have

$$u_x = -\frac{1}{r^2} r_x = -\frac{x}{r^3}.$$

Differentiating once more and using the quotient and the chain rule again we have

$$u_{xx} = -\frac{r^3 - 3r^2 r_x x}{r^6} = -\frac{r^2 - 3x^2}{r^5}.$$

By symmetry, we have

$$u_{yy} = -\frac{r^2 - 3y^2}{r^5}, \quad u_{zz} = -\frac{r^2 - 3z^2}{r^5}.$$

Therefore,

$$
\begin{aligned}
u_{xx} + u_{yy} + u_{zz} &= -\frac{r^2 - 3x^2}{r^5} - \frac{r^2 - 3y^2}{r^5} - \frac{r^2 - 3z^2}{r^5} \\
&= -\frac{3r^2 - 3(x^2 + y^2 + z^2)}{r^5} = -\frac{3r^2 - 3r^2}{r^5} = 0.
\end{aligned}
$$

Example 4.1.3. Find the general solution $u = u(x, y)$ of the equation

$$u_{xy} + u_x = 0.$$

Solution. We can consider this partial differential equation as an ordinary differential equation. Indeed, if we introduce a new function $w = w(x, y) = u_x$, then the given equation takes the form

$$w_y + w = 0.$$

The general solution of the above equation is easily found and it is given by

$$w(x, y)) = a(x)e^{-y},$$

where $a(x)$ is any differentiable function. Therefore,

$$u_x = e^{-y} a(x).$$

Integrating the above equation with respect to x (keeping y constant), we obtain that the general solution of the given partial differential equation is

$$u(x, y) = e^{-y} f(x) + g(y),$$

where f is any differentiable function of a single variable and g an arbitrary function of a single variable.

Example 4.1.4. The equations

$$u_x + xu_y + u = 0, \quad u_y - uu_xu_y + u^2 = 0, \quad u_x - yu_y + zu_z = x - y - z$$

are all of the first order, while the equations

$$u_{xx} + u_{yy} = 0, \quad u_y - xyu_{xy} + (x - y)u_{yy} = 0, \quad u_{xx} + u_{yy} + u_{zz} = 0$$

are of the second order.

Exercises for Section 4.1.

1. Show that the function $u = u(x, y)$ defined by

$$u(x, y) = f\left(\sqrt{x^2 + y^2}\right),$$

where f is any differentiable function on $[a, b]$, $a \geq 0$, satisfies the differential equation

$$yu_x - xu_y = 0.$$

2. Let $u(x, y) = f(x)g(y)$, where f and g are any differentiable functions of one variable. Verify that the function $u = u(x, y)$ is a solution of the partial differential equation

$$uu_{xy} = u_x u_y.$$

3. Let $u(x, y) = e^x f(2x - y)$, where f is any differentiable function of one variable. Show that $u = u(x, y)$ is a solution of the partial differential equation

$$u_x + 2u_y = 0.$$

4. Find the general solution $u(x, y)$ of the following differential equations.

   (a) $u_x = f(x)$,

   (b) $u_x = f(y)$,
   where f is a continuous function, defined on an open interval.

5. Find the general solution $u(x, y)$ of the following differential equations.

   (a) $u_{yy} = 0$.

   (b) $u_{xy} = 0$.

(c) $u_{xx} + u = 0 = 0$.

(d) $u_{yy} + u = 0 = 0$.

6. Show that the function

$$u(x,t) = \frac{1}{\sqrt{t}} e^{-\frac{x^2}{16t}}, \quad t > 0$$

satisfies the differential equation

$$u_t = 4u_{xx}.$$

7. Show that the function

$$u(x,t) = \phi(x - at) + \psi(x + at),$$

where a is any constant, and ϕ and ψ are any differentiable functions, each of a single variable, satisfies the differential equation

$$u_{tt} = a^2 u_{xx}.$$

8 Find a partial differential equation which is satisfied by each of the following families of functions.

(a) $u = x + y + f(xy)$, f is any differentiable function of one variable.

(b) $u = (x + a)(y + b)$, a and b are any constants.

(c) $ax^2 + by^2 + cz^2 = 1$, a, b and c are any constants.

(d) $u = xy + f(x^2 + y^2)$, f is any differentiable function of one variable.

(e) $u = f(\frac{xy}{u})$, f is any differentiable function of one variable.

9. Show that the function $u = u(x,y)$ defined by

$$u(x,y) = \begin{cases} \frac{xy}{x^2+y^2} & \text{if } (x,y) \neq (0,0) \\ 0 & \text{if } (x,y) = (0,0) \end{cases}$$

is discontinuous only at the point $(0,0)$, but still satisfies the equation

$$xu_x + yu_y = 0$$

in the entire plane.

10. Show that all surfaces of revolution around the u-axis satisfy the partial differential equation

$$u(x,y) = f(x^2 + y^2),$$

where f is some differentiable function of one variable.

4.2 Partial Differential Equations of the First Order.

In this section we will study partial differential equations of the first order. We will restrict our discussion only to equations in $\mathbb{R}^2$. These types of equations are used in many physical problems, such as traffic flow, shock waves, convection–diffusion processes in chemistry and chemical engineering.

We start with the linear equations.

Linear Partial Differential Equations of the First Order. A differential equation of the form

$$(4.2.1) \qquad a(x,y)u_x(x,y) + b(x,y)u_y(x,y) = c(x,y)u(x,y)$$

is called a *linear partial differential equation of the first order.* We suppose that the functions $a = a(x,y)$, $b = b(x,y)$ and $c = c(x,y)$ are continuous in some domain in the plane.

Let $u = u(x,y)$ be a solution of (4.2.1). Geometrically, the set (x,y,u) represents a surface S in $\mathbb{R}^3$. We know from calculus that a normal vector $\mathbf{n}$ to the surface S at a point (x,y,u) of the surface is given by

$$\mathbf{n} = (u_x,\, u_y,\, -1).$$

From Equation (4.2.1) it follows that the vector (a, b, cu) is perpendicular to the vector $\mathbf{n}$, and therefore the vector (a, b, cu) must be in the tangent plane to the surface S. Therefore, in order to solve Equation (4.2.1) we need to find the surface S. It is obvious that S is the union of all curves which lie on the surface S with the property that the vector

$$(4.2.2) \qquad \big(a(x,y),\, b(x,y),\, c(x,y)u(x,y)\big)$$

is tangent to each curve. Let us consider one such curve $\mathcal{C}$, whose parametric equations are given by

$$(\mathcal{C}) \qquad \begin{cases} x = x(t), \\ y = y(t), \\ u = u(t). \end{cases}$$

A tangent vector to the curve $\mathcal{C}$ is given by

$$(4.2.3) \qquad \left(\frac{dx}{dt},\, \frac{dy}{dt},\, \frac{du}{dt}\right).$$

If we compare (4.2.2) and (4.2.3), then we obtain

$$(4.2.4) \qquad \begin{cases} \dfrac{dx}{dt} = a(x,y), \\[2mm] \dfrac{dy}{dt} = b(x,y), \\[2mm] \dfrac{du}{dt} = c(x,y)u(x,y). \end{cases}$$

The curves $\mathcal{C}$ given by (4.2.4) are called the *characteristic curves* for Equation (4.2.1). The system (4.2.4) is called the *characteristic equations* for Equation (4.2.1). There are two linearly independent solutions of the system (4.2.4), and any solution of Equation (4.2.1) is constant on such curves. To summarize, we have the following theorem.

Theorem 4.2.1. *The general solution* $u = u(x, y)$ *of the partial differential equation* (4.2.1) *is given by*

$$f\big(F(x, y, u), G(x, y, u)\big) = 0,$$

where f *is any differentiable function of two variables and* $F(x, y, u) = c_1$ *and* $G(x, y, u) = c_2$ *are two linearly independent solutions of the characteristic equations*

$$(4.2.5) \qquad \frac{dx}{a(x, y)} = \frac{dy}{b(x, y)} = \frac{du}{c(x, y)u(x, y)}.$$

Example 4.2.1. Solve the partial differential equation $u_x + 3x^2 y^2 u_y = 0$.

Solution. The characteristic curves of this equation are given by the general solution of

$$3x^2 y^3 \, dx - dy = 0.$$

After the separation of the variables in the above equation we obtain that its general solution is given by

$$x^3 = -\frac{1}{2y^2} + c.$$

Therefore

$$c = x^3 + \frac{1}{2y^2},$$

and so the general solution of the given partial differential equation is

$$u(x, y) = f\left(x^3 + \frac{1}{2y^2}\right),$$

where f is any differentiable function of a single variable.

Example 4.2.2. Find the particular solution $u = u(x, y)$ of the differential equation

$$-y u_x + x u_y = 0$$

which satisfies the condition $u(x, x) = 2x^2$.

Solution. The characteristic curves of this equation are given by the general solution of

$$x \, dx + y \, dy = 0.$$

Integrating the above equation we obtain

$$x^2 + y^2 = c.$$

Therefore, the general solution of the given partial differential equation is

$$u(x, y) = f(x^2 + y^2),$$

where f is any differentiable function of one variable. From the condition $u(x, x) = 2x^2$ we have $f(2x^2) = 2x^2$ and so $f(x) = x$. Therefore, the particular solution of the given equation that satisfies the given condition is

$$u(x, y) = x^2 + y^2.$$

Example 4.2.3. Find the general solution of

$$xu_x + yu_y = nu.$$

Solution. The characteristic equations (4.2.5) of the given differential equation are

$$\frac{dx}{x} = \frac{dy}{y} = \frac{du}{nu}.$$

From the first equation we obtain

$$\frac{y}{x} = c_1, \quad c_1 \text{ is any constant.}$$

From the other equation

$$\frac{dy}{y} = \frac{du}{nu}$$

we find

$$\frac{u}{y^n} = c_2, \quad c_2 \text{ is any constant.}$$

Therefore, the general solution of the equation is given by

$$f\left(\frac{y}{x}, \frac{u}{y^n}\right) = 0,$$

where f is any differentiable function. Notice that the solution can also be expressed as

$$u(x, y) = y^n g\left(\frac{y}{x}\right),$$

where g is any differentiable function of a single variable.

Example 4.2.4. Find the particular solution $u = u(x, y)$ of the equation

$$(4.2.6) \qquad \sqrt{x}\,u_x - \sqrt{y}\,u_y = \frac{u}{\sqrt{x} - \sqrt{y}},$$

which satisfies the condition $u(4x, x) = \sqrt{x}$.

Solution. The characteristic equations (4.2.5) in this case are

$$(4.2.7) \qquad \frac{dx}{\sqrt{x}} = -\frac{dy}{\sqrt{y}} = (\sqrt{x} - \sqrt{y})\frac{du}{u}.$$

The first equation in (4.2.7) implies

$$\sqrt{x} + \sqrt{y} = c_1, \text{ where } c_1 \text{ is any constant,}$$

and so $F(x, y)$ in Theorem 4.2.1 is given by

$$(4.2.8) \qquad F(x, y) = \sqrt{x} + \sqrt{y}.$$

Further, from the characteristic equations (4.2.7) it follows that

$$dx = \sqrt{x}(\sqrt{x} - \sqrt{y})\frac{du}{u} \quad dy = -\sqrt{y}(\sqrt{x} - \sqrt{y})\frac{du}{u}.$$

Thus,

$$\frac{dx - dy}{\sqrt{x} - \sqrt{y}} = (\sqrt{x} + \sqrt{y})\frac{du}{u},$$

and so

$$\frac{d(x - y)}{x - y} = \frac{du}{u}.$$

Integrating the last equation we obtain

$$\frac{u}{x - y} = c_2, \text{ where } c_2 \text{ is any constant,}$$

and so $G(x, y)$ in Theorem 4.2.1 is given by

$$(4.2.9) \qquad G(x, y) = \frac{u}{x - y}.$$

Therefore, from (4.2.7) and (4.2.8) we obtain that the general solution of Equation (4.2.6) is given by

$$f\left(\sqrt{x} + \sqrt{y}, \frac{u}{x - y}\right) = 0,$$

where f is any differentiable function. The last equation implies that the general solution can be expressed as

$$u(x, y) = (x - y)g(\sqrt{x} + \sqrt{y}),$$

where g is any differentiable function of a single variable. From the given condition $u(4x, x) = \sqrt{x}$ we have $3xg(\sqrt{x}) = 3\sqrt{x}$, and so $g(x) = 1/x$. Therefore,

$$u(x, y) = (x - y)\frac{1}{\sqrt{x} + \sqrt{y}} = \sqrt{x} - \sqrt{y}$$

is the required particular solution.

Now, we will describe another method for solving Equation (4.2.1). The idea is to replace the variables x and y with new variables ξ and η and transform the given equation into a new equation which can be solved. We will find a transformation to new variables

(4.2.10) $$\xi = \xi(x, y), \ \eta = \eta(x, y),$$

which transforms Equation (4.2.1) into an equation of the form

(4.2.11) $$w_\xi + g(\xi, \eta)w = h(\xi, \eta),$$

where $w = w(\xi, \eta) = u\big(x(\xi, \eta), y(\xi, \eta)\big)$. Equation (4.2.11) is a linear ordinary differential equation of the first order (with respect to ξ), and as such it can be solved.

So, the problem is to find such functions ξ and η with the above properties. By the chain rule we have

$$u_x = w_\xi \xi_x + w_\eta \eta_x, \quad u_y = w_\xi \xi_y + w_\eta \eta_y.$$

Substituting the above partial derivatives into Equation (4.2.1), after some rearrangements, we obtain

$$\big(a\xi_x + b\xi_y\big)w_\xi + \big(a\eta_x + b\eta_y\big)w_\eta = c\big(x(\xi, \eta), y(\xi, \eta)\big).$$

In order for the last equation to be of the form (4.2.11) we need to choose $\xi = \xi(x, y)$ such that

$$a(x, y)\xi_x + b(x, y)\xi_y = 0.$$

But the last equation is homogeneous and therefore along its characteristic curves

(4.2.12) $$-b(x, y)\, dx + a(x, y)\, dy = 0$$

the function $\xi = \xi(x, y)$ is constant. Thus,

$$\xi(x, y) = F(x, y),$$

where $F(x, y) = c$, c is any constant, is the general solution of Equation (4.2.12). For the function $\eta = \eta(x, y)$ we can choose $\eta(x, y) = y$.

Example 4.2.5. Find the particular solution $u = u(x, y)$ of the equation

$$\frac{1}{x}u_x + \frac{1}{y}u_y = \frac{1}{y}, \quad x \neq 0, \ y \neq 0,$$

for which $u(x, 1) = 3 - x^2$.

Solution. Equation (4.2.12) in this case is

$$-\frac{1}{y}dx + \frac{1}{x}dy = 0,$$

whose general solution is given by $x^2 - y^2 = c$. If $w(\xi, \eta) = u(x(\xi, \eta), y(\xi, \eta))$, then the new variables ξ and η are given by

$$\begin{cases} \xi = x^2 - y^2, \\ \eta = y. \end{cases}$$

Now, if we substitute the partial derivatives $u_x = w_\xi \xi_x + w_\eta t_x = 2x v_\xi$ and $u_y = w_\xi \xi_y + w_\eta \eta_y = -2y w_\xi + w_\eta$ in the given partial differential equation, we obtain the simple equation $w_t = 1$. The general solution of this equation is given by $w(\xi, \eta) = \eta + f(\xi)$, where f is any differentiable function. Returning to the original variables x and y we have

$$u(x, y) = y + f(x^2 - y^2).$$

From the condition $u(x, 1) = 3 - x^2$ it follows that $3 - x^2 = 1 + f(x^2 - 1)$, i.e., $f(x^2 - 1) = 2 - x^2$. If we let $s = x^2 - 1$, then $f(s) = 2 - (s + 1) = 1 - s$. Therefore, the particular solution of the given equation is

$$u(x, y) = y + 1 + y^2 - x^2.$$

The plot of the particular solution $u(x, y)$ is given in Figure 4.2.1.

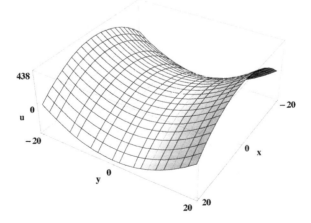

Figure 4.2.1

Example 4.2.6. Find the solution $u = u(x, y)$ of the equation

$$u_x + yu_y + u = 0, \quad y > 0,$$

for which

$$u(x, e^{2x}) = e^{2x}.$$

Solution. Equation (4.2.12) in this case is

$$-y\,dx + dy = 0,$$

whose general solution is given by $y = ce^x$. If $w(\xi, \eta) = u(x(\xi, \eta), y(\xi, \eta))$, then the new variables ξ and η are given by

$$\begin{cases} \xi = ye^{-x}, \\ \eta = y. \end{cases}$$

Now, substituting the partial derivatives $u_x = w_\xi \xi_x + w_\eta \eta_x = -ye^{-x}w_\xi$ and $u_y = w_\xi \xi_y + w_\eta \eta_y = e^{-x}w_\xi + w_\eta$ in the given partial differential equation we obtain

$$\eta w_\eta + w = 0.$$

This equation can be solved by the method of separation of variables:

$$\frac{dw(\xi, \eta)}{v} = -\frac{d\eta}{\eta}.$$

The general solution of the last equation is

$$w(\xi, \eta) = \frac{f(\xi)}{\eta},$$

where f is an arbitrary differentiable function. Therefore, the general solution of the given partial differential equation is

$$u(x, y) = \frac{f(ye^{-x})}{y}.$$

Now, from $u(x, e^{2x}) = 1$ it follows that $f(e^x) = e^{2x}$, i.e., $f(x) = x^2$. Therefore, the particular solution of the given partial differential equation is

$$u(x, y) = ye^{-2x}.$$

The last two examples are only particular examples of a much more general problem, the so-called *Cauchy Problem*. We will state this problem without any further discussion.

Cauchy Problem. Consider a curve $\mathcal{C}$ in a plane domain Ω. The Cauchy problem is to find a solution $u = u(x, y)$ of a partial differential equation of the first order

$$F(x, y, u, u_x, u_y) = 0, \quad (x, y) \in \Omega,$$

such that u takes prescribed values u_0 on $\mathcal{C}$:

$$u(x, y)\bigg|_{\mathcal{C}} = u_0(x, y).$$

In applications, a linear partial differential equation of the first order of a function $u = u(t, x)$ of the form

$$u_t(t, x) + cu_x(t, x) = f(t, x)$$

is called the *transport equation*. This equation can be applied to model pollution (contaminant) or dye dispersion, in which case $u(t, x)$ represents the density of the pollutant or the dye at time t and position x. Also it can be applied to traffic flow in which case $u(x, t)$ represents the density of the traffic at moment t and position x. In an initial value problem for the transport equation, we seek the function $u(t, x)$ which satisfies the transport equation and satisfies the initial condition $u(0, x) = u_0(x)$ for some given initial density $u_0(x)$.

Let us illustrate one of these applications with a concrete example.

Example 4.2.7. Fluid flows with constant speed v in a thin, uniform, straight tube (cylinder) whose cross sections have constant area A. Suppose that the fluid contains a pollutant whose concentration at position x and time t is denoted by $u(t, x)$. If there are no other sources of pollution in the tube and there is no loss of pollutant through the walls of the tube, derive a partial differential equation for the function $u(t, x)$.

Solution. Consider a part of the tube between any two points x_1 and x_2. At moment t, the amount of the pollutant in this part is equal to

$$\int_{x_1}^{x_2} Au(t, x) \, dx.$$

Similarly, the amount of the pollutant at a fixed position x during any time interval (t_1, t_2) is given by

$$\int_{t_1}^{t_2} u(t, x)v \, dt.$$

Now we apply the mass *conservation principle*, which states that the total amount of the pollutant in the section (x_1, x_2) at time t_2 is equal to the total amount of the pollutant in the section (x_1, x_2) at time t_1 plus the amount of contaminant that flowed through the position x_1 during the time interval (t_1, t_2) minus the amount of pollutant that flowed through the position x_2 during the time interval (t_1, t_2):

$$(4.2.13) \quad \int_{x_1}^{x_2} u(t_2, x) A \, dx = \int_{x_1}^{x_2} u(t_1, x) A \, dx + \int_{t_1}^{t_2} u(t, x_1) v \, dt - \int_{t_1}^{t_2} u(t, x_2) v \, dt.$$

From the fundamental theorem of calculus we have

$$\int_{x_1}^{x_2} u(t_2, x) A \, dx - \int_{x_1}^{x_2} u(t_1, x) A \, dx = \int_{x_1}^{x_2} \left(u(t_2, x) A - u(t_1, x) A \right) dx$$

$$(4.2.14) \qquad = \int_{x_1}^{x_2} \int_{t_1}^{t_2} \partial_t \left(u(t, x) A \right) dt \, dx.$$

Similarly,

$$\int_{t_1}^{t_2} u(t, x_2) A v \, dt - \int_{t_1}^{t_2} u(t, x_1) A v \, dt$$

$$(4.2.15)$$

$$= \int_{t_1}^{t_2} \left(u(t, x_2) A - u(t, x_1) A v \right) dt = \int_{t_1}^{t_2} \int_{x_1}^{x_2} \partial_x \left(u(t, x) A v \right) dx \, dt.$$

If we substitute (4.2.14) and (4.2.15) into the mass conservation law (4.2.13) and interchange the order of integration, then we obtain that

$$\int_{x_1}^{x_2} \int_{t_1}^{t_2} \left[\partial_t \left(u(t, x) A \right) + \partial_x \left(u(t, x) A v \right) \right] dx \, dt = 0.$$

Since the last equation holds for every x_1, x_2, t_1 and t_2 (under the assumption that $u_t(x, y)$ and $u_x(t, x)$ are continuous functions), we have

$$u_t(t, x) + v u_x(t, x) = 0.$$

Example 4.2.8. Find the density function $u = u(t, x)$ which satisfies the differential equation

$$u_t + xu_x = 0,$$

if the initial density $u_0(x) = u(0, x)$ is given by

$$u_0(x) = e^{-(x-3)^2}.$$

Solution. The characteristic curves of this partial differential equation are given by the general solution of the equation

$$-x\, dt + dx = 0.$$

The general solution of the above ordinary differential equation is given by

$$xe^{-t} = c,$$

where c is any constant. Therefore the general solution of the partial differential equation is given by

$$u(t, x) = f(xe^{-t}),$$

where f is any differentiable function of one variable. Using the initial density u_0 we have

$$f(x) = e^{-(x-3)^2}.$$

Therefore

$$u(t, x) = e^{-\left(xe^{-t}-3\right)^2}.$$

We can use Mathematica to check that really this is the solution:
$In[2] := DSolve[\{D[u[t, x], t] + xD[u[t, x], x] == 0,$
$u[0, x] == e^{-(x-3)^2}\},\ u[t, x], \{t, x\}]$
$Out[2] = \{\{u[t, x]- > e^{-(-3+e^{-t}x)^2}\}\}$

Using Mathematica plots of the solution $u(t, x)$ and several characteristic curves are given in Figure 4.2.2 (a) and (b), respectively

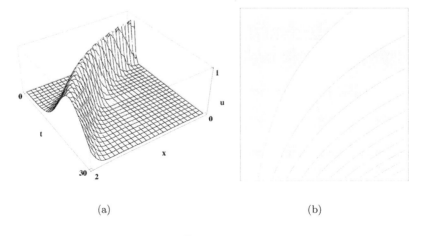

(a)   (b)

Figure 4.2.2

The method of characteristics, described for linear partial differential equations of the first order, can be applied also for some nonlinear equations.

Quasi-linear Equation of the First Order. A differential equation of the form

$$(4.2.16) \qquad a(x, y, u)u_x(x, y, u) + b(x, y, u)u_y(x, y) = c(x, y, u),$$

where $u = u(x, y)$, $(x, y) \in \Omega \subseteq \mathbb{R}^2$, is called a *quasi-linear differential equation of the first order*. We suppose that $a(x, y, u)$, $b(x, y, u)$ and $c(x, y, u)$ are continuous functions in Ω.

A similar discussion as for the linear equation gives the following theorem.

Theorem 4.2.2. *The general solution* $u = u(x, y)$ *of Equation (4.2.16) is given by*

$$f(F(x, y, u), G(x, y, u)) = 0,$$

where f *is any differentiable function of two variables and* $F(x, y, u) = c_1$ *and* $G(x, y, u) = c_2$ *are two linearly independent solutions of the characteristic equations*

$$\frac{dx}{a(x, y, u)} = \frac{dy}{b(x, y, u)} = \frac{du}{c(x, y, u)}.$$

Example 4.2.9. Solve the partial differential equation $uu_x + 2xu_y = 2xu^2$.

Solution. It is obvious that $u \equiv 0$ is a solution of the above equation. If $u \neq 0$, then the characteristic equations are

$$\frac{dx}{u} = \frac{dy}{2x} = \frac{du}{2xu^2}.$$

From $2xu^2\, dx = u\, du$ we have $u = c_1 e^{x^2}$. Using this solution and the other differential equation $2x\, dx = u\, du$ we have $c_1 x + e^{-x^2} = c_2$. Therefore, the general solution of the given nonlinear equation is given implicitly

$$f\left(ue^{-x^2},\, xue^{-x^2} + e^{-x^2}\right) = 0,$$

where f is any differentiable function of two variables.

Exercises for Section 4.2.

1. Find the general solution of each of the following partial differential equations.

(a) $xu_x + y^2 u_y = 0$, $x > 0$, $y > 0$.

(b) $xu_x + yu_y = xy \sin(xy)$, $y > 0$.

(c) $(1+x^2)u_x + u_y = 0$.

(d) $u_x + 2xy^2u_y = 0$.

2. Find the general solution of each of the following partial differential equations.

   (a) $au_x + bu_y + cu = 0$, $a \neq 0$, b and c are numerical constants.

   (b) $y^3u_x - xy^2u_y = cxu$, $c \neq 0$ is a constant, $y > 0$.

   (c) $u_x + (x+y)u_y = xu$.

   (d) $xu_x - yu_y + y^2u = y^2$.

   (e) $xu_x + yu_y = u$, $x \neq 0$.

   (f) $x^2u_x + y^2u_y = (x+y)u$, $x \neq 0$ and $y \neq 0$.

   (g) $3u_x + 4u_y + 14(x+y)u = 6xe^{-(x+y)^2}$.

   (h) $x^2u_x + u_y = xu$.

3. Find the general solution $u(x, y)$ of each of the following partial differential equations. Find the particular solution $u_p(x, y)$ which satisfies the given condition.

   (a) $3u_x - 5u_y = 0$, $u(0, y) = \sin y$.

   (b) $2u_x - 3u_y = \cos x$, $u(x, x) = x^2$.

   (c) $xu_x + yu_y = 2xy$, $u(x, x^2) = 2$.

   (d) $u_x + xu_y = \left(y - \frac{1}{2}x^2\right)^2$, $u(0, y) = e^y$.

   (e) $yu_x - 2xyu_y = 2xu$, $u(0, y) = y^3$.

   (f) $2xu_x + yu_y - x - u = 0$, $u(1, y) = \dfrac{1}{y}$.

4. Find the particular solution of each of the following partial differential equations which satisfies the given condition.

   (a) $\dfrac{1}{x}u_x + \dfrac{1}{y}u_y = 0$, $u(x, 0) = cx^4$, where c is a constant.

(b) $\dfrac{1}{x}u_x + \dfrac{1}{y}u_y = \dfrac{1}{y}$, $u(x,1) = \dfrac{1}{2}(3 - x^2)$.

(c) $xu_x + (x+y)u_y = u + 1$, $u(x,0) = x^2$.

(d) $2xyu_x + (x^2 + y^2)u_y = 0$, $u(x,y) = e^{\frac{x}{x-y}}$ if $x + y = 1$.

(e) $xu_x - yu_y = u$, $u(y,y) = y^2$.

(f) $xu_x + yu_y = 2xyu$, $u(x,y) = f(x,y)$ on the circle $x^2 + y^2 = 1$, f is a given differentiable function.

(g) $e^x u_x + u_y = y$, $u(x,0) = g(x)$, g is a given differentiable function.

5. Solve the following equations with the given initial conditions.

   (a) $u_t + 2xtu_x = u$, $u(0,x) = x$, where c is a constant.

   (b) $u_t + xu_x = 0$, $u(0,x) = f(x)$, where f is a given differentiable function.

   (c) $u_t = x^2 u_x$, $u(x,x) = f(x)$, where f is a given differentiable function.

   (d) $u_t + cu_x = 0$, $u(0,x) = f(x)$, where f is a given differentiable function.

   (e) $u_t + u_x = x \cos t$, $u(0,x) = \sin x$.

   (f) $u_t + u_x - u = t$, $u(0,x) = e^x$.

6. Solve the transport equation $u_t + u_x = 0$, $0 < x < \infty$, $t > 0$, subject to the boundary condition $u(t,0) = t$, and the initial condition

$$u(0,x) = \begin{cases} x^2, & 0 \le x \le 2 \\ x, & x > 2. \end{cases}$$

4.3 Linear Partial Differential Equations of the Second Order.

In this section we will study linear equations of the second order in one unknown function. The following are examples of this type of equations:

$$\begin{aligned} u_t &= c^2 u_{xx}, & \text{the Heat equation.} \\ u_{tt} &= c^2 u_{xx}, & \text{the Wave equation.} \\ u_{xx} + u_{yy} &= 0, & \text{the Laplace equation.} \end{aligned}$$

A fairly large class of important problems of physics and engineering can be written in the form of the above listed equations. The heat equation is used to model heat distribution (flow) along a rod or wire. It is easily generalized to describe heat flow in planar or three dimensional objects. This equation can also be used to describe many diffusion processes. The wave equation is used to describe many physical processes, such as the vibrations of a guitar string, vibrations of a stretched membrane, waves in water, sound waves, light waves and radio waves. Laplace's equation is applied to problems involving electrostatic and gravitational potentials.

4.3.1 Important Equations of Mathematical Physics.

In this part we study several physical problems which lead to second order linear partial differential equations.

The Wave Equation. Chapter 5 is entirely devoted to the wave equation. Now, we will only derive this important equation.

Example 4.3.1. Very thin, uniform and perfectly flexible string of length l is placed on the Ox-axis. Let us assume that the string is fastened at its end points $x = 0$ and $x = l$. We also assume that the string is free to vibrate only in the vertical direction. Initially, the string is in a horizontal position. At the initial moment $t = 0$, we displace the string slightly and the string will start to vibrate. By $u(t, x)$ we denote the vertical displacement of the point x of the string from its equilibrium position at moment t. We assume that the function $u(x, t)$ is twice differentiable with respect to both variables x and t, i.e., $u \in C^2\big((0, l) \times (0, \infty)\big)$. We will also assume that the density ρ of the string is constant. Let us consider a small segment of the string between the points A and B (see Figure 4.3.1). This segment has mass which is equal to ρh and acceleration u_{tt}. The only forces acting along this string segment are the string tension T. We ignore gravity g since it is very small relative to the tension forces. From physics we know that tension is the magnitude of the pulling force exerted by the string, and the tension force is in the direction of the tangent line. According to Hooke's law, the tension of the string at the point x and time t is equal to $u_x(x, t)$. The vertical components of the tension forces at the points A and B are

$$-T_1 \sin \alpha_1 \quad \text{and} \quad T_2 \sin \alpha_2$$

respectively.

Since there is no horizontal motion, the horizontal components of the tension forces at the points A and B must be the same:

(4.3.1) $$T_1 \cos \alpha_1 = T_2 \cos \alpha_2 = T.$$

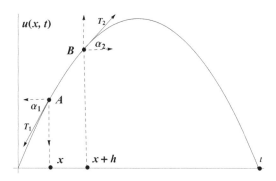

Figure 4.3.1

Now if we apply Newton's Second Law of motion, which states that mass times acceleration equals force, then we have

$$\rho h u_{tt} = T_2 \sin \alpha_2 - T_1 \sin \alpha_1.$$

If we divide the last equation by T and use Equation (4.3.1), we obtain that

$$\frac{\rho h}{T} u_{tt}(x,t) = \frac{T_2 \sin \alpha_2}{T_2 \cos \alpha_2} - \frac{T_1 \sin \alpha_1}{T_1 \cos \alpha_1}.$$

$$= \tan \alpha_2 - \tan \alpha_1.$$

Now since

$$\tan \alpha_1 = u_x(x,t), \quad \tan \alpha_2 = u_x(x+h,t)$$

we have

$$\frac{\rho}{T} u_{tt}(x,t) = \frac{u_x(x+h,t) - u_x(x,t)}{h}.$$

Letting $h \to 0$ and $a^2 = \dfrac{T}{\rho}$ we obtain

$$u_{tt}(x,t) = a^2 u_{xx}(x,t).$$

This equation is called the *wave equation*.

We can consider the vibration of a two dimensional elastic membrane (drumhead) over a plane domain Ω. In a similar way as for the vibrating string we find that the equation of the deviation $u(x,y,t)$ at a point $(x,y) \in \Omega$ and at time t has the form

$$u_{tt}(x,y,t) = a^2 \big[u_{xx}(x,y,t) + u_{yy}(x,y,t) \big], \quad (x,y) \in \Omega, \ t > 0,$$

where a is a constant which depends on the physical characteristics of the membrane.

Three dimensional vibrations can be treated in the same way, resulting in the three dimensional wave equation

$$u_{tt} = a^2 \left(u_{xx} + u_{yy} + u_{zz} \right).$$

Remark. If we introduce the *Laplace operator* or *Laplacian* Δ by

(4.3.2) $$\Delta = \partial_x^2 + \partial_y^2 + \partial_z^2,$$

then the wave equation in any dimension can be written in the form

$$u_{tt} = a^2 \Delta u.$$

Example 4.3.2. The Telegraph Equation. A variation of the wave equation is the telegraph equation

$$u_{tt} - c^2 u_{xx} + a u_t + b u = 0, \quad 0 < x < l, \; t > 0,$$

where a, b and c are constants. This equation is applied in problems related to transmission of electrical signals in telephone or electrical lines (cables). A mathematical derivation for the telegraph equation in terms of voltage and current for a section of a transmission line will be investigated.

Consider a small element of a telegraph cable as an electrical circuit of length h at a point x, where x is the distance from one of the ends of the cable. Let $i(x,t)$ be the current in the cable, v the voltage in the cable, $e(x,t)$ the potential at any point x of the cable. Further, let R be the coefficient of resistance of the cable, L the coefficient of inductance of the cable, G the coefficient of conductance to the ground and C the capacitance to the ground. According to Ohm's Law, the voltage change in the cable is given by

(4.3.3) $$v = Ri(x,t).$$

According to another Ohm's Laws the voltage decrease across the capacitor is given by

(4.3.4) $$v = \frac{1}{C} \int i(x,t) \, dt$$

and the voltage decrease across the inductor is given by

(4.3.5) $$v = L\frac{\partial i(x,t)}{\partial t}.$$

The electrical potential $e(x+h,t)$ at the point $x+h$ is equal to the potential $e(x,t)$ at x minus the decrease in potential along the cable element $[x, x+h]$. Therefore, from (4.3.3), (4.3.4) and (4.3.5) it follows that

$$e(x + h, t) - e(x, t) = -Rhi(x, t) - Lhi_t(x, t).$$

If we divide the last equation by h and let $h \to 0$ we obtain that

(4.3.6) $$e_x(x, t) = -Ri(x, t) - Li_t(x, t).$$

For the current i we have the following. The current $i(x + h, t)$ at $x + h$ is equal to the current $i(x, t)$ at x minus the current lost through leakage to the ground. Using the formula for current through the capacitor,

$$i(x, t) = Ce_t(x, t),$$

therefore we have

(4.3.7) $$i(x + h, t) - i(x, t) = -Ghi(x, t) - Che_t(x, t).$$

Dividing both sides of (4.3.7) by h and letting $h \to 0$ we obtain

(4.3.8) $$i_x(x, t) = -Ghi(x, t) - Ce_t(x, t).$$

If now we differentiate (4.3.6) with respect to x and (4.3.8) with respect to t, the following equations are obtained.

(4.3.9) $$.e_{xx}(x, t) = -Ri_x(x, t) - Li_{xt}(x, t),$$

and

(4.3.10) $$i_{xt}(x, t) = -Ge_x(x, t) - Ce_{tt}(x, t).$$

From (4.3.9) and (4.3.10) it easily follows that

(4.3.11) $$e_{xx}(x, t) - LCe_{tt}(x, t) - (RC + GL)e_t(x, t) - GRe(x, t) = 0,$$

and

(4.3.12) $$i_{xx}(x, t) - LCi_{tt}(x, t) - (RC + GL)i_t(x, t) - GRi(x, t) = 0.$$

Equations (4.3.11) and (4.3.12) are the telegraphic equations for $e(x, t)$ and $i(x, t)$.

Example 4.3.3. The Heat/Diffusion Equation. In this example we will derive the heat equation using some well known physical laws and later, in Chapter 6, we will solve this equation.

Let us consider a homogeneous cylindrical metal rod of length l which has constant cross sectional area A and constant density ρ. We place the rod along the x-axis from $x = 0$ to $x = l$. We assume that the lateral surface of the rod is heat insulated, that is, we assume that the heat can escape or enter the rod only at either end. Let $u(t, x)$ denote the temperature of the rod at position x at time t. We also assume that the temperature is constant at any point of the cross section. For derivation of the equation we will use the following physical laws of heat conduction:

1. *The Fourier/Ficks Law.* When a body is heated the heat flows in the direction of temperature decrease from points of higher temperature to places of lower temperature. The amount of heat which flows through the section at a position x during time interval $(t, t + \Delta t)$ is equal to

$$(4.3.13) \qquad Q = -A \int_t^{t+\Delta t} \gamma u_x(x, t)\, dt,$$

where the constant γ is the *thermal conductivity*.

2. The rate of change of heat in a section of the rod between x and $x + h$ is given by

$$(4.3.14) \qquad \int_x^{x+h} c\rho A u_t(x, t)\, dx,$$

where the constant c is the *specific heat*, which is the amount of heat that it takes to raise one unit mass by one unit temperature.

3. *Conservation Law of Energy.* This law states that the rate of change of heat in the element of the rod between the points x and $x + h$ is equal to the sum of the rate of change of heat that flows in and the rate of change of heat that flows out, that is,

$$(4.3.15) \qquad c\rho h u_t(x, t) = -\gamma A u_x(x + h, t) - \gamma A u_x(x, t).$$

From (4.3.13), (4.3.14) and (4.3.15) it follows that

$$(4.3.16) \quad \int_t^{t+\Delta t} \left[\gamma u_x(x+h, \tau) - \gamma u_x(x, \tau)\right] d\tau = \int_x^{x+h} c\rho\left[u(\xi, t+\Delta t) - u(\xi, t)\right] d\xi.$$

Since we assumed that u_{xx} and u_t are continuous, from the mean value theorem of integral calculus we obtain that

$$(4.3.17) \quad \left[\gamma u_x(x + h, \tau_*) - \gamma u_x(x, \tau_*)\right]\Delta t = c\rho\left[u(\xi_*, t + \Delta t) - u(\xi_*, t)\right]h,$$

for some $\tau_* \in (t, t + \Delta t)$ and $\xi_* \in (x, x + h)$. If we apply the mean value theorem of differential calculus to (4.3.17), then we have

$$\frac{\partial}{\partial x}\left[\gamma u_x(\xi_1, \tau_*)\right]\Delta t h = c\rho u_t(\xi_*, t_1)h\Delta t,$$

for some $\xi_1 \in (x, x + h)$ and $t_1 \in (t, t + \Delta t)$. If we divide the last equation by $h\rho$, and we let $\Delta t \to 0$ and $h \to 0$, we obtain that

$$\frac{\partial}{\partial x}\left(\gamma u_x\right) = c\rho u_t.$$

Since the rod is homogeneous, γ, c and ρ are constants, and the last equation takes the form

(4.3.18)
$$u_t = a^2 u_{xx}, \quad a^2 = \frac{\gamma}{c\rho}.$$

Equation (4.3.18) is called the *one dimensional heat equation* or *one dimensional heat-conduction equation*.

As in the case of the wave equation, we can consider a two dimensional heat equation for a two dimensional plate, or three dimensional object:

$$u_t = a^2 \Delta u.$$

The Laplace Equation. There are many physical problems which lead to the Laplace equation. A detailed study of this equation will be taken up in Chapter 7.

Let us consider some applications of this equation with a few examples.

Example 4.3.4. It was shown that the temperature of a nonstationary heat flow without the presence of heat sources satisfies the heat differential equation

$$u_t = a^2 \Delta u.$$

If a stationary heat process occurs (the heat state does not change with time), then the temperature distribution is constant with time; thus it is only a function of position. Consequently, the heat equation in this case reduces to

$$\Delta u = 0.$$

This equation is called the *Laplace equation*.

However, if heat sources are present, the heat equation in a stationary situation takes the form

(4.3.19)
$$\Delta u = -f, \quad f = \frac{F}{k},$$

where F is the heat density of the sources, and k is the heat conductivity coefficient. The nonhomogeneous Laplace equation (4.3.19) is usually called the *Poisson equation*.

Example 4.3.5. As a second example, we will investigate the work in a conservative vector field. Gravitational force, spring force, electrical and magnetic force are several examples of conservative forces.

Let Ω be the whole plane $\mathbb{R}^2$ except the origin O. Suppose that a particle starts at any point $M(x,y)$ of the plane Ω and it is attracted toward the origin O by a force $\mathbf{F} = (f,g)$ of magnitude

$$\frac{1}{r} = \frac{1}{\sqrt{x^2 + y^2}}.$$

We will show that the work w performed only by the attraction force $\mathbf{F}$ toward O as the particle moves from M to $N(1,0)$ is path independent, i.e., w does not depend on the curve along which the particle moves, but only on the initial and terminal points M and N (see Figure 4.3.2.)

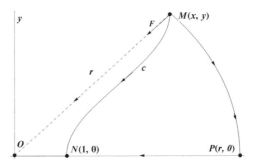

Figure 4.3.2

Since the magnitude of $\mathbf{F} = (f,g)$ is $1/\sqrt{x^2+y^2}$ we have

$$f = f(x,y) = -\frac{x}{x^2+y^2}, \quad g = g(x,y) = -\frac{y}{x^2+y^2}.$$

Consider the function

(4.3.20) $$v(x,y) = -\frac{1}{2}\ln\left(x^2+y^2\right).$$

It is easily seen that

(4.3.21) $$f(x,y) = v_x(x,y), \quad g(x,y) = v_y(x,y).$$

Let c be any continuous and piecewise smooth curve in Ω from the point M to the point N and let

$$x = x(t), \quad y = y(t), \ a \le t \le b,$$

be the parametric equations of the curve c. The work w_c done to move the particle from the point M to point N along the curve c under the force $\mathbf{F}$ is given by the line integral

$$(4.3.22) \qquad w_c = \int_a^b \left[f(x(t), y(t)) x'(t) + g(x(t), y(t)) y'(t) \right] dt.$$

From (4.3.21) and (4.3.22) it follows that

$$w_c = \int_a^b \left[f(x(t), y(t)) x'(t) + g(x(t), y(t)) y'(t) \right] dt$$

$$= \int_a^b \left[v_x x'(t) + v_y y'(t) \right] dt = \int_a^b \frac{d}{dt} \left(v(x(t), y(t)) \right) dt$$

$$= v(x(b), y(b)) - v(x(a), y(a)) = v(N) - v(M).$$

Therefore, the work w_c does not depend on c (it depends only on the initial point $M(x, y)$ and terminal point $N(1, 0)$). Thus

$$w_c = u(x, y)$$

for some function $u(x, y)$.

Now, since the work w_c is path independent we have

$$w_c = w_{\overset{\frown}{MP}} + w_{\overline{PN}}.$$

For the work $w_{\overset{\frown}{MP}}$ along the arc $\overset{\frown}{MP}$, from (4.3.21) and the path independence of the work we have

$$w_{\overset{\frown}{MP}} = v(P) - v(M) = -\frac{1}{2} \ln(r^2) + \frac{1}{2} \ln(r^2) = 0.$$

For the work $w_{\overline{PN}}$ along the line segment $\overline{PN}$ we have

$$w_{\overline{PN}} = \int_{\overline{PN}} f(x, y)\, dx + g(x, y)\, dy = - \int_{\overline{PN}} f(x, y)\, dx$$

$$= - \int_r^1 \frac{1}{x}\, dx = \ln r = \frac{1}{2} \ln \left(x^2 + y^2 \right).$$

Therefore,

$$u(x, y) = \frac{1}{2} \ln \left(x^2 + y^2 \right).$$

Differentiating the above function $u(x, y)$ twice with respect to x and y we obtain

$$u_{xx} = \frac{y^2 - x^2}{(x^2 + y^2)^2}, \quad u_{yy} = \frac{x^2 - y^2}{(x^2 + y^2)^2}.$$

Thus,

$$u_{xx} + u_{yy} = 0.$$

4.3.2 Classification of Linear PDEs of the Second Order.

In this section we will consider the most general linear partial differential equation of the second order of a function $u(x, y)$ of two independent variables x and y in a domain $\Omega \subseteq \mathbb{R}^2$:

(4.3.23) $L u \equiv a u_{xx} + 2 b u_{xy} + c u_{yy} + a_1 u_x + c_1 u_y + d u = f(x, y), \ (x, y) \in \Omega.$

We assume that the coefficients $a = a(x, y)$, $b = b(x, y)$, $c = c(x, y)$, $a_1 = a_1(x, y)$, $c_1 = c_1(x, y)$ and $d = d(x, y)$ are functions of x and y, and the functions $a(x, y)$, $b(x, y)$ and $c(x, y)$ are twice continuously differentiable in Ω. We assume that the function $u(x, y)$ is also twice continuously differentiable. If the function $f(x, y)$ in (4.3.23) is zero in Ω, then the equation

(4.3.24) $a u_{xx} + 2 b u_{xy} + c u_{yy} + a_1 u_x + c_1 u_y + d u = 0, \ (x, y) \in \Omega$

is called *homogeneous*. We will find an invertible transformation of the variables x and y to new variables ξ and η such that the transformed differential equation has a much simpler form than (4.3.23) or (4.3.24). Let the functions

(4.3.25) $\xi = \xi(x, y), \quad \eta = \eta(x, y)$

be twice differentiable, and let the Jacobian

$$\begin{vmatrix} \xi_x & \xi_y \\ \eta_x & \eta_y \end{vmatrix}$$

be different from zero in the domain Ω of consideration. Then the transformation defined by (4.3.25) is invertible in some domain. Using the chain rule we compute the first partial derivatives

$$u_x = u_\xi \xi_x + u_\eta \eta_x, \quad u_y = u_\xi \xi_y + u_\eta \eta_y,$$

and after that the second order partial derivatives

$$u_{xx} = u_{\xi\xi} \xi_x^2 + 2 u_{\xi\eta} \xi_x \eta_x + u_{\eta\eta} \eta_x^2 + u_\xi \xi_{xx} + u_\eta \eta_{xx},$$
$$u_{xy} = u_{\xi\xi} \xi_x \xi_y + u_{\xi\eta} (\xi_x \eta_y + \xi_y \eta_x) + u_{\eta\eta} \eta_x \eta_y + u_\xi \xi_{xy} + u_\eta \eta_{xy},$$
$$u_{yy} = u_{\xi\xi} \xi_y^2 + 2 u_{\xi\eta} \xi_y \eta_y + u_{\eta\eta} \eta_y^2 + u_\xi \xi_{yy} + u_\eta \eta_{yy}.$$

If we substitute the above derivatives into Equation (4.3.23), we obtain

(4.3.26) $A u_{\xi\xi} + 2 B u_{\xi\eta} + C u_{\eta\eta} + \tilde{A} u_\xi + \tilde{C} u_\eta + d u = f,$

where

(4.3.27)
$$\begin{cases} A = a \xi_x^2 + 2 b \xi_x \xi_y + c \xi_y^2, \\ B = a \xi_x \eta_y + b (\xi_x \eta_y + \eta_x \xi_y) + c \xi_y \eta_y, \\ C = a \eta_x^2 + 2 b \eta_x \eta_y + c \eta_y^2, \\ \tilde{A} = a_1 \xi_x + c_1 \xi_y, \\ \tilde{C} = a_1 \eta_x + c_1 \eta_y. \end{cases}$$

We can simplify Equation (4.3.26) if we can select ξ and η such that at least one of the coefficients A, B and C is zero.

Consider the family of functions $y = y(x)$ defined by $\xi(x, y) = c_1$, where c_1 is any numerical constant. Differentiating this equation with respect to x we have

$$\xi_x + \xi_y y'(x) = 0,$$

from which it follows that

$$\xi_x = -\xi_y y'(x).$$

If we substitute ξ_x from the last equation into the expression for A in (4.3.27) it follows that

$$(4.3.28) \qquad A = \xi_y^2 \left(a y'^2(x) - 2b y'(x) + c \right).$$

Working similarly, if we consider the family of functions $y = y(x)$ defined by $\eta(x, y) = $ numerical constant, we obtain

$$(4.3.29) \qquad B = \eta_y^2 \left(a y'^2(x) - 2b y'(x) + c \right).$$

Equations (4.3.28) and (4.3.29) suggest that we need to consider the nonlinear ordinary differential equation

$$(4.3.30) \qquad a y'^2(x) - 2b y'(x) + c = 0,$$

Equation (4.3.30) is called the *characteristic equation* of the given partial differential equation. If $a \neq 0$, from (4.3.30) we obtain

$$(4.3.31) \qquad y'(x) = \frac{b \pm \sqrt{b^2 - ac}}{a}.$$

From (4.3.31) it follows that we have to consider the following three cases: $b^2 - ac > 0$, $b^2 - ac < 0$ and $b^2 - ac = 0$.

Case 1^0. $b^2 - ac > 0$. In this case the linear partial differential equation (4.3.23) is called *hyperbolic*. The wave equation is a prototype of a hyperbolic equation.

Let $\varphi(x, y) = C_1$ and $\psi(x, y) = C_2$ be the two (linearly independent) general solutions of Equations (4.3.31). If we select now

$$\xi = \varphi(x, y) \quad \text{and} \quad \eta = \psi(x, y),$$

then from (4.3.28) and (4.3.29) it follows that both coefficients A and C will be zero, and so the transformed equation (4.3.26) takes the form

$$(4.3.32) \qquad u_{\xi\eta} + \frac{\tilde{A}}{2B} u_\xi + \frac{\tilde{C}}{2B} u_\eta + \frac{du}{2B} = \frac{f}{2B}.$$

Equation (4.3.32) is the *canonical form* of the equation of hyperbolic type. Very often, another canonical form is used. If we introduce new variables α and β by

$$\xi = \alpha + \beta, \quad \eta = \alpha - \beta,$$

then from

$$u_\xi = \frac{1}{2}(u_\alpha + u_\beta), \; u_\eta = \frac{1}{2}(u_\alpha - u_\beta), \; u_{\xi\eta} = \frac{1}{2}(u_{\alpha\alpha} - u_{\beta\beta})$$

Equation (4.3.32) takes the form

$$u_{\alpha\alpha} - u_{\beta\beta} = F(u, u_\alpha, u_\beta).$$

Case 2^0. $b^2 - ac = 0$. An equation for which $b^2 - ac = 0$ is said to be of *parabolic type* or simply *parabolic equation*. A prototype of this type of equation is the heat equation.

In this case, the two equations (4.3.31) coincide and we have a single differential equation. Let the general solution of this equation be given by $\xi(x, y) = $ constant. Now, select the variables ξ and η by

$$\xi = \xi(x, y), \quad \eta = \eta(x, y),$$

where for $\eta(x, y)$ we can take to be any function, linearly independent of ξ. Because of this choice of the variable ξ, from $b^2 = ac$ and (4.3.31) we have

$$\sqrt{a}\, \xi_x + \sqrt{c}\, \xi_y = \frac{1}{\sqrt{a}}\left(a\xi_x + b\xi_y\right) = 0.$$

Therefore,

$$A = a\xi_x^2 + 2b\xi_x\xi_y + c\xi_y^2 = \left(\sqrt{a}\xi_x + \sqrt{c}\xi_y\right)^2 = 0,$$

and

$$B = a\xi_x\eta_y + b\left(\xi_x\eta_y + \eta_x\xi_y\right) + c\xi_y\eta_y$$
$$= \left(\sqrt{a}\, \xi_x + \sqrt{c}\xi_y\right)\left(\sqrt{a}\, \eta_x + \sqrt{c}\, \eta_y\right) = 0,$$

and so, the *canonical form* of the given partial differential equation (after dividing (4.3.26) by C) is

$$u_{\eta\eta} = F\left(\xi, \eta, u, u_\xi, u_\eta\right).$$

Case 3^0. $b^2 - ac < 0$. In this case, the linear partial differential equation is called *elliptic*. The Laplace and Poisson equations are prototypes of this type of equation. The right hand sides of Equations (4.3.31) in this case will be complex functions. Let the general solution of one of Equations (4.3.31) be given by $\phi(x, y) = $ constant, where

$$\phi(x, y) = \varphi(x, y) + i\psi(x, y).$$

It is easy to check that the function $\phi(x, y)$ satisfies the equation

$$a(\phi_x)^2 + 2b\phi_x\phi_y + c(\phi_x)^2 = 0.$$

If we separate the real and imaginary parts in the above equation, we obtain

$$(4.3.33) \qquad a(\varphi_x)^2 + 2b\varphi_x\varphi_y + c(\varphi_x)^2 = a(\psi_x)^2 + 2b\psi_x\psi_y + c(\psi_x)^2 = 0,$$

and

$$(4.3.34) \qquad a\varphi_x\psi_x + b(\varphi_x\psi_y + \varphi_y\psi_x) + c\varphi_y\psi_y = 0.$$

From (4.3.32) it follows that Equations (4.3.33) and (4.3.32) can be written in the form $A = C$ and $B = 0$, where A, B and C are the coefficients that appear in Equation (4.3.33). Dividing both sides of Equation (4.3.26) by A (provided that $A \neq 0$) we obtain

$$(4.3.35) \qquad u_{\xi\xi} + u_{\eta\eta} = F(\xi, \eta, u, u_\xi, u_\eta).$$

Equation (4.3.35) is called the *canonical form* of equation of the elliptic type.

Let us take a few examples.

Example 4.3.6. Find the domains in which the equation

$$yu_{xx} - 2xu_{xy} + yu_{yy} = 0$$

is hyperbolic, parabolic or elliptic.

Solution. The discriminant for this equation is $D = b^2 - ac = x^2 - y^2$. Therefore the equation is hyperbolic in the region Ω of points (x, y) for which $y^2 < x^2$, i.e., for which $|y| < |x|$. The equation is parabolic when $D = 0$, i.e., when $y = \pm x$. The equation is elliptic when $D < 0$, i.e., when $|y| > |x|$.

Example 4.3.7. Solve the equation

$$y^2 u_{xx} + 2xyu_{xy} + x^2 u_{yy} - \frac{y^2}{x}u_x - \frac{x^2}{y}u_y = 0$$

by reducing it to its canonical form.

Solution. For this equation, $b^2 - ac = x^2y^2 - x^2y^2 = 0$ and so the equation is parabolic in the whole plane. The characteristic equation (4.3.30) for this equation is

$$y^2 y'^2(x) - 2xyby'(x) + 3x^2 = 0.$$

Solving for $y'(x)$ we obtain

$$y'(x) = \frac{x}{y}.$$

The general solution of the above equation is given by $x^2 - y^2 = c$, where c is an arbitrary constant. Therefore, we introduce the new variable η to be an arbitrary function (for example, we can take $\eta = x$) and the variable ξ is introduced by

$$\xi = x^2 - y^2, \quad \eta = x.$$

Then, using the chain rule we obtain

$$u_x = 2xu_\xi + u_\eta, \ u_y = -2yu_\xi,$$
$$u_{xx} = 2u_\xi + 4x^2u_{\xi\xi} + 4xu_{\xi\eta} + u_{\eta\eta},$$
$$u_{xy} = -4xyu_{\xi\xi} - 2yu_{\xi\eta}, \ u_{yy} = -2u_\xi + 4y^2u_{\xi\xi}.$$

Substituting the above partial derivatives into the given equation we obtain

$$\eta u_{\eta\eta} = u_\eta.$$

This equation can be treated as an ordinary differential equation. By separation of variables we find that its general solution is

$$u(\xi, \eta) = \eta^2 f(\xi) + g(\xi),$$

where f and g are arbitrary twice differentiable functions. Returning to the variables x and y it follows that the solution of the equation is

$$u(x, y) = x^2 f(x^2 - y^2) + g(x^2 - y^2).$$

Example 4.3.8. Solve the equation

$$y^2 u_{xx} - 4xy u_{xy} + 3x^2 u_{yy} - \frac{y^2}{x} u_x - \frac{3x^2}{y} u_y = 0$$

by reducing it to its canonical form.

Solution. For this equation, $b^2 - ac = 16x^2y^2 - 12x^2y^2 = 4x^2y^2 > 0$ for every $x \neq 0$ and $y \neq 0$. Therefore, the equation is hyperbolic in $\mathbb{R}^2 \setminus \{(0,0)\}$. It is parabolic on the coordinate axes Ox and Oy. For the hyperbolic case, the characteristic equation (4.3.30) is

$$y^2 {y'}^2(x) + 4xy y'(x) + x^2 = 0.$$

Solving for $y'(x)$ we obtain

$$y'(x) = -\frac{x}{y}, \quad y'(x) = -3\frac{x}{y}.$$

The general solution of the last two ordinary differential equations is given by

$$x^2 + y^2 = C_1, \quad \text{and} \quad 3x^2 + y^2 = C_2,$$

where C_1 and C_2 are arbitrary numerical constants. Therefore, we introduce new variables ξ and η by

$$\xi = 3x^2 + y^2 \quad \text{and} \quad \eta = x^2 + y^2.$$

Therefore, using the chain rule we obtain

$$u_x = 6xu_\xi + 2xu_\eta, \; u_y = 2yu_\xi + 2yu_\eta,$$
$$u_{xx} = 6u_\xi + 2u_\eta + 36x^2u_{\xi\xi} + 24x^2u_{\xi\eta} + 4x^2u_{\eta\eta},$$
$$u_{xy} = 12xyu_{\xi\xi} + 16xu_{\xi\eta} + 4xyu_{\eta\eta}, \; u_{yy} = 2u_\xi + 2u_\eta + 4y^2u_{\xi\xi} + 8y^2u_{\xi\eta} + 4y^2u_{\eta\eta}.$$

If we substitute the above partial derivatives into the given equation we obtain

$$u_{\xi\eta} = 0.$$

It is easy to find that the general solution of the above equation is

$$u(\xi, \eta) = f(\xi) + g(\xi),$$

where f and g are arbitrary twice differentiable functions. Returning to the variables x and y it follows that the solution of the equation is

$$u(x, y) = f(x^2 + y^2) + g(3x^2 + y^2).$$

Example 4.3.9. Reduce to a canonical form the equation

$$5u_{xx} + 4u_{xy} + 4u_{yy} = 0.$$

Solution. For this equation, $b^2 - ac = -4 < 0$ for every $(x, y) \in \mathbb{R}^2$ and therefore, the equation is elliptic in the whole plane $\mathbb{R}^2$. The characteristic equation (4.3.30) in this case is

$$5y'^2(x) - 4y'(x) + 4 = 0,$$

and solving for $y'(x)$ we obtain

$$y'(x) = \frac{2}{5} \pm \frac{4}{5}i.$$

The general solutions of the last two ordinary differential equations are given by

$$y - \left(\frac{2}{5} + \frac{4}{5}i\right)x = C_1, \quad \text{and} \quad y - \left(\frac{2}{5} - \frac{4}{5}i\right)x = C_2,$$

where C_1 and C_2 are arbitrary numerical constants. Therefore, we introduce new variables ξ and η by

$$\xi = 5y - (2 + 4i)x \quad \text{and} \quad \eta = y - (2 - 4i)x.$$

In order to avoid working with complex numbers we introduce two new variables α and β by

$$\alpha = \xi + \eta \quad \text{and} \quad \beta = i(\xi - \eta).$$

Then

$$\alpha = 10y - 4x \quad \text{and} \quad \beta = 8x.$$

Using the chain rule we obtain

$$u_x = -4u_\alpha + 8u_\beta, \; u_y = 10u_\alpha, \; u_{xx} = 16u_{\alpha\alpha} - 64u_{\alpha\beta} + 8u_{\beta\beta},$$
$$u_{xy} = -40u_{\alpha\alpha} + 80u_{\beta\beta}, \; u_{yy} = 100u_{\alpha\alpha}.$$

If we substitute the above partial derivatives in the given equation, then its canonical form is

$$u_{\alpha\alpha} + u_{\beta\beta} = 0.$$

Example 4.3.10. Find the general solution of the wave equation

$$u_{tt} - c^2 u_{xx} = 0, \quad c > 0.$$

Solution. The characteristic equation is given by

$$x'^2(t) - c^2 = 0,$$

whose solutions are $x + ct = c_1$ and $x - ct = c_2$. Therefore, we introduce the new variable ξ and η by $\xi = x + ct$ and $\eta = x - ct$. From

$$u_{xx} = u_{\xi\xi} + 2u_{\xi\eta} + u_{\eta\eta} \quad u_{tt} = c^2 u_{\xi\xi} - 2c^2 u_{\xi\eta} + c^2 u_{\eta\eta}$$

the wave equation takes the form

$$u_{\xi\eta} = 0.$$

The general solution of the last equation is given by

$$u(\xi, \eta) = f(\xi) + g(\eta),$$

where f and g are twice differentiable functions, each of a single variable. Therefore, the general solution of the wave equation is given by

$$u(x, y) = f(x + ct) + g(x - ct).$$

The classification of linear partial differential equations of the second order in more than two variables is much more complicated and involves elements of multi-variable differential calculus and linear algebra. For these reasons this topic is not discussed in this book and the interested reader can consult more advanced books.

Exercises for Section 4.3.

1. Suppose that f and g are functions of class $C^2(-\infty, \infty)$. Find a second order partial differential equation that is satisfied by all functions of the form $u(x, y) = y^3 + f(xy) + g(x)$.

2. Classify the following linear partial differential equations with constant coefficients as hyperbolic, parabolic or elliptic.

    (a) $u_{xx} - 6u_{xy} = 0$.

    (b) $4u_{xx} - 12u_{xy} + u_x - u_y = 0$.

    (c) $4u_{xx} + 6u_{xy} + 9u_{yy} + u_y = 0$.

    (d) $u_{xx} + 4u_{xy} + 5u_{yy} - u_x + u_y = 0$.

    (e) $u_{xx} - 4u_{xy} + 4u_{yy} + 3u_x - 5u_y = 0$.

    (f) $u_{xx} + 2u_{xy} - 3u_{yy} + u_y = 0$.

3. Determine the domains in the Oxy plane where each of the following equations is hyperbolic, parabolic or elliptic.

    (a) $u_{xx} + 2yu_{xy} + u_y = 0$.

    (b) $(1 + x)yu_{xx} + 2xyu_{xy} - y^2u_{yy} = 0$.

    (c) $yu_{xx} + yu_{yy} + u_x - u_y = 0$.

    (d) $x^2u_{xx} + 4yu_{xy} + u_{yy} + 2u_x - 3u_y = 0$.

    (e) $yu_{xx} - 2u_{xy} + e^xu_{yy} + x^2u_y - u = 0$.

    (f) $3yu_{xx} - xu_{xy} + u = 0$.

    (g) $u_{xx} + 2xyu_{xy} + a^2u_{yy} + u = 0$.

4. Classify each of the following equations as hyperbolic, parabolic or elliptic, and reduce it to its canonical form equations.

    (a) $u_{xx} + yu_{xy} + u_{yy} = 0$.

    (b) $xu_{xx} + u_{yy} = x^2$.

    (c) $4u_{xx} + 5u_{xy} + u_{yy} + u_x + u_y = 0$.

(d) $2u_{xx} - 3yu_{xy} + u_{yy} = y$.

(e) $u_{xx} + yu_{yy} = 0$.

(f) $y^2 u_{xx} + x^2 u_{yy} = 0$.

(g) $(a^2 + x^2)u_{xx} + (a^2 + x^2)u_{yy} + xu_x + yu_y = 0$.   Hint: $\xi = \ln\left(x + \sqrt{a^2 + x^2}\right)$, $\eta = \ln\left(x + \sqrt{a^2 + y^2}\right)$.

5. Classify each of the following partial differential equations as hyperbolic or parabolic and find its general solution.

(a) $9u_{xx} + 12u_{xy} + 4u_{yy} = 0$.

(b) $u_{xx} + 8u_{xy} + 16u_{yy} = 0$.

(c) $u_{xx} + 2u_{xy} - 3u_{yy} = 0$.

(d) $2xyu_{xx} + x^2 u_{xy} - u_x = 0$.

(e) $x^2 u_{xx} + 2xyu_{xy} + y^2 u_{yy} = 4x^2$.

(f) $(1 + \sin x)u_{xx} - 2\cos x + (1 - \sin x)u_{yy} + \dfrac{(1 + \sin x)^2}{2\cos x}u_x$
$+ \dfrac{1}{2}(1 - \sin x)u_y = 0$.

6. Classify each of the following partial differential equations as hyperbolic or parabolic and find the indicated particular solution.

(a) $u_{xx} + u_{xy} - 2u_{yy} + 1 = 0$, $u(0, y) = u_y(0, y) = x$.

(b) $e^{2x}u_{xx} - 2e^{x+y}u_{xy} + e^{2y}u_{yy} + e^{2x}u_x + e^{2y}u_y = 0$, $u(0, y) = e^{-y}$,
$u_y(0, y) = -e^{-y}$.

(c) $u_{xx} - 2u_{xy} + u_{yy} = 4e^{3y} + \cos x$, $u(x, 0) = 1 + \cos x - \dfrac{5}{9}$,
$u(0, y) = \dfrac{4}{9}e^{3y}$.

4.4 Boundary and Initial Conditions.

In the previous sections we have seen that partial differential equations have infinitely many solutions. Therefore, additional conditions are required in order for a partial differential equation to have a unique solution. In general, we have two types of conditions, boundary value conditions and initial conditions. Boundary conditions are constraints on the function and the space variable, while initial conditions are constraints on the unknown function and the time variable.

Consider a linear partial differential equation of the second order

$$(4.4.1) \qquad L\,u = G(t, \mathbf{x}), \quad \mathbf{x} \in \Omega$$

where L is a linear differential operator of the second order (in $L\,u$ partial derivatives up to the second order are involved). An example of such an operator was the operator given by Equation (4.3.23). Ω is a domain in $\mathbb{R}^n$, $n = 1, 2, 3$. The boundary of Ω is denoted by $\partial\Omega$.

Related to differential equation (4.4.1) we introduce the following terms:

Boundary conditions are a set of constraints that describe the nature of the unknown function $u(t, \mathbf{x})$, $\mathbf{x} \in \Omega$, on the boundary $\partial\Omega$ of the domain Ω. There are three important types of boundary conditions:

Dirichlet Conditions. These conditions specify prescribed values $f(t, \mathbf{x})$, $\mathbf{x} \in \partial\Omega$, of the unknown function $u(t, \mathbf{x})$ on the boundary $\partial\Omega$. We usually write these conditions in the form

$$(4.4.2) \qquad u(t, \mathbf{x})\Big|_{\partial\Omega} = f(t, \mathbf{x}).$$

Neumann Conditions. With these conditions, the value of the normal derivative $\frac{\partial u(t,\mathbf{x})}{\mathbf{n}}$ on the boundary $\partial\Omega$ is specified. Symbolically we write this as

$$(4.4.3) \qquad \frac{\partial u(t, \mathbf{x})}{\mathbf{n}}\bigg|_{\partial\Omega} = g(t, \mathbf{x}).$$

Robin (Mixed) Conditions. These conditions are linear combinations of the Dirichlet and Neumann Conditions:

$$(4.4.4) \qquad a\,u(t, \mathbf{x})\Big|_{\partial\Omega} + b\frac{\partial u(t, \mathbf{x})}{\mathbf{n}}\bigg|_{\partial\Omega} = h(t, \mathbf{x}),$$

for some nonzero constants or functions a and b and a given function h, defined on $\partial\Omega$.

Let us note that different portions of the boundary $\partial\Omega$ can have different types of boundary conditions.

In cases when we have a partial differential equation that involves the time variable t, then we have to consider *initial (Cauchy) conditions*. These conditions specify the value of the unknown function and its higher order derivatives at the initial time $t = t_0$.

If the functions $f(t, \mathbf{x})$, $g(t, \mathbf{x})$ and $h(t, \mathbf{x})$ in Equations (4.4.2), (4.4.3) and (4.4.4), respectively, are identically zero in the domain, then we have so called *homogeneous* boundary conditions; otherwise, the boundary conditions are *nonhomogeneous*.

The choice of boundary and initial conditions depends on the given partial differential equation and the physical problem that the equation describes. For example, in the case of a vibrating string, described by the one space dimensional wave equation

$$u_{tt} - u_{xx} = 0$$

in the domain $(0, \infty) \times (0, l)$, the initial conditions are given by specifying the initial position and velocity of the string. If, for example, we impose the boundary conditions

$$u(0) = u(l) = 0,$$

then it means that the two ends $x = 0$ and $x = l$ of the string are fixed.

If we consider heat distribution problems, then the Dirichlet conditions specify the temperature of the body (a bar, a rectangular plate or a ball) on the boundary. For this type of problem the Neumann conditions specify the heat flux across the boundary.

A similar discussion can be applied to the Laplace/Poisson equation. The questions of choice of boundary and initial conditions for concrete partial differential equations (the heat, the wave and the Laplace equation) will be studied in more detail in the next chapters.

4.5 Projects Using Mathematica.

In this section we will see how Mathematica can be used to solve many partial differential equations. Mathematica can find symbolic (analytic) solutions of many types of partial differential equations (linear of the first and second order, semi linear and quasi linear of the first order). This can be done by the command

$In[] := \text{Dsolve}[\text{expression}, u, \{x, y, \ldots\}];$

Mathematica can solve many partial differential equations. The command for solving (numerically) differential equations is

$In[] := \text{NDsolve}[\text{expression}, u, \{x, y, \ldots\}];$

Project 4.5.1. In this project we solve several types of partial differential equations.

Example 1. Solve the linear partial differential equation of the first order

$$x^3 u_x(x, y) + x^2 y u_y(x, y) - (x + y)u(x, y) = 0.$$

Solution. The solution of the given partial differential is obtained by the commands

$$In[1] := e1 = x^3 * D[u[x, y], x] + x^2 * y * D[u[x, y], y] - (x + y) * u[x, y] == 0;$$

$In[2] := s1 = \text{DSolve}[e1, u, \{x, y\}]$

$Out[2] = \{\{u \to \text{Function}\ [\{x, y\}, e^{-\frac{1}{x} - \frac{y}{x^2}}\ C[1]\ [\frac{y}{x}]]\}\}$

By specifying the constant $C[1]$, we can find a particular solution.

$In[3] := u[x, y]/.s[[1]]/.\{C[1][x_-] \to x^2\}$

$Out[3] = \dfrac{e^{-\frac{1}{x} - \frac{y}{x^2}}\ y^2}{x^2}$

Example 2. Solve the quasi linear partial differential equation of the first order

$$xu_x(x, y) + yu_y(x, y) - u(x, y) - u^2(x, y) = 0.$$

Solution. The solution of the given partial differential is obtained by the commands

$In[4] := e2 = x * D[u[x, y], x] + y * D[u[x, y], y] - u[x, y] - u[x, y]^2 == 0;$

$In[5] := s2 = \text{DSolve}[e1, u, \{x, y\}]$

$Out[5] = \{\{u \to \text{Function}\ [\{x, y\}, -\dfrac{e^{C[1][\frac{y}{x}]}\ x}{-1 + e^{C[1][\frac{y}{x}]}\ x}]\}\}$

Example 3. Solve the nonlinear partial differential equation of the first order

$$u_x^2(x, y) + u_y^2(x, y) - u(x, y) = u^2(x, y).$$

Solution. The solution of the given partial differential is obtained by the commands

$In[6] := e3 = D[u[x, y], x]^2 + y * D[u[x, y], y]^2 - u[x, y]^2 == 0;$

$In[7] := s3 = \text{DSolve}[e3, u, \{x, y\}]$

$Out[7] = \{\{u \to \text{Function}\ [\{x, y\}, e^{-\frac{y}{\sqrt{1+C[1]^2}} - \frac{C[2]}{\sqrt{1+C[1]^2}} - \frac{y}{\sqrt{1+C[1]^2}}}]\}\}$

Example 4. Solve the second order linear partial differential equation

$$u_{xx}(x, y) + 2u_{xy}(x, y) - 5u_{yy}(x, y) = 0.$$

Solution. The solution of the given partial differential is obtained by the commands

$In[8] := e4 = D[u[x, y], x, 2] + 2 * D[u[x, y], x, y] - *D[u[x, y], y, 2] == 0;$

$In[9] := \text{DSolve}\ [e4, u, \{x, y\}]$

$Out[9] = \{\{u \to \text{Function}\ [\{x, y\}, C[1][-\frac{1}{2}(2 + 2\sqrt{6})x + y]$
$+ C[2][-\frac{1}{2}(2 - 2\sqrt{6})x + y]]\}\}$

Example 5. Plot the solution of the initial boundary value problem

$$\begin{cases} u_t(t,x) = 9u_{xxx}(t,x), & 0 < x < 5, \ t > 0 \\ u(0,x) = 0, & 0 < x < 5 \\ u(t,0) = 9\sin^4 t, & u(t,5) = 0, \ t > 0. \end{cases}$$

Solution. We use NDSolve to plot the solution of the given problem.

$In[10] := \text{NDSolve} \left[\left\{ u^{(1,0)}[t,x] == 9\, u^{(0,2)}[t,x], \ u[0,x] == 0, \right. \right.$
$\left. u[t,0] = 9 \ Sin[t]^4, \ u[t,5] == 0 \right\}, \ u, \ \{t,0,10\}, \{x,0,5\} \right]$

$Out[10] = \{\{u \to \text{InterpolatingFunction}\, [\{\{0.,10\}, \{0.,5.\}\}, <>]\}\}$

$In[11] = \text{Plot3D}[\text{Evaluate}[u[t,x]/.\%], \{t,0,10\}, \{x,0,5\}, \text{PlotRange} - > \text{All}]$

The plot of the solution of the given problem is displayed in Figure 4.5.1.

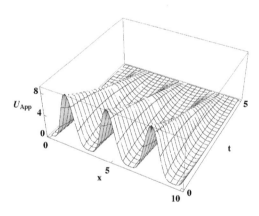

Figure 4.5.1

CHAPTER 5

THE WAVE EQUATION

The purpose of this chapter is to study the one dimensional *wave equation*

$$u_{tt}(x,t) = a^2 u_{xx}(x,t) + h(x,t), \quad x \in (a,b) \subseteq \mathbb{R}, \ t > 0,$$

also known as the *vibrating string equation*, and its higher dimensional version

$$u_{tt}(\mathbf{x},t) = a^2 \Delta u(\mathbf{x},t) + h(\mathbf{x},t), \quad \mathbf{x} \in \Omega \subseteq \mathbb{R}^n, \ n = 2,3.$$

This equation is important in many applications and describes many physical phenomena, e.g., sound waves, ocean waves and mechanical and electromagnetic waves. As mentioned in the previous chapter, the wave equation is an important representative of the very large class of hyperbolic partial differential equations.

In the first section of this chapter we will derive d'Alembert's formula, a general method for the solution of the one dimensional homogeneous and nonhomogeneous wave equation. In the next several sections we will apply the *Fourier Method*, or so called *Separation of Variables Method*, for constructing the solution of the one and higher dimensional wave equation in rectangular, polar and spherical coordinates. In the last section of this chapter we will apply the Laplace and Fourier transforms to solve the wave equation.

5.1 d'Alembert's Method.

In this section we will use d'Alembert's formula to solve the wave equation on an infinite, semi-infinite and finite interval.

Case 1^0. *Infinite String. Homogeneous Equation.* Let us consider first the homogeneous wave equation for an infinite string

(5.1.1) $$u_{tt}(x,t) = a^2 u_{xx}(x,t), \quad -\infty < x < \infty, \ t > 0,$$

which satisfies the initial conditions

(5.1.2) $$u(x,0) = f(x), \quad u_t(x,0) = g(x), \quad -\infty < x < \infty.$$

We introduce new variables ξ and η by

$$\xi = x + at, \quad \text{and} \quad \eta = x - at.$$

Recall from Chapter 4 that the variables ξ and η are called the *characteristics* of the wave equation. If $u(x,t) = w(\xi,\eta)$, then using the chain rule, the wave equation (5.1.1) is transformed into the simple equation

$$w_{\xi\eta} = 0,$$

whose general solution is given by

$$w(\xi,\eta) = F(\xi) + G(\eta),$$

where F and G are any twice differentiable functions, each of a single variable. Therefore, the general solution of the wave equation (5.1.1) is

(5.1.3) $u(x,t) = F(x + at) + G(x - at).$

Using the initial condition $u(x,0) = f(x)$ from (5.1.2) and setting $t = 0$ in (5.1.3) we obtain

(5.1.4) $F(x) + G(x) = f(x).$

If we differentiate (5.1.3) with respect to t we obtain

(5.1.5) $u_t(x,t) = aF'(x + at) - aG'(x - at).$

If we substitute $t = 0$ in (5.1.5) and use the initial condition $u_t(x,0) = g(x)$, then we obtain

(5.1.6) $aF'(x) - aG'(x) = g(x).$

From (5.1.6), by integration it follows that

(5.1.7) $$aF(x) - aG(x) = \int_0^x g(s)\,ds + C,$$

where C is any integration constant. From (5.1.4) and (5.1.7) we obtain

(5.1.8) $$F(x) = \frac{f(x)}{2} + \frac{1}{2a}\int_0^x g(s)\,ds + \frac{C}{2},$$

and

$$G(x) = \frac{f(x)}{2} - \frac{1}{2a}\int_0^x g(s)\,ds - \frac{C}{2}.$$

If we substitute the obtained formulas for F and G into (5.1.3) we obtain that the solution $u(x,t)$ of the wave equation (5.1.1) subject to the initial conditions (5.1.2) is given by

$$(5.1.9) \qquad u(x,t) = \frac{f(x+at) + f(x-at)}{2} + \frac{1}{2a} \int_{x-at}^{x+at} g(s)\,ds.$$

Formula (5.1.9) is known as *d'Alembert's formula.*

Remark. The first term

$$\frac{f(x+at) + f(x-at)}{2}$$

in Equation (5.1.9) represents the propagation of the initial displacement without the initial velocity, i.e., when $g(x) = 0$. The second term

$$\frac{1}{2a} \int_{x-at}^{x+at} g(s)\,ds$$

is the initial velocity (impulse) when we have zero initial displacement, i.e., when $f(x) = 0$.

A function of type $f(x-at)$ in physics is known as a *propagation forward wave* with velocity a, and $f(x+at)$ as a *propagation backward wave* with velocity a.

The straight lines $u = x + at$ and $u = x - at$ (known as *characteristics*) give the propagation paths along which the wave function $f(x)$ propagates.

Remark. It is relatively easy to show that if $f \in C^2(\mathbb{R})$ and $g \in C(\mathbb{R})$, then the C^2 function $u(x,t)$, given by d'Alembert's formula (5.1.9), is the unique solution of the problem (5.1.1), (5.1.2). (See Exercise 1 of this section.)

Example 5.1.1. Find the solution of the wave equation (5.1.1) satisfying the initial conditions

$$u(x,0) = \cos x, \quad u_t(x,0) = 0, \quad -\infty < x < \infty.$$

Solution. Using d'Alembert's formula (5.1.9) we obtain

$$u(x,t) = \frac{\cos(x+at) + \cos(x-at)}{2} = 2\cos x \cos at.$$

Example 5.1.2. Find the solution of the wave equation (5.1.1) satisfying the initial conditions

$$u(x,0) = 0 \quad u_t(x,0) = \sin x, \quad -\infty < x < \infty.$$

Solution. Using d'Alembert's formula (5.1.9) we obtain

$$u(x,t) = \frac{1}{2a} \int\limits_{x-at}^{x+at} \sin s \, ds = \frac{1}{a} \sin x \sin at.$$

Example 5.1.3. Illustrate the solution of the wave equation (5.1.1) taking $a = 2$, subject to the initial velocity $g(x) = 0$ and initial displacement $f(x)$ given by

$$f(x) = \begin{cases} 2 - |x|, & -2 \le x \le 2 \\ 0, & \text{otherwise.} \end{cases}$$

Solution. Figure 5.1.1 shows the behavior of the string and the propagation of the forward and backward waves at the moments $t = 0$, $t = 1/2$, $t = 1$, $t = 1.5$, $t = 2$ and $t = 2.5$.

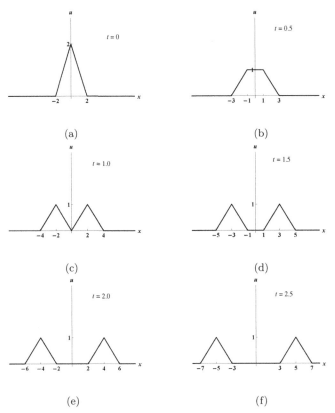

Figure 5.1.1

The solution $u(x,t)$ of this wave equation is given by

$$u(x,t) = \frac{f(x+2t) + f(x-2t)}{2}.$$

Example 5.1.4. Using d'Alembert's method find the solution of the wave equation (5.1.1), subject to the initial displacement $f(x) = 0$ and the initial velocity $u_t(x,0) = g(x)$, given by

$$g(x) = \cos(2x), \quad -\infty < x < \infty.$$

Solution. From d'Alembert's formula (5.1.9) it follows that

$$u(x,t) = \frac{1}{2a} \int\limits_{x-at}^{x+at} g(s)\, ds = \frac{1}{2a} \int\limits_{x-at}^{x+at} \cos 2s\, ds$$

$$= \frac{1}{4a} \sin 2s \Big|_{s=x-at}^{s=x+at} = -\frac{1}{2a} \cos 2x \sin 2at.$$

Example 5.1.5. Show that if both functions $f(x)$ and $g(x)$ in d'Alembert's solution of the wave equation are odd functions, then the solution $u(x,t)$ is an odd function of the spatial variable x. Also, if both functions $f(x)$ and $g(x)$ in d'Alembert's solution of the wave equation are even functions, then the solution $u(x,t)$ is an even function of the spatial variable x.

Solution. Let us check the case when both functions are odd. The other case is left as an exercise. So, assume that

$$f(-x) = -f(x), \quad g(-x) = -g(x), \quad x \in \mathbb{R}.$$

Then by d'Alembert's formula (5.1.9) it follows that

$$u(-x,t) = \frac{f(-x+at) + f(-x-at)}{2} + \frac{1}{2a} \int\limits_{-x-at}^{-x+at} g(s)\, ds \quad (\; s = -v\;)$$

$$= -\frac{f(x-at) + f(x+at)}{2} - \frac{1}{2a} \int\limits_{x+at}^{x-at} g(-v)\, dv$$

$$= -\frac{f(x-at) + f(x+at)}{2} - \frac{1}{2a} \int\limits_{x-at}^{x+at} g(v)\, dv = -u(x,t).$$

Case 2^0. Semi-Infinite String. Homogeneous Equation. Let us consider now the homogeneous wave equation for a semi-infinite string

(5.1.10) $$u_{tt}(x,t) = a^2 u_{xx}(x,t), \quad 0 < x < \infty, \ t > 0,$$

which satisfies the initial conditions

(5.1.11) $u(x, 0) = f(x), \quad u_t(x, 0) = g(x), \; 0 < x < \infty$

and the boundary condition

$$u(0, t) = 0, \quad t > 0.$$

In order to solve this problem we will use the result of Example 5.1.5. First, we extend the functions f, g and $u(x, t)$ to the odd functions f_o, f_o and $w(x, t)$, respectively, by

$$f_o(x) = \begin{cases} -f(-x), & -\infty < x < 0 \\ f(x), & 0 < x < \infty, \end{cases} \quad g_o(x) = \begin{cases} -g(-x), & -\infty < x < 0 \\ g(x), & 0 < x < \infty, \end{cases}$$

and

$$w(x, t) = \begin{cases} -u(-x, t), & -\infty < x < 0, \; t > 0 \\ u(x, t), & 0 < x < \infty, \; t > 0. \end{cases}$$

Next, we consider the initial boundary value problem

$$\begin{cases} w_{tt}(x, t) = a^2 w_{xx}(x, t), & -\infty < x < \infty, \; t > 0 \\ w(x, 0) = f_o(x), \quad w_t(x, 0) = g_o(x), & -\infty < x < \infty. \end{cases}$$

By d'Alembert's formula (5.1.9), the solution of the last problem is given by

(5.1.12) $w(x, t) = \dfrac{f_o(x + at) + f_o(x - at)}{2} + \dfrac{1}{2a} \displaystyle\int_{x-at}^{x+at} g_o(s) \, ds.$

Since f_o and g_o are odd functions from (5.1.12) it follows that

$$w(0, t) = \frac{f_o(at) + f_o(-at)}{2} + \frac{1}{2a} \int_{at}^{at} g_o(s) \, ds = 0,$$

and so,

$$w(x, t) = u(x, t), \text{ for } x > 0 \text{ and } t > 0.$$

Therefore, the solution of the initial boundary value problem (5.1.11) on the semi-infinite interval $0 < x < \infty$ is given by

(5.1.13) $u(x, t) = \begin{cases} \dfrac{f(x+at) - f(x-at)}{2} + \dfrac{1}{2a} \displaystyle\int_{x-at}^{x+at} g(s) \, ds, & 0 < x < at \\[4mm] \dfrac{f(x+at) + f(x-at)}{2} + \dfrac{1}{2a} \displaystyle\int_{x-at}^{x+at} g(s) \, ds, & at \leq x < \infty. \end{cases}$

Example 5.1.6. Find the solution of the wave equation

$$u_{tt}(x,t) = 4u_{xx}(x,t), \quad x > 0, \ t > 0,$$

subject to the conditions

$$\begin{cases} u(x,0) = 3e^{-x}, & u_t(x,0) = 0, \ x > 0 \\ u(0,t) = 0, & t > 0. \end{cases}$$

Solution. Apply Equation (5.1.13) to this problem.

$$u(x,t) = \begin{cases} \frac{3}{2}\left(e^{-x-2t} - e^{-x+2t}\right), & 0 < x < 2t \\[2mm] \frac{3}{2}\left(e^{-x-2t} + e^{-x+2t}\right), & 2t \le x < \infty. \end{cases}$$

Case 3^0. Homogeneous Wave Equation for a Bounded Interval. We will consider now a bounded interval $(0,l)$ and find the solution of the homogeneous problem

$$\begin{cases} u_{tt}(x,t) = a^2 u_{xx}(x,t), & 0 < x < l, \ t > 0 \\ u(x,0) = f(x), & u_t(x,0) = g(x), \ 0 < x < l \\ u(0,t) = u(l,t) = 0, & t > 0. \end{cases}$$

As in the case of the semi-infinite interval, we extend both functions f and g to the odd functions f_o and g_o, respectively, on the interval $(-l,l)$. After that, we periodically extend the functions f_o and g_o on the whole real line $\mathbb{R}$ to new, $2l$-periodic functions f_p and g_p, defined by

$$f_p(x) = \begin{cases} f_o(x), & -l < x < l \\ f_p(x+2l), & \text{otherwise,} \end{cases} \quad g_p(x) = \begin{cases} g_o(x), & -l < x < l \\ g_p(x+2l), & \text{otherwise.} \end{cases}$$

Notice that the functions f_p and g_p have the following properties.

$$(5.1.14) \quad \begin{cases} f_p(-x) = -f_p(x), & f_p(l-x) = -f_p(l+x) \\ g_p(-x) = -g_p(x), & g_p(l-x) = -g_p(l+x). \end{cases}$$

Now, we consider the following problem for the wave equation on $(-\infty, \infty)$.

$$\begin{cases} w_{tt}(x,t) = a^2 w_{xx}(x,t), & -\infty < x < \infty, \ t > 0 \\ w(x,0) = f_p(x), & w_t(x,0) = g_p(x), \ -\infty < x < \infty. \end{cases}$$

By d'Alembert's formula (5.1.9) the solution of the above problem is given by

$$(5.1.15) \quad w(x,t) = \frac{f_p(x+at) + f_p(x-at)}{2} + \frac{1}{2a} \int_{x-at}^{x+at} g_p(s)\, ds.$$

From (5.1.14) and properties (5.1.14) for the functions f_p and g_p it follows that

$$w(0,t) = w(l,t) = 0, \ t > 0.$$

Therefore,

$$w(x,t) = u(x,t), \quad 0 < x < l, \ t > 0,$$

and so, the solution $u(x,t)$ of the wave equation on the finite interval $0 < x < l$ is given by

(5.1.16) $$u(x,t) = \frac{f_p(x+at) + f_p(x-at)}{2} + \frac{1}{2a} \int_{x-at}^{x+at} g_p(s)\,ds.$$

Example 5.1.7. Using d'Alemebert's formula (5.1.16) find the solution of the problem

$$u_{tt}(x,t) = u_{xx}(x,t), \quad 0 < x < \pi, \ t > 0,$$

subject to the initial conditions

$$u(x,0) = f(x) = 2\sin x + 4\sin 2x, \ u_t(x,0) = g(x) = \sin x, \ 0 < x < \pi,$$

and the boundary conditions

$$u(0,t) = u(\pi,t) = 0, \quad t > 0.$$

Solution. The odd periodical extensions f_p and g_p of f and g, respectively, of period 2π, are given by

$$f_p(x) = 2\sin x + 4\sin 2x, \quad g_p(x) = \sin x, \ x \in \mathbb{R}.$$

Therefore, by (5.1.16) the solution of the given problem is

$$u(x,t) = \frac{1}{2}\left[2\sin(x+t) + 4\sin 2(x+t) + 2\sin(x-t) + 4\sin 2(x-t)\right]$$

$$+ \frac{1}{2}\int_{x-t}^{x+t} \sin s\,ds = 6\sin x \cos t - \frac{1}{2}\left[\cos(x+t) - \cos(x-t)\right]$$

$$= 6\sin x \cos t + \sin x \sin t.$$

Example 5.1.8. Using d'Alemebert's formula (5.1.16) find the solution of the string problem

$$u_{tt}(x,t) = u_{xx}(x,t), \quad 0 < x < \pi, \ t > 0,$$

subject to the initial conditions

$$\begin{cases} u(x,0) = f(x) = 0, & 0 < x < \pi, \\ u_t(x,0) = g(x) = 2x, & 0 < x < \pi, \end{cases}$$

and the boundary conditions

$$u(0,t) = u(\pi,t) = 0, \quad t > 0.$$

At what instances does the string return to its initial shape $u(x,0) = 0$?

Solution. By d'Alembert's formula (5.1.16) the solution of the given initial boundary value problem is

(5.1.17) $$u(x,t) = \frac{1}{2} \int_{x-t}^{x+t} g_p(s) \, ds.$$

Let G be an antiderivative of g_p on $\mathbb{R}$, i.e., let $G'(x) = g_p(x)$, $x \in \mathbb{R}$. Then

$$G(x) = \int_0^x g_p(s) \, ds.$$

From the above formula for the antiderivative G and from (5.1.17) we have

(5.1.18) $$u(x,t) = \frac{1}{2}\big[G(x+t) - G(x-t)\big].$$

Now, we will find explicitly the function G. Since g_p is an odd and 2π-periodic function it follows that G is also a 2π-periodic function. Indeed,

$$G(x+2\pi) = \int_0^{x+2\pi} g_p(s) \, ds = \int_0^x g_p(s) \, ds + \int_0^{x+2\pi} g_p(s) \, ds$$

$$= \int_0^x g_p(s) \, ds + \int_{-\pi}^{\pi} g_p(s) \, ds = \int_0^x g_p(s) \, ds + 0 = G(x).$$

For $-x \le x \le 1$ we have $g_p(x) = 2x$ and thus, for $-x \le x \le 1$ we have

$$G(x) = x^2.$$

Let $t_r > 0$ be the time moments when the string returns to its initial position $u(x, 0) = 0$. From (5.1.18) it follows that this will happen only when

$$G(x + t_r) = G(x - t_r).$$

Since G is 2π periodic, from the last equation we obtain

$$x + t_r = x - t_r + 2n\pi, \ n = 1, 2, \ldots.$$

Therefore,

$$t_r = n\pi, \ n = 1, 2, \ldots$$

are the moments when the string will come back to its original position.

In the next example we discuss the important notion of *energy of the string* in the wave equation.

Example 5.1.9. Consider the string equation on the finite interval $0 < x < l$:

$$\begin{cases} u_{tt}(x, t) = a^2 u_{xx}(x, t), & 0 < x < l, \ t > 0. \\ u(x, 0) = f(x), & u_t(x, 0) = g(x), \ 0 < x < l \\ u(0, t) = u(l, t) = 0, & t > 0. \end{cases}$$

The *energy of the string* at moment t, denoted by $E(t)$, is defined by

$$(5.1.19) \qquad E(t) = \frac{1}{2} \int_0^l \left[u_t^2(x, t) + a^2 u_x^2(x, t) \right] dx.$$

The terms

$$E_k(t) = \frac{1}{2} \int_0^l u_t^2(x, t) \, dx, \quad E_p(t) = \frac{1}{2} \int_0^l a^2 u_x^2(x, t) \, dx$$

are called *kinetic* and *potential energy* of the string, respectively.
    The energy $E(t)$ of the wave equation is a constant function.

Solution. Since $u(x, t)$ is a twice differentiable function from (5.1.19) it follows that

$$(5.1.20) \qquad E'(t) = \int_0^l u_t(x, t) u_{tt}(x, t) \, dx + a^2 \int_0^l u_x(x, t) u_{xt}(x, t) \, dx.$$

Using the integration by parts formula for the second integral in Equation (5.1.20) and the fact that $u_{tt}(x,t) = a^2 u_{xx}(x,t)$, from (5.1.20) we obtain

$$E'(t) = \int_0^l u_t(x,t)u_{tt}(x,t)\,dx + a^2\big[u_x(l,t)u_t(l,t) - u_x(0,t)u_t(0,t)\big]$$

$$- a^2 \int_0^l u_t(x,t)u_{xx}(x,t)\,dx = a^2\big[u_t(l,t)u_x(l,t) - u_t(0,t)u_x(0,t)\big].$$

Therefore

$$(5.1.21) \qquad E'(t) = a^2\big[u_t(l,t)u_x(l,t) - u_t(0,t)u_x(0,t)\big].$$

From the given boundary conditions $u(0,t) = u(l,t) = 0$ for all $t > 0$ it follows that

$$u_t(0,t) = u_t(l,t) = 0, \ t > 0,$$

and hence from (5.1.21) we obtain that $E'(t) = 0$. Therefore $E(t)$ is a constant function. In fact, we can evaluate this constant. Indeed,

$$E(t) = E(0) = \frac{1}{2}\int_0^l \big[u_t^2(x,0) + a^2 u_x^2(x,0)\big]\,dx$$

$$= \frac{1}{2}\int_0^l \big[f^2(x) + a^2\big(g'(x)\big)^2\big]\,dx.$$

The result in Example 5.1.9. can be used to prove the following theorem.

Theorem 5.1.1. *The initial boundary value problem*

$$\begin{cases} u_{tt}(x,t) = a^2 u_{xx}(x,t), & 0 < x < l, \ t > 0, \\ u(x,0) = f(x), & u_t(x,0) = g(x), \ 0 < x < l \\ u(0,t) = u(l,t) = 0, & t > 0 \end{cases}$$

has only one solution.

Proof. If $v(x,t)$ and $w(x,t)$ are solutions of the above problem, then the function $U(x,t) = v(x,t) - w(x,t)$ satisfies the problem

$$\begin{cases} U_{tt}(x,t) = a^2 U_{xx}(x,t), & 0 < x < l, \ t > 0 \\ U(x,0) = 0, & U_t(x,0) = 0, \ 0 < x < l \\ U(0,t) = U(l,t) = 0, & t > 0. \end{cases}$$

If $E(t)$ is the energy of $U(x,t)$, then by the result in Example 5.1.9 it follows that $E(t) = 0$ for every $t > 0$. Therefore,

$$\int_0^l \left[U_t^2(x,t) + a^2 U_x^2(x,t) \right] dx = 0,$$

from which it follows that

$$U_t^2(x,t) + a^2 U_x^2(x,t) = 0, \quad 0 < x < l, \ t > 0.$$

Thus, $U(x,t) = U_t(x,t) = 0$ for every $0 < x < l$ and every $t > 0$ and so,

$$U(x,t) = U(0,t) + \int_0^x U_s(s,t)\, ds = 0,$$

for every $0 < x < l$ and every $t > 0$. Hence, $u(x,t) = v(x,t)$ for every $0 < x < l$ and every $t > 0$. ∎

Case 4^0. Nonhomogeneous Wave Equation. Duhamel's Principle. Consider now the nonhomogeneous problem

(5.1.22) $\quad \begin{cases} u_{tt}(x,t) = a^2 u_{xx}(x,t) + h(x,t), & -\infty < x < \infty, \ t > 0 \\ u(x,0) = f(x), \quad u_t(x,0) = g(x), & -\infty < x < \infty. \end{cases}$

First, we know already how to solve the homogeneous problem

(5.1.23) $\quad \begin{cases} v_{tt}(x,t) = a^2 v_{xx}(x,t), & -\infty < x < \infty, \ t > 0 \\ v(x,0) = f(x), \quad v_t(x,0) = g(x), & -\infty < x < \infty. \end{cases}$

Next, we will solve the problem

(5.1.24) $\quad \begin{cases} w_{tt}(x,t) = a^2 w_{xx}(x,t) + h(x,t), & -\infty < x < \infty, \ t > 0 \\ w(x,0) = 0, \quad w_t(x,0) = 0, & -\infty < x < \infty. \end{cases}$

Now, if $v(x,t)$ is the solution of problem (5.1.23) and $w(x,t)$ is the solution of problem (5.1.24), then $u(x,t) = v(x,t) + w(x,t)$ will be the solution of our nonhomogeneous problem (5.1.22).

From Case 1^0 it follows that the solution $v(x,t)$ of the homogeneous problem (5.1.23) is given by

$$v(x,t) = \frac{f(x+at) + f(x-at)}{2} + \frac{1}{2a} \int_{x-at}^{x+at} g(s)\, ds.$$

Next, let us consider problem (5.1.24). If we introduce new variables ξ and η by

$$\xi = x + at, \quad \eta = x - at,$$

then problem (5.1.24) is reduced to the differential equation

(5.1.25) $$\widetilde{w}_{\xi\eta}(\xi, \eta) = -\frac{1}{4a^2} H(\xi, \eta),$$

subject to the conditions

(5.1.26) $$\widetilde{w}(\xi, \xi) = 0, \quad \widetilde{w}_\xi(\xi, \xi) = 0, \quad \widetilde{w}_\eta(\xi, \xi) = 0,$$

where

$$\widetilde{w}(\xi, \eta) = u\left(\frac{\xi + \eta}{2}, \frac{\xi - \eta}{2a}\right) \quad H(\xi, \eta) = h\left(\frac{\xi + \eta}{2}, \frac{\xi - \eta}{2a}\right).$$

If we integrate (5.1.25) with respect to η we obtain

$$\widetilde{w}_\xi(\xi, \xi) - \widetilde{w}_\xi(\xi, \eta) = -\frac{1}{4a^2} \int_\eta^\xi H(\xi, s)\, ds.$$

From the last equation and conditions (5.1.26) it follows that

$$\widetilde{w}_\xi(\xi, \eta) = \frac{1}{4a^2} \int_\eta^\xi H(\xi, s)\, ds.$$

Integrating the last equation with respect to ξ we have

$$\widetilde{w}(\xi, \eta) = \frac{1}{4a^2} \int_\eta^\xi \int_\eta^y H(y, s)\, ds\, dy.$$

If in the last iterated integrals we introduce new variables μ and ν by the transformation

$$s = \mu - a\nu, \quad y = \mu + a\nu,$$

then we obtain that the solution $w(x, t)$ of problem (5.1.24) is given by

$$w(x, t) = \frac{1}{2a} \int_0^t \int_{x-a(t-\nu)}^{x+a(t-\nu)} f(\mu, \nu)\, d\mu\, d\nu.$$

Therefore, the solution $u(x, t)$ of the original problem (5.1.22) is given by

$$u(x, t) = \frac{f(x + at) + f(x - at)}{2} + \frac{1}{2a} \int_{x-at}^{x+at} g(s)\, ds + \frac{1}{2a} \int_0^t \int_{x-a(t-\nu)}^{x+a(t-\nu)} f(\mu, \nu)\, d\mu\, d\nu.$$

Example 5.1.10. Find the solution of the problem

$$\begin{cases} u_{tt}(x,t) = u_{xx}(x,t) + e^{x-t}, & -\infty < x < \infty, \ t > 0 \\ u(x,0) = 0, \quad u_t(x,0) = 0, & -\infty < x < \infty. \end{cases}$$

Solution. Using the formula in Duhamel's principle we find that the solution of the problem is

$$u(x,t) = \frac{1}{2} \int_0^t \int_{x-(t-\nu)}^{x+(t-\nu)} e^{\mu-\nu} \, d\mu \, d\nu = \frac{1}{2} \int_0^t \left(e^{x+t-\nu} - e^{x-t+\nu} \right) e^{-\nu} \, d\nu$$

$$= \frac{1}{2} \int_0^t \left(e^{x+t-2\nu} - e^{x-t} \right) d\nu = \frac{1}{4} e^{x-t} + \frac{1}{2} e^{x+t} - te^{-x-t}.$$

d'Alemberts method can be used to solve other types of equations.

Example 5.1.11. Consider the partial differential equation

$$(5.1.27) \qquad\qquad u_{tt}(x,t) - 2u_{xt}(x,t) + u_{tt}(x,t) = 0.$$

    (a) Find the general solution of (5.1.27).

    (b) Find the particular solution of (5.1.27) in the domain $x \geq 0$, $t \geq 0$, subject to the following initial and boundary conditions

$$(5.1.28) \qquad\qquad u(x,0) = 2x^2, \quad u(0,t) = 0, \ x \geq 0, \ t \geq 0.$$

Solution. (a) If $u(x,t)$ is a solution of (5.1.27), then it belongs to the class $C^2(\mathbb{R}^2)$. Let us introduce new variables ξ and η by

$$\xi = x + t, \ \eta = x - t.$$

With these new variables we have a new function

$$w(\xi,\eta) = u\left(\frac{\xi+\eta}{2}, \frac{\xi-\eta}{2} \right).$$

Using the chain rule we have

$$u_{xx} = w_{\xi\xi} + 2w_{\xi\eta} + w_{\eta\eta}, \quad u_{tt} = w_{\xi\xi} - 2w_{\xi\eta} + w_{\eta\eta}, \quad u_{xt} = w_{\xi\xi} - w_{\eta\eta}.$$

If we substitute the above partial derivatives in Equation (5.1.27), then we obtain the simple equation

$$w_{\eta\eta} = 0.$$

From the last equation it follows that

$$w_\eta = f(\xi)$$

where f is any twice differentiable function on $\mathbb{R}$. Integrating the last equation with respect to η we obtain

$$w(\xi, \eta) = \eta f(\xi) + g(\xi),$$

where g is any twice differentiable function on $\mathbb{R}$.

Therefore, the general solution of Equation (5.1.27) is given by

$$u(x, t) = (x - t)f(x + t) + g(x + t).$$

(b) From the initial and boundary conditions (5.1.28) it follows that

$$x f(x) + g(x) = 2x^2, \quad -t f(t) + g(t) = 0.$$

From the last two equations (taking $t = x$ in the second equation) it follows that

$$g(x) = x^2 \text{ and } f(x) = x.$$

Therefore, the solution of the given initial boundary value problem is

$$u(x, t) = (x - t)(x + t) + (x + t)^2 = 2x^2 + 2xt.$$

Exercises for Section 5.1.

1. Show that if $f \in C^2(\mathbb{R})$ and $g \in C(\mathbb{R})$, then the C^2 function $u(x, t)$, defined by

$$u(x, t) = \frac{f(x + at) + f(x - at)}{2} + \frac{1}{2a} \int_{x-at}^{x+at} g(s)\, ds$$

    is the unique solution of the problem

$$u_{tt}(x, t) = a^2 u_{xx}(x, t), \quad -\infty < x < \infty, \ t > 0,$$

    subject to the initial conditions

$$u(x, 0) = f(x), \quad u_t(x, 0) = g(x), \quad -\infty < x < \infty.$$

2. Show that the wave equation

$$u_{tt}(x, t) = a^2 u_{xx}(x, t)$$

    satisfies the following properties:
    (a) If $u(x, t)$ is a solution of the equation, then $u(x - y, t)$ is also a solution for any constant y.
    (b) If $u(x, t)$ is a solution of the equation, then the derivative $u_x(x, t)$ is also a solution.
    (c) If $u(x, t)$ is a solution of the equation, then $u(kx, kt)$ is also a solution for any constant k.

In Exercises 3–10 using d'Alembert's formula solve the wave equation

$$u_{tt}(x,t) = a^2 u_{xx}(x,t), \quad -\infty < x < \infty, \; t > 0,$$

subject to the indicated initial conditions.

3. $a = 5$, $u(x,0) = \sin x$, $u_t(x,0) = \sin 3x$, $-\infty < x < \infty$.

4. $u(x,0) = \frac{1}{1+x^2}$, $u_t(x,0) = \sin x$, $-\infty < x < \infty$.

5. $u(x,0) = 0$, $u_t(x,0) = 2xe^{-x^2}$, $-\infty < x < \infty$.

6. $u(x,0) = 2\sin x \cos x$, $u_t(x,0) = \cos x$, $-\infty < x < \infty$.

7. $u(x,0) = \frac{1}{1+x^2}$, $u_t(x,0) = e^{-x}$, $-\infty < x < \infty$.

8. $u(x,0) = \cos \pi x2$, $u_t(x,0) = \frac{e^{kx} - e^{-kx}}{2}$, k is a constant.

9. $u(x,0) = e^x$, $u_t(x,0) = \sin x$, $-\infty < x < \infty$.

10. $u(x,0) = e^{-x^2}$, $u_t(x,0) = \cos^2 x$, $-\infty < x < \infty$.

11. Solve the wave equation on the semi-infinite interval

$$u_{tt}(x,t) = 25u_{xx}(x,t), \quad -\infty < x < \infty, \; t > 0,$$

subject to the initial conditions

$$u(x,0) = 0, \quad u_t(x,0) = \begin{cases} 1, & x < 0 \\ 0, & x \geq 0. \end{cases}$$

12. Solve the wave equation

$$u_{tt}(x,t) = a^2 u_{xx}(x,t), \quad 0 < x < \infty, \; t > 0,$$

subject to the initial conditions

$$u(x,0) = \sin x, \quad u_t(x,0) = \frac{1}{1+x^2}, \quad 0 < x < \infty.$$

13. Solve the wave equation

$$u_{tt}(x,t) = a^2 u_{xx}(x,t), \quad 0 < x < \infty, \; t > 0,$$

subject to the initial conditions

$$u(x,0) = f(x) = \begin{cases} 0, & 0 < x < 2 \\ 1, & 2 < x < 3, \\ 0, & x > 3 \end{cases}$$

$$u_t(x,0) = 0, \; x > 0, \quad u_x(0,t) = 0, \; t > 0.$$

14. Using d'Alembert's method solve the wave equation

$$u_{tt}(x,t) = u_{xx}(x,t), \ 0 < x < 2, \ t > 0,$$

subject to the initial conditions

$$u(x,0) = \sin 2\pi x, \ u_t(x,0) = \sin \pi x, \ 0 < x < 2.$$

15. Using d'Alembert's method solve the wave equation

$$u_{tt}(x,t) = u_{xx}(x,t), \ 0 < x < 1, \ t > 0,$$

subject to the initial conditions

$$u(x,0) = 0, \ u_t(x,0) = 1, \ 0 < x < 1.$$

When does the string for the first time return to its initial position $u(x,0) = 0$?

16. Solve the initial value problem

$$u_{tt}(x,t) = a^2 u_{xx}(x,t) + e^{-t} \sin x, \ -\infty < x < \infty, \ t > 0$$
$$u(x,0) = 0, \ u_t(x,0) = 0, \ -\infty < x < \infty.$$

17. Compute the potential, kinetic and total energy of the string equation

$$u_{tt}(x,t) = u_{xx}(x,t), \ 0 < x < \pi, \ t > 0,$$

subject to the conditions

$$\begin{cases} u(x,0) = \sin x, & u_t(x,0) = 0, \ 0 < x < \pi, \\ u(0,t) = u(\pi,t) = 0, & t > 0. \end{cases}$$

5.2 Separation of Variables Method for the Wave Equation.

In this section we will discuss the *Method of Separation of Variables*, also known as the *Fourier Method* or the *Eigenfunctions Expansion Method* for solving the one dimensional wave equation. We will discuss homogeneous and non-homogeneous wave equations with homogeneous and nonhomogeneous boundary conditions on a bounded interval. This method can be applied to wave equations in higher dimensions and it can be used to solve some other partial differential equations, as we will see in the next sections of this chapter and the following chapters.

Case 1⁰. Homogeneous Wave Equation. Homogeneous Boundary Conditions.
Consider the initial boundary value problem

(5.2.1) $u_{tt}(x,t) = a^2 u_{xx}(x,t), \quad 0 < x < l, \ t > 0,$

(5.2.2) $u(x,0) = f(x), \quad u_t(x,0) = g(x), \ 0 < x < l$

(5.2.3) $u(0,t) = u(l,t) = 0,$

where a is a positive constant.

We will look for a nontrivial solution $u(x,t)$ of the above problem of the form

(5.2.4) $u(x,t) = X(x)T(t),$

where $X(x)$ and $T(t)$ are functions of single variables x and t, respectively.

Differentiating (5.2.4) with respect to x and t and substituting the partial derivatives in Equation (5.2.1) we obtain

(5.2.5) $\dfrac{X''(x)}{X(x)} = \dfrac{1}{a^2}\dfrac{T''(t)}{T(t)}.$

Equation (5.2.5) holds identically for every $0 < x < l$ and every $t > 0$. Notice that the left side of this equation is a function which depends only on x, while the right side is a function which depends on t. Since x and t are independent variables this can happen only if each function in both sides of (5.2.4) is equal to the same constant λ:

$$\frac{X''(x)}{X(x)} = \frac{1}{a^2}\frac{T''(t)}{T(t)} = \lambda.$$

From the last equation we obtain the two ordinary differential equations

(5.2.6) $X''(x) - \lambda X(x) = 0,$

and

(5.2.7) $T''(t) - a^2\lambda T(t) = 0.$

From the boundary conditions

$$u(0,t) = u(l,t) = 0, \quad t > 0$$

it follows that

$$X(0)T(t) = X(l)T(t) = 0, \ t > 0.$$

From the last equations we obtain the boundary conditions

$$X(0) = X(l) = 0,$$

since $T(t) \equiv 0$ would imply $u(x, t) \equiv 0$. Solving the eigenvalue problem (5.2.6) with the above boundary conditions, just as in Chapter 3, we find that the eigenvalues λ_n and corresponding eigenfunctions $X_n(x)$ are given by

$$\lambda_n = -\left(\frac{n\pi}{l}\right)^2, \quad X_n(x) = \sin\frac{n\pi x}{l}, \quad n = 1, 2, \ldots.$$

The solution of the differential equation (5.2.7), corresponding to the above found λ_n is given by

$$T_n(t) = a_n \cos\frac{n\pi a t}{l} + b_n \sin\frac{n\pi a t}{l}, \quad n = 1, 2, \ldots,$$

where a_n and b_n are constants which will be determined. Therefore, we obtain a sequence of functions

$$u_n(x, t) = \left(a_n \cos\frac{n\pi a t}{l} + b_n \sin\frac{n\pi a t}{l}\right)\sin\frac{n\pi x}{l}, \quad n = 1, 2, \ldots$$

each of which satisfies the wave equation (5.2.1) and the boundary conditions (5.2.3). Since the wave equation and the boundary conditions are linear and homogeneous, a function $u(x, t)$ of the form

$$(5.2.8) \quad u(x, t) = \sum_{n=1}^{\infty} u_n(x, t) = \sum_{n=1}^{\infty}\left(a_n \cos\frac{n\pi a t}{l} + b_n \sin\frac{n\pi a t}{l}\right)\sin\frac{n\pi x}{l}$$

also will satisfy the wave equation and the boundary conditions. If we assume that the above series is convergent and that it can be differentiated term by term with respect to t, from (5.2.8) and the initial conditions (5.2.2) we obtain

$$(5.2.9) \qquad f(x) = u(x, 0) = \sum_{n=1}^{\infty} a_n \sin\frac{n\pi x}{l}, \quad 0 < x < l$$

and

$$(5.2.10) \qquad g(x) = u_t(x, 0) = \sum_{n=1}^{\infty} \frac{n\pi a}{l} b_n \sin\frac{n\pi x}{l}, \quad 0 < x < l$$

Using the Fourier sine series (from Chapter 1) for the functions $f(x)$ and $g(x)$, from (5.2.9) and (5.2.10) we obtain

$$(5.2.11) \quad a_n = \frac{2}{l}\int_0^l f(x)\sin\frac{n\pi x}{l}\, dx, \quad b_n = \frac{2}{n\pi a}\int_0^l g(x)\sin\frac{n\pi x}{l}\, dx, \quad n \in \mathbb{N}.$$

A formal justification that the obtained function $u(x, t)$ is the solution of the wave equation is given by the following theorem, stated without a proof.

Theorem 5.2.1. *Suppose that the functions f and g are of class $C^2[0, l]$ and that $f''(x)$ and $g''(x)$ are piecewise continuous on $[0, l]$. If*

$$f(0) = f''(0) = f(l) = f''(l) = 0$$
$$g(0) = g(l),$$

then the function $u(x, t)$ given by (5.2.8), where a_n and b_n are given by (5.2.11), is the unique solution of the problem (5.2.1), (5.2.2), (5.2.3).

Example 5.2.1. Solve the following initial boundary value problem.

$$
\begin{cases}
u_{tt}(x, t) = a^2 u_{xx}(x, t), & 0 < x < 4, \ t > 0, \\
u(x, 0) = f(x) = \begin{cases} x, & 0 < x < 2, \\ 4 - x, & 2 < x < 4, \end{cases} \\
u_t(x, 0) = g(x) = 0, & 0 < x < 4, \\
u(0, t) = u(4, t) = 0, & t > 0.
\end{cases}
$$

Solution. Since $g(x) \equiv 0$ on $(0, 4)$, $b_n = 0$ for every $n = 1, 2, \ldots$. For the coefficients a_n, by formulas (5.2.11), using the integration by parts formula we have

$$a_n = \frac{1}{2} \int_0^2 x \sin \frac{n\pi x}{4} \, dx + \frac{1}{2} \int_2^4 (4 - x) \sin \frac{n\pi x}{4} \, dx = \frac{16}{\pi^2 n^2} \sin \frac{n\pi}{2}.$$

Therefore, the solution $u(x, t)$ of the problem is given by

$$(5.2.12) \qquad u(x, t) = \frac{16}{\pi^2} \sum_{n=1}^{\infty} \frac{\sin\left(\frac{n\pi}{2}\right)}{n^2} \sin \frac{n\pi x}{4} \cos \frac{n\pi a t}{4}.$$

The solution $u(x, t)$ of the given problem, using d'Alembert's method, is given by

$$u(x, t) = \frac{f_p(x + at) + f_p(x - at)}{2},$$

where $f_p(x)$ is the odd, 4-periodic extension of the function $f(x)$. This solution can be easily derived from the solution $u(x, t)$, given by (5.2.12). Indeed, using the trigonometric formula

$$\sin \alpha \cos \beta = \frac{1}{2} [\sin(\alpha + \beta) + \sin(\alpha - \beta)]$$

in (5.2.12), we can write the function $u(x, t)$, given by (5.2.12), as

$$(5.2.13) \qquad
\begin{aligned}
u(x, t) = & \frac{1}{2} \cdot \frac{16}{\pi^2} \sum_{n=1}^{\infty} \frac{\sin\left(\frac{n\pi}{2}\right)}{n^2} \sin \frac{n\pi(x + at)}{4} \\
& + \frac{1}{2} \cdot \frac{16}{\pi^2} \sum_{n=1}^{\infty} \frac{\sin\left(\frac{n\pi}{2}\right)}{n^2} \sin \frac{n\pi(x - at)}{4}.
\end{aligned}
$$

Now, since $f_p(x)$ is a continuous function on the whole real line $\mathbb{R}$, its Fourier series, given by

$$\frac{16}{\pi^2} \sum_{n=1}^{\infty} \frac{\sin \frac{n\pi}{2}}{n^2} \sin \frac{n\pi x}{4},$$

converges to $f_p(x)$. Therefore, the first and second terms in (5.2.13) are equal to

$$\frac{1}{2} f_p(x + at) \quad \text{and} \quad \frac{1}{2} f_p(x - at),$$

respectively, which was to be shown.

Case 2^0. *Nonhomogeneous Wave Equation. Homogeneous Boundary Conditions.* Consider the initial boundary value problem

(5.2.14)
$$\begin{cases} u_{tt}(x,t) = a^2 u_{xx}(x,t) + F(x,t), \quad 0 < x < l, \ t > 0, \\ u(x,0) = f(x), \quad u_t(x,0) = g(x), \ 0 < x < l \\ u(0,t) = u(l,t) = 0. \end{cases}$$

In order to find the solution to problem (5.2.14) we split the problem into the following two problems:

(5.2.15)
$$\begin{cases} v_{tt}(x,t) = a^2 v_{xx}(x,t), \quad 0 < x < l, \ t > 0, \\ v(x,0) = f(x), \quad v_t(x,0) = g(x), \ 0 < x < l \\ v(0,t) = v(l,t) = 0, \end{cases}$$

and

(5.2.16)
$$\begin{cases} w_{tt}(x,t) = a^2 w_{xx}(x,t) + F(x,t), \ 0 < x < l, \ t > 0, \\ w(x,0) = 0, \ w_t(x,0) = 0, \ 0 < x < l \\ u(0,t) = u(l,t) = 0. \end{cases}$$

Problem (5.2.15) has been considered in *Case* 1^0 and we already know how to solve it. Let its solution be $v(x,t)$. If $w(x,t)$ is the solution of problem (5.2.16), then

$$u(x,t) = v(x,t) + w(x,t)$$

will be the solution of the given problem (5.2.14). Therefore, we need to solve only problem (5.2.16).

There are several approaches to solving problem (5.2.16). One of them, presented without a rigorous justification, is the following.

For each fixed $t > 0$, we expand the function $F(x,t)$ in the Fourier sine series

(5.2.17)
$$F(x,t) = \sum_{n=1}^{\infty} F_n(t) \sin \frac{n\pi x}{l}, \quad 0 < x < l,$$

where

(5.2.18) $$F_n(t) = \frac{2}{l} \int_0^l F(x,t) \sin \frac{n\pi x}{l} \, dx, \; n = 1, 2, \dots.$$

Next, again for each fixed $t > 0$, we expand the unknown function $w(x,t)$ in the Fourier sine series

(5.2.19) $$w(x,t) = \sum_{n=1}^{\infty} w_n(t) \sin \frac{n\pi x}{l}, \; 0 < x < l,$$

where

(5.2.20) $$w_n(t) = \frac{2}{l} \int_0^l w(x,t) \sin \frac{n\pi x}{l} \, dx, \; n = 1, 2, \dots.$$

From the initial conditions $w(x,0) = w_t(x,0) = 0$ it follows that

(5.2.21) $$w_n(0) = w_n'(0) = 0.$$

If we substitute (5.2.17) and (5.2.19) into the wave equation (5.2.16) and compare the Fourier coefficients we obtain

(5.2.22) $$w_n''(t) + \frac{n^2 \pi^2 a^2}{l^2} 2 w_n(t) = F_n(t).$$

The solution of the second order, linear differential equation (5.2.22), in view of the initial conditions (5.2.21), is given by

(5.2.23) $$w_n(t) = \frac{1}{n\pi a} \int_0^t F_n(s) \sin\left(\frac{n\pi a}{l}(t-s)\right) ds.$$

Let us take an example.

Example 5.2.2. Solve the wave equation

$$\begin{cases} u_{tt}(x,t) = a^2 u_{xx}(x,t) + xt, & 0 < x < 4, \; t > 0, \\ u(x,0) = f(x) = \begin{cases} x, & 0 < x < 2, \\ 4-x, & 2 < x < 4, \end{cases} \\ u_t(x,0) = g(x) = 0, & 0 < x < 4, \\ u(0,t) = u(4,t) = 0, & t > 0. \end{cases}$$

Take $a = 4$ in the wave equation and plot the displacements $u(x,t)$ of the string at the moments $t = 0$, $t = 0.15$, $t = 0.85$ and $t = 1$.

Solution. The corresponding homogeneous wave equation with homogeneous boundary conditions was solved in Example 5.2.1. and its solution is given by

$$(5.2.24) \qquad v(x,t) = \frac{16}{\pi^2} \sum_{n=1}^{\infty} \frac{\sin \frac{n\pi}{2}}{n^2} \sin \frac{n\pi x}{4} \cos \frac{n\pi a t}{4}.$$

Now, we solve the nonhomogeneous problem

$$\begin{cases} w_{tt}(x,t) = a^2 w_{xx}(x,t) + xt, \ 0 < x < 4, \ t > 0, \\ w(x,0) = 0, \ w_t(x,0) = 0, \ 0 < x < 4 \\ w(0,t) = w(4,t) = 0. \end{cases}$$

First, expand the function xt (for each fixed $t > 0$) in the Fourier sine series on the interval $(0,4)$:

$$xt = \sum_{n=1}^{\infty} F_n(t) \sin \frac{n\pi x}{4}.$$

The coefficients $F_n(t)$ in this expansion are given by

$$F_n(t) = \frac{2}{4} \int_0^4 xt \sin \frac{n\pi x}{4} \, dx = -\frac{8t \cos n\pi}{n\pi}.$$

Next, again for each fixed $t > 0$, we expand the unknown function $w(x,t)$ in the Fourier sine series

$$w(x,t) = \sum_{n=1}^{\infty} w_n(t) \sin \frac{n\pi x}{4}, \ 0 < x < l,$$

where the coefficients $w_n(t)$ are determined using Equation (5.2.23).

$$w_n(t) = \frac{1}{n\pi a} \int_0^t F_n(s) \sin \left(\frac{n\pi a}{4}(t-s) \right) ds$$

$$= -\frac{8 \cos n\pi}{an^2\pi^2} \int_0^t s \sin \left(\frac{n\pi a}{4}(t-s) \right) ds$$

$$= -\frac{32 \cos n\pi}{a^2 n^3 \pi^3} \left(n\pi a t - 4 \sin \frac{n\pi a}{4} t \right).$$

Thus,

$$(5.2.25) \qquad w(x,t) = -\frac{32}{a^2\pi^3} \sum_{n=1}^{\infty} \frac{\cos n\pi}{n^3} \left(n\pi a t - 4 \sin \frac{n\pi a}{4} t \right) \sin \frac{n\pi x}{4}.$$

Therefore, the solution $u(x,t)$ of the given problem is $u(x,t) = v(x,t) + w(x,t)$, where $v(x,t)$ and $w(x,t)$ are given by (5.2.24) and (5.2.25), respectively.

The plots of $u(x,t)$ at the given time instances are displayed in Figure 5.2.1.

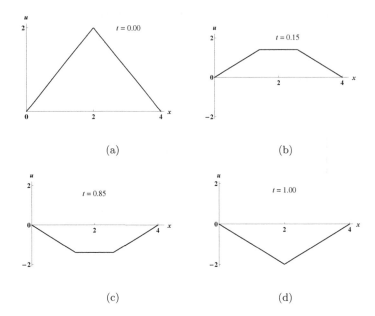

Figure 5.2.1

Case 3^0. *Nonhomogeneous Wave Equation with Nonhomogeneous Dirichlet Boundary Conditions.* Consider the initial boundary value problem

(5.2.26)
$$\begin{cases} u_{tt}(x,t) = a^2 u_{xx}(x,t) + F(x,t), & 0 < x < l,\ t > 0, \\ u(x,0) = f(x), \quad u_t(x,0) = g(x),\ 0 < x < l \\ u(0,t) = \varphi(t), \quad u(l,t) = \psi(t). \end{cases}$$

The solution of this problem can be found by superposition of the solution $v(x,t)$ of the homogeneous problem (5.2.15) with the solution $w(x,t)$ of the problem

(5.2.27)
$$\begin{cases} w_{tt}(x,t) = a^2 w_{xx}(x,t), & 0 < x < l,\ t > 0, \\ w(x,0) = 0, \quad u_t(x,0) = 0,\ 0 < x < l, \\ w(0,t) = \varphi(t), \quad w(l,t) = \psi(t). \end{cases}$$

If we introduce a new function $\widetilde{w}(x,t)$ by

$$\widetilde{w}(x,t) = w(x,t) - x\psi(t) + (x - l)\varphi(t),$$

then problem (5.2.27) is transformed to the problem:

(5.2.28)
$$\begin{cases} \widetilde{w}_{tt}(x,t) = a^2 \widetilde{w}_{xx}(x,t) + \widetilde{F}(x,t), & 0 < x < l,\ t > 0, \\ \widetilde{w}(0,t) = \widetilde{w}(l,t) = 0, \quad t > 0 \\ \widetilde{w}(x,0) = \widetilde{f}(x), \quad \widetilde{w}_t(x,0) = \widetilde{g}(x), \end{cases}$$

where

$$\widetilde{F}(x,t) = (x - l)\varphi''(t) - x\psi''(t),$$
$$\widetilde{f}(x) = (x - l)\varphi(0) - x\psi(0),$$
$$\widetilde{g}(x) = (x - l)\varphi'(0) - x\psi'(0).$$

Notice that problem (5.2.28) has homogeneous boundary conditions and it was already considered in Case 1^0; therefore we know how to solve it.

Case 4^0. Homogeneous Wave Equation. Neumann Boundary Conditions. Consider the initial boundary value problem

(5.2.29)
$$\begin{cases} u_{tt}(x,t) = a^2 u_{xx}(x,t), & 0 < x < l, \ t > 0, \\ u(x,0) = f(x), & u_t(x,0) = g(x), \ 0 < x < l \\ u_x(0,t) = u_x(l,t) = 0, & t > 0. \end{cases}$$

We will solve this problem by the separation of variables method. Let the solution $u(x,t)$ of the above problem be of the form

(5.2.30)
$$u(x,t) = X(x)T(t),$$

where $X(x)$ and $T(t)$ are functions of single variables x and t, respectively.

Differentiating (5.2.30) with respect to x and t and substituting the partial derivatives in the wave equation in problem (5.2.29) we obtain

(5.2.31)
$$\frac{X''(x)}{X(x)} = \frac{1}{a^2}\frac{T''(t)}{T(t)}.$$

Equation (5.2.31) holds identically for every $0 < x < l$ and every $t > 0$. Notice that the left side of this equation is a function which depends only on x, while the right side is a function which depends on t. Since x and t are independent variables this can happen only if each function in both sides of (5.2.31) is equal to the same constant λ.

$$\frac{X''(x)}{X(x)} = \frac{1}{a^2}\frac{T''(t)}{T(t)} = \lambda$$

From the last equation we obtain the two ordinary differential equations

(5.2.32)
$$X''(x) - \lambda X(x) = 0,$$

and

(5.2.33)
$$T''(t) - a^2\lambda T(t) = 0.$$

From the boundary conditions

$$u_x(0,t) = u_x(l,t) = 0, \quad t > 0$$

it follows that

$$X'(0)T(t) = X'(l)T(t) = 0, \ t > 0.$$

To avoid the trivial solution, from the last equations we obtain

(5.2.34) $$X'(0) = X'(l) = 0.$$

Solving the eigenvalue problem (5.2.32), (5.2.34) (see Chapter 3), we obtain that its eigenvalues $\lambda = \lambda_n$ and the corresponding eigenfunctions $X_n(x)$ are

$$\lambda_0 = 0, \quad X_0(x) = 1, \ 0 < x < l$$

and

$$\lambda_n = -\left(\frac{n\pi}{l}\right)^2, \quad X_n(x) = \cos\frac{n\pi x}{l}, \ n = 0, 1, 2, \ldots, \ 0 < x < l.$$

The solution of the differential equation (5.2.33), corresponding to the above found λ_n, is given by

$$T_0(t) = a_0 + b_0 t$$

$$T_n(t) = a_n \cos\frac{n\pi a t}{l} + b_n \sin\frac{n\pi a t}{l}, \ n = 1, 2, \ldots,$$

where a_n and b_n are constants which will be determined.

Therefore we obtain a sequence of functions

$$\{u_n(x,t) = X_n(x)T_n(t), \ n = 0, 1, \ldots\},$$

given by

$$\begin{cases} u_0(x,t) = a_0 + b_0 t, \\ u_n(x,t) = \left(a_n \cos\dfrac{n\pi a t}{l} + b_n \sin\dfrac{n\pi a t}{l}\right) \cos\dfrac{n\pi x}{l}, \ n \in \mathbb{N} \end{cases}$$

each of which satisfies the wave equation and the Neumann boundary conditions in problem (5.2.29). Since the given wave equation and the boundary conditions are linear and homogeneous, a function $u(x,t)$ of the form

(5.2.35) $$u(x,t) = a_0 + b_0 t + \sum_{n=1}^{\infty} \left(a_n \cos\frac{n\pi a t}{l} + b_n \sin\frac{n\pi a t}{l}\right) \cos\frac{n\pi x}{l}$$

also will satisfy the wave equation and the boundary conditions. If we assume that the above series is convergent and it can be differentiated term by term with respect to t, then from (5.2.35) and the initial conditions in problem (5.2.29) we obtain

(5.2.36) $$f(x) = u(x,0) = a_0 + \sum_{n=1}^{\infty} a_n \cos\frac{n\pi x}{l}, \quad 0 < x < l$$

and

$$\text{(5.2.37)} \qquad g(x) = u_t(x, 0) = \sum_{n=1}^{\infty} \frac{n\pi a}{l} b_n \cos \frac{n\pi x}{l}, \quad 0 < x < l.$$

Using the Fourier cosine series (from Chapter 1) for the functions $f(x)$ and $g(x)$, or the fact, from Chapter 2, that the eigenfunctions

$$1, \cos \frac{\pi x}{l}, \cos \frac{2\pi x}{l}, \ldots, \cos \frac{n\pi x}{l}, \ldots$$

are pairwise orthogonal on the interval $[0, l]$, from (5.2.36) and (5.2.37) we obtain

$$\text{(5.2.38)} \qquad \begin{cases} a_n = \dfrac{2}{l} \displaystyle\int_0^l f(x) \cos \dfrac{n\pi x}{l} \, dx, \quad n = 0, 1, 2, \ldots \\[4mm] b_n = \dfrac{2}{n\pi a} \displaystyle\int_0^l g(x) \cos \dfrac{n\pi x}{l} \, dx, \quad n = 1, 2, \ldots \end{cases}$$

Remark. The problem (5.2.29) can be solved in a similar way to problem (5.2.2), except we use the Fourier cosine series instead of the Fourier sine series (see Exercise 9 of this section.)

Example 5.2.3. Using the separation of variables method solve the following problem:
$$\begin{cases} u_{tt}(x, t) = u_{xx}(x, t), \quad 0 < x < \pi, \ t > 0, \\ u(x, 0) = \pi^2 - x^2, \quad u_t(x, 0) = 0, \ 0 < x < \pi, \\ u_x(0, t) = u_x(\pi, t) = 0, \quad t > 0. \end{cases}$$

Solution. Let the solution $u(x, t)$ of the problem be of the form

$$u(x, t) = X(x)T(t).$$

From the wave equation and the given boundary conditions we obtain the eigenvalue problem

$$\text{(5.2.39)} \qquad \begin{cases} X''(x) - \lambda X(x) = 0, \quad 0 < x < \pi \\ X'(0) = X'(\pi) = 0, \end{cases}$$

and the ordinary differential equation

$$\text{(5.2.40)} \qquad T''(t) - \lambda T(t) = 0, \quad t > 0.$$

Solving the eigenvalue problem (5.2.39) we obtain that its eigenvalues $\lambda = \lambda_n$ and the corresponding eigenfunctions $X_n(x)$ are

$$\lambda_0 = 0, \quad X_0(x) = 1, \ 0 < x < \pi$$

and

$$\lambda_n = -n^2, \quad X_n(x) = \cos nx, \ n = 1, 2, \ldots, \ 0 < x < \pi.$$

The solutions of the differential equation (5.2.40), corresponding to the above found λ_n, are given by

$$T_0(t) = a_0 + b_0 t$$
$$T_n(t) = a_n \cos nt + b_n \sin nt, \quad n = 1, 2, \ldots,$$

where a_n and b_n are constants which will be determined.

Hence, the solution of the given problem will be of the form

$$u(x, t) = a_0 + b_0 t + \sum_{n=1}^{\infty} \left(a_n \cos nt + b_n \sin nt \right) \cos nx, \ 0 < x < \pi, \ t > 0.$$

From the initial condition $u_t(x, 0) = 0$, $0 < x < \pi$, and the orthogonality property of the eigenfunctions

$$\{1, \cos x, \cos 2x, \ldots, \cos nx, \ldots\}$$

on the interval $[0, \pi]$ we obtain $b_n = 0$ for every $n = 0, 1, 2, \ldots$. From the other initial condition $u(x, 0) = \pi^2 - x^2$, $0 < x < \pi$, and again from the orthogonality of the eigenfunctions, we obtain

$$a_0 = \frac{1}{\pi} \int_0^{\pi} \left(\pi^2 - x^2 \right) dx = \frac{2\pi^2}{3}.$$

For $n = 1, 2, \ldots$ we have

$$a_n = \frac{1}{\int_0^{\pi} \cos^2 nx \, dx} \int_0^{\pi} \left(\pi^2 - x^2 \right) \cos nx \, dx = -\frac{2}{\pi} \int_0^{\pi} x^2 \cos nx \, dx$$

$$= -\frac{2}{\pi} \cdot \frac{2\pi}{n^2} \cos n\pi = \frac{4}{n^2} (-1)^{n+1}.$$

Therefore, the solution $u(x, t)$ of the problem is given by

$$u(x, t) = \frac{2\pi^2}{3} + \sum_{n=1}^{\infty} \frac{4}{n^2} (-1)^{n+1} \cos nt \cos nx, \ 0 < x < \pi, \ t > 0.$$

The next example describes vibrations of a string in a medium (air or water).

Example 5.2.4. Using the separation of variables method solve the following problem:

$$\begin{cases} u_{tt}(x,t) = a^2 u_{xx}(x,t) - c^2 u, & 0 < x < l, \ t > 0, \\ u(x,0) = f(x), \quad u_t(x,0) = g(x), \ 0 < x < l \\ u(0,t) = u(l,t) = 0, \quad t > 0. \end{cases}$$

The term $-c^2 u$ represents the force of reaction of the medium.

Solution. Let the solution $u(x,t)$ of the problem be of the form

$$u(x,t) = X(x)T(t).$$

If we substitute the partial derivatives of u into the string equation, then we obtain

$$X(x)T''(t) = a^2 X''(x)T(t) - c^2 X(x)T(t), \quad 0 < x < l, \ t > 0.$$

From the last equation it follows that

$$a^2 \frac{X''(x)}{X(x)} - c^2 = \frac{T''(t)}{T(t)}.$$

Since x and t are independent, from the above equation we have

$$a^2 \frac{X''(x)}{X(x)} - c^2 = -\lambda,$$

i.e.,

(5.2.41) $$X''(x) - \frac{c^2 - \lambda}{a^2} X(x) = 0$$

and

(5.2.42) $$\frac{T''(t)}{T(t)} = -\lambda,$$

where λ is a constant to be determined. From $u(0,t) = u(l,t) = 0$ it follows that $X(x)$ satisfies the boundary conditions

(5.2.43) $$X(0) = X(l) = 0.$$

Now, we will solve the eigenvalue problem (5.2.41), (5.2.42). If $c^2 - \lambda = 0$, then the general solution of the differential equation (5.2.41) is given by

$$X(x) = A + Bx$$

where A and B are arbitrary constants. From the conditions (5.2.43) it follows that $A = B = 0$. Thus, $X(x) \equiv 0$, and so $\lambda = c^2$ can't be an eigenvalue of the problem.

If $c^2 - \lambda > 0$, i.e., if $\lambda < c^2$, then the general solution of (5.2.41) is

$$X(x) = A e^{\frac{1}{a}\sqrt{c^2 - \lambda}\,x} + B e^{-\frac{1}{a}\sqrt{c^2 - \lambda}\,x},$$

where A and B are arbitrary constants. From the conditions (5.2.43) again it follows that $A = B = 0$. Thus, $X(x) \equiv 0$, and so any $\lambda > -c^2$ cannot be an eigenvalue of the problem.

Finally, if $c^2 - \lambda < 0$, i.e., $\lambda > c^2$, then the general solution of (5.2.41) is

$$X(x) = A \sin\left(\frac{1}{a}\sqrt{\lambda - c^2}\,x\right) + B \cos\left(\frac{1}{a}\sqrt{\lambda - c^2}\,x\right),$$

where A and B are arbitrary constants. From the boundary condition $X(0) = 0$ it follows that $B = 0$. From the other boundary condition $X(l) = 0$ it follows that

$$\sin\left(\frac{1}{a}\sqrt{\lambda - c^2}\,l\right) = 0.$$

From the last equation we have

$$\frac{1}{a}\sqrt{\lambda - c^2}\,l = n\pi, \quad n \in \mathbb{N}.$$

Therefore, the eigenvalues of the eigenvalue problem are

$$(5.2.44) \qquad\qquad \lambda_n = c^2 + \frac{a^2 n^2 \pi^2}{l^2}, \quad n \in \mathbb{N}.$$

The corresponding eigenfunctions are

$$X_n(x) = \sin\frac{n\pi x}{l}, \quad n \in \mathbb{N}.$$

The general solution of the differential equation (5.2.42), which corresponds to the above found eigenvalues λ_n, is

$$T_n(t) = a_n \cos\left(\sqrt{\lambda_n}\,t\right) + b_n \sin\left(\sqrt{\lambda_n}\,t\right).$$

Therefore, the solution of the given initial value problem is

$$u(x, t) = \sum_{n=1}^{\infty} \left(a_n \cos\left(\sqrt{\lambda_n}\,t\right) + b_n \sin\left(\sqrt{\lambda_n}\,t\right)\right) \sin\frac{n\pi x}{l},$$

where a_n and b_n can be found using the orthogonality property of the functions

$$\sin\frac{\pi x}{l}, \sin\frac{2\pi x}{l}, \ldots, \sin\frac{n\pi x}{l}, \ldots$$

and the initial conditions for the function $u(x, t)$.

$$a_n = \frac{2}{l}\int_0^l f(x)\sin\frac{n\pi x}{l}\,dx, \quad b_n = \frac{2}{l\lambda_n}\int_0^l g(x)\sin\frac{n\pi x}{l}\,dx, \quad n \in \mathbb{N}.$$

Exercises for Section 5.2.

1. Using separation of variables solve the equation

$$u_{tt}(x,t) = u_{xx}(x,t), \quad 0 < x < \pi, \; t > 0,$$

subject to the Dirichlet boundary conditions

$$u(0,t) = u(\pi,t) = 0, \quad t > 0$$

and the following initial conditions:

(a) $u(x,0) = 0, \; u_t(x,0) = 1, \; 0 < x < \pi.$

(b) $u(x,0) = \pi x - x^2, \; u_t(x,0) = 0, \; 0 < x < \pi.$

(c) $u(x,0) = \begin{cases} \frac{3}{2}x, & 0 < x < \frac{2\pi}{3} \\ 3(\pi - x), & \frac{2\pi}{3} < x < \pi, \end{cases}$ $\quad u_t(x,0) = 0, \; 0 < x < \pi.$

(d) $u(x,0) = \sin x, \quad u_t(x,0) = \begin{cases} 0, & 0 < x < \frac{\pi}{4} \\ 1, & \frac{\pi}{4} < x < \frac{3\pi}{4} \\ 0, & \frac{3\pi}{4} < x < \pi, \end{cases} \quad 0 < x < \pi.$

(e) $u(x,0) = \begin{cases} x, & 0 < x < \frac{\pi}{2} \\ \pi - x, & \frac{\pi}{2} < x < \pi, \end{cases}$ $\quad u_t(x,0) = 0, \; 0 < x < \pi.$

2. Use separation of variables method to solve the damped wave equation

$$\begin{cases} u_{tt}(x,t) = a^2 u_{xx}(x,t) - 2k u_t(x,t), \quad 0 < x < l, \; t > 0, \\ u(0,t) = u(l,t) = 0, \; t > 0, \\ u(x,0) = f(x), \; u_t(x,0) = g(x), \; 0 < x < l. \end{cases}$$

The term $-2k u_t(x,t)$ is the frictional forces that cause the damping.
Consider the following three cases:

$$\text{(a) } k < \frac{\pi a}{l}; \quad \text{(b) } k = \frac{\pi a}{l}; \quad \text{(c) } k > \frac{\pi a}{l}.$$

3. Solve the damped wave equation

$$\begin{cases} u_{tt}(x,t) = u_{xx}(x,t) - 4u_t(x,t), \quad 0 < x < \pi, \; t > 0, \\ u(0,t) = u(\pi,t) = 0, \quad t > 0, \\ u(x,0) = 1, \quad u_t(x,0) = 0, \; 0 < x < \pi. \end{cases}$$

4. Solve the damped wave equation

$$\begin{cases} u_{tt}(x,t) = u_{xx}(x,t) - 2u_t(x,t), \quad 0 < x < \pi, \; t > 0, \\ u(0,t) = u(\pi,t) = 0, \quad t > 0, \\ u(x,0) = 0, \quad u_t(x,0) = \sin x + \sin 2x, \; 0 < x < \pi. \end{cases}$$

5. Solve the initial boundary value problem

$$\begin{cases} u_{tt}(x,t) = a^2 u_{xx}(x,t), \quad 0 < x < l, \; t > 0, \\ u(x,0) = f(x), \quad u_t(x,0) = g(x), \; 0 < x < l, \\ u(0,t) = u_x(0,t) = 0, \; t > 0. \end{cases}$$

6. Solve the problem

$$\begin{cases} u_{tt}(x,t) = 4u_{xx}(x,t), \quad 0 < x < 1, \; t > 0, \\ u(x,0) = x(1-x), \quad u_t(x,0) = \cos \dfrac{\pi x}{2}, \; 0 < x < 1, \\ u(0,t) = u(1,t) = 0, \quad t > 0. \end{cases}$$

7. Solve the following wave equation with Neumann boundary conditions:

$$\begin{cases} u_{tt}(x,t) = 4u_{xx}(x,t), \; 0 < x < \pi, \; t > 0, \\ u_x(0,t) = u_x(\pi,t) = 0, \; t > 0, \\ u(x,0) = \sin x, \; u_t(x,0) = 0, \; 0 < x < \pi. \end{cases}$$

8. Solve the following wave equation with Neumann boundary conditions:

$$\begin{cases} u_{tt}(x,t) = u_{xx}(x,t), \; 0 < x < 1, \; t > 0, \\ u_x(0,t) = u_x(1,t) = 0, \; t > 0, \\ u(x,0) = 0, \; u_t(x,0) = 2x - 1, \; 0 < x < 1. \end{cases}$$

9. Solve the initial boundary value problem

$$\begin{cases} u_{tt}(x,t) = a^2 u_{xx}(x,t), \; 0 < x < l, \; t > 0, \\ u(x,0) = f(x), \; u_t(x,0) = g(x), \; 0 < x < l \\ u_x(0,t) = u_x(l,t) = 0, \; t > 0 \end{cases}$$

by expanding the functions $f(x)$, $g(x)$ and $u(x,t)$ in the Fourier cosine series

$$f(x) = \frac{1}{2}a_0 + \sum_{n=1}^{\infty} a_n \cos \frac{n\pi x}{l},$$

$$g(x) = \frac{1}{2} + \sum_{n=1}^{\infty} a'_n \cos \frac{n\pi x}{l},$$

$$u(x,t) = \frac{1}{2}A_0(t) + \sum_{n=1}^{\infty} A_n(t) \cos \frac{n\pi x}{l}.$$

In Exercises 10–13 solve the wave equation

$$u_{tt}(x,t) = a^2 u_{xx}(x,t) + F(x,t), \quad 0 < x < l, \ t > 0,$$

subject to the Dirichlet initial conditions

$$u(x,0) = u_t(x,0) = 0, \quad 0 < x < l$$

and the following forcing functions $F(x,t)$ and boundary conditions.

10. $F(x,t) = Ae^{-t}\sin\frac{\pi x}{l}, \ \ u(0,t) = u(l,t) = 0.$

11. $F(x,t) = Axe^{-t}, \ \ u(0,t) = u(l,t) = 0.$

12. $F(x,t) = A\sin t, \ \ u(0,t) = u_x(l,t) = 0.$

13. $F(x,t) = Ae^{-t}\cos\frac{\pi x}{2l}, \ \ u_x(0,t) = u(l,t) = 0.$

In Exercises 14–16 solve the wave equation

$$u_{tt}(x,t) = u_{xx}(x,t), \ 0 < x < \pi, \ t > 0,$$

subject to the the following initial and boundary conditions.

14. $u(0,t) = t^2, \ \ u(\pi,t) = t^3,$

  $u(x,0) = \sin x,$

  $u_t(x,0) = 0, \ \ 0 < x < \pi, \ t > 0.$

15. $u(0,t) = e^{-t}, \ u(\pi,t) = t,$

  $u(x,0) = \sin x \cos x,$

  $u_t(x,0) = 1, \ 0 < x < \pi, \ t > 0.$

16. $u(0,t) = t, \ u_x(\pi,t) = 1,$

  $u(x,0) = \sin \frac{x}{2},$

  $u_t(x,0) = 1, \ 0 < x < \pi, \ t > 0.$

5.3 The Wave Equation on Rectangular Domains.

   In this section we will study the two and three dimensional wave equation on rectangular domains. We begin by considering the homogeneous wave equation on a rectangle.

5.3.1 Homogeneous Wave Equation on a Rectangle

   Consider an infinitely thin, perfectly elastic membrane stretched across a rectangular frame. Let

$$D = \{(x,y) : 0 < x < a, \ 0 < y < b\}$$

be the rectangle, whose boundary ∂D is the frame.
   The following initial boundary value problem describes the vibration of the membrane:

(5.3.1)
$$\begin{cases} u_{tt} = c^2 \Delta u \equiv c^2 (u_{xx} + u_{yy}), & (x,y) \in D, \ t > 0 \\ u(x,y,0) = f(x,y), & (x,y) \in D \\ u_t(x,y,0) = g(x,y), & (x,y) \in D \\ u(x,y,t) = 0, & (x,y) \in \partial D, \ t > 0. \end{cases}$$

   If a solution $u(x,y,t)$ of the wave equation is of the form

$$u(x,y,t) = W(x,y)T(t), \quad (x,y) \in D, \ t > 0,$$

then the wave equation becomes

$$W(x,y)T''(t) = c^2 \Delta W(x,y) T(t),$$

where Δ is the Laplace operator. The last equation can be written in the form

$$\frac{\Delta W(x,y)}{W(x,y)} = \frac{T''(t)}{c^2 T(t)}, \quad (x,y) \in D, \ t > 0.$$

The above equation is possible only when both of its sides are equal to the same constant, denoted by $-\lambda$:

$$\begin{cases} \dfrac{T''(t)}{c^2 T(t)} = -\lambda, \\[4mm] \dfrac{\Delta W(x,y)}{W(x,y)} = -\lambda. \end{cases}$$

From the second equations it follows that

$$(5.3.2) \qquad \Delta W(x,y) + \lambda W(x,y) = 0, \quad (x,y) \in D,$$

and from the first equation we have

$$(5.3.3) \qquad T''(t) + c^2 \lambda T(t) = 0, \quad t > 0.$$

To avoid the trivial solution, from the boundary conditions for $u(x,y,t)$, the following boundary conditions for the function $W(x,y)$ are obtained:

$$(5.3.4) \qquad W(x,y) = 0, \quad (x,y) \in \partial D,$$

Equation (5.3.2) is called the *Helmholtz Equation*.

In order to find the eigenvalues λ of Equation (5.3.2) (*eigenvalues of the Laplace operator Δ*), subject to the boundary conditions (5.3.4), we separate the variables x and y. We write $W(x,y)$ in the form

$$W(x,y) = X(x)Y(y).$$

If we substitute $W(x,y)$ in Equation (5.3.2) we obtain

$$X''(x)X(x) + Y''(y)Y(y) + \lambda X(x)Y(y) = 0, \quad (x,y) \in D,$$

which can be written in the form

$$\frac{X''(x)}{X(x)} = -\frac{Y''(y)}{Y(y)} - \lambda, \quad (x,y) \in D.$$

The last equation is possible only when both sides are equal to the same constant, denoted by $-\mu$:

$$\frac{X''(x)}{X(x)} = -\mu, \quad -\frac{Y''(y)}{Y(y)} - \lambda = -\mu, \; (x,y) \in D.$$

The boundary conditions (5.3.4) for the function $W(x,y)$ imply $X(0) = X(a) = 0$ and $Y(0) = Y(b) = 0$. Therefore, we have the following boundary value problems:

$$(5.3.5) \qquad \begin{cases} X''(x) + \mu X(x) = 0, \quad X(0) = X(a) = 0, \; 0 < x < a, \\ Y''(y) + (\lambda - \mu)Y(y) = 0, \quad Y(0) = Y(b) = 0, \; 0 < y < b. \end{cases}$$

We have already solved the eigenvalue problems (5.3.5) (see Chapter 2). Their eigenvalues and corresponding eigenfunctions are

$$\mu = \frac{m^2\pi^2}{a^2}, \quad X(x) = \sin\frac{m\pi x}{a}, \quad m = 1, 2, \ldots$$

$$\lambda - \mu = \frac{n^2\pi^2}{b^2}, \quad Y(y) = \sin\frac{n\pi y}{b}, \quad n = 1, 2, \ldots.$$

Therefore, the eigenvalues λ_{mn} and corresponding eigenfunctions $W_{mn}(x, y)$ of the eigenvalue problem (5.3.2), (5.3.4) are given by

$$\lambda_{mn} = \frac{m^2\pi^2}{a^2} + \frac{n^2\pi^2}{b^2}$$

and

$$W_{mn}(x, y) = \sin\frac{m\pi x}{a} \sin\frac{n\pi y}{b}.$$

A general solution of Equation (5.3.3), corresponding to the above found λ_{mn}, is given by

$$T_{mn}(t) = a_{mn} \cos\left(\sqrt{\lambda_{mn}}\, ct\right) + b_{mn} \sin\left(\sqrt{\lambda_{mn}}\, ct\right).$$

Therefore, the solution $u = u(x, y, t)$ of the problem (5.3.1) will be of the form

$$(5.3.6) \quad \sum_{m=1}^{\infty}\sum_{n=1}^{\infty} \sin\frac{m\pi x}{a} \sin\frac{n\pi y}{b}\left(a_{mn}\cos\sqrt{\lambda_{mn}}\, ct + b_{mn}\sin\sqrt{\lambda_{mn}}\, ct\right),$$

where the coefficients a_{mn} will be found from the initial conditions for the function $u(x, y, t)$ and the following orthogonality property of the eigenfunctions

$$f_{m,n}(x, y) \equiv \sin\left(\frac{m\pi x}{a}\right) \sin\left(\frac{n\pi y}{b}\right), \quad m, n = 1, 2, \ldots$$

on the rectangle $[0, a] \times [0, b]$:

$$\int_0^a \int_0^b f_{m,n}(x, y)\, f_{p,q}(x, y)\, dx\, dy = 0, \quad \text{for every } (m, n) \neq (p, q).$$

Using this property, we find

$$(5.3.7) \qquad a_{mn} = \frac{4}{ab}\int_0^a \int_0^b f(x, y)\sin\frac{m\pi x}{a}\sin\frac{n\pi y}{b}\, dy\, dx$$

and

$$(5.3.8) \qquad b_{mn} = \frac{4}{ab\pi\sqrt{\frac{m^2}{a^2} + \frac{n^2}{b^2}}} \int_0^a \int_0^b g(x,y) \sin \frac{m\pi x}{a} \sin \frac{n\pi y}{b} \, dy \, dx.$$

Remark. For a given function $f(x,y)$ on a rectangle $[0,a] \times [0,b]$, the series of the form

$$\sum_{m=1}^{\infty} \sum_{n=1}^{\infty} A_{mn} \sin\left(\frac{m\pi x}{a}\right) \sin\left(\frac{n\pi y}{b}\right),$$

$$\sum_{m=0}^{\infty} \sum_{n=0}^{\infty} B_{mn} \cos\left(\frac{m\pi x}{a}\right) \cos\left(\frac{n\pi y}{b}\right),$$

where

$$A_{mn} = \frac{4}{ab} \int_0^a \int_0^b f(x,y) \sin\left(\frac{m\pi x}{a}\right) \sin\left(\frac{n\pi y}{b}\right) dy \, dx,$$

$$B_{mn} = \frac{4}{ab} \int_0^a \int_0^b f(x,y) \cos\left(\frac{m\pi x}{a}\right) \cos\left(\frac{n\pi y}{b}\right) dy \, dx$$

are called the *Double Fourier Sine Series* and the *Double Fourier Cosine Series*, respectively, of the function f(x,y) on the rectangle $[0,a] \times [0,b]$.

Example 5.3.1. Solve the following membrane problem.

$$\begin{cases} u_{tt}(x,y,t) = \dfrac{1}{\pi^2}\left(u_{xx}(x,y,t) + u_{yy}(x,y,t)\right), & (x,y) \in S, \ t > 0, \\ u(x,y,0) = \sin 3\pi x \sin \pi y, & (x,y) \in S, \\ u_t(x,y,0) = 0, & t > 0, \\ u(x,y,t) = 0, & (x,y) \in \partial S, \ t > 0, \end{cases}$$

where S is the unit square

$$S = \{(x,y) : 0 < x < 1, \ 0 < y < 1\}$$

and ∂S is its boundary.

Display the shape of the membrane at the moment $t = 0.7$.

Solution. From $g(x,y) = 0$ and (5.3.8) we have $b_{mn} = 0$ for all $m, n = 1, 2, \ldots$. From $f(x,y) = \sin 3\pi x \sin \pi y$ and (5.3.7) we have

$$a_{mn} = 4 \int_0^1 \int_0^1 \sin 3\pi x \sin \pi y \sin m\pi x \sin n\pi y \, dy \, dx$$

$$= 4 \int_0^1 \sin 3\pi x \sin m\pi x \, dx \cdot \int_0^1 \sin \pi y \sin n\pi y \, dy$$

$$= \begin{cases} 1, & m = 3, \ n = 1 \\ 0, & \text{otherwise.} \end{cases}$$

Therefore, the solution of this problem, by formula (5.3.6), is given by

$$u(x, y, t) = \sin 3\pi x \, \sin \pi y \, \cos \sqrt{10}\, t.$$

Notice that when $\sqrt{10}\, t = \frac{\pi}{2}$, i.e., when $t = \frac{\pi}{2\sqrt{10}} \approx 0.4966$, for the first time the vertical displacement of the membrane from the Oxy plane is zero.

The shape of the membrane at several time instances is displayed in Figure 5.3.1.

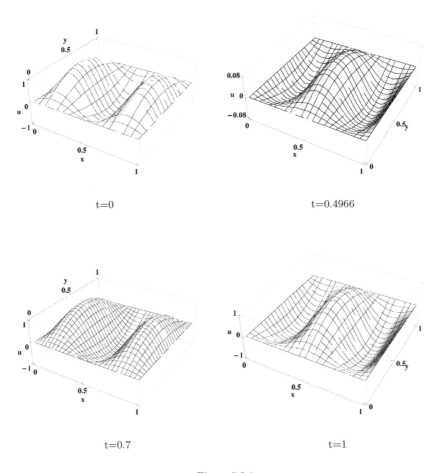

t=0   t=0.4966

t=0.7   t=1

Figure 5.3.1

5.3.2 Nonhomogeneous Wave Equation on a Rectangle.

Let R be the rectangle given by

$$R = \{(x, y) : 0 < x < a, \ 0 < y < b\}$$

whose boundary is ∂R.

Consider the following two dimensional initial boundary value problem for the wave equation

$$(5.3.9) \quad \begin{cases} u_{tt} = c^2 (u_{xx} + u_{yy}) + F(x, y, t), & (x, y) \in R, \ t > 0, \\ u(x, y, 0) = f(x, y), \quad u_t(x, y, 0) = g(x, y), \ (x, y) \in R \\ u(x, y, t) = 0, \quad (x, y) \in \partial R, \ t > 0. \end{cases}$$

As in the one dimensional case, we split problem $(5.3.9)$ into the following two problems:

$$(5.3.10) \quad \begin{cases} v_{tt} = c^2 (v_{xx} + v_{yy}), & (x, y) \in R, \ t > 0, \\ v(x, y, 0) = f(x, y), \quad v_t(x, y, 0) = g(x, y), \ (x, y) \in R, \\ v(x, y, t) = 0, \quad (x, y) \in \partial R, \ t > 0. \end{cases}$$

and

$$(5.3.11) \quad \begin{cases} w_{tt} = c^2 (w_{xx} + w_{yy}) + F(x, y, t), & (x, y) \in R, \ t > 0, \\ w(x, y, 0) = 0, \quad w_t(x, y, 0) = 0, \ (x, y) \in R, \\ w(x, y, t) = 0, \quad (x, y) \in \partial R, \ t > 0. \end{cases}$$

Problem $(5.3.10)$ was considered in *Case* 1^0. Let its solution be $v(x, y, t)$. If $w(x, y, t)$ is the solution of problem $(5.3.11)$, then

$$u(x, y, t) = v(x, y, t) + w(x, y, t)$$

will be the solution of the given problem $(5.3.9)$. So, it remains to solve problem $(5.3.11)$.

For each fixed $t > 0$, we expand the function $F(x, y, t)$ in the double Fourier sine series

$$(5.3.12) \quad F(x, y, t) = \sum_{m=1}^{\infty} \sum_{n=1}^{\infty} F_{mn}(t) \sin \frac{m\pi x}{a} \sin \frac{n\pi y}{b}, \ (x, y) \in R, \ t > 0,$$

where

$$(5.3.13) \quad F_{mn}(t) = \frac{4}{ab} \int_0^a \int_0^b F(x, y, t) \sin \frac{m\pi x}{a} \sin \frac{n\pi y}{b} \, dy \, dx.$$

Next, for each fixed $t > 0$, we expand the unknown function $w(x, y, t)$ in the double Fourier sine series

$$(5.3.14) \quad w(x, y, t) = \sum_{m=1}^{\infty} \sum_{n=1}^{\infty} w_{mn}(t) \sin \frac{m\pi x}{a} \sin \frac{n\pi y}{b}, \quad (x, y) \in R, \ t > 0,$$

where

$$w_{mn}(t) = \frac{4}{ab} \int_0^a \int_0^b w(x, y, t) \sin \frac{m\pi x}{a} \sin \frac{n\pi y}{b} \, dy \, dx.$$

From the initial conditions $w(x, y, 0) = w_t(x, y, 0) = 0$ it follows that

$$(5.3.15) \quad\quad\quad\quad w_{mn}(0) = w'_{mn}(0) = 0.$$

If we substitute (5.3.12) and (5.3.14) into the wave equation (5.3.11) and compare the Fourier coefficients we obtain

$$w''_{mn}(t) + c^2 \left(\frac{m^2 \pi^2}{a^2} + \frac{n^2 \pi^2}{b^2} \right) w_{mn}(t) = F_{mn}(t).$$

The last equation is a linear, second order, nonhomogeneous ordinary differential equation with constant coefficients and given initial conditions (5.3.15) and it can be solved.

Example 5.3.2. Solve the following damped membrane problem

$$(5.3.16) \quad \begin{cases} u_{tt} = \dfrac{1}{\pi^2} (u_{xx} + u_{yy}) + xyt, & (x, y) \in S, \ t > 0, \\ u(x, y, 0) = \sin 3\pi x \sin \pi y, \quad u_t(x, y, 0) = 0, \ (x, y) \in S, \\ u(x, y, t) = 0, \quad (x, y) \in \partial S, \ t > 0, \end{cases}$$

where

$$S = \{(x, y) : 0 < x < 1, \ 0 < y < 1\}$$

is the unit square and ∂S is its boundary.

Solution. The corresponding homogeneous problem is

$$(5.3.17) \quad \begin{cases} v_{tt} = \dfrac{1}{\pi^2} (u_{xx} + u_{yy}), & (x, y) \in S, \ t > 0, \\ v(x, y, 0) = \sin 3\pi x \sin \pi y, \quad v_t(x, y, 0) = 0, \ (x, y) \in S, \\ v(x, y, t) = 0, \quad (x, y) \in \partial S, \ t > 0, \end{cases}$$

and its solution $v(x, t)$, found in Example 5.3.1, is given by

$$(5.3.18) \quad\quad\quad v(x, y, t) = \sin 3\pi x \sin \pi y \cos \sqrt{10}\, t.$$

The corresponding nonhomogeneous wave equation is

(5.3.19)
$$\begin{cases} w_{tt} = \dfrac{1}{\pi^2}(u_{xx} + u_{yy}) + xyt, & (x,y) \in S,\ t > 0, \\ w(x,y,0) = \sin 3\pi x \sin \pi y, & w_t(x,y,0) = 0,\ (x,y) \in S, \\ w(x,y,t) = 0, & (x,y) \in \partial S,\ t > 0. \end{cases}$$

For each fixed $t > 0$, expand the functions xyt and $w(x,y,t)$ in the double Fourier sine series

(5.3.20)
$$xyt = \sum_{m=1}^{\infty} \sum_{n=1}^{\infty} F_{mn}(t) \sin m\pi x \sin n\pi y, \quad (x,y) \in S,\ t > 0,$$

where

(5.3.21)
$$F_{mn}(t) = 4 \int_0^1 \int_0^1 xyt \sin m\pi x \sin n\pi y \, dy \, dx$$

$$= 4t \left(\int_0^1 x \sin m\pi x \, dx \right) \left(\int_0^1 y \sin n\pi y \, dy \right)$$

$$= 4t \frac{(-1)^m (-1)^n}{mn\pi^2}, \quad m, n = 1, 2, \ldots,$$

and

(5.3.22) $\quad w(x,y,t) = \displaystyle\sum_{m=1}^{\infty} \sum_{n=1}^{\infty} w_{mn}(t) \sin m\pi x \sin n\pi y, \quad (x,y) \in S,\ t > 0,$

where

$$w_{mn}(t) = 4 \int_0^1 \int_0^1 w(x,y,t) \sin m\pi x \sin n\pi y \, dy \, dx, \quad m, n = 1, 2, \ldots$$

From the initial conditions $w(x,y,0) = w_t(x,y,0) = 0$ it follows that

(5.3.23)
$$w_{mn}(0) = w'_{mn}(0) = 0.$$

If we substitute (5.3.20) and (5.3.22) into the wave equation (5.3.19) and compare the Fourier coefficients we obtain

$$w''_{mn}(t) + (m^2 + n^2) w_{mn}(t) = 4t \frac{(-1)^{m+n}}{mn\pi^2}.$$

The solution of the last equation, subject to the initial conditions (5.3.23), is

$$w_{mn}(t) = \frac{4(-1)^m(-1)^n}{mn(m^2 + n^2)\pi^2} t + \frac{4(-1)^m(-1)^n}{mn(m^2 + n^2)\sqrt{m^2 + n^2}\,\pi^2} \sin \sqrt{m^2 + n^2}\, t$$

and so, from (5.3.22), it follows that the solution $w = w(x, y, t)$ is given by

$$(5.3.24) \quad w = \sum_{m=1}^{\infty} \sum_{n=1}^{\infty} \frac{4(-1)^m(-1)^n}{mn\lambda_{mn}^2 \pi^2} \left[t + \frac{\sin(\lambda_{mn} t)}{\lambda_{mn}} \right] \sin m\pi x \sin n\pi y,$$

where

$$\lambda_{mn} = \sqrt{m^2 + n^2}.$$

Hence, the solution $u(x, y, t)$ of the given problem (5.3.16) is

$$u(x, y, t) = v(x, y, t) + w(x, y, t),$$

where $u(x, y, t)$ and $w(x, y, t)$ are given by (5.3.18) and (5.3.24), respectively.

5.3.3 The Wave Equation on a Rectangular Solid.

Let V be the three dimensional solid given by

$$V = \{(x, y, z) : 0 < x < a, \, 0 < y < b, \, 0 < z < c\}$$

whose boundary is ∂V.

Consider the initial boundary value problem

$$(5.3.25) \quad \begin{cases} u_{tt} = A^2 \left(u_{xx} + u_{yy} + u_{zz} \right), & (x, y, z) \in V, \; t > 0, \\ u(x, y, z, 0) = f(x, y, z), & (x, y, z) \in V, \\ u_t(x, y, z, 0) = g(x, y, z), & (x, y, z) \in V, \\ u(x, y, z, t) = 0, & (x, y, z) \in \partial V, \; t > 0. \end{cases}$$

This problem can be solved by the separation of variables method. Let the solution $u(x, y, z, t)$ of the problem be of the form

$$u(x, y, z, t) = X(x)Y(y)Z(z)T(t), \quad (x, y, z) \in V, \; t > 0.$$

From the boundary conditions for the function $u(x, y, z, t)$ we have

$$(5.3.26) \qquad X(0) = Y(0) = Z(0) = X(a) = Y(b) = Z(c) = 0.$$

If we differentiate this function twice with respect to the variables and substitute the partial derivatives in the wave equation, after a rearrangement we obtain

$$\frac{T''(t)}{A^2 T(t)} = \frac{X''(x)}{X(x)} + \frac{Y''(y)}{Y(y)} + \frac{Z''(z)}{Z(z)}.$$

From the last equation it follows that

$$(5.3.27) \qquad T''(t) + \lambda A^2 T(t) = 0, \quad t > 0$$

and

$$\frac{X''(x)}{X(x)} = -\frac{Y''(y)}{Y(y)} - \frac{Z''(z)}{Z(z)} - \lambda$$

for some constant λ.

The above equation is possible only if

$$\frac{X''(x)}{X(x)} = -\mu,$$

i.e.,

(5.3.28) $X''(x) + \mu X(x) = 0, \quad 0 < x < a$

and

$$-\frac{Y''(y)}{Y(y)} - \frac{Z''(z)}{Z(z)} - \lambda = -\mu,$$

where μ is a constant

The last equation can be written in the form

$$-\frac{Y''(y)}{Y(y)} = \frac{Z''(z)}{Z(z)} + \lambda - \mu,$$

which is possible only if

$$-\frac{Y''(y)}{Y(y)} = \frac{Z''(z)}{Z(z)} + \lambda - \mu = \nu,$$

where ν is a constant.

From the last equations we obtain

(5.3.29) $Y''(y) + \nu Y(y) = 0, \quad 0 < y < b$

and

(5.3.30) $Z''(z) + (\lambda - \mu - \nu)Z(z) = 0, \quad 0 < z < c.$

Solving the eigenvalue problems (5.3.28), (5.3.29), (5.3.30), (5.3.26) we find

$$\begin{cases} \mu_i = \dfrac{i^2\pi^2}{a^2}, & X_i(x) = \sin\dfrac{i\pi x}{a}, & 0 < x < a, \quad i \in \mathbb{N} \\[2mm] \nu_j = \dfrac{j^2\pi^2}{b^2}, & Y_j(y) = \sin\dfrac{j\pi y}{b}, & 0 < y < b, \quad j \in \mathbb{N} \\[2mm] \lambda_{ijk} - \mu_j - \nu_k = \dfrac{k^2\pi^2}{c^2}, & Z_k(z) = \sin\dfrac{k\pi z}{c}, & 0 < z < c, \quad i,j,k \in \mathbb{N}. \end{cases}$$

If we solve the differential equation (5.3.27) for the above obtained λ_{ijk} we have

$$T_k(t) = a_{ijk}\cos A\sqrt{\lambda_{ijk}}\,t + b_{ijk}\sin A\sqrt{\lambda_{ijk}}\,t.$$

Therefore, the solution $u = u(x, y, z, t)$ of the three dimensional problem (5.3.25) is given by

$$u = \sum_{i=1}^{\infty}\sum_{j=1}^{\infty}\sum_{k=1}^{\infty} \sin\frac{i\pi x}{a} \sin\frac{j\pi y}{b} \sin\frac{k\pi z}{c} \left(a_{ijk}\cos A\sqrt{\lambda_{ijk}}t + b_{ijk}\sin A\sqrt{\lambda_{ijk}}t\right),$$

where the coefficients a_{ijk} and b_{ijk} are determined using the initial conditions

$$f(x, y, z) = \sum_{i=1}^{\infty}\sum_{j=1}^{\infty}\sum_{k=1}^{\infty} a_{ijk}\sin\frac{i\pi x}{a} \sin\frac{j\pi y}{b} \sin\frac{k\pi z}{c},$$

$$g(x, y, z) = A\sum_{i=1}^{\infty}\sum_{j=1}^{\infty}\sum_{k=1}^{\infty} b_{ijk}\sqrt{\lambda_{ijk}}\sin\frac{i\pi x}{a} \sin\frac{j\pi y}{b} \sin\frac{k\pi z}{c}.$$

Using the orthogonality property of the sine functions

$$\left\{\sin\frac{i\pi x}{a}, \sin\frac{j\pi y}{b}, \sin\frac{k\pi z}{c}, \quad i, j, k = 1, 2, \ldots\right\}$$

on $0 \le x \le a$, $0 \le y \le b$, $0 \le z \le c$ we find

$$\begin{cases} a_{ijk} = \dfrac{8}{abc}\displaystyle\int_0^a\int_0^b\int_0^c f(x, y, z)\sin\frac{i\pi x}{a} \sin\frac{j\pi y}{b} \sin\frac{k\pi z}{c}\, dz\, dy\, dx, \\[3mm] b_{ijk} = \dfrac{8}{abcA\sqrt{\lambda_{ijk}}}\displaystyle\int_0^a\int_0^b\int_0^c g(x, y, z)\sin\frac{i\pi x}{a} \sin\frac{j\pi y}{b} \sin\frac{k\pi z}{c}\, dz\, dy\, dx. \end{cases}$$

The coefficients λ_{ijk} in the above formulas are given by

$$\lambda_{ijk} = \frac{i^2\pi^2}{a^2} + \frac{j^2\pi^2}{b^2} + \frac{k^2\pi^2}{c^2}, \quad i, j, k \in \mathbb{N}.$$

Exercises for Section 5.3.

In Exercises 1–5 solve the two dimensional wave equation

$$u_{tt} = \frac{1}{\pi^2}\left(u_{xx}(x, y, t) + u_{yy}(x, y, t)\right), \quad 0 < x < 1,\ 0 < y < 1,\ t > 0,$$

subject to the boundary conditions

$$u(0, y, t) = u(1, y, t) = u(x, 0, t) = u(x, 1, t) = 0,\ 0 < x < 1,\ 0 < y < 1,\ t > 0$$

and the following initial conditions.

1. $u(x, y, 0) = x(1 - x)y(1 - y)$, $u_t(x, y, 0) = 0$, $0 < x < 1$, $0 < y < 1$.

2. $u(x, y, 0) = \sin \pi x \sin \pi y$, $u_t(x, y, 0) = \sin \pi x$, $0 < x < 1$, $0 < y < 1$.

3. $u(x, y, 0) = x(1 - x)y(1 - y)$, $u_t(x, y, 0) = 2 \sin \pi x \sin 2\pi x$, $0 < x < 1$, $0 < y < 1$.

4. $u(x, y, 0) = 0$, $u_t(x, y, 0) = 1$, $0 < x < 1$, $0 < y < 1$.

5. $u(x, y, 0) = 0$, $u_t(x, y, 0) = x(1 - x)y(1 - y)$, $0 < x < 1$, $0 < y < 1$.

6. Let $S = \{(x, y) : 0 < x < a,\ 0 < y < b\}$ be a rectangle with boundary ∂S. Find the general solution of the two dimensional wave equation

$$u_{tt} = u_{xx} + u_{yy}, \quad (x, y) \in S,\ t > 0,$$

subject to the following boundary conditions.

(a) $u(x, y, t) = 0$, $(x, y) \in S$, $t > 0$.

(b) $u_x(x, y, t) = u(x, y, t) = 0$, $(x, y) \in \partial S$, $t > 0$.

7. Solve the two dimensional wave initial boundary value problem
$$\begin{cases} u_{tt} = u_{xx} + u_{yy}, \quad (x, y) \in S,\ t > 0, \\ u(x, y, 0) = 3 \sin 4y - 5 \cos 2x \sin y, \quad (x, y) \in S, \\ u_t(x, y, 0) = 7 \cos x \sin 3y, \quad (x, y) \in S, \\ u_x(x, y, t) = u_x(x, y, t) = 0, \quad (x, y) \in \partial S,\ t > 0, \\ u(x, y, t) = u(x, y, t) = 0, \quad (x, y) \in \partial S,\ t > 0, \end{cases}$$

where $S = \{(x, y) : 0 < x < \pi,\ 0 < y < \pi\}$ is a square and ∂S its boundary.

8. Let R be the rectangle $R = \{(x, y) : 0 < x < a,\ 0 < y < b\}$ and ∂R be its boundary. Use the separation of variables method to solve the following two dimensional damped membrane vibration problem.
$$\begin{cases} u_{tt} = c^2 (u_{xx} + u_{yy}) - 2k^2 u_t, \quad (x, y) \in R,\ t > 0, \\ u(x, y, 0) = f(x, y), \quad u_t(x, y, 0) = g(x, y),\ (x, y) \in R, \\ u(x, y, t) = 0, \quad (x, y) \in \partial R,\ t > 0. \end{cases}$$

9. Let R be the rectangle $R = \{(x, y) : 0 < x < a,\ 0 < y < b\}$ and ∂R be its boundary. Solve the following two dimensional damped membrane vibration resonance problem.

$$\begin{cases} u_{tt} = c^2\left(u_{xx} + u_{yy}\right) + F(x, y)\sin \omega t, \quad (x, y) \in R,\ t > 0, \\ u(x, y, 0) = 0, \quad u_t(x, y, 0) = 0,\ (x, y) \in R, \\ u(x, y, t) = u(a, y, t) = 0, \quad (x, y) \in \partial R,\ t > 0. \end{cases}$$

10. Solve the previous problem taking

    (a) $a = b = \pi$, $c = 1$, and $F(x, y, t) = xy \sin x \sin y \sin 5t$.

    (b) $a = b = \pi$, $c = 1$, and $F(x, y, t) = xy \sin x \sin y \sin \sqrt{5}t$.

11. Let R be the rectangle $R = \{(x, y) : 0 < x < a,\ 0 < y < b\}$ and ∂R be its boundary. Solve the following two dimensional damped membrane vibration resonance problem.

$$\begin{cases} u_{tt} = c^2\left(u_{xx} + u_{yy}\right) - 2ku_t + A\sin \omega t, \quad (x, y) \in R,\ t > 0, \\ u(x, y, 0) = 0, \quad u_t(x, y, 0) = 0,\ (x, y) \in R, \\ u(x, y, t) = 0, \quad \in \partial R,\ t > 0. \end{cases}$$

5.4 The Wave Equation on Circular Domains.

In this section we will discuss and solve problems of vibrations of a circular drum and a three dimensional ball.

5.4.1 The Wave Equation in Polar Coordinates.

Consider a membrane stretched over a circular frame of radius a. Let D be the disc inside the frame:

$$D = \{(x, y) : x^2 + y^2 < a^2\},$$

whose boundary is ∂D.

We assume that the membrane has uniform tension. If $u(x, y, t)$ is the displacement of the point (x, y, u) of the membrane from the Oxy plane at moment t, then the vibrations of the membrane are modeled by the wave equation

(5.4.1) $\qquad u_{tt} = c^2 \Delta_{x,y}\, u \equiv c^2\left(u_{xx} + u_{yy}\right), \quad (x, y) \in D,\ t > 0,$

subject to the Dirichlet boundary condition

(5.4.2) $$u(x, y, t) = 0, \quad (x, y) \in \partial D, \ t > 0.$$

In order to find the solution of this boundary value problem we take the solution $u(x, y, t)$ to be of the form $u(x, y, t) = W(x, y)T(t)$. With this new function $W(x, y)$ the given membrane problem is reduced to solving the following eigenvalue problem:

(5.4.3) $$\begin{cases} \Delta_{x,y} W(x, y, t) + \lambda W(x, y, t) = 0, \quad (x, y) \in D, \\ W(x, y, t) = 0, \quad (x, y) \in \partial D, \ t > 0, \\ W(x, y, t) \text{ is continuous for } (x, y) \in D \cup \partial D. \end{cases}$$

It is difficult to find the eigenvalues λ and corresponding eigenfunctions $W(x, y)$ of the above eigenvalue problem stated in this form. But, it is relatively easy to show that all eigenvalues are positive. To prove this we will use the Green's formula (see Appendix E):

$$\iint\limits_{D} p\Delta q \, dx \, dy = \oint\limits_{\partial D} p\nabla q \cdot \mathbf{n} \, ds - \iint\limits_{D} \nabla p \cdot \nabla q \, dx \, dy,$$

and if, in particular, p and q vanish on the boundary ∂D, then

(5.4.4) $$\iint\limits_{D} p\Delta q \, dx \, dy = -\iint\limits_{D} \nabla p \cdot \nabla q \, dx \, dy.$$

If we take $p = q = W$ in formula (5.4.4), then we obtain

$$0 \leq \iint\limits_{D} \nabla W \cdot \nabla W \, dx \, dy = -\iint\limits_{D} W\Delta W \, dx \, dy$$

$$= \lambda \iint\limits_{D} W^2 \, dx \, dy.$$

Thus $\lambda \geq 0$.

The parameter λ cannot be zero, since otherwise, from the above we would have

$$\iint\limits_{D} \nabla W \cdot \nabla W \, dx \, dy = -\iint\limits_{D} W\Delta W \, dx \, dy$$

$$= \lambda \iint\limits_{D} W^2 \, dx \, dy = 0,$$

which would imply $\nabla W = 0$ on $D \cup \partial D$. So, W would be a constant function on $D \cup \partial D$. But, from $W = 0$ on ∂D and the continuity of W on

$D \cup \partial D$ we would have $W = 0$ on $D \cup \partial D$, which is impossible since W is an eigenfunction.

Instead of considering cartesian coordinates when solving the circular membrane problem, it is much more convenient to use polar coordinates.

Let us recall that the polar coordinates are given by

$$x = r \cos \varphi, \quad y = r \sin \varphi, \quad -\pi \leq \varphi < \pi, \ 0 \leq r < \infty,$$

and the Laplace operator in polar coordinates is given by

$$c^2 \Delta_{r,\varphi} \, u \equiv c^2 \left(u_{rr} + \frac{1}{r} u_r + \frac{1}{r^2} u_{\varphi\varphi} \right).$$

(See Appendix E for the Laplace operator in polar coordinates.)

Let us have a circular membrane fixed along its boundary, which occupies the disc of radius a, centered at the origin. If the displacement of the membrane is denoted by $u(r, \varphi, t)$, then the wave equation in polar coordinates is given by

$$(5.4.5) \qquad u_{tt}(r, \varphi) = c^2 \Delta_{r,\varphi} \, u(r, \varphi), \quad 0 < r \leq a, \ -\pi \leq \varphi < \pi, \ t > 0,$$

Since the membrane is fixed for the frame, we impose Dirichlet boundary conditions:

$$(5.4.6) \qquad u(a, \varphi, t) = 0, \quad -\pi \leq \varphi < \pi, \ t > 0.$$

Since $u(r, \varphi, t)$ is a single valued and differentiable function, we have to require that $u(r, \varphi, t)$ is a periodical function of φ, i.e., we have to have the following periodic condition:

$$(5.4.7) \qquad u(r, \varphi + 2\pi, t) = u(r, \varphi, t), \quad 0 < r \leq a, \ -\pi \leq \varphi < \pi, \ t > 0.$$

Further, since the solution $u(r, \varphi, t)$ is a continuous function on the whole disc D we have that the solution is bounded (finite) at the origin. Therefore we have the boundary condition

$$(5.4.8) \qquad \mid u(0, \varphi, t) \mid = \mid \lim_{r \to 0^+} u(r, \varphi, t) \mid < \infty, \qquad -\pi \leq \varphi < \pi, \ t > 0.$$

Also we impose the following initial conditions:

$$(5.4.9) \qquad \begin{cases} u(r, \varphi, 0) = f(r, \varphi), & 0 < r < a, \ -\pi \leq \varphi < \pi, \\ u_t(r, \varphi, 0) = g(r, \varphi), & 0 < r < a, \ -\pi \leq \varphi < \pi. \end{cases}$$

Now we can separate out the variables. Let

$$(5.4.10) \qquad u(r, \varphi, t) = w(r, \varphi)T(t), \quad 0 < r \leq a, \ -\pi \leq \varphi < \pi.$$

If we substitute (5.4.10) into (5.4.5) we obtain

$$\frac{\Delta_{r,\varphi}\, w(r,\varphi)}{w(r,\varphi)} = \frac{T''(t)}{c^2 T(t)}.$$

The last equation is possible only when both sides are equal to the same constant, which will be denoted by $-\lambda^2$ (we already established that the eigenvalues of the Helmholtz equation are positive):

$$(5.4.11) \quad w_{rr} + \frac{1}{r}w_r + \frac{1}{r^2}w_{\varphi\varphi} + \lambda^2 w(r,\varphi) = 0, \quad 0 < r \le a, \ -\pi \le \varphi < \pi,$$

and

$$(5.4.12) \qquad\qquad T''(t) + \lambda^2 c^2 T(t) = 0, \quad t > 0.$$

Now we separate the Helmholtz equation (5.4.11) by letting

$$(5.4.13) \qquad\qquad w(r,\varphi) = R(r)\Phi(\varphi).$$

From the boundary conditions (5.4.6), (5.4.7), and (5.4.8) for the function $u(r,\varphi,t)$ it follows that the functions $R(r)$ and $\Phi(\varphi)$ satisfy the boundary conditions

$$(5.4.14) \qquad \begin{cases} \Phi(\varphi) = \Phi(\varphi + 2\pi), & -\pi \le \varphi < \pi, \\ R(a) = 0, & |\lim_{r\to 0^+} R(r)| < \infty. \end{cases}$$

Substituting (5.4.13) into Equation (5.4.5) it follows that

$$\frac{r^2 R''(r)}{R(r)} + r\frac{R'(r)}{R(r)} + \lambda^2 r^2 = -\frac{\Phi''(\varphi)}{\Phi(\varphi)},$$

which is possible only if both sides are equal to the same constant μ. Therefore, in view of the conditions (5.4.14), we obtain the eigenvalue problems

$$(5.4.15) \qquad \Phi''(\varphi) + \mu\Phi(\varphi) = 0, \quad \Phi(\varphi) = \Phi(\varphi + 2\pi), \ -\pi \le \varphi < \pi$$

and

$$(5.4.16) \quad r^2 R'' + rR' + (\lambda^2 r^2 - \mu)R = 0, \quad R(a) = 0, \ |\lim_{r\to 0^+} R(r)| < \infty.$$

Let us consider first the problem (5.4.15). If $\mu = 0$, then the general solution of the differential equation in (5.4.15) is

$$\Phi(\varphi) = A\varphi + B.$$

The periodicity of $\Phi(\varphi)$ implies $A = 0$, and so $\mu = 0$ is an eigenvalue of (5.4.15) with corresponding eigenfunction $\Phi(\varphi) = 1$.

If $\mu > 0$, then the general solution of the differential equation in (5.4.15) is

$$\Phi(\varphi) = A \cos \sqrt{\mu}\, \varphi + B \sin \sqrt{\mu}\, \varphi.$$

Using the periodic condition $\Phi(\pi) = \Phi(-\pi)$, from the above equation it follows that

$$A \cos \sqrt{\mu}\, \pi + B \sin \sqrt{\mu}\, \pi = A \cos \sqrt{\mu}\, \pi - B \sin \sqrt{\mu}\, \pi,$$

from which we have

$$\sin \sqrt{\mu}\, \pi = 0,$$

that is, $\sqrt{\mu}\, \pi = m\pi$, $m = 1, 2, \ldots$. Therefore,

(5.4.17) $\mu_m = m^2, \quad \Phi_m(\varphi) = a_m \cos m\varphi + b_m \sin m\varphi, \quad m = 0, 1, 2, \ldots$

are the eigenvalues and corresponding eigenfunctions of problem (5.4.15).

As before, for these found $\mu_m = m^2$ we will solve problem (5.4.16). The equation in (5.4.16) is the Bessel equation of order m, discussed and solved in Section 3.3. of Chapter 3, and its general solution is given by

$$R(r) = C J_m(\lambda r) + D Y_m(\lambda r),$$

where $J_m(\cdot)$ and $Y(\cdot)$ are the Bessel functions of the first and second kind, respectively. Since $R(r)$ is bounded at $r = 0$, and because Bessel functions of the second kind have singularity at $r = 0$, i.e.,

$$\lim_{r \to 0} Y_m(r) = \infty,$$

we have to choose $D = 0$. From the other boundary condition $R(a) = 0$ it follows that the eigenvalue λ must be chosen such that

$$J_m(\lambda a) = 0.$$

Since each Bessel function has infinitely many positive zeroes, the last equation has infinitely many zeroes which will be denoted by z_{mn}. Using these zeroes we have that the eigenfunctions of the problem (5.4.16) are given by n

$$R_{mn}(r) = J_m(\lambda_{mn}\, r) \quad m = 0, 1, 2, \ldots; \; n = 1, 2, \ldots.$$

For the above found eigenvalues λ_{mn}, the general solution of Equation (5.4.12) is given by

$$T_{mn}(t) = a_{mn} \cos \frac{z_{mn}}{a} t + b_{mn} \sin \frac{z_{mn}}{a} t,$$

and the general solution $w(r, \theta)$ of Equation (5.4.11) is a linear combination of the following, called *modes* of the membrane:

(5.4.18)
$$\begin{cases} w_{mn}^{(c)}(r, \varphi) = J_m\left(\frac{z_{mn}}{a}r\right) \cos m\varphi, \\ w_{mn}^{(s)}(r, \varphi) = J_m\left(\frac{z_{mn}}{a}r\right) \sin m\varphi. \end{cases}$$

Therefore, the solution of our original problem (5.4.5), which was to find the vibration of the circular membrane with given initial displacement and initial velocity, is given by

(5.4.19)
$$u(r, \varphi, t) = \sum_{m=0}^{\infty} \sum_{n=1}^{\infty} J_m\left(\frac{z_{mn}}{a}r\right) \left[a_m \cos m\varphi + b_m \sin m\varphi\right] \cdot$$
$$\left[A_{mn} \cos \frac{z_{mn}}{a}ct + B_{mn} \sin \frac{z_{mn}}{a}ct\right].$$

The coefficients in (5.4.19) are found by using the initial conditions (5.4.9):

$$f(r, \varphi) = u(r, \varphi, 0) = \sum_{m=0}^{\infty} \sum_{n=1}^{\infty} \left[a_m \cos m\varphi + b_m \sin m\varphi\right] A_{mn} J_m\left(\frac{z_{mn}}{a}r\right)$$

and

$$g(r, \varphi) = u_t(r, \varphi, 0) = \sum_{m=0}^{\infty} \sum_{n=1}^{\infty} \left[a_m \cos m\varphi + b_m \sin m\varphi\right] cB_{mn} \frac{z_{mn}}{a} J_m\left(\frac{z_{mn}}{a}r\right).$$

Using the orthogonality property of the Bessel eigenfunctions $\{J_m(\cdot), \; m = 0, 1, \ldots\}$, as well as the orthogonality property of

$$\{\cos m\varphi, \; \sin m\varphi, \; m = 0, 1, \ldots\}$$

we find

(5.4.20)
$$\begin{cases} a_m A_{mn} = \dfrac{\int_0^{2\pi}\int_0^a f(r, \varphi) J_m\left(\frac{z_{mn}}{a}r\right) r \cos m\varphi \, dr \, d\varphi}{\int_0^{2\pi}\int_0^a J_m^2\left(\frac{z_{mn}}{a}r\right) r \cos^2 m\varphi \, dr \, d\varphi}, \\[4ex] c\dfrac{z_{mn}}{a} a_m B_{mn} = \dfrac{\int_0^{2\pi}\int_0^a g(r, \varphi) J_m\left(\frac{z_{mn}}{a}r\right) r \cos m\varphi \, dr \, d\varphi}{\int_0^{2\pi}\int_0^a J_m^2\left(\frac{z_{mn}}{a}r\right) r \cos^2 m\varphi \, dr \, d\varphi}, \\[4ex] b_m A_{mn} = \dfrac{\int_0^{2\pi}\int_0^a f(r, \varphi) J_m\left(\frac{z_{mn}}{a}r\right) r \sin m\varphi \, dr \, d\varphi}{\int_0^{2\pi}\int_0^a J_m^2\left(\frac{z_{mn}}{a}r\right) r \sin^2 m\varphi \, dr \, d\varphi}, \\[4ex] c\dfrac{z_{mn}}{a} b_m B_{mn} = \dfrac{\int_0^{2\pi}\int_0^a g(r, \varphi) J_m\left(\frac{z_{mn}}{a}r\right) r \sin m\varphi \, dr \, d\varphi}{\int_0^{2\pi}\int_0^a J_m^2\left(\frac{z_{mn}}{a}r\right) r \sin^2 m\varphi \, dr \, d\varphi}. \end{cases}$$

Remark. As mentioned in Chapter 3, the zeroes z_{mn}, $n = 1, 2, \ldots$ of the Bessel function $J_m(\cdot)$ are not easy to find. In Table 5.4.1. we list the first n zeroes for the Bessel functions of order m of the first kind.

Table 5.4.1

$m \setminus n$	n=1	n=2	n=3	n=4	n=5	n=6	n=7
m=0	2.40482	5.52007	8.65372	11.7915	14.9309	18.0710	21.2116
m=1	3.83170	7.01558	10.1734	13.3236	16.4706	19.6158	22.7600
m=2	5.13562	8.41724	11.6198	14.7959	17.9598	21.1170	24.2701
m=3	6.38016	9.76102	13.0152	16.2234	19.4094	22.5827	25.7481
m=4	7.58834	11.0647	14.3725	17.6159	20.8263	24.0190	27.1990
m=5	8.77148	12.3386	15.7001	18.9801	22.2178	25.4303	28.6266
m=6	9.93611	13.5892	17.0038	20.3207	23.5860	26.8201	30.0337
m=7	11.0863	14.8212	18.2875	21.6415	24.9349	28.1911	31.4227
m=8	12.2250	16.0377	19.5545	22.9451	26.2668	29.5456	32.7958
m=9	13.3543	17.2412	20.8070	24.2338	27.5837	30.8853	34.1543

Example 5.4.1. Solve the vibrating membrane problem (5.4.5), (5.4.9), (5.4.10) if $a = c = 1$ and the initial displacement $f(r, \varphi)$ is given by

$$f(r, \varphi) = (1 - r^2) r^2 \sin 2\varphi, \quad 0 \le r \le 1, \ 0 \le \varphi < 2\pi$$

and the initial velocity is

$$g(r, \varphi) = 0, \quad 0 \le r \le 1, \ 0 \le \varphi < 2\pi.$$

Display the shapes of the membrane at the time instances $t = 0$ and $t = 0.7$.

Solution. Since $g(r, \varphi) = 0$, from (5.4.20) it follows that the coefficients $a_m B_{mn}$ and $b_m B_{mn}$ are all zero.

Since

$$\int_0^{2\pi} \sin 2\varphi \cos m\varphi \, d\varphi = 0, \quad m = 0, 1, 2, \ldots,$$

$$\int_0^{2\pi} \sin 2\varphi \sin m\varphi \, d\varphi = 0, \quad \text{for every } m \ne 2,$$

from (5.4.20) we have $a_m A_{mn} = 0$ for every $m = 0, 1, 2, \ldots$ and every $n = 1, 2, \ldots$; and $b_m A_{mn} = 0$ for every $m \ne 2$ and every $n = 1, 2, \ldots$.

For $m = 2$ and $n \in \mathbb{N}$, from (5.4.20) we have

(5.4.21)
$$b_2 A_{2n} = \frac{\int_0^{2\pi} \int_0^1 (1 - r^2) r^2 \sin 2\varphi \, J_2(z_{2n}r) r \sin 2\varphi \, dr \, d\varphi}{\int_0^{2\pi} \int_0^1 J_2^2(z_{2n}r) r \sin^2 2\varphi \, dr \, d\varphi}$$

$$= \frac{\int_0^1 (1 - r^2) r^3 J_2(z_{2n}r) \, dr}{\int_0^1 r J_2^2(z_{2n}r) \, dr}.$$

The numbers z_{2n} in (5.4.21) are the zeroes of the Bessel function $J_2(x)$.

Taking $a = 1$, $p = 2$ and $\lambda = z_{2n}$ in Example 3.2.12 of Chapter 3, for the integral in the numerator in (5.4.21) we have

(5.4.22)
$$\int_0^1 (1 - r^2) r^3 J_2(z_{2n}r) \, dr = \frac{2}{z_{2n}^2} J_4(z_{2n}).$$

In Theorem 3.3.11 of Chapter 3 we proved the following result.

If λ_k are the zeroes of the Bessel function $J_\mu(x)$, then

$$\int_0^1 x J_\mu^2(\lambda_k x) \, dx = \frac{1}{2} J_{\mu+1}^2(\lambda_k).$$

If we take $\mu = 2$ in this formula, then for the integral in the denominator in (5.4.21) we have

(5.4.23)
$$\int_0^1 r J_2^2(z_{2n}r) \, dr = \frac{1}{2} J_3^2(z_{2n}).$$

If we substitute (5.4.22) and (5.4.23) into (5.4.21), then we obtain

$$b_2 A_{2n} = \frac{4 J_4(z_{2n})}{z_{2n}^2 J_3^2(z_{2n})}.$$

The right hand side of the last equation can be simplified using the following recurrence formula from Chapter 3 for the Bessel functions.

$$J_{\mu+1}(x) + J_{\mu-1}(x) = \frac{2\mu}{x} J_\mu(x).$$

If we take $\mu = 3$ and $x = z_{2n}$ in this formula and use the fact that $J_2(z_{2n}) = 0$, then it follows that

$$z_{2n} J_4(z_{2n}) = 6 J_3(z_{2n})$$

and thus

$$b_2 A_{2n} = \frac{24}{z_{2n}^3 J_3(z_{2n})}.$$

Therefore, from (5.4.19), the solution $u(r, \varphi, t)$ of the vibration membrane problem is given by

$$u(r, \varphi, t) = 24 \sum_{n=1}^{\infty} \frac{1}{z_{2n}^3 J_3(z_{2n})} J_2(z_{2n} r) \sin 2\varphi \cos(z_{2n} t).$$

The shapes of the membrane at $t = 0$ and $t = 0.7$ are displayed in Figure 5.4.1.

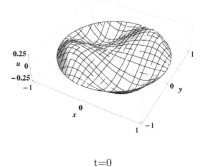

t=0   t=0.7

Figure 5.4.1

Example 5.4.2. Solve the vibrating circular membrane problem (5.4.5), (5.4.9), (5.4.10) if $a = 2$, $c = 1$ and

$$f(r, \varphi) = (4 - r^2) r \sin \varphi, \quad 0 \le r \le 2,\ 0 \le \varphi < 2\pi,$$
$$g(r, \varphi) = 1, \quad 0 \le r \le 2,\ 0 \le \varphi < 2\pi.$$

Solution. From (5.4.20) and the orthogonality of the sine and cosine functions we have

$$a_m A_{mn} = b_m B_{mn} = 0$$

for every $m \ge 0$, $n \ge 1$. For the same reason, we have

$$b_m A_{mn} = 0$$

for every $m \ne 1$, $n \ge 1$, and $a_m B_{mn} = 0$ for every $m \ne 0$, $n \ge 1$.

For $m = 1$ we have

$$b_1 A_{1n} = \frac{\int_0^{2\pi} \int_0^2 (4 - r^2) r^2 \sin^2 \varphi J_1\left(\frac{z_{1n}}{2}r\right) dr \, d\varphi}{\int_0^{2\pi} \int_0^2 J_1^2\left(\frac{z_{1n}}{2}r\right) r \sin^2 \varphi \, dr \, d\varphi} = \frac{\int_0^2 (4 - r^2) r^2 J_1\left(\frac{z_{1n}}{2}r\right) dr}{\int_0^2 J_1^2\left(\frac{z_{1n}}{2}r\right) r \, dr}.$$

Using the same argument as in the previous example we obtain

$$b_1 A_{1n} = 128 \frac{J_1\left(\frac{z_{1n}}{2}r\right)}{z_{1n}^3 J_2(z_{1n})}.$$

Similarly, for $m = 0$ we have

$$\frac{z_{0n}}{2} a_0 B_{0n} = \frac{\int_0^{2\pi} \int_0^2 J_0\left(\frac{z_{0n}}{2}r\right) r \, dr \, d\varphi}{\int_0^{2\pi} \int_0^2 J_0^2\left(\frac{z_{0n}}{2}r\right) r \, dr \, d\varphi}$$

$$= \frac{\int_0^2 J_0\left(\frac{z_{0n}}{2}r\right) r \, dr}{\int_0^2 J_0^2\left(\frac{z_{0n}}{2}r\right) r \, dr} = 2 \frac{J_0\left(\frac{z_{0n}}{2}r\right)}{z_{0n} J_1(z_{0n})}.$$

Therefore, the solution $u(r, \varphi, t)$ of the vibration problem is given by

$$u(r, \varphi, t) = 128 \sum_{n=1}^{\infty} \frac{J_1\left(\frac{z_{1n}}{2}r\right)}{z_{1n}^3 J_2(z_{1n})} \sin n\varphi \, \cos\left(\frac{z_{1n}}{2}t\right)$$

$$+ 4 \sum_{n=1}^{\infty} \frac{J_0\left(\frac{z_{0n}}{2}r\right)}{z_{0n}^2 J_1(z_{0n})} \sin\left(\frac{z_{0n}}{2}t\right).$$

5.4.2 The Wave Equation in Spherical Coordinates.

Let B be the three dimensional ball

$$B = \{(x, y, z) \in \mathbb{R}^3 : x^2 + y^2 + z^2 < a^2\},$$

whose boundary will be denoted by S. The wave equation for a function $u = u(x, y, z, t)$, $(x, y, z) \in B$, $t > 0$ is given by

$$u_{tt} = c^2 \left(u_{xx} + u_{yy} + u_{zz}\right) = c^2 \Delta_{x,y,z} \, u(x, y, x, t), \quad (x, y, z) \in B, \ t > 0,$$

subject to a zero Dirichlet boundary condition and given initial conditions. This equation describes many physical phenomena, such as the propagation of light, magnetic and X-ray waves. Also, in electromagnetism, the components of electric and magnetic fields satisfy the wave equation. Vibrations of the ball are also modeled by the wave equation. The solution $u(x, y, z, t)$ of the wave equation represents the radial displacement of the ball at position $(x, y, z) \in B$ and time t.

It is much more convenient if we write the wave equation in spherical coordinates:

$$(5.4.24) \qquad u_{tt} = c^2 \, \Delta \, u, \quad 0 \leq r < a,\ 0 \leq \theta < \pi,\ -\pi \leq \varphi < \pi,\ t > 0,$$

where

$$(5.4.25) \qquad \begin{aligned} \Delta u &= \frac{1}{r^2} \frac{\partial}{\partial r} \left(r^2 \frac{\partial u}{\partial r} \right) + \frac{1}{r^2 \sin \theta} \frac{\partial}{\partial \theta} \left(\sin \theta \frac{\partial u}{\partial \theta} \right) + \frac{1}{r^2 \sin^2 \theta} \frac{\partial^2 u}{\partial \varphi^2} \\ &\equiv \frac{\partial^2 u}{\partial r^2} + \frac{2}{r} \frac{\partial u}{\partial r} + \frac{1}{r^2} \left(\frac{\partial^2 u}{\partial \theta^2} + \cot \theta \frac{\partial u}{\partial \theta} + \csc^2 \theta \frac{\partial^2 u}{\partial \varphi^2} \right). \end{aligned}$$

(See Appendix E for the Laplacian in spherical coordinates.)

Along with this equation, we incorporate the boundary and initial conditions

$$(5.4.26) \qquad \begin{cases} u(a, \varphi, \theta, t) = 0, & 0 \leq \theta < \pi,\ 0 \leq \varphi < 2\pi, \\ u(r, \varphi, \theta, 0) = f(r, \theta, \varphi), & 0 \leq r < a,\ 0 \leq \theta < \pi,\ -\pi \leq \varphi < \pi, \\ u_t(r, \varphi, \theta, 0) = g(r, \varphi, \theta), & 0 \leq r < a,\ 0 \leq \theta < \pi,\ -\pi \leq \varphi < \pi. \end{cases}$$

In order to solve equation (5.4.24), subject to the boundary and initial conditions (5.4.26), we work similarly as in the wave equation, describing the vibrations of a circular membrane.

First we write the solution $u(r, \theta, \varphi, t)$ of the above initial boundary value problem in the form

$$(5.4.27) \qquad u(r, \varphi, \theta, t) = F(r, \varphi, \theta) T(t).$$

From the boundary condition $u(a, \varphi, \theta, t) = 0$ we obtain that the function $F(r, \varphi, \theta)$ satisfies the boundary condition

$$(5.4.28) \qquad F(a, \varphi, \theta) = 0, \quad 0 \leq \theta < \pi,\ -\pi \leq \varphi < \pi.$$

If we substitute (5.4.27) in the wave equation (5.4.24), after separating the variables, we obtain

$$(5.4.29) \qquad T''(t) + \lambda c^2 T(t) = 0, \quad t > 0$$

and again, the Helmholtz equation

$$(5.4.30) \qquad \Delta F(r, \varphi, \theta) + \lambda^2 \, F = 0, \quad 0 \leq r < a,\ 0 \leq \theta < \pi,\ -\pi \leq \varphi < \pi,$$

where $\Delta F(r, \varphi, \theta)$ is given by (5.4.25) and λ is a positive parameter to be determined.

The fact that λ^2 was used in (5.4.30) follows from the following result.

Theorem 5.4.1. *Suppose that* $P = P(x, y, z)$ *is a twice differentiable function in the ball* B, *continuous on the closed ball* $\overline{B}$ *and vanishes on the boundary* S. *If* P *is not identically zero in the ball* B *and satisfies the Helmholtz equation*

$$\Delta P(x, y, z) + \mu P(x, y, z) = 0, \quad (x, y, z) \in B,$$

then $\mu > 0$.

Proof. We use the Divergence Theorem (see Appendix E). If $\mathbf{F} = \mathbf{F}(x, y, z)$ is a twice differentiable vector field on the closed ball $\overline{B}$, then

$$\iiint\limits_{B} \nabla \cdot \mathbf{F} \, dx \, dy \, dz = \iint\limits_{S} \mathbf{F} \cdot \mathbf{n} \, d\sigma,$$

where ∇ is the gradient vector operator and $\mathbf{n}$ is the outward unit normal vector on the sphere S. If we take $\mathbf{F} = P\nabla P$ in the divergence formula and use the fact that $P = 0$ on the boundary S we obtain that

$$\iiint\limits_{B} \nabla \cdot (P\nabla P) \, dx \, dy \, dz = \iint\limits_{S} P\nabla P \cdot \mathbf{n} \, d\sigma = 0.$$

From this equation and from

$$\nabla \cdot (P\nabla P) = \nabla P \cdot \nabla P + P\nabla \cdot \nabla P = \nabla P \cdot \nabla P + P\Delta P$$

it follows that

$$\iiint\limits_{B} \nabla P \cdot \nabla P \, dx \, dy \, dz = - \iiint\limits_{B} P\Delta P, \, dx \, dy \, dz.$$

Using the above formula and Helmholtz equation we have

$$0 \leq \iiint\limits_{B} \nabla P \cdot \nabla P \, dx \, dy \, dz = - \iiint\limits_{B} P\Delta P \, dx \, dy \, dz = \mu \iiint\limits_{B} P^2 \, dx \, dy \, dz.$$

Therefore, $\mu \geq 0$. If $\mu = 0$, then using the same argument as above, the continuity of P on $\overline{B}$ and the fact that P vanishes on S would imply that $P \equiv 0$ on B, contrary to the assumption that P is not identically zero. Therefore $\mu > 0$. ∎

In order to solve (5.4.30) we assume a solution of the form

$$F(r, \varphi, \theta) = R(r)\Theta(\theta)\Phi(\varphi).$$

If we substitute this function into (5.4.30) we obtain (after rearrangement)

$$\frac{1}{R}\frac{d}{dr}\left(r^2\frac{dR}{dr}\right) + \lambda^2 r^2 = -\frac{1}{\Theta\sin\theta}\frac{d}{d\theta}\left(\sin\theta\frac{d\Theta}{d\theta}\right) - \frac{1}{\Phi\sin^2\theta}\frac{d^2\Phi}{d\varphi^2}.$$

The last equation is possible only when both sides are equal to the same constant ν.

(5.4.31) $$\frac{d}{dr}\left(r^2\frac{dR}{dr}\right) + (\lambda^2 r^2 - \nu)R = 0,$$

and

$$-\frac{1}{\Theta\sin\theta}\frac{d}{d\theta}\left(\sin\theta\frac{d\Theta}{d\theta}\right) - \frac{1}{\Phi\sin^2\theta}\frac{d^2\Phi}{d\varphi^2} = \nu.$$

From the last equation it follows that

$$-\frac{\sin\theta}{\theta}\frac{d}{d\theta}\left(\sin\theta\frac{d\Theta}{d\theta}\right) - \nu\sin^2\theta = \frac{1}{\Phi}\frac{d^2\Phi}{d\varphi^2}.$$

And again, this equation is possible only when both sides are equal to the same constant $-\mu$:

(5.4.32) $$\frac{d^2\Phi}{d\varphi^2} + \mu\Phi = 0,$$

and

(5.4.33) $$\frac{d^2\Theta}{d\theta^2} + \cot\varphi\frac{d\Theta}{d\theta} + \left(\nu - \frac{\mu}{\sin^2\theta}\right)\Theta = 0.$$

Because of physical reasons, the function $\Phi(\varphi)$ must be 2π periodic, i.e., it should satisfy the periodic condition

(5.4.34) $$\Phi(\varphi + 2\pi) = \Phi(\varphi).$$

We already solved the eigenvalue problem (5.4.32), (5.4.34) when we studied the vibrations of a circular membrane. We found that the eigenvalues and corresponding eigenfunctions of this problem are

(5.4.35) $\mu_m = m^2,$ $\Phi_m(\varphi) = a_m\cos m\varphi + b_m\sin m\varphi,$ $m = 0, 1, \ldots.$

Now, we turn our attention to Equation (5.4.33). This equation has two singular points at $\theta = 0$ and $\theta = \pi$. If we introduce a new variable θ by $x = \cos\theta$, then we obtain the differential equation

(5.4.36) $(1 - x^2)\dfrac{d^2\Theta}{dx^2} - 2x\dfrac{d\Theta}{dx} + \left[\nu - \dfrac{m^2}{1 - x^2}\right]\Theta(x) = 0,$ $-1 < x < 1.$

Equation (5.4.36) has two singularities at the points $x = -1$ and $x = 1$. Let us recall that this equation was discussed in Section 3.3 of Chapter 3, where we found out that in order for the solutions of this equation to remain bounded for all x we are required to have

$$(5.4.37) \qquad \nu = n(n+1), \quad n = 0, 1, 2, \ldots.$$

The corresponding bounded solutions of Equation (5.4.36) are the associated Legendre functions $P_n^{(m)}(x)$ of order m, given by

$$P_n^{(m)}(x) = (1 - x^2)^{\frac{m}{2}} \frac{d^m}{dx^m}(P_n(x)),$$

where $P_n(x)$ are the Legendre polynomials of order n.

Therefore, the bounded solutions of Equation (5.4.33) are given (up to a multiplicative constant) by

$$(5.4.38) \qquad \Theta_{nm}(\theta) = P_n^{(m)}(\cos \theta).$$

Finally, for the above obtained ν we are ready to solve Equation (5.4.31). This equation has singularity at the point $r = 0$. First, we rewrite the equation in the form

$$(5.4.39) \qquad r^2 \frac{d^2 R}{dr^2} + 2r \frac{dR}{dr} + \left[\lambda r^2 - n(n+1) \right] R = 0.$$

From the boundary condition $F(a, \theta, \varphi) = 0$, given in (5.4.28), and from the fact that $R(r)$ remains bounded for all $0 \le r < a$, it follows that the function $R(r)$ should satisfy the conditions:

$$(5.4.40) \qquad R(a) = 0, \quad \mid \lim_{r \to 0+} R(r) \mid < \infty.$$

If we introduce a new function $y = y(r)$ by the substitution

$$R(r) = \frac{y}{\sqrt{r}},$$

then Equation (5.4.39) is transformed into the following equation

$$(5.4.41) \qquad r^2 \frac{d^2 y}{dr^2} + r \frac{dy}{dr} + \left[\lambda r^2 - (n + \frac{1}{2})^2 \right] y = 0.$$

Equation (5.4.41) is the Bessel equation of order $n + \frac{1}{2}$ and its solution is

$$y(r) = A J_{n+\frac{1}{2}}(\sqrt{\lambda} r) + B Y_{n+\frac{1}{2}}(\sqrt{\lambda} r),$$

where J and Y are the Bessel functions of the first and second type, respectively. From the boundary conditions (5.4.40) it follows that $B = 0$ and

$$(5.4.42) \qquad\qquad\qquad J_{n+\frac{1}{2}}(\sqrt{\lambda}\, a) = 0.$$

If z_{nj} are the zeroes of the Bessel function $J_{n+\frac{1}{2}}(x)$, then from (5.4.42) it follows that the eigenvalues λ_{mn} are given by

$$(5.4.43) \qquad\qquad \lambda_{nj} = \left(\frac{z_{jn}}{a}\right)^2, \quad n = 0, 1, 2, \dots; \ j = 1, 2, \dots.$$

So, the eigenfunctions of (5.4.31) are

$$(5.4.44) \qquad\qquad\qquad R_{nj}(r) = J_{n+\frac{1}{2}}\left(\frac{z_{nj}}{a} r\right).$$

For the above found λ_{nj} we have that the general solution of Equation (5.4.29) is given by

$$(5.4.45) \qquad\qquad T_{nj}(t) = a_{nj} \cos\left(\sqrt{\lambda_{nj}}\, ct\right) + b_{nj} \sin\left(\sqrt{\lambda_{nj}}\, ct\right).$$

Therefore, the solution $u(r, \varphi, \theta, t)$ of the vibration ball problem (5.4.25) is given by

$$(5.4.46) \qquad u(r, \varphi, \theta, t) = \sum_{m=0}^{\infty} \sum_{n=0}^{\infty} \sum_{j=1}^{\infty} A_{mnj} \Phi_m(\varphi) \Theta_{mn}(\theta) R_{nj}(r) T_{nj}(t),$$

where $\Phi_m(\varphi)$, $\Theta_{mn}(\theta)$, $R_{nj}(r)$ and $T_{nj}(t)$ are the functions given by the formulas (5.4.35), (5.4.38), (5.4.44) and (5.4.45).

The coefficients A_{mnj} in (5.4.46) are determined using the initial conditions given in (5.4.26) and the orthogonality properties of the eigenfunctions involved in (5.4.46).

___

Exercises for Section 5.4.

In Exercises 1–7 solve the following circular membrane problem:

$$\begin{cases} u_{tt}(r, \varphi, t) = c^2 \left(u_{rr} + \dfrac{1}{r} u_r + \dfrac{1}{r^2} u_{\varphi\varphi}\right), & 0 \le r < a, \ -\pi \le \varphi < \pi, \ t > 0, \\ u(a, \varphi, t) = 0, \quad -\pi \le \varphi < \pi, \ t > 0, \\ u(r, \varphi, 0) = f(r, \varphi), \quad u_t(r, \varphi, 0) = g(r, \varphi), \ 0 < r \le a, \ -\pi \le \varphi < \pi, \end{cases}$$

if it is given that

1. $c = a = 1$, $f(r, \varphi) = 1 - r^2$, $g(r, \varphi) = 0$.

2. $c = a = 1$, $f(r, \varphi) = 0$, $g(r, \varphi) = \begin{cases} 1, & 0 < r < \frac{1}{2} \\ 0, & \frac{1}{2} < r < 1. \end{cases}$

3. $c = a = 1$, $f(r, \varphi) = (1 - r^2)r \sin \varphi$, $g(r, \varphi) = 0$.

4. $c = a = 1$, $f(r, \varphi) = (1 - r^2)r \sin \varphi$, $g(r, \varphi) = (1 - r^2)r^2 \sin 2\varphi$.

5. $c = a = 1$, $f(r, \varphi) = 5J_4(z_{4,1}r) \cos 4\varphi - J_2(z_{2,3}r) \sin 2\varphi$,

   $g(r, \varphi) = 0$ (z_{mn} is the nth zero of $J_m(x)$).

6. $c = a = 1$, $f(r, \varphi) = J_0(z_{03}r)$, $g(r, \varphi) = 1 - r^2$ (z_{mn} is the nth zero of $J_m(x)$).

7. $c = 1$, $a = 2$, $f(r, \varphi) = 0$, $g(r, \varphi) = 1$.

8. Use the separation of variables to find the solution $u = u(r, t)$ of the following boundary value problems:

$$\begin{cases} u_{tt}(r, t) = c^2 \left(\dfrac{\partial^2 u}{\partial r^2} + \dfrac{1}{r} \dfrac{\partial u}{\partial r} \right), & 0 < r < a, \ t > 0 \\[2mm] u(r, 0) = f(r), \quad \dfrac{\partial u(r, t)}{\partial r} \Big|_{t=0} = g(r), \ 0 < r < a \\[2mm] u(a, t) = 0, \quad t > 0. \end{cases}$$

9. Use the separation of variables to find the solution $u = u(r, t)$ of the following boundary value problems:

$$\begin{cases} u_{tt}(r, t) = c^2 \left(\dfrac{\partial^2 u}{\partial r^2} + \dfrac{1}{r} \dfrac{\partial u}{\partial r} \right), & 0 < r < a, \ t > 0 \\[2mm] u(r, 0) = f(r), \quad \dfrac{\partial u(r, t)}{\partial r} \Big|_{t=0} = g(r), \ 0 < r < a \\[2mm] |u(0, t)| < \infty, \quad \dfrac{\partial u(r, t)}{\partial r} \Big|_{r=a} = 0, \ t > 0. \end{cases}$$

10. Use the separation of variables to find the solution $u = u(r,t)$ of the following boundary value problems:

$$\begin{cases} u_{tt}(r,t) = c^2 \left(\dfrac{\partial^2 u}{\partial r^2} + \dfrac{1}{r} \dfrac{\partial u}{\partial r} \right) + A, & 0 < r < a,\ t > 0 \\[2mm] u(r,0) = 0, & \dfrac{\partial u(r,t)}{\partial r}\Big|_{t=0},\ 0 < r < a \\[2mm] u(a,t) = 0, & t > 0. \end{cases}$$

11. Use the separation of variables to find the solution $u = u(r,t)$ of the following boundary value problems:

$$\begin{cases} u_{tt}(r,t) = c^2 \left(\dfrac{\partial^2 u}{\partial r^2} + \dfrac{1}{r} \dfrac{\partial u}{\partial r} \right) + A f(r,t), & 0 < r < a,\ t > 0 \\[2mm] u(r,0) = 0, & \dfrac{\partial u(r,t)}{\partial r}\Big|_{t=0},\ 0 < r < a \\[2mm] u(a,t) = 0, & t > 0. \end{cases}$$

12. Find the solution $u = u(r,t)$ of the following boundary value problems:

$$\begin{cases} u_{tt}(r,t) = c^2 \left(\dfrac{\partial^2 u}{\partial r^2} + \dfrac{2}{r} \dfrac{\partial u}{\partial r} \right), & r_1 < r < r_2,\ t > 0 \\[2mm] u(r,0) = 0, & \dfrac{\partial u(r,t)}{\partial t}\Big|_{t=0} = f(r),\ r_1 < r < r_2 \\[2mm] \dfrac{\partial u(r,t)}{\partial r}\Big|_{r=r_1} = 0, & \dfrac{\partial u(r,t)}{\partial r}\Big|_{r=r_2} = 0,\ t > 0. \end{cases}$$

13. Find the solution $u = u(r,\varphi,t)$ of the following boundary value problems:

$$\begin{cases} u_{tt}(r,\varphi,t) = c^2 \left(\dfrac{\partial^2 u}{\partial r^2} + \dfrac{1}{r} \dfrac{\partial u}{\partial r} + \dfrac{1}{r^2} \dfrac{\partial^2 u}{\partial \varphi^2} \right), & 0 < r < a,\ 0 \le \varphi < 2\pi,\ t > 0 \\[2mm] u(r,\varphi,0) = A r \cos\varphi, & \dfrac{\partial u_t(r,\varphi,t)}{\partial t}\Big|_{t=0} = 0,\ 0 < r < a,\ 0 \le \varphi < 2\pi \\[2mm] \dfrac{\partial u(r,t)}{\partial r}\Big|_{r=a} = 0, & 0 \le \varphi < 2\pi,\ t > 0. \end{cases}$$

5.5 Integral Transform Methods for the Wave Equation.

In this section we will apply the Laplace and Fourier transforms to solve the wave equation and similar partial differential equations. We begin with the Laplace transform method.

5.5.1 The Laplace Transform Method for the Wave Equation.

Besides the separation of variables method, the Laplace transform method is used very often to solve linear partial differential equations.

As usual, we use capital letters for the Laplace transform with respect to t of a given function $g(x,t)$. For example, we write

$$\mathcal{L}\, u(x,t) = U(x,s), \quad \mathcal{L}\, y(x,t) = Y(x,s), \ \mathcal{L}\, f(x,t) = F(x,s).$$

Let us recall from Chapter 2 several properties of the Laplace transform.

$$(5.5.1) \qquad \mathcal{L}\left(u(x,t)\right) = U(x,s) = \int_0^\infty u(x,t)e^{-st}\,dt.$$

$$(5.5.2) \qquad \mathcal{L}\left(u_t(x,t)\right) = sU(x,s) - U(x,0).$$

$$(5.5.3) \qquad \mathcal{L}\left(u_{tt}(x,t)\right) = s^2 U(x,s) - su(x,0) - u_t(x,0).$$

$$(5.5.4) \qquad \mathcal{L}\left(u_x(x,t)\right) = \frac{d}{dx}\left(\mathcal{L}\left(u(x,t)\right)\right)$$

$$(5.5.5) \qquad \mathcal{L}\left(u_{xx}(x,t)\right) = \frac{d^2}{dx^2}\left(\mathcal{L}\left(u(x,t)\right)\right) = \frac{d^2 U(x,s)}{dx^2}.$$

Let us take a few examples.

Example 5.5.1. Using the Laplace transform method solve the following boundary value problem.

$$\begin{cases} u_{tt}(x,t) = u_{xx}(x,t), & 0 < x < 1, \ t > 0, \\ u(0,t) = u(1,t) = 0, & t > 0, \\ u(x,0) = 0, \quad u_t(x,0) = \cos \pi x, \ 0 < x < 1. \end{cases}$$

Solution. If $U = U(x,s) = \mathcal{L}\left(u(x,t)\right)$, then from (5.5.3), (5.5.5) and the initial conditions we have

$$\frac{d^2 U}{dx^2} - s^2 U = -\cos \pi x.$$

The general solution of the above ordinary differential equation is

$$U(x,s) = \frac{\cos \pi x}{s^2 + \pi^2} + c_1 e^{sx} + c_2 e^{-sx}.$$

Using the boundary conditions for the function $u(x,t)$ we obtain that the function $U(x,s)$ satisfies the conditions

$$U(0,s) = \mathcal{L}\big(u(0,t)\big) = 0, \quad U(1,s) = \mathcal{L}\big(u(1,t)\big) = 0.$$

These conditions for $U(x,s)$ imply $c_1 = c_2 = 0$ and so

$$U(x,s) = \frac{\cos \pi x}{s^2 + \pi^2}.$$

Therefore,

$$u(x,t) = \mathcal{L}^{-1}\big(U(x,s)\big) = \mathcal{L}^{-1}\left(\frac{\cos \pi x}{s^2 + \pi^2}\right)$$

$$= \cos \pi x \, \mathcal{L}^{-1}\left(\frac{1}{s^2 + \pi^2}\right) = \frac{1}{\pi} \cos \pi x \sin \pi t.$$

Example 5.5.2. Using the Laplace transform method solve the following boundary value problem.

$$\begin{cases} u_{tt}(x,t) = u_{xx}(x,t), & x > 0, \ t > 0, \\ u(0,t) = 0, \quad \lim_{x \to \infty} u_x(x,t) = 0, \ t > 0, \\ u(x,0) = xe^{-x}, \quad u_t(x,0) = 0, \ x > 0. \end{cases}$$

Solution. If $U = U(x,s) = \mathcal{L}\big(u(x,t)\big)$, then from (5.5.3) and (5.5.5) we have

$$s^2 U - su(x,0) - u_t(x,0) = \frac{d^2 U}{dx^2}.$$

The last equation in view of the initial conditions becomes

$$\frac{d^2 U}{dx^2} - s^2 U = -sxe^{-x},$$

whose general solution is

$$U(x,s) = -2\frac{se^{-sx}}{(s^2-1)^2} + x\frac{se^{-sx}}{s^2-1} + c_1 e^{sx} + c_2 e^{-xs}.$$

The boundary condition $\lim\limits_{x \to \infty} u_x(x,t) = 0$ implies $\lim\limits_{x \to \infty} U_x(x,s) = 0$ and so, $c_1 = 0$. From the other boundary conditions $u(0,t) = 0$ we have $U(0,s) = \mathcal{L}\big(u(0,t)\big) = 0$, which implies

$$c_2 = \frac{2s}{(s^2-1)^2}.$$

Therefore,

$$U(x,s) = 2\frac{se^{-sx}}{(s^2-1)^2} - 2e^{-x}\frac{s}{(s^2-1)^2} + x\frac{e^{-sx}}{s^2-1}.$$

Using the linearity and translation property of the inverse Laplace transform, together with the decomposition into partial fractions, we have

$$u(x,t) = 2\mathcal{L}^{-1}\left(\frac{se^{-sx}}{(s^2-1)^2}\right) - 2e^{-x}\mathcal{L}^{-1}\left(\frac{s}{(s^2-1)^2}\right) + xe^{-x}\mathcal{L}^{-1}\left(\frac{1}{s^2-1}\right)$$

$$= \frac{1}{2}(t-x)e^{-t-x}\left[1 - e^{2t} + (e^{2t} - e^{2x})H(t-x)\right],$$

where

$$H(t) = \begin{cases} 1, & \text{if } t \geq 0 \\ 0, & \text{if } t < 0 \end{cases}$$

is the Heaviside unit step function.

Example 5.5.3. Find the solution of the following problem (the telegraphic equation).

$$\begin{cases} u_{tt}(x,t) = \dfrac{1}{\pi^2}u_{xx}(x,t), & 0 < x < 1, \ t > 0, \\ u(0,t) = \sin \pi t, & t > 0, \\ u_t(1,t) = 0, & t > 0, \\ u(x,0) = 0, & u_t(x,0) = 0, \ 0 < x < 1. \end{cases}$$

Solution. If $U = U(x,s) = \mathcal{L}\big(u(x,t)\big)$, then from (5.5.3), (5.5.5), in view of the initial conditions, we have

$$\frac{d^2U}{dx^2} - \pi^2 s^2 U = 0.$$

The general solution of the last equation is

$$U(x,s) = c_1 e^{\pi sx} + c_2 e^{-\pi sx}.$$

From the boundary condition $u(1,t) = 0$ it follows that $U(1,s) = 0$. From the other boundary condition $u(0,t) = \sin \pi t$ we have

$$U(0,s) = \mathcal{L}\big(\sin \pi t\big)\big) = \frac{\pi}{s^2 + \pi^2}.$$

Using these boundary conditions for $U(x,s)$ we find

$$U(x,s) = \pi\frac{e^{s\pi(1-x)} - e^{-s\pi(1-x)}}{\left(e^{\pi s} - e^{-\pi s}\right)\left(s^2 + \pi^2\right)}.$$

In order to find the inverse Laplace transform of $U(x, s)$ we use the formula for the inverse Laplace transform by contour integration

$$u(x,t) = \frac{1}{2\pi i} \int_{b-i\infty}^{b+i\infty} e^{st} U(x, s)\, ds = \frac{1}{2\pi i} \int_{b-i\infty}^{b+i\infty} F(x, s)\, ds,$$

where

$$F(x, s) = \pi \frac{e^{s\pi(1-x)} - e^{-s\pi(1-x)}}{\left(e^{\pi s} - e^{-\pi s}\right)\left(s^2 + \pi^2\right)}.$$

(See the section the Inverse Laplace Transform by Contour Integration in Chapter 2). One way to compute the above contour integral is by the Cauchy Theorem of Residues (see Appendix F).

The singularities of the function $F(x, s)$ are found by solving the equation

$$\left(e^{\pi s} - e^{-\pi s}\right)\left(s^2 + \pi^2\right) = 0,$$

or equivalently,

$$\left(e^{2\pi s} - 1\right)\left(s^2 + \pi^2\right) = 0.$$

The solutions of the above equation are $s = \pm ni$, $n = 0, 1, 2, \ldots$. We exclude $s = 0$ since it is a removable singularity for the function $F(z)$. Notice that all these singularities are simple poles. Using the formula for residues (see Appendix F) we find

$$Res\big(F(x, s), s = \pi i\big) = \lim_{s \to \pi i} \big[(s - \pi i)F(x, s)\big] = \frac{e^{\pi t i}}{2i} \frac{\sin\left(\pi^2(1 - x)\right)}{\sin \pi^2},$$

$$Res\big(F(x, s), s = -\pi i\big) = \lim_{s \to -\pi i} \big[(s + \pi i)F(x, s)\big] = -\frac{e^{-\pi t i}}{2i} \frac{\sin\left(\pi^2(1 - x)\right)}{\sin \pi^2},$$

$$Res\big(F(x, s), s = ni\big) = \lim_{s \to ni} \big[(s - ni)F(x, s)\big] = ie^{n\pi t i} \frac{\sin\left(n\pi(1 - x)\right)}{(\pi^2 - n^2)\cos n\pi},$$

$$Res\big(F(x, s), s = -ni\big) = \lim_{s \to -ni} \big[(s + ni)F(x, s)\big] = -\frac{i}{e^{n\pi t i}} \frac{\sin\left(n\pi(1 - x)\right)}{(\pi^2 - n^2)\cos n\pi}.$$

Therefore, by the Residue Theorem we have

$$u(x,t) = Res\big(F(x, s), \pi i\big) + Res\big(F(x, s), -\pi i\big) + \sum_{\substack{n=-\infty \\ n \neq 0}}^{\infty} Res\big(F(x, s), ni\big)$$

$$= \frac{e^{\pi t i}}{2i} \frac{\sin\left(\pi^2(1 - x)\right)}{\sin \pi^2} - \frac{e^{-\pi t i}}{2i} \frac{\sin\left(\pi^2(1 - x)\right)}{\sin \pi^2}$$

$$+ \sum_{n=1}^{\infty} \left(ie^{n\pi t i} \frac{\sin\left(n\pi(1 - x)\right)}{(\pi^2 - n^2)\cos n\pi} - ie^{-n\pi t i} \frac{\sin\left(n\pi(1 - x)\right)}{(\pi^2 - n^2)\cos n\pi} \right)$$

$$= \frac{\sin \pi t \, \sin\left(\pi^2(1 - x)\right)}{\sin \pi^2} - 2\sum_{n=1}^{\infty} \frac{\sin n\pi t \, \sin\left(n\pi(1 - x)\right)}{(\pi^2 - n^2)\cos n\pi}.$$

The Convolution Theorem for the Laplace transform can be used to solve some boundary value problems. The next example illustrates this method.

Example 5.5.4. Find the solution of the following problem.

$$\begin{cases} u_{tt}(x,t) = u_{xx}(x,t) + \sin t, & x > 0, \ t > 0, \\ u(0,t) = 0, & \lim_{x \to \infty} u_x(x,t) = 0, \ t > 0, \\ u(x,0) = 0, & u_t(x,0) = 0, \ 0 < x < 1. \end{cases}$$

Solution. If $U = U(x,s) = \mathcal{L}\big(u(x,t)\big)$, then from (5.5.3) and (5.5.5) we have

$$s^2 U - su(x,0) - u_t(x,0) = \frac{d^2 U}{dx^2} + \frac{1}{s^2 + 1}.$$

The last equation, in view of the initial conditions, becomes

$$\frac{d^2 U}{dx^2} - s^2 U = -\frac{1}{s^2 + 1},$$

whose general solution is

$$U(x,s) = c_1 e^{sx} + c_2 e^{-sx} + \frac{1}{s^2(s^2 + 1)}.$$

From the condition $\lim_{x \to \infty} u_x(x,t) = 0$ it follows that $\lim_{x \to \infty} U_x(x,s) = 0$ and so $c_1 = 0$. From the other boundary conditions $u(0,t) = 0$ it follows that $U(0,s) = \mathcal{L}\big(u(0,t)\big) = 0$, which implies that

$$c_2 = -\frac{1}{s^2(s^2 + 1)}.$$

Therefore,

$$U(x,s) = \frac{1}{s^2 + 1} \frac{1 - e^{-sx}}{s^2}.$$

From the last equation, by the Convolution Theorem for the Laplace transform we have

$$u(x,t) = (\sin t) * \left(\mathcal{L}^{-1}\left(\frac{1 - e^{-sx}}{s^2} \right) \right) = (\sin t) * \big(t - (t - x) H(t - x) \big)$$

$$= \int_0^t \sin(t - y) \big(y - (y - x) H(y - x) \big) \, dy$$

$$= \int_0^t y \sin(t - y) \, dy - \int_0^t (y - x) H(y - x) \, dy$$

$$= t \sin t - \int_0^t (y - x) H(y - x) \, dy.$$

For the last integral we consider two cases. If $t < x$, then $y - x < 0$, and so $H(x - y) = 0$. Thus, the integral is zero in this case. If $t > x$, then

$$\int_0^t (y - x)H(y - x)\,dy = \int_0^x (y - x)H(y - x)\,dy + \int_x^t (y - x)H(y - x)\,dy$$

$$= 0 + \int_x^t (y - x)\,dy = \frac{1}{2}(t - x)^2.$$

Therefore,

$$u(x,t) = \begin{cases} t\sin t, & \text{if } 0 < t < x \\ t\sin t - \frac{1}{2}(t - x)^2, & \text{if } t > x. \end{cases}$$

5.5.2 The Fourier Transform Method for the Wave Equation.

The Fourier transform, like the Laplace transform, can be used to solve certain classes of partial differential equations. The wave equation, and like equations, in one and higher spatial dimensions can be solved using the Fourier transform.

The fundamentals of the Fourier transform were developed in Chapter 2. We use the same, but capital letters for the Fourier transform of a given function $u(x,t)$ with respect to the variable $x \in \mathbb{R}$. For example, we write

$$\mathcal{F}u(x,t) = U(\omega,t), \quad \mathcal{F}y(x,t) = Y(\omega,x), \quad \mathcal{F}f(x,t) = F(\omega,t).$$

Let us recall several properties of the Fourier transform.

$$(5.5.6) \qquad \mathcal{F}\big(u(x,t)\big) = U(\omega,t) = \int_{-\infty}^{\infty} u(x,t)e^{-i\omega x}\,dx.$$

$$(5.5.7) \qquad \mathcal{F}\big(u_x(x,t)\big) = i\omega U(\omega,t).$$

$$(5.5.8) \qquad \mathcal{F}\big(u_{xx}(x,t)\big) = (i\omega)^2 U(\omega,t) = -\omega^2 U(\omega,t).$$

$$(5.5.9) \qquad u(x,t) = \frac{1}{2\pi}\int_{-\infty}^{\infty} U(\omega,t)e^{i\omega x}\,d\omega.$$

$$(5.5.10) \qquad \mathcal{F}\big(u_t(x,t)\big) = \frac{d}{dt}\Big(\mathcal{F}\big(u(x,t)\big)\Big).$$

$$(5.5.11) \qquad \mathcal{F}\big(u_{tt}(x,t)\big) = \frac{d^2}{dt^2}\Big(\mathcal{F}\big(u(x,t)\big)\Big).$$

Let us take first an easy example.

Example 5.5.5. Using the Fourier transform solve the following transport boundary value problem.

$$\begin{cases} u_t(x,t) + au_x(x,t) = 0, & -\infty < x < \infty, \ t > 0 \\ u(x,0) = e^{-x^2}, & -\infty < x < \infty. \end{cases}$$

Solution. If $U = U(\omega,t) = \mathcal{F}\left(u(x,t)\right)$, then from (5.5.7) and (5.5.10) we have

$$\frac{dU(\omega,t)}{dt} + ai\omega U(\omega,t) = 0.$$

The general solution of this equation (keeping ω constant) is

$$U(\omega,t) = C(\omega)e^{-i\omega at}.$$

From the initial condition

$$u(x,0) = e^{-x^2}$$

it follows that

$$U(\omega,0) = \mathcal{F}\left(u(x,0)\right) = \mathcal{F}\left(e^{-x^2}\right) = \sqrt{\pi}e^{-\frac{\omega^2}{4}} \quad \text{(see Table B in Appendix B)}.$$

Therefore, $C(\omega) = \sqrt{\pi}\,e^{-\frac{\omega^2}{4}}$ and so,

$$U(\omega,t) = \sqrt{\pi}\,e^{-\frac{\omega^2}{4}}e^{-i\omega at}.$$

Using the modulation property of the Fourier transform (see Chapter 2) we have

$$u(x,t) = e^{-(x-at)^2}.$$

Example 5.5.6. Using the Fourier transform solve the following transport boundary value problem.

$$\begin{cases} u_{tt}(x,t) = c^2 u_{xx}(x,t) + f(x,t), & -\infty < x < \infty, \ t > 0, \\ u(x,0) = 0, \quad u_t(x,0) = 0, & -\infty < x < \infty. \end{cases}$$

Solution. If $U = U(\omega,t) = \mathcal{F}\left(u(x,t)\right)$ and $F = F(\omega,t) = \mathcal{F}\left(f(x,t)\right)$, then from (5.5.6), (5.5.8) and (5.5.11) the equation becomes

(5.5.12)
$$\frac{d^2U(\omega,t)}{dt^2} + c^2\omega^2 U(\omega,t) = F(\omega,t).$$

From the initial conditions $u(x,0) = 0$ and $u_t(x,0) = 0$ it follows that $U(\omega,0) = \mathcal{F}\left(u(x,0)\right) = 0$ and $U_t(\omega,0) = \mathcal{F}\left(u_t(x,0)\right) = 0$. Using these initial conditions and the method of variation of parameters (see the Appendix D) the solution of (5.5.12) is

$$U(\omega, t) = \frac{1}{c\omega} \int_0^t F(\omega, \tau) \sin c\omega(t - \tau) \, d\tau.$$

Therefore, by the inversion formula we have

$$(5.5.13) \qquad u(x, t) = \frac{1}{2\pi} \int_{-\infty}^{\infty} \left(\int_0^t F(\omega, \tau) \frac{\sin c\omega(t - \tau)}{c\omega} \, d\tau \right) e^{i\omega x} \, d\omega.$$

To complete the problem, let us recall the following properties of the Fourier transform.

$$\mathcal{F} f(x - a) = e^{-i\omega} (\mathcal{F} f)(\omega), \quad \mathcal{F} \left(e^{iax} f(x) \right) = (\mathcal{F} f)(\omega - a),$$

$$\mathcal{F} \left(\chi_a(x) \right)(\omega) = 2 \frac{\sin a\omega}{\omega},$$

where χ_a is the *characteristic function* on the interval $(-a, a)$, defined by

$$\chi_a(x) = \begin{cases} 1, & |x| \leq 1 \\ 0, & |x| > 1, \end{cases}$$

Based on these properties, from (5.5.13) it follows that

$$u(x, t) = \frac{1}{2\pi} \int_0^t \left(\int_{-\infty}^{\infty} F(\omega, \tau) \frac{\sin c\omega(t - \tau)}{c\omega} e^{i\omega x} \, d\omega \right) d\tau$$

$$= \frac{1}{2\pi} \int_0^t \left(\int_{x-c(t-\tau)}^{x+c(t-\tau)} f(\omega, \tau) \, d\omega \right) d\tau.$$

In the next example we consider the important *telegraph equation*.

Example 5.5.7. Using the Fourier transform solve the following problem.

$$\begin{cases} u_{tt}(x, t) + 2(a + b)u_t(x, t) + 4ab\,u(x, t) = c^2 u_{xx}(x, t) = 0, & x \in \mathbb{R}, \ t > 0, \\ u(x, 0) = f(x), & x \in \mathbb{R}, \\ u_t(x, 0) = g(x), & x \in \mathbb{R}, \end{cases}$$

where a and b are positive given constants.

Solution. The current and voltage in a transmission line are governed by equations of this type. If $a = b = 0$, then the wave equation is a particular case of the telegraph equation. This equation has many other applications in fluid mechanics, acoustics and elasticity.

If $U = U(\omega, t) = \mathcal{F}\big(u(x,t)\big)$, then from (5.5.6), (5.5.8), (5.5.10) and (5.5.11) the telegraph equation becomes

$$(5.5.14) \qquad \frac{d^2 U}{dt^2} + 2(a+b)\frac{dU}{dt} + (4ab + c^2\omega^2)U = 0.$$

The solutions of the characteristic equation of the above ordinary differential equation (keeping ω constant) are

$$r_{1,2} = -(a+b) \pm \sqrt{(a-b)^2 - c^2\omega^2}.$$

Case 1^0. If $D = (a-b)^2 - c^2\omega^2 > 0$, then r_1 and r_2 are real and distinct, and so the general solution of (5.5.14) is given by

$$U(\omega, t) = C_1(\omega)e^{r_1 t} + C_2(\omega)e^{r_2 t}.$$

Notice that $r_1 < 0$ and $r_2 < 0$ in this case. From the given initial conditions $u(x,0) = f(x)$ and $u_t(x,0) = g(x)$ it follows that $U(\omega, 0) = F(\omega)$ and $U_t(\omega, 0) = G(\omega)$, where $F = \mathcal{F}f$ and $G = \mathcal{F}g$. Using these initial conditions we have

$$C_1(\omega) = \frac{G - r_2 F}{r_2 - r_1}, \qquad C_2(\omega) = \frac{G - r_1 F}{r_2 - r_1}.$$

Therefore,

$$U(\omega, t) = \frac{e^{r_1 t} + e^{r_2 t}}{r_2 - r_1} G(\omega, t) - \frac{r_2 e^{r_1 t} + r_1 e^{r_2 t}}{r_2 - r_1} F(\omega, t).$$

In general, it is not very easy to find the inverse transform from the last equation.

Case 2^0. If $D = (a-b)^2 - c^2\omega^2 < 0$, then

$$U(\omega, t) = e^{-(a+b)t}\left(C_1(\omega)e^{\sqrt{-D}\,it} + C_2(\omega)e^{-\sqrt{-D}\,it} \right).$$

In some particular cases we can find explicitly the inverse. For example, if $a = b$, then $r_{1,2} = (a+b) \pm c\omega i$, and so,

$$U(\omega, t) = e^{-2at}\left(C_1(\omega)e^{c\omega t i} + C_2(\omega)e^{-c\omega t i} \right).$$

From the last equation, by the formula for the inverse Fourier transform and the translation property we obtain

$$u(x,t) = \frac{1}{2\pi} e^{-2at} \int_{-\infty}^{\infty} \left[C_1(\omega)e^{c\omega t i} + C_2(\omega)e^{-c\omega t i} \right] e^{i\omega x}\, d\omega$$

$$= e^{-2at}\left(c_1(x - ct) + c_2(x + ct) \right),$$

where $c_1(x) = \mathcal{F}^{-1}\left(C_1(\omega)\right)$ and $c_2(x) = \mathcal{F}^{-1}\left(C_2(\omega)\right)$.

If the wave equation is considered on the interval $(0, \infty)$, then we use the Fourier sine or Fourier cosine transform. Let us recall from Chapter 2 a few properties of these transforms.

$$(5.5.15) \qquad \mathcal{F}_s\, u(x,t) = \int_0^\infty u(x,t)\sin \omega x\, dx.$$

$$(5.5.16) \qquad \mathcal{F}_c\, u(x,t) = \int_0^\infty u(x,t)\cos \omega x\, dx.$$

$$(5.5.17) \qquad \mathcal{F}_s\left(u_{xx}(x,t)\right) = -\omega^2 \mathcal{F}_s\left(u(x,t)\right) + \omega\, u(0,t).$$

$$(5.5.18) \qquad \mathcal{F}_c\left(u_{xx}(x,t)\right) = -\omega^2 \mathcal{F}_s\left(u(x,t)\right) - u_x(0,t).$$

$$(5.5.19) \qquad \mathcal{F}_s\left(u_{tt}(x,t)\right) = \frac{d^2}{dt^2}\left(\mathcal{F}_s\left(u(x,t)\right)\right).$$

$$(5.5.20) \qquad \mathcal{F}_c\left(u_{tt}(x,t)\right) = \frac{d^2}{dt^2}\left(\mathcal{F}_c\left(u(x,t)\right)\right).$$

$$(5.5.21) \qquad u(x,t) = \frac{2}{\pi}\int_0^\infty \mathcal{F}_s\left(u(x,t)\right)(\omega,t)\sin \omega x\, d\omega.$$

$$(5.5.22) \qquad u(x,t) = \frac{2}{\pi}\int_0^\infty \mathcal{F}_c\left(u(x,t)\right)(\omega,t)\cos \omega x\, d\omega.$$

The choice of whether to use the Fourier sine or Fourier cosine transform depends on the nature of the boundary conditions at zero. Let us illustrate this remark with a few examples of a wave equation on the semi-infinite interval $(0,\infty)$.

Example 5.5.8. Using the Fourier cosine or Fourier sine transform solve the following boundary value problem.

$$\begin{cases} u_{tt}(x,t) = c^2 u_{xx}(x,t), & 0 < x < \infty,\ t > 0, \\ u(x,0) = f(x), & u_t(x,0) = g(x),\ 0 < x < \infty, \\ u(0,t) = 0, & t > 0. \end{cases}$$

Solution. Let us apply the Fourier sine transform to solve this problem. If $U = U(\omega,t) = \mathcal{F}_s\left(u(x,t)\right)$, then from (5.5.17) and (5.5.19) the wave equation becomes

$$(5.5.23) \qquad \frac{d^2 U(\omega,t)}{dt^2} + c^2\omega^2 U(\omega,t) = 0.$$

From the initial conditions $u(x,0) = f(x)$ and $u_t(x,0) = g(x)$ it follows that $U(\omega,0) = \mathcal{F}_s\big(u(x,0)\big) = F(\omega)$ and $U_t(\omega,0) = \mathcal{F}_s\big(u_t(x,0)\big) = G(\omega)$. Using these initial conditions, the solution of (5.5.23) is

$$U(\omega,t) = F(\omega)\cos c\omega t + G(\omega)\frac{\sin c\omega t}{c\omega}.$$

To find the inverse Fourier transform we consider first the case when $x - ct > 0$. In this case we have $x + ct > 0$ and so from the inversion formula (5.5.21) it follows that

$$u(x,t) = \frac{2}{\pi}\int_0^\infty \left(F(\omega)\cos c\omega t + G(\omega)\frac{\sin c\omega t}{c\omega}\right)\sin \omega x\, d\omega$$

$$= \frac{2}{\pi}\int_0^\infty F(\omega)\cos c\omega t \sin \omega x\, d\omega + \frac{2}{\pi}\int_0^\infty G(\omega)\frac{\sin c\omega t \sin \omega x}{c\omega}\, d\omega$$

$$= \frac{1}{\pi}\int_0^\infty F(\omega)\big[\sin (x + ct)\omega + \sin (x - ct)\omega\big]\, d\omega$$

$$+ \frac{1}{\pi c}\int_0^\infty F(\omega)\frac{\cos (x - ct)\omega - \cos (x + ct)\omega}{\omega}\, d\omega.$$

For the first integral in the above equation, by the inversion formula (5.5.21) we have

$$\frac{1}{\pi}\int_0^\infty F(\omega)\big[\sin (x + ct)\omega + \sin (x - ct)\omega\big]\, d\omega = \frac{1}{\pi}\int_0^\infty F(\omega)\sin (x + ct)\omega\, d\omega$$

$$+ \frac{1}{\pi}\int_0^\infty F(\omega)\sin (x - ct)\omega\, d\omega = \frac{1}{2}f(x + ct) + \frac{1}{2}f(x - ct).$$

For the second integral in the equation, again by the inversion formula (5.5.21) we have

$$\frac{1}{\pi c}\int_0^\infty G(\omega)\frac{\cos (x-ct)\omega - \cos (x+ct)\omega}{\omega}\, d\omega = \frac{1}{\pi c}\int_0^\infty \left(G(\omega)\int_{x-ct}^{x+ct}\sin \omega s\, ds\right)d\omega$$

$$= \int_{x-ct}^{x+ct}\left(\frac{1}{\pi c}\int_0^\infty G(\omega)\sin \omega s\, d\omega\right)ds = \frac{1}{2c}\int_{x-ct}^{x+ct} g(s)\, ds.$$

Thus, $u(x,t)$, in this case, is given by

$$u(x,t) = \frac{f(x + ct) + f(x - ct)}{2} + \frac{1}{2c}\int_{x-ct}^{x+ct} g(s)\, ds.$$

If $x - ct < 0$, then $ct - x > 0$ and working as above and using the facts that $\sin(x - ct)\omega = -\sin(ct - x)\omega$ and $\cos(x - ct)\omega = \cos(ct - x)\omega$ we obtain

$$u(x,t) = \frac{f(x+ct) - f(ct-x)}{2} + \frac{1}{2c}\int_{ct-x}^{x+ct} g(s)\,ds.$$

Therefore,

$$u(x,t) = \frac{f(x+ct) + f(|x-ct|)sign(x-ct)}{2} + \frac{1}{2c}\int_{|x-ct|}^{x+ct} g(s)\,ds,$$

where the function $sign(\cdot)$ is defined by

$$sign(x) = \begin{cases} -1, & x < 0 \\ 1, & x > 0. \end{cases}$$

Example 5.5.9. Using the Fourier cosine or Fourier sine transform solve the following boundary value problem.

$$\begin{cases} u_{tt}(x,t) = c^2 u_{xx}(x,t), & 0 < x < \infty, \ t > 0, \\ u(x,0) = f(x), & u_t(x,0) = g(x), \ 0 < x < \infty, \\ u_x(0,t) = 0, & t > 0. \end{cases}$$

Solution. We apply the Fourier cosine transform to solve this problem. If $U = U(\omega,t) = \mathcal{F}_c(u(x,t))$, then from (5.5.17) and (5.5.19) the wave equation becomes

(5.5.24) $$\frac{d^2 U(\omega,t)}{dt^2} + c^2\omega^2 U(\omega,t) = 0.$$

From the initial conditions $u(x,0) = f(x)$ and $u_t(x,0) = g(x)$ it follows that $U(\omega,0) = \mathcal{F}_s(u(x,0)) = F(\omega)$ and $U_t(\omega,0) = \mathcal{F}_s(u_t(x,0)) = G(\omega)$. Using these initial conditions, the solution of (5.5.24) is

$$U(\omega,t) = F(\omega)\cos c\omega t + G(\omega)\frac{\sin c\omega t}{c\omega}.$$

If $x - ct > 0$, then $x + ct > 0$ and so from the inversion formula (5.5.22) it follows that

$$u(x,t) = \frac{2}{\pi} \int_0^\infty \left(F(\omega)\cos c\omega t + G(\omega)\frac{\sin c\omega t}{c\omega} \right) \cos \omega x \, d\omega$$

$$= \frac{2}{\pi} \int_0^\infty F(\omega)\cos c\omega t \cos \omega x \, d\omega + \frac{2}{\pi} \int_0^\infty G(\omega)\frac{\sin c\omega t \cos \omega x}{c\omega} \, d\omega$$

$$= \frac{1}{\pi} \int_0^\infty F(\omega)\big[\cos (x+ct)\omega + \cos (x-ct)\omega\big] \, d\omega$$

$$+ \frac{1}{\pi c} \int_0^\infty G(\omega)\frac{\sin (x-ct)\omega - \sin (x+ct)\omega}{\omega} \, d\omega.$$

For the first integral in the above equation, by the inversion formula (5.5.22) we have

$$\frac{1}{\pi} \int_0^\infty F(\omega)\big[\cos (x+ct)\omega + \cos (x-ct)\omega\big] \, d\omega = \frac{1}{\pi} \int_0^\infty F(\omega)\cos (x+ct)\omega \, d\omega$$

$$+ \frac{1}{\pi} \int_0^\infty F(\omega)\cos (x-ct)\omega \, d\omega = \frac{1}{2}f(x+ct) + \frac{1}{2}f(x-ct).$$

For the second integral in the equation, again by the inversion formula (5.5.22) we have

$$\frac{1}{\pi c} \int_0^\infty G(\omega)\frac{\sin (x-ct)\omega - \sin (x+ct)\omega}{\omega} \, d\omega = -\frac{1}{\pi c} \int_0^\infty G(\omega) \int_{x-ct}^{x+ct} \cos \omega s \, ds \, d\omega$$

$$= - \int_{x-ct}^{x+ct} \left(\frac{1}{\pi c} \int_0^\infty G(\omega)\cos \omega s \, d\omega \right) ds = -\frac{1}{2c} \int_{x-ct}^{x+ct} g(s) \, ds.$$

Thus, $u(x,t)$ in this case is given by

$$u(x,t) = \frac{f(x+ct) + f(x-ct)}{2} + \frac{1}{2c} \int_{x-ct}^{x+ct} g(s) \, ds.$$

If $x - ct < 0$, then $ct - x > 0$, and working as above, we obtain $u(x,t)$ in this case is given by

$$u(x,t) = \frac{f(x+ct) + f(x-ct)}{2} + \frac{1}{2c} \left(\int_0^{x+ct} g(s) \, ds + \int_0^{ct-x} g(s) \, ds \right).$$

Therefore,

$$u(x,t) = \frac{f(x+ct) + f(|\, x - ct \,|)}{2} + \frac{1}{2c}\left(\int_0^{x+ct} g(s)\, ds - sign(x-ct)\int_0^{|x-ct|} g(s)\, ds\right),$$

Exercises for Section 5.5.

In Problems 1–9, using the Laplace transform solve the indicated initial boundary value problem on the interval $(0, \infty)$, subject to the given conditions.

1. $u_t(x,t) + 2u_x(x,t) = 0$, $x > 0$, $t > 0$, $u(x,0) = 3$, $u(0,t) = 5$.

2. $u_t(x,t) + u_x(x,t) = 0$, $x > 0$, $t > 0$, $u(x,0) = \sin x$, $u(0,t) = 0$.

3. $u_t(x,t) + u_x(x,t) = -u$, $x > 0$, $t > 0$, $u(x,0) = \sin x$, $u(0,t) = 0$.

4. $u_t(x,t) - u_x(x,t) = u$, $x > 0$, $t > 0$, $u(x,0) = e^{-5x}$, $u(0,t) = 0$.

5. $u_t(x,t) + u_x(x,t) = t$, $x > 0$, $t > 0$, $u(x,0) = 0$, $u(0,t) = t^2$.

6. $u_{tt}(x,t) = c^2 u_{xx}(x,t)$, $x > 0$, $t > 0$ $u(0,t) = f(t)$, $\lim_{x\to\infty} u(x,t) = 0$
   $t > 0$, $u(x,0) = 0$, $u_t(x,0) = 0$, $x > 0$.

7. $u_{tt}(x,t) = u_{xx}(x,t)$, $x > 0$, $t > 0$ $u(0,t) = 0$, $\lim_{x\to\infty} u(x,t) = 0$ $t > 0$,
   $u(x,0) = xe^{-x}$, $u_t(x,0) = 0$, $x > 0$.

8. $u_{tt}(x,t) = u_{xx}(x,t) + t$, $x > 0$, $t > 0$ $u(0,t) = 0, t > 0$, $u(x,0) = 0$,
   $u_t(x,0) = 0$, $x > 0$.

9. $u_{tt}(x,t) = u_{xx}(x,t)$, $x > 0$, $t > 0$ $u(0,t) = \sin t$, $t > 0$, $u(x,0) = 0$,
   $u_t(x,0) = 0$, $x > 0$.

In Problems 10–15, use the Laplace transform to solve the initial boundary value problem on the indicated interval, subject to the given conditions.

10. $u_{tt}(x,t) = c^2 u_{xx}(x,t) + kc^2 \sin \frac{\pi x}{a}$, $0 < x < a$, $t > 0$, $u(x,0) = u_t(x,0) = 0$, $u(0,t) = u(a,t) = 0$, $t > 0$.

11. $u_{tt}(x,t) = u_{xx}(x,t)$, $0 < x < 1$, $t > 0$, $u(x,0) = 0$, $u_t(x,0) = 0$ $u(0,t) = 0$, $u(0,t) = 0$.

12. $u_{tt}(x,t) = u_{xx}(x,t)$, $0 < x < 1$, $t > 0$, with initial conditions $u(x,0) = 0$, $u_t(x,0) = \sin \pi x$, and boundary conditions $u(0,t) = 0$, $u_x(1,t) = 1$.

$\left[\text{Hint: Expand } \dfrac{1}{1 + e^{-2s}} \text{ in a geometric series.} \right]$

13. $u_{tt}(x,t) = u_{xx}(x,t)$, $0 < x < 1$, $t > 0$, with initial conditions $u(x,0) = 0$, $u_t(x,0) = 1$, and boundary conditions $u(0,t) = 0$, $u(1,t) = 0$.

14. $u_{tt}(x,t) = u_{xx}(x,t)$, $0 < x < 1$, $t > 0$, with initial conditions $u(x,0) = \sin \pi x$, $u_t(x,0) = -\sin \pi x$ and boundary conditions $u(0,t) = u(1,t) = 0$.

In Problems 15–19, use a Fourier transform to solve the given boundary value problem on the indicated interval, subject to the given conditions.

15. $tu_x(x,t) + u_t(x,t) = 0$, $-\infty < x < \infty$, $t > 0$ subject to the initial condition $u(x,0) = f(x)$.

16. $u_x(x,t) + 3u_t(x,t) = 0$, $-\infty < x < \infty$, $t > 0$ subject to the initial condition $u(x,0) = f(x)$.

17. $t^2 u_x(x,t) - u_t(x,t) = 0$, $-\infty < x < \infty$, $t > 0$ subject to the initial condition $u(x,0) = 3 \cos x$.

18. $u_{tt}(x,t) + u_t(x,t) = -u(x,t)$, $-\infty < x < \infty$, $t > 0$ subject to the initial conditions $u(x,0) = f(x)$ and $u_t(x,0) = g(x)$.

19. $u_{xt}(x,t) = u_{xx}(x,t)$, $-\infty < x < \infty$, $t > 0$ subject to the initial condition $u(x,0) = \sqrt{\dfrac{\pi}{2}}\, e^{-|x|}$.

In Problems 20–25, use one of the Fourier transforms to solve the indicated wave equation, subject to the given initial and boundary conditions. Find explicitly the inverse Fourier transform where it is possible.

20. $u_{tt}(x,t) = u_{xx}(x,t)$, $-\infty < x < \infty$, $t > 0$ subject to the initial conditions $u(x,0) = \dfrac{1}{1+x^2}$ and $u_t(x,0) = 0$.

21. $u_{tt}(x,t) = c^2 u_{xx}(x,t)$, $0 < x < \infty$, $t > 0$ subject to the initial conditions $u(x,0) = 0$, $u_t(x,0) = 0$ and the boundary condition $u_x(0,t) = f(t)$, $0 < t < \infty$.

    $\left[\text{Hint: Use the Fourier cosine transform.}\right]$

22. $u_{tt}(x,t) = c^2 u_{xx}(x,t) + f(x,t)$, $0 < x < \infty$, $t > 0$ subject to the initial conditions $u(x,0) = 0$, $u_t(x,0) = 0$ and the boundary condition $u(0,t) = 0$, $0 < t < \infty$.

    $\left[\text{Hint: Use the Fourier sine transform.}\right]$

23. $u_{tt}(x,t) = c^2 u_{xx}(x,t) + f(x,t)$, $0 < x < \infty$, $t > 0$ subject to the initial conditions $u(x,0) = 0$, $u_t(x,0) = 0$ and the condition $u_x(0,t) = 0$, $0 < t < \infty$.

    $\left[\text{Hint: Use the Fourier cosine transform.}\right]$

24. $u_{tt}(x,t) = c^2 u_{xx}(x,t) + 2a u_t(x,t) + b^2 u(x,t)$, $0 < x < \infty$, $t > 0$, subject to the initial conditions $u(x,0) = u_t(x,0) = 0$, $0 < x < \infty$ and the boundary conditions $u(0,t) = f(t)$, $t > 0$ and $\lim\limits_{x\to\infty} |\,u(x,t)\,| < \infty$, $t > 0$.
    $\left[\text{Hint: Use the Fourier sine transform.}\right]$

25. $u_{tt}(x,t) = c^2 u_{xx}(x,t)$, $0 < x < \infty$, $t > 0$, subject to the initial conditions $u(x,0) = u_t(x,0) = 0$, $0 < x < \infty$, and the boundary conditions $u_x(0,t) - ku(0,t) = f(t)$, $\lim\limits_{x\to\infty} |\,u(x,t)\,| = 0$, $t > 0$.
    $\left[\text{Hint: Use the Fourier cosine transform.}\right]$

5.6 Projects Using Mathematica.

In this section we will see how Mathematica can be used to solve several problems involving the wave equation. In particular, we will develop several Mathematica notebooks which automate the computation, the solution of this equation. For a brief overview of the computer software Mathematica consult Appendix H.

Project 5.6.1. Using d'Alembert's formula solve the wave equation

$$\begin{cases} u_{tt}(x,t) = u_{xx}(x,t), & -\infty < x < \infty, \ t > 0, \\ u(x,0) = \dfrac{1}{x^2+1}, & u_t(x,0) = e^x. \end{cases}$$

All calculation should be done using Mathematica. Display the plot of $u(x,t)$ at several time moments t.

Solution. Use the DSolve command to find the general solution of the given partial differential equation.

$In[1] := DSolve[D[u[x,t],\{t,2\}] == DS[u[x,t],\{x,2\}], u[x,t],\{t,x\}];$

$Out[2] = u[x,t] -> C[1][-t+x] + C[2][t+x];$

Next, define the functions $f(x)$ and $g(x)$ and apply the initial conditions.

$In[3] := f[x\_] := 1/(1+x^2);$

$In[4] := g[x\_] := xExp[-x^2];$

$In[5] : In[5] := C[1][x] + C[2][x] == f[x];$

$In[6] := -C[1]'[x] + C[2]'[x] == g[x]:$

$In[7] := Solve[C[1]'[x] + C[2]'[x] == f'[x]\&\&$
$-C[1]'[x] + C[2]'[x] == g[x], \{[C[1]'][x], C[2]'[x]\}]$

$Out[7] = \{\{C[1]'[x] -> \frac{1}{2}(-g[x] + f'[x]), C[2]'[x] -> \frac{1}{2}(g[x] + f'[x])\}\}$

Next, find $C1[x]$ and $C2[x]$.

$In[8] := DSolve[C[1]'[x] == (1/2)(-g[x] + f'[x]), C[1][x], x];$

$In[8] := DSolve[C[2]'[x] == (1/2)(g[x] + f'[x]), C[2][x], x];$

The obtained $C[1][x]$ and $C2[x]$ substitute in the solution and obtain

$In[9] := u[x,t] = (f[x-t] + f[x+t])/2 + (1/2)Integrate[g[u], u, x-t, x+t]$

$Out[10] = \dfrac{f[-t+x] + f[t+x] + f[t+x]}{2} + \dfrac{1}{2}\int\limits_{x-t}^{x+t} g[u]\, du.$

Plots of $u[x,t]$ at several time instances t are displayed in Figure 5.6.1.

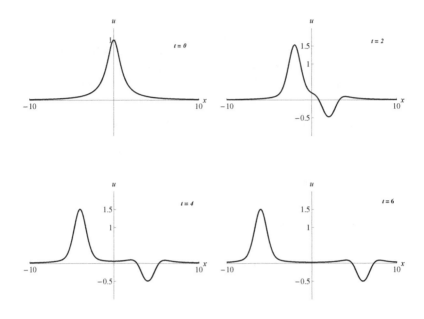

Figure 5.6.1

Project 5.6.2. Use the Laplace transform to solve the wave equation

$$\begin{cases} u_{tt}(x,t) = u_{xx}(x,t), & 0 < x < 1, \ t > 0, \\ u(x,0) = \sin \pi x, & u_t(x,0) = -\sin \pi x, \ 0 < x < 1, \\ u(0,t) = \sin t, & u_t(0,t) = 0, \ t > 0. \end{cases}$$

Do all calculation using Mathematica. Display the plots of $u(x,t)$ at several time instances t.

Solution. First we define the wave expression:

$In[1] := wave = D[u[x,t], \{t,2\}] - c^2 D[u[x,t], \{x,2\}];$

$Out[2] = u^{(0,2)}[x,t] - c^2 u^{(2,0)}[x,t];$

Next, take the Laplace transform

$In[3] := LaplaceTransform[wave,t,s];$

$Out[4] = s^2 LaplaceTransform[u[x,t],t,s]$
$-c^2 LaplaceTransform[[u^{(2,0)}[x,t],t,s] - s\, u[x,0] - u^{(0,1)}[x,0];$

Further, define

$In[5] := U[x_-,s_-] := LaplaceTransform[u[x,t],t,s];$

Next, define the initial conditions

$In[6] := f[x\_] = Sin[Pi\, x];$
$In[7] := g[x\_] = -Sin[Pi\, x];$

Using the initial conditions define the expression

$In[8] := eq = s^2\, U[x, s] - c^2 D[U[x, s], \{x, 2\}] - sf[x] - g[x];$

Next find the general solution of

$$\frac{d^2 U(x, s)}{dx^2} - s^2 U(x, s) + sf(x) + g(x) = 0$$

$In[9] := gensol = DSolve[eq == 0, U[x, s], x]$

$Out[10] := \{\{U[x, s]-> e^{\frac{sx}{c}} C[1] + e^{-\frac{sx}{c}} C[2] + \frac{-Sin[\pi\, x] + sSin[\pi x]}{c^2 \pi^2 + s^2}\}\}$

$In[11] := Solution = U[x, s]/.gensol[[1]];$

Define the boundary condition

$In[12] := b[t\_] = Sin[t];$

From the boundary conditions and the bounded property of the Laplace transform as $s \to \infty$ we find the constants $C[1]$ and $C[2]$:

$In[13] := BC = LaplaceTransform[b[t], t, s];$

$Out[14] = \frac{1}{1+s^2}$

$In[15] := BC = (Solution/.\{x \to 0, C[1]-> 0\})$

$Out[16] = \frac{1}{1+s^2} = C[2]$

$In[17] := FinalExpression = Solution/.\{x \to 0, C[1] \to 0\}$

$Out[18] = \frac{e^{-\frac{sx}{c}}}{1+s^2} + \frac{-Sin[\pi x] + s\, Sin[\pi x]}{c^2 \pi^2 + s^2};$

Find the inverse Laplace of the last expression:

$In[19] := InverseLaplace[\frac{e^{-\frac{sx}{c}}}{1+s^2} + \frac{-Sin[\pi\, x] + s\, Sin[\pi\, x]}{c^2 \pi^2 + s^2}, s, t]$

$Out[20] = \frac{\left(c\pi Cos[c\,\pi\, t] - Sin[c\,\pi\, t]\right) Sin[\pi\, x]}{c\pi} + HeavisideTheta[t - \frac{x}{c}] Sin[t - \frac{x}{c}].$

Plots of $u[x,t]$ at several time instances t are displayed in Figure 5.6.2.

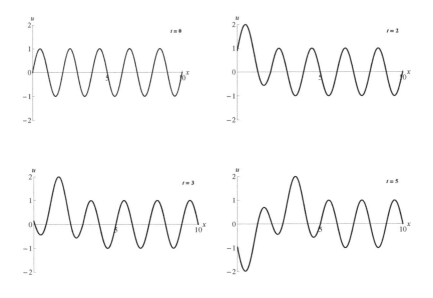

Figure 5.6.2

Project 5.6.3. In Example 5.4.1 we solved the circular membrane problem

$$u_{tt}(r,\varphi,t) = c^2\Big(u_{rr} + \frac{1}{r}u_r + \frac{1}{r^2}u_{\varphi\varphi}\Big), \quad 0 < r \le a,\ 0 \le \varphi < 2\pi,\ t > 0$$

for $a = c = 1$, subject to the initial conditions

$$\begin{cases} u(r,\varphi,0) = f(r,\varphi) = (1-r^2)r^2 \sin 2\varphi, & 0 < r \le a,\ 0 \le \varphi < 2\pi, \\ u_t(r,\varphi,0) = 0, & 0 < r \le a,\ 0 \le \varphi < 2\pi \end{cases}$$

and the boundary condition

$$u(a,\varphi) = 0, \quad 0 \le \varphi < 2\pi.$$

In Example 5.4.2 we solved the same wave equation for $a = 2$, $c = 1$, subject to the boundary conditions

$$u(r,\varphi,0) = f(r,\varphi) = (4-r^2)r \sin \varphi, \quad u_t(r,\varphi,0) = 1,\ 0 \le r \le 1,\ 0 \le \varphi < 2\pi.$$

Mathematica generates the plots for the solutions $u(r,\varphi)$ of these problems, displayed in Figure 5.4.1 and Figure 5.4.2.

Solution. The solution $v(r, \varphi, t)$ of the problem found in Example 5.4.1 is given by

$$v(r, \varphi, t) = 24 \sum_{n=1}^{\infty} \frac{1}{z_{2n}^3 J_3(z_{2n})} J_2(z_{2n}r) \sin 2\varphi \cos (z_{2n}t)$$

and the solution $w(r, \varphi, t)$ of the problem found in Example 5.4.2 is given by

$$w(r, \varphi, t) = 128 \sum_{n=1}^{\infty} \frac{J_1\left(\frac{z_{1n}}{2}r\right)}{z_{1n}^3 J_2(z_{1n})} \sin \varphi \cos \left(\frac{z_{1n}}{2}t\right) + 4 \sum_{n=1}^{\infty} \frac{J_0\left(\frac{z_{0n}}{2}r\right)}{z_{0n}^2 J_1(z_{0n})} \sin \left(\frac{z_{0n}}{2}t\right)$$

In the above sums, z_{mn} $(m = 0, 1, 2)$ denotes the nth zero of the Bessel function $J_m(\cdot)$.

First generate a list of zeroes of the mth Bessel function.

$In[1] := BZ[m\_, n\_] = N[BesselJZero[m, k, 0]];$
(the nth positive zero of $J_m(\cdot)$)

Next define the Nth partial sums of the solutions $u_1(r, \varphi, t)$ and $u_2(r, \varphi, t)$.

$In[2] := u_1[r\_, \varphi\_, t\_, N\_] := 24 \, Sum[(BesselJ[2, Bz[2, n] \, r])/((Bz[2, n])^3$
$BesselJ[3, Bz[2, n]]) \, Sin[2 \, \varphi]] \, Cos[Bz[2, n] \, t], \{n, 1, N\}];$

$In[3] := u_2[r\_, \varphi\_, t\_, N\_] := 128 \, Sum[(BesselJ[1, Bz[1, n]r/2])$
$/((Bz[1, n])^3 \, BesselJ[2, Bz[1, n]]) \, Sin[\varphi]] \, Cos[Bz[1, n] \, t/2], \{n, 1, N\}]$
$+4 \, Sum[(BesselJ[0, Bz[0, n] \, r/2])/((Bz[0, n])^2 \, BesselJ[1, Bz[0, n]])$
$Sin[Bz[0, n] \, t/2], \{n, 1, N\}];$

$In[4] := ParametricPlot3D[\{rCos[\varphi], rSin[\varphi], u_1[r, \varphi, 1, 3]\}, \{\varphi, 0, 2 \, Pi]\},$
$\{r, 0, 1\}, Ticks- > \{\{-1, 0, 1\}, \{-1, 0, 1\}, \{-.26, 0, 0.26\}\}, \text{RegionFunction}$
$- > Function[\{r, \varphi, u\}, r <= 1], BoxRatios- > Automatic,$
$AxesLabel- > \{x, y, u\}];$

$In[5] := ParametricPlot3D[\{rCos[\varphi], rSin[\varphi], u_2[r, \varphi, 1, 3]\}, \{\varphi, 0, 2 \, Pi]\},$
$\{r, 0, 1\}, Ticks- > \{\{-1, 0, 1\}, \{-1, 0, 1\}, \{-.26, 0, 0.26\}\}, \text{RegionFunction}$
$- > Function[\{r, \varphi, u\}, r <= 1], BoxRatios- > Automatic,$
$AxesLabel- > \{x, y, u\}];$

CHAPTER 6

THE HEAT EQUATION

The purpose of this chapter is to study the one dimensional *heat equation*

$$u_t(x,t) = c^2 u_{xx}(x,t) + F(x,t), \ x \in (a,b) \subseteq \mathbb{R}, \ t > 0,$$

also known as the *diffusion equation*, and its higher dimensional version

$$u_t(\mathbf{x},t) = a^2 \Delta u(\mathbf{x},t) + F(\mathbf{x},t), \ \mathbf{x} \in \Omega \subseteq \mathbb{R}^n, \ n = 2,3.$$

In the first section we will find the *fundamental solution* of the initial value problem for the heat equation. In the next sections we will apply the *separation of variables method* for constructing the solution of the one and higher dimensional heat equation in rectangular, polar and spherical coordinates. In the last section of this chapter we will apply the Laplace and Fourier transforms to solve the heat equation.

6.1 The Fundamental Solution of the Heat Equation.

In this section we will solve the homogeneous and nonhomogeneous heat equation on the whole real line and on the half line.

1^0. *The Homogeneous Heat Equation on* $\mathbb{R}$. Consider the following homogeneous *Cauchy problem*.

(6.1.1) $\qquad u_t(x,t) = c^2 u_{xx}(x,t), \quad -\infty < x < \infty, \ t > 0$

(6.1.2) $\qquad u(x,0) = f(x), \quad -\infty < x < \infty.$

First we will show that the Cauchy problem (6.1.1), (6.1.2) has a unique solution under the assumption that $u \in C^2(\mathbb{R} \times [0,\infty))$ and

(6.1.3) $\qquad\qquad\qquad \lim_{|x| \to \infty} u_x(x,t) = 0, \quad t > 0.$

As for the wave equation we define the notion of *energy* $E(t)$ of the heat equation (6.1.1) by

(6.1.4) $$E(t) = \frac{1}{2} \int\limits_{-\infty}^{\infty} u^2(x,t)\, dx.$$

326

Let $v(x,t)$ and $w(x,t)$ be two solutions of the problem (6.1.1), (6.1.2) and assume that both v and w satisfy the boundary condition (6.1.3). If $u = v - w$, then u satisfies the heat equation (6.1.1) and the initial and boundary conditions

$$\begin{cases} u(x,0) = 0, & -\infty < x < \infty \\ \lim_{x \to \pm\infty} u_x(x,t) = 0, & t > 0. \end{cases}$$

From (6.1.4), using the heat equation (6.1.1) and the above boundary and initial conditions, by the integration by parts formula we obtain that

$$E'(t) = \int_{-\infty}^{\infty} u(x,t)u_t(x,t)\,dx = c^2 \int_{-\infty}^{\infty} u_{xx}(x,t)u(x,t)\,dx$$

$$= c^2 \int_{-\infty}^{\infty} u(x,t)\frac{\partial}{\partial x}\big(u_x(x,t)\big)\,dx = c^2 u(x,t)u_x(x,t)\Big|_{x=-\infty}^{x=\infty}$$

$$- c^2 \int_{-\infty}^{\infty} u_x^2(x,t)\,dx = -c^2 \int_{-\infty}^{\infty} u_x^2(x,t)\,dx \le 0.$$

Thus, $E(t)$ is a decreasing function. Therefore, since $E(t) \ge 0$ and $E(0) = 0$ we have $E(t) = 0$ for every $t > 0$. Hence, $u(x,t) = 0$ for every $t > 0$ and so $v(x,t) = w(x,t)$.   ∎

Now we will solve the heat equation (6.1.1). We look for a solution $G(x,t)$ of the form

$$G(x,t) = \frac{1}{\sqrt{t}} g\left(\frac{x}{c\sqrt{t}}\right)$$

for some function g of a single variable. If we introduce the new variable ξ by

$$\xi = \frac{x}{c\sqrt{t}},$$

then by the chain rule, the heat equation (6.1.1) is transformed into the ordinary differential equation

$$2cg''(\xi) + \xi g'(\xi) + g(\xi) = 0.$$

This equation can be written in the form

$$2cg''(\xi) + \big(\xi g(\xi)\big)' = 0.$$

One solution of the last equation is

$$g(\xi) = e^{-\frac{\xi^2}{4c}}.$$

Therefore, one solution of the heat equation (6.1.1) is given by

$$(6.1.5) \qquad\qquad G(x,t) = \frac{1}{\sqrt{4\pi ct}}\, e^{-\frac{x^2}{4ct}}.$$

The function $G(x,t)$, defined by (6.1.5), is called the *fundamental solution* or *Green function* of the heat equation (6.1.1).

Some properties of the function $G(x,t)$ are given by the following theorem.

Theorem 6.1.1. *The fundamental solution $G(x,t)$ satisfies the following properties.*

(a) $G(x,t) > 0$ *for all $x \in \mathbb{R}$ and all $t > 0$.*

(b) $G(x,t)$ *is infinitely differentiable with respect to x and t.*

(c) $G(x-a,t)$ *is a solution of the heat equation (6.1.1) for every a.*

(d) $G(\sqrt{k}\,x, kt)$ *is a solution of the heat equation (6.1.1) for every $k > 0$.*

(e) $\int\limits_{-\infty}^{\infty} G(x,t)\,dx = 1$ *for every $t > 0$.*

(f) $u(x,t) = G(x,t) * f(x) = \int\limits_{-\infty}^{\infty} G(x-y,t)f(y)\,dy$ *is the solution of the Cauchy problem (6.1.1), (6.1.2) for every bounded and continuous function f on $\mathbb{R}$.*

Proof. Properties (a) and (b) are obvious. Property (d) can be checked by direct differentiation. For property (e) we have

$$\int\limits_{-\infty}^{\infty} G(x,t)\,dx = \frac{1}{\sqrt{4\pi ct}} \int\limits_{-\infty}^{\infty} e^{-\frac{x^2}{4ct}}\,dx \qquad (\text{change of variable } x = \sqrt{4ct}\,y)$$

$$= \frac{1}{\sqrt{\pi}} \int\limits_{-\infty}^{\infty} e^{-y^2}\,dy = \frac{1}{\sqrt{\pi}}\sqrt{\pi} = 1.$$

To prove (f) let

$$u(x,t) = G(x,t) * f(x) = \frac{1}{\sqrt{4\pi ct}} \int\limits_{-\infty}^{\infty} e^{-\frac{(x-y)^2}{4ct}} f(y)\,dy.$$

By direct calculation of the partial derivatives we can check (see Exercise 1 of this section) that the above function $u(x,t)$ satisfies the heat equation (6.1.1). We omit the proof of the fact that the function $u(x,t)$ satisfies the initial condition $u(x,0) = f(x)$. We cannot simply substitute $t = 0$ directly in the definition of $u(x,t)$ since the kernel function $G(x,t)$ has singularity (is not defined) at $t = 0$). For the proof of this omitted fact the reader is referred to the book by G. B. Folland [6].   ∎

Example 6.1.1. Solve the Cauchy problem.

$$\begin{cases} u_t(x,t) = \dfrac{1}{16} u_{xx}(x,t), & -\infty < x < \infty, \ t > 0 \\[2mm] u(x,0) = e^{-x^2}, & -\infty < x < \infty. \end{cases}$$

Solution. From Theorem 6.1.1 it follows that the solution of the given problem is

$$u(x,t) = G(x,t) * f(x) = \frac{1}{\sqrt{\pi t}} \int_{-\infty}^{\infty} e^{-\frac{s^2}{t}} e^{-(x-s)^2} \, ds$$

$$= \frac{e^{-x^2}}{\sqrt{\pi t}} \int_{-\infty}^{\infty} e^{-\frac{(t+1)s^2}{t} + 2xs} \, ds = \frac{e^{-x^2}}{\sqrt{\pi t}} \int_{-\infty}^{\infty} e^{-\left(\sqrt{\frac{t+1}{t}}\, s - \sqrt{\frac{t}{t+1}}\, x\right)^2 + \frac{tx^2}{t+1}} \, ds$$

$$= \frac{e^{-x^2}}{\sqrt{\pi t}} \, e^{\frac{tx^2}{t+1}} \int_{-\infty}^{\infty} e^{-\left(\sqrt{\frac{t+1}{t}}\, s - \sqrt{\frac{t}{t+1}}\, x\right)^2} \, ds = \frac{e^{-\frac{x^2}{1+t}}}{\sqrt{1+t}}.$$

The plots of the heat distribution at several time instances t are displayed in Figure 6.1.1.

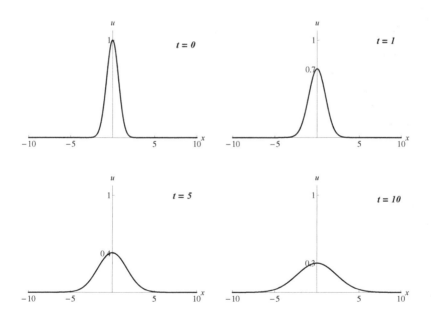

Figure 6.1.1

2^0. *Nonhomogeneous Heat Equation on* $\mathbb{R}$. Consider the nonhomogeneous problem

(6.1.6)
$$\begin{cases} u_t(x,t) = c^2 u_{xx}(x,t) + F(x,t), & -\infty < x < \infty, \ t > 0 \\ u(x,0) = f(x), & -\infty < x < \infty. \end{cases}$$

As in the wave equation, the solution $u(x,t)$ of Problem (6.1.6) is given by

$$u(x,t) = v(x,t) + w(x,t),$$

where $v(x,t)$ and $w(x,t)$ are the solutions of the following two problems.

(6.1.7)
$$\begin{cases} v_t(x,t) = c^2 u_{xx}(x,t), & -\infty < x < \infty, \ t > 0 \\ v(x,0) = f(x), & -\infty < x < \infty \end{cases}$$

and

(6.1.8)
$$\begin{cases} w_t(x,t) = c^2 w_{xx}(x,t) + F(x,t), & -\infty < x < \infty, \ t > 0 \\ w(x,0) = 0, & -\infty < x < \infty. \end{cases}$$

From Case 1^0, the solution of Problem (6.1.7) is given by

$$v(x,t) = \int_{-\infty}^{\infty} G(x-s,t) f(s) \, ds,$$

where $G(x,t)$ is the Green function, given by (6.1.5).

We will "show" that the solution $w(x,t)$ of problem (6.1.8) is given by

(6.1.9)
$$w(x,t) = \int_0^t \int_{-\infty}^{\infty} G(x-s, t-s) F(s,\tau) \, ds \, d\tau.$$

First, let us recall the following result from Chapter 2. If $g(x)$ is a continuous function on $\mathbb{R}$ which vanishes outside a bounded interval, then

(6.1.10)
$$\int_{-\infty}^{\infty} g(x) \delta_a(x) \, dx = g(a),$$

where $\delta_a(x)$ is the Dirac impulse function concentrated at the point $x = a$.

If we ignore the question about differentiation under the sign of integration, then from (6.1.9) it follows that

(6.1.11)
$$w_t(x,t) = \int_0^t \int_{-\infty}^{\infty} G_t(x-s, t-s) F(s,\tau) \, ds \, d\tau$$
$$+ \lim_{\tau \to t} \int_{-\infty}^{\infty} G(x-s, t-s) F(s,\tau) \, ds.$$

For the first term in Equation (6.1.11), using the fact that the Green function $G(x - s, t - s)$ satisfies

$$G_t((x - s, t - s) = c^2 G_{xx}(x - s, t - s),$$

we have

$$\int_0^t \int_{-\infty}^\infty G_t(x - s, t - s)F(s, \tau)ds\, d\tau$$

$$= c^2 \frac{\partial^2}{\partial x^2} \left(\int_0^t \int_{-\infty}^\infty G(x - s, t - s)F(s, \tau)\, ds\, d\tau \right)$$

$$= c^2 w_{xx}(x, t).$$

For the second term in (6.1.11), from (6.1.10) and the fact that

$$\lim_{t \to 0} G(x, t) = \delta_0(x),$$

we have

$$\lim_{\tau \to t} \int_{-\infty}^\infty G(x - s, t - s)F(s, \tau)\, ds = F(x, t).$$

Therefore, Equation (6.1.11) becomes

$$w_t(x, t) = a^2 w_{xx}(x, t) + F(x, t).$$

Thus, the function $w(x, t)$, defined by (6.1.9) satisfies the heat equation in (6.1.8). It remains to show that $w(x, t)$ satisfies the initial condition in (6.1.8). Ignoring some details we have

$$\lim_{t \to 0} w(x, t) = \lim_{t \to 0} \int_0^t \int_{-\infty}^\infty G(x - s, t - s)F(s, \tau)\, ds\, d\tau = 0. \qquad \blacksquare$$

3^0. *Homogeneous Heat Equation on* $(0, \infty)$. Consider the homogeneous problem

(6.1.12)
$$\begin{cases} u_t(x, t) = c^2 u_{xx}(x, t), & 0 < x < \infty, \ t > 0 \\ u(x, 0) = f(x), & 0 < x < \infty \\ u(0, t) = 0, & t > 0. \end{cases}$$

As in d'Alembert's method for the wave equation, we will convert the initial boundary value problem (6.1.12) into a boundary value problem on the whole real line $(-\infty, \infty)$. In d'Alembert's method for the wave equation we have used either the odd or even extension of the function $u(x,t)$. For the heat equation let us work with an odd extension. Let $\widetilde{f}(x)$ be the odd extension of the initial function $f(x)$.

Now consider the following initial value problem on $(-\infty, \infty)$.

(6.1.13)
$$\begin{cases} v_t(x,t) = c^2 v_{xx}(x,t), & -\infty < x < \infty, \ t > 0 \\ v(x,0) = \widetilde{f}(x), & -\infty < x < \infty. \end{cases}$$

From case 1^0 it follows that the solution $v(x,t)$ of problem (6.1.13) is given by

$$v(x,t) = \int_{-\infty}^{\infty} G(x-s)\widetilde{f}(s)\,ds.$$

Since $\widetilde{f}$ is an odd function, the function $v(x,t)$ is also an odd function with respect to x, i.e., $v(-x,t) = -v(x,t)$ for all $x \in \mathbb{R}$ and all $t > 0$ (see Exercise 5 of this section). Therefore, $v(0,t) = 0$ and so $v(x,t)$ satisfies the same boundary condition as the function $u(x,t)$. Now, since problems (6.1.12) and (6.1.13), coupled with the condition $v(0,t) = 0$, have unique solutions, for $x > 0$ we have

$$u(x,t) = v(x,t) = \int_{-\infty}^{\infty} G(x-s)\widetilde{f}(s)\,ds$$

$$= \frac{1}{\sqrt{4c\pi t}} \int_{-\infty}^{0} e^{-\frac{(x-s)^2}{4c\pi t}} \widetilde{f}(s)\,ds + \frac{1}{\sqrt{4c\pi t}} \int_{0}^{\infty} e^{-\frac{(x-s)^2}{4c\pi t}} \widetilde{f}(s)\,ds$$

$$= \frac{1}{\sqrt{4c\pi t}} \int_{\infty}^{0} e^{-\frac{(x+\tau)^2}{4c\pi t}} \widetilde{f}(-\tau)\,d\tau + \frac{1}{\sqrt{4c\pi t}} \int_{0}^{\infty} e^{-\frac{(x-s)^2}{4c\pi t}} f(s)\,ds$$

$$= \frac{1}{\sqrt{4c\pi t}} \int_{0}^{\infty} \left(e^{-\frac{(x-s)^2}{4c\pi t}} + e^{-\frac{(x+s)^2}{4c\pi t}} \right) f(s)\,ds.$$

Remark. The heat equation on a bounded interval will be solved in the next section using the separation of variables method and in the following section of this chapter by the Laplace and Fourier transforms.

Exercises for Section 6.1.

1. Complete the proof of Theorem 6.1.1.

2. Show that the the fundamental solution $G(x, t)$ satisfies the following.

    (a) $G(x, t) \to +\infty$ as $t \to 0+$ with $x = 0$.

    (b) $G(x, t) \to 0$ as $t \to 0+$ with $x \neq 0$.

    (c) $G(x, t) \to 0$ as $t \to \infty$.

3. Solve the heat equation (6.1.1) subject to the initial condition

$$u(x, 0) = f(x) = \begin{cases} 0, & x < 0 \\ 10 - x, & x \geq 0. \end{cases}$$

4. Solve the heat equation

$$\begin{cases} u_t(x, t) = c^2 u_{xx}(x, t) + u(x, t), & -\infty < x < \infty, \ t > 0, \\ u(x, 0) = f(x), & -\infty < x < \infty. \end{cases}$$

5. Let $u(x, t)$ be the solution of the Cauchy problem

$$\begin{cases} u_t(x, t) = c^2 u_{xx}(x, t), & \infty < x < \infty, \ t > 0, \\ u(x, 0) = f(x), & -\infty < x < \infty. \end{cases}$$

    Show that

    (a) If $f(x)$ is an odd function, then $u(-x, t) = -u(x, t)$ for all x and all $t > 0$. Conclude that $u(0, t) = 0$.

    (b) If $f(x)$ is an even function, then $u(-x, t) = u(x, t)$ for all x and all $t > 0$.

6.2 Separation of Variables Method for the Heat Equation.

In this section we will discuss the *separation of variables method* for solving the one dimensional heat equation. We will consider homogeneous and nonhomogeneous heat equations with homogeneous and nonhomogeneous boundary conditions on a bounded interval. This method can be applied to the heat equation in higher dimensions and it can be used to solve some other partial differential equations, as we will see in the next sections of this chapter and the following chapters.

1^0. *Homogeneous Heat Equation with Homogeneous Boundary Conditions.* Consider the homogeneous heat equation

$$(6.2.1) \qquad u_t(x,t) = c^2 u_{xx}(x,t), \quad 0 < x < l, \ t > 0,$$

which satisfies the initial condition

$$(6.2.2) \qquad u(x,0) = f(x), \quad 0 < x < l$$

and the homogeneous Dirichlet boundary conditions

$$(6.2.3) \qquad u(0,t) = u(l,t) = 0.$$

We will find a nontrivial solution $u(x,t)$ of the above initial boundary value problem of the form

$$(6.2.4) \qquad u(x,t) = X(x)T(t),$$

where $X(x)$ and $T(t)$ are functions of single variables x and t, respectively.

Differentiating (6.2.4) with respect to x and t and substituting the partial derivatives in Equation (6.2.1) we obtain

$$(6.2.5) \qquad \frac{X''(x)}{X(x)} = \frac{1}{c^2}\frac{T'(t)}{T(t)}.$$

Equation (6.2.5) holds identically for every $0 < x < l$ and every $t > 0$. Since x and t are independent variables, Equation (6.2.5) is possible only if each function on both sides is equal to the same constant λ:

$$\frac{X''(x)}{X(x)} = \frac{1}{c^2}\frac{T'(t)}{T(t)} = \lambda$$

From the last equations we obtain two ordinary differential equations.

$$(6.2.6) \qquad X''(x) - \lambda X(x) = 0,$$

and

$$(6.2.7) \qquad T'(t) - c^2 \lambda T(t) = 0.$$

From the boundary conditions

$$u(0,t) = u(l,t) = 0, \ t > 0$$

it follows that

$$X(0)T(t) = X(l)T(t) = 0, \ t > 0.$$

From the last equations we have

(6.2.8) $$X(0) = X(l) = 0,$$

since $T(t) = 0$ for every $t > 0$ would imply $u(x,t) \equiv 0$. Solving the eigenvalue problem (6.2.6), (6.2.8), as in Chapter 3, it follows that the eigenvalues λ_n and the corresponding eigenfunctions $X_n(x)$ are given by

$$\lambda_n = -\left(\frac{n\pi}{l}\right)^2, \quad X_n(x) = \sin\frac{n\pi x}{l}, \ n = 1, 2, \ldots.$$

The solution of the differential equation (6.2.7) corresponding to the above found λ_n is given by

$$T_n(t) = a_n e^{-\frac{n^2\pi^2 c^2}{l^2} t}, \quad n = 1, 2, \ldots$$

where a_n are constants which will be determined. Therefore we obtain a sequence of functions

$$u_n(x,t) = a_n e^{-\frac{n^2\pi^2 c^2}{l^2} t} \sin\frac{n\pi x}{l}, \quad n = 1, 2, \ldots,$$

each of which satisfies the heat equation (6.2.1) and the boundary conditions (6.2.3). Since the heat equation and the boundary conditions are linear and homogeneous, a function $u(x,t)$ of the form

(6.2.9) $$u(x,t) = \sum_{n=1}^{\infty} u_n(x,t) = \sum_{n=1}^{\infty} a_n e^{-\frac{n^2\pi^2 c^2}{l^2} t} \sin\frac{n\pi x}{l}$$

also will satisfy the heat equation and the boundary conditions. If we assume that the above series is convergent, from (6.2.9) and the initial condition (6.2.2) we obtain

(6.2.10) $$f(x) = u(x,0) = \sum_{n=1}^{\infty} a_n \sin\frac{n\pi x}{l}, \quad 0 < x < l.$$

Using the orthogonality of sine functions, from (6.2.10) we obtain

(6.2.11) $$a_n = \frac{2}{l} \int_0^l f(x) \sin\frac{n\pi x}{l} \, dx.$$

A formal justification of the above solution of the heat equation is given by the following theorem, stated without a proof.

Theorem 6.2.1. *Suppose that the function f is continuous and its derivative f' is a piecewise continuous function on $[0, l]$. If $f(0) = f(l) = 0$, then the function $u(x, t)$ given in (6.2.8), where the coefficients a_n are given by (6.2.10), is the unique solution of the problem defined by (6.2.1), (6.2.2), (6.2.3).*

Example 6.2.1. Solve the initial boundary value problem

$$\begin{cases} u_t(x,t) = c^2 u_{xx}(x,t), & 0 < x < \pi, \ t > 0, \\ u(x,0) = f(x) = x(x - \pi), & 0 < x < \pi, \\ u(0,t) = u(\pi,t) = 0, & t > 0. \end{cases}$$

Solution. For the coefficients a_n, applying Equation (6.2.11) and using the integration by parts formula we have

$$a_n = \frac{2}{\pi} \int_0^\pi f(x) \sin nx \, dx = \frac{2}{\pi} \int_0^\pi x(\pi - x) \sin nx \, dx$$

$$= 2 \int_0^\pi x \sin nx \, dx - \frac{2}{\pi} \int_0^\pi x^2 \sin nx \, dx = \frac{4}{\pi} \frac{1 - (-1)^n}{n^3}.$$

Therefore, the solution $u(x, t)$ of the given heat problem is

$$u(x,t) = \frac{8}{\pi} \sum_{n=1}^\infty \frac{1}{(2n-1)^3} e^{-(2n-1)^2 c^2 t} \sin (2n - 1)x.$$

Case 2^0. Nonhomogeneous Heat Equation. Homogeneous Boundary Conditions. Consider the initial boundary value problem

(6.2.12) $$\begin{cases} u_t(x,t) = c^2 u_{xx}(x,t) + F(x,t), & 0 < x < l, \ t > 0, \\ u(x,0) = f(x), & 0 < x < l \\ u(0,t) = u(l,t) = 0. \end{cases}$$

In order to find the solution of problem (6.2.12) we split the problem into the following two problems:

(6.2.13) $$\begin{cases} v_t(x,t) = c^2 u_{xx}(x,t), & 0 < x < l, \ t > 0, \\ v(x,0) = f(x), & 0 < x < l \\ v(0,t) = v(l,t) = 0, \end{cases}$$

and

$$
(6.2.14) \qquad
\begin{cases}
w_t(x,t) = c^2 w_{xx}(x,t) + F(x,t), & 0 < x < l,\ t > 0, \\
w(x,0) = 0, & 0 < x < l \\
w(0,t) = w(l,t) = 0.
\end{cases}
$$

Problem (6.2.13) was considered in *Case* 1^0 and it has been solved. Let $v(x,t)$ be its solution. If $w(x,t)$ is the solution of problem (6.2.14), then

$$
u(x,t) = v(x,t) + w(x,t)
$$

will be the solution of the given problem (6.2.12). So, the remaining problem to be solved is (6.2.14).

In order to solve problem (6.2.14) we proceed exactly as in the nonhomogeneous wave equation of Chapter 5. For each fixed $t > 0$, expand the nonhomogeneous term $F(x,t)$ in the Fourier sine series

$$
(6.2.15) \qquad F(x,t) = \sum_{n=1}^{\infty} F_n(t) \sin \frac{n\pi x}{l}, \quad 0 < x < l,
$$

where

$$
(6.2.16) \qquad F_n(t) = \frac{2}{l} \int_0^l F(\xi,t) \sin \frac{n\pi\xi}{l}\, d\xi, \quad n = 1, 2, \ldots.
$$

Next, again for each fixed $t > 0$, expand the unknown function $w(x,t)$ in the Fourier sine series

$$
(6.2.17) \qquad w(x,t) = \sum_{n=1}^{\infty} w_n(t) \sin \frac{n\pi x}{l}, \quad 0 < x < l,
$$

where

$$
(6.2.18) \qquad w_n(t) = \frac{2}{l} \int_0^l w(\xi,t) \sin \frac{n\pi\xi}{l}\, d\xi, \quad n = 1, 2, \ldots.
$$

From the initial condition $w(x,0) = 0$ it follows that

$$
(6.2.19) \qquad w_n(0) = 0.
$$

If we substitute (6.2.15) and (6.2.17) into the heat equation (6.2.14) and compare the Fourier coefficients we obtain

$$
(6.2.20) \qquad w_n'(t) + \frac{n^2 \pi^2 c^2}{l^2} w_n(t) = F_n(t).
$$

The solution of Equation (6.2.20) in view of the initial condition (6.2.19) is

$$
(6.2.21) \qquad w_n(t) = \int_0^t F_n(\tau) e^{-\frac{c^2 n^2 \pi^2}{l^2}(t-\tau)} \, d\tau.
$$

Therefore, from (6.2.15) and (6.2.16) the solution of problem (6.2.14) is given by

$$
(6.2.22) \qquad w(x,t) = \int_0^l \int_0^t G(x,\xi,t-\tau) F(\xi,\tau) \, d\xi \, d\tau,
$$

where

$$
(6.2.23) \qquad G(x,\xi,t-\tau) = \frac{2}{l} \sum_{n=1}^{\infty} e^{-\frac{c^2 n^2 \pi^2}{l^2}(t-\tau)} \sin \frac{n\pi}{l} x \sin \frac{n\pi}{l}\xi.
$$

Remark. The function

$$
(6.2.24) \qquad G(x,\xi,t) = \frac{2}{l} \sum_{n=1}^{\infty} e^{-\frac{c^2 n^2 \pi^2}{l^2} t} \sin \frac{n\pi}{l} x \sin \frac{n\pi}{l}\xi
$$

is called the *Green function* of the heat equation on the interval $[0,l]$.

Example 6.2.2. Solve the initial boundary value problem for the heat equation

$$
\begin{cases}
u_t(x,t) = c^2 u_{xx}(x,t) + xt, & 0 < x < \pi, \ t > 0, \\
u(x,0) = x(x - \pi), & 0 < x < \pi, \\
u(0,t) = u(\pi,t) = 0, & t > 0.
\end{cases}
$$

For $c = 1$ plot the heat distribution $u(x,t)$ at several time instances.

Solution. The corresponding homogeneous heat equation was solved in Example 6.2.1 and its solution is given by

$$
(6.2.25) \qquad v(x,t) = \frac{8}{\pi} \sum_{n=1}^{\infty} \frac{1}{(2n-1)^3} e^{-(2n-1)^2 c^2 t} \sin (2n-1)x.
$$

Now, we solve the nonhomogeneous heat equation with homogeneous boundary conditions:

$$
\begin{cases}
w_t(x,t) = c^2 w_{xx}(x,t) + xt, & 0 < x < \pi, \ t > 0, \\
w(x,0) = 0, & 0 < x < \pi \\
w(0,t) = w(\pi,t) = 0.
\end{cases}
$$

For each fixed $t > 0$, expand the function xt and the unknown function $w(x,t)$ in the Fourier sine series on the interval $(0, \pi)$:

$$xt = \sum_{n=1}^{\infty} F_n(t) \sin nx, \quad w(x,t) = \sum_{n=1}^{\infty} w_n(t) \sin nx,$$

where the coefficients $F_n(t)$ and $w_n(t)$ in these expansions are given by

$$F_n(t) = \frac{2}{\pi} \int_0^\pi x \sin nx \, dx = -\frac{2t \cos n\pi}{n},$$

$$w_n(t) = \int_0^t F_n(\tau) e^{-c^2 n^2 (t-\tau)} \, d\tau = -\frac{2 \cos n\pi}{n} \int_0^t \tau e^{-c^2 n^2 (t-\tau)} \, d\tau$$

$$= \frac{2 \cos n\pi}{c^4 n^5} \left[-1 + e^{-c^2 n^2 t} + c^2 n^2 t \right].$$

Thus,

$$(6.2.26) \qquad w(x,t) = \frac{2}{c^4} \sum_{n=1}^{\infty} \frac{\cos n\pi}{n^5} \left[-1 + e^{-c^2 n^2 t} + c^2 n^2 t \right] \sin nx,$$

and so, the solution $u(x,t)$ of the problem is given by $u(x,t) = v(x,t) + w(x,t)$, where v and w are given by (6.2.25) and (6.2.26), respectively.

For $c = 1$ the plots of $u(x,t)$ at 4 instances are displayed in Figure 6.2.1.

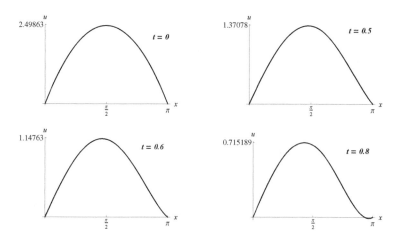

Figure 6.2.1

3^0 *Nonhomogeneous Heat Equation. Nonhomogeneous Dirichlet Boundary Conditions.* Consider the initial boundary value problem

(6.2.27)
$$\begin{cases} u_t(x,t) = c^2 u_{xx}(x,t) + F(x,t), & 0 < x < l, \ t > 0, \\ u(x,0) = f(x), & 0 < x < l \\ u(0,t) = \alpha(t), & u(l,t) = \beta(t) \ t \geq 0. \end{cases}$$

The solution of this problem can be found by superposition of the solution $v(x,t)$ of the homogeneous problem (6.2.12) (considered in *Case* 2^0) with the solution $w(x,t)$ of the problem

(6.2.28)
$$\begin{cases} w_t(x,t) = c^2 w_{xx}(x,t), & 0 < x < l, \ t > 0, \\ w(x,0) = 0, & 0 < x < l \\ w(0,t) = \alpha(t), & w(l,t) = \beta(t) \ . \end{cases}$$

If we introduce a new function $\widetilde{w}(x,t)$ by

$$\widetilde{w}(x,t) = w(x,t) - x\beta(t) + (x-l)\alpha(t),$$

then problem (6.2.28) is transformed into the problem

(6.2.29)
$$\begin{cases} \widetilde{w}_t(x,t) = c^2 \widetilde{w}_{xx}(x,t) + \widetilde{F}(x,t), & 0 < x < l, \ t > 0, \\ \widetilde{w}(0,t) = \widetilde{w}(l,t) = 0, & t > 0 \\ \widetilde{w}(x,0) = \widetilde{f}(x), & 0 < x < l, \end{cases}$$

where

$$\widetilde{F}(x,t) = (x-l)\alpha'(t) - x\beta'(t),$$
$$\widetilde{f}(x) = (x-l)\alpha(0) - x\beta(0).$$

Notice that problem (6.2.29) has homogeneous boundary conditions and it was considered in *Case* 1^0 and therefore we know how to solve it.

4^0. *Homogeneous Heat Equation. Neumann Boundary Conditions.* Consider the initial boundary value problem

(6.2.30)
$$\begin{cases} u_t(x,t) = c^2 u_{xx}(x,t), & 0 < x < l, \ t > 0, \\ u(x,0) = f(x), & 0 < x < l, \\ u_x(0,t) = u_x(l,t) = 0, & t > 0. \end{cases}$$

We will solve this problem by the separation of variables method. Let the solution $u(x,t)$ (not identically zero) of the above problem be of the form

(6.2.31)
$$u(x,t) = X(x)T(t),$$

where $X(x)$ and $T(t)$ are functions of single variables x and t, respectively.

Differentiating (6.2.31) with respect to x and t and substituting the partial derivatives in the heat equation in problem (6.2.30) we obtain that

$$\frac{X''(x)}{X(x)} = \frac{1}{c^2}\frac{T'(t)}{T(t)}.$$

The last equation is possible only if

$$\frac{X''(x)}{X(x)} = \frac{1}{c^2}\frac{T'(t)}{T(t)} = \lambda$$

for some constant λ. From the last equation we obtain the two ordinary differential equations

(6.2.32) $$X''(x) - \lambda X(x) = 0,$$

and

(6.2.33) $$T'(t) - c^2\lambda T(t) = 0.$$

From the boundary conditions

$$u_x(0,t) = u_x(l,t) = 0, \ t > 0$$

it follows that

$$X'(0)T(t) = X'(l)T(t) = 0, \ t > 0.$$

From the last equations we obtain

(6.2.34) $$X'(0) = X'(l) = 0.$$

Solving the eigenvalue problem (6.2.32), (6.2.34) (see Chapter 3) we obtain that its eigenvalues $\lambda = \lambda_n$ and the corresponding eigenfunctions $X_n(x)$ are

$$\lambda_0 = 0, \quad X_0(x) = 1, \ 0 < x < l$$

and

$$\lambda_n = -\left(\frac{n\pi}{l}\right)^2, \quad X_n(x) = \cos\frac{n\pi x}{l}, \ n = 0, 1, 2, \ldots, \ 0 < x < l.$$

The solution of the differential equation (6.2.33) corresponding to the above found λ_n is given by

$$T_n(t) = a_n e^{c^2\lambda_n t}, \quad n = 0, 1, 2, \ldots,$$

where a_n are constants which will be determined.

From (6.2.31) we obtain a sequence $\{u_n(x,t) = X_n(x)T_n(t),\ n = 0, 1, \ldots\}$ given by

$$u_n(x,t) = a_n e^{c^2 \lambda_n t} \cos \frac{n\pi x}{l}, \qquad n = 0, 1, 2, \ldots$$

each of which satisfies the heat equation and the Neumann boundary conditions involved in problem (6.2.30). Since the given heat equation and the boundary conditions are linear and homogeneous, a function $u(x,t)$ of the form

$$(6.2.35) \qquad u(x,t) = \sum_{n=0}^{\infty} u_n(x,t) = \sum_{n=0}^{\infty} a_n e^{c^2 \lambda_n t} \cos \frac{n\pi x}{l}$$

also will satisfy the heat equation and the boundary conditions. If we assume that the above series is convergent, from (6.2.35) and the initial condition in problem (6.2.30) we obtain

$$(6.2.36) \qquad f(x) = u(x,0) = \sum_{n=0}^{\infty} a_n \cos \frac{n\pi x}{l}, \qquad 0 < x < l.$$

Using the Fourier cosine series (from Chapter 1) for the function $f(x)$ or the fact, from Chapter 2, that the eigenfunctions

$$1, \ \cos \frac{2\pi x}{l}, \ \cos \frac{3\pi x}{l}, \ \ldots, \cos \frac{n\pi x}{l}, \ \ldots$$

are pairwise orthogonal on the interval $[0, l]$, from (6.2.36) we obtain that

$$(6.2.37) \qquad a_n = \frac{2}{l} \int_0^l f(x) \cos \frac{n\pi x}{l}\, dx \quad n = 0, 1, 2, \ldots$$

Remark. Problem (6.2.29) can be solved in a similar way to Example 6.2.2., except we use the Fourier cosine series instead of the Fourier sine series.

Example 6.2.3. Using the separation of variables method solve the problem

$$\begin{cases} u_t(x,t) = u_{xx}(x,t), & 0 < x < \pi, \ t > 0, \\ u(x,0) = \pi^2 - x^2, & 0 < x < \pi \\ u_x(0,t) = u_x(\pi,t) = 0, & t > 0. \end{cases}$$

Solution. Let the solution $u(x,t)$ of the problem be of the form

$$u(x,t) = X(x)T(t).$$

From the heat equation and the given boundary conditions we obtain the eigenvalue problem

$$(6.2.38) \qquad \begin{aligned} X''(x) - \lambda X(x) &= 0, \quad 0 < x < \pi \\ X'(0) = X'(\pi) &= 0, \end{aligned}$$

and the ordinary differential equation

(6.2.39) $$T'(t) - \lambda T(t) = 0, \quad t > 0.$$

Solving the eigenvalue problem (6.2.38) we obtain that its eigenvalues $\lambda = \lambda_n$ and the corresponding eigenfunctions $X_n(x)$ are

$$\lambda_0 = 0, \quad X_0(x) = 1, \ 0 < x < \pi$$

and

$$\lambda_n = -n^2, \quad X_n(x) = \cos nx, \ n = 1, 2, \ldots, \ 0 < x < \pi.$$

The solution of the differential equation (6.2.39) corresponding to the above found λ_n is given by

$$T_n(t) = a_n e^{-n^2 t}, \quad n = 0, 1, 2, \ldots.$$

Hence, the solution of the given problem will be of the form

$$u(x, t) = \sum_{n=0}^{\infty} a_n e^{-n^2 t} \cos nx, \quad 0 < x < \pi, \ t > 0,$$

where a_n are coefficients to be determined.

From the other initial condition $u(x, 0) = \pi^2 - x^2, 0 < x < \pi$, and the orthogonality property of the eigenfunctions

$$\{1, \cos x, \cos 2x, \ldots, \cos nx, \ldots\}$$

on the interval $[0, \pi]$ we obtain

$$a_0 = \frac{1}{\pi} \int_0^{\pi} (\pi^2 - x^2) \, dx = \frac{2\pi^2}{3}$$

and

$$a_n = \frac{1}{\int_0^{\pi} \cos^2 nx \, dx} \int_0^{\pi} (\pi^2 - x^2) \cos nx \, dx = -\frac{2}{\pi} \int_0^{\pi} x^2 \cos nx \, dx$$

$$= -\frac{2}{\pi} \cdot \frac{2\pi}{n^2} \cos n\pi = \frac{4}{n^2}(-1)^{n+1},$$

for $n = 1, 2, \ldots$. Therefore, the solution $u(x, t)$ of the problem is given by

$$u(x, t) = \frac{2\pi^2}{3} + \sum_{n=1}^{\infty} \frac{4}{n^2}(-1)^{n+1} e^{-n^2 t} \cos nx, \quad 0 < x < \pi, \ t > 0.$$

Example 6.2.4. Let the temperature $u(x,t)$ of a rod, composed of two different metals, be defined by

$$u(x,t) = \begin{cases} v(x,t), & 0 < x < \frac{1}{2}, \ t > 0 \\ w(x,t), & \frac{1}{2} < x < 1, \ t > 0. \end{cases}$$

Find the temperature $u(x,t)$ of the rod by solving the following system of partial differential equations.

(6.2.40) $v_t(x,t) = v_{xx}(x,t), \quad 0 < x < \dfrac{1}{2}, \ t > 0,$

(6.2.41) $w_t(x,t) = 4w_{xx}(x,t), \quad \dfrac{1}{2} < x < 1, \ t > 0,$

(6.2.42) $v(0,t) = w(1,t) = 0, \quad t > 0$

(6.2.43) $v\Big(\dfrac{1}{2},t\Big) = w\Big(\dfrac{1}{2},t\Big), \quad v_x\Big(\dfrac{1}{2},t\Big) = w_x\Big(\dfrac{1}{2},t\Big),$

(6.2.44) $u(x,0) = f(x) = x(1-x), \quad 0 < x < 1.$

Solution. If $v(x,t) = T_1(t)Y_1(x)$ and $w(x,t) = T_2(t)Y_2(x)$, then using the separation of variables we obtain the following system of ordinary differential equations.

(6.2.45) $Y_1''(x) + \lambda Y_1(x) = 0, \quad 0 < x < \dfrac{1}{2}$

(6.2.46) $4Y_2''(x) + \lambda Y_2(x) = 0, \quad \dfrac{1}{2} < x < 1$

(6.2.47) $T_k'(t) + \lambda T_k(t) = 0, \quad t > 0, \ k = 1,2.$

Solving the differential equations (6.2.45) and (6.2.46) and using the boundary conditions $Y_1(0) = Y_2(\pi) = 0$ we obtain

$$Y_1(x) = A \sin\left(\sqrt{\lambda}\, x\right), \quad Y_2(x) = B \sin\left(\tfrac{1}{2}\sqrt{\lambda}\,(x-1)\right),$$

for some constants A and B. From the conditions (6.2.43) we obtain the following conditions for the functions $Y_1(x)$ and $Y_2(x)$:

(6.2.48)
$$A \sin\left(\frac{\sqrt{\lambda}}{2}\right) = -B \sin\left(\frac{\sqrt{\lambda}}{2}\right),$$
$$A\lambda \cos\left(\frac{\sqrt{\lambda}}{2}\right) = 2B\lambda \cos\left(\frac{\sqrt{\lambda}}{2}\right).$$

One solution of the above system is $\lambda = 0$. If we eliminate A and B from (6.2.48), then we find the other solution:

(6.2.49) $\cot\left(\dfrac{\sqrt{\lambda}}{2}\right) = 0.$

If λ_n, $n = 1, 2, \ldots$ are the solutions of Equation (6.2.49), then

$$\lambda_n = \pi^2(2n - 1)^2, \quad n = 1, 2, \ldots.$$

For the found λ_n, the coefficients A and B in (6.2.48) can be chosen to be

$$A = -\frac{1}{\sin(2n-1)\frac{\pi}{2}} = (-1)^n, \quad B = \frac{1}{\sin(2n-1)\frac{\pi}{2}} = -(-1)^n.$$

Therefore the eigenfunctions $u_n(x)$, corresponding to the eigenvalues λ_n, are given by

$$(6.2.50) \qquad u_n(x) = \begin{cases} (-1)^n \sin(2n-1)\pi x, & 0 < x < \frac{1}{2} \\ (-1)^n \sin\big((2n-1)\pi(1-x)\big), & \frac{1}{2} < x < 1. \end{cases}$$

The solution of Equation (6.2.47) for the above found λ_n is given by

$$T_n(t) = e^{-(2n-1)^2\pi^2 t},$$

and so the solution of the given problem is

$$u(x,t) = \sum_{n=1}^{\infty} c_n e^{-(2n-1)^2\pi^2 t} w_n(x),$$

where $u_n(x)$ are given by (6.2.50).

The coefficients c_n are found as usual from the initial condition $u(x, 0) = f(x)$ and the orthogonality of the functions $u_n(x)$ on the interval $(0, 1)$:

$$c_n = \frac{1}{\left(\int\limits_0^1 u_n^2(x)\,dx\right)^2} \int\limits_0^1 u_n(x) f(x)\,dx.$$

For the integral in the denominator, we have

$$\int\limits_0^1 u_n^2(x)\,dx = \int\limits_0^{\frac{1}{2}} \sin^2\big((2n-1)\pi x\big)\,dx + \int\limits_{\frac{1}{2}}^1 \sin^2\big((2n-1)\pi(1-x)\big)\,dx$$

$$= \frac{1}{4} + \frac{1}{4} = \frac{1}{2}.$$

For the integral in the numerator of c_n, we have

$$\int\limits_0^1 u_n(x) f(x)\,dx = (-1)^n \int\limits_0^{\frac{1}{2}} x(1-x) \sin\big((2n-1)\pi x\big)\,dx$$

$$+ (-1)^n \int\limits_{\frac{1}{2}}^1 x(1-x) \sin\big((2n-1)\pi(1-x)\big)\,dx$$

$$= (-1)^n \left[\frac{2}{\pi^3(2n-1)^3} + \frac{2}{\pi^3(2n-1)^3}\right] = \frac{4(-1)^n}{\pi^3(2n-1)^3}.$$

Therefore,

$$c_n = \frac{8(-1)^n}{\pi^3(2n-1)^3},$$

and so the solution of the problem is given by

$$u(x,t) = \frac{8}{\pi^3} \sum_{n=1}^{\infty} \frac{(-1)^n}{(2n-1)^3} e^{-(2n-1)^2\pi^2 t} u_n(x),$$

where $u_n(x)$ are given by (6.2.50).

The plots of the heat distribution $w(x,t)$ of the rod at several time instances t are displayed in Figure 6.2.2.

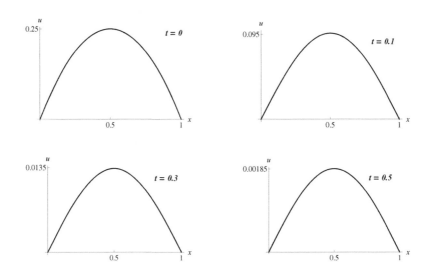

Figure 6.2.2

Exercises for Section 6.2.

In Exercises 1–12 use the separation of variables method to solve the heat equation

$$u_t(x,t) = a^2 u_{xx}(x,t), \quad 0 < x < l, \, t > 0,$$

subject to the following boundary conditions and the following initial conditions:

1. $a = \sqrt{3}$, $l = \pi$, $u(x,0) = x(\pi - x)$, $u(0,t) = u(\pi,t) = 0$.

2. $a = 1$, $l = 3$, $u(x,0) = x$, $u_x(0,t) = u_x(3,t) = 0$.

3. $a = \sqrt{2}$, $l = 4$, $u(x,0) = x + 1$, $u(0,t) = u_x(4,t) = 0$.

4. $a = \sqrt{2}$, $l = \pi$, $u(x,0) = 20$, $u(0,t) = u(\pi,t) = 0$.

5. $a = \sqrt{2}$, $l = 2$, $u(0,t) = u(2,t) = 0$, and

$$u(x,0) = \begin{cases} 20, & 0 \leq x < 1 \\ 0, & 1 \leq x \leq 2. \end{cases}$$

6. $a = 2$, $l = \pi$, $u(x,0) = x^2$, $u(0,t) = u(\pi,t) = 0$.

7. $a = 1$, $l = \pi$, $u(x,0) = 100$, $u(0,t) = u_x(\pi,t) = 0$.

8. $l = \pi$, $u(0,t) = u_x(\pi,t) = 0$, $u(x,0) = \begin{cases} x, & 0 < x < \frac{\pi}{2} \\ \pi - x, & \frac{\pi}{2} < x < \pi. \end{cases}$

9. $l = \pi$, $u(x,0) = \pi - x$, $u(0,t) = u_x(\pi,t) = 0$.

10. $l = \pi$, $u(x,0) = T_0$, $u(0,t) = 0$, $u(\pi,t) = T_0$.

11. $a = 1$, $l = \pi$, $u(x,0) = x \sin x$, $u(0,t) = u(\pi,t) = 0$.

12. $a = 1$, $l = \pi$, $u(x,0) = \sin^2 x$, $u(0,t) = u_x(\pi,t) = 0$.

13. Suppose that a rod of length l is radioactive and produces heat at a constant rate r. Solve the rod heat distribution equation

$$u_t(x,t) = a^2 u_{xx}(x,t) + r, \quad u(x,0) = 0, \ u(0,t) = u(l,t) = 0.$$

In Exercises 14–17 solve the nonhomogeneous heat equation

$$u_t(x,t) = a^2 u_{xx}(x,t) + F(x,t), \ 0 < x < l, \ t > 0,$$

subject to the given initial condition $u(x,0) = f(x)$ and the following source functions $F(x,t)$ and boundary conditions.

14. $l = 1$, $F(x,t) = -1$, $u_x(0,t) = u_x(1,t) = 0$, $f(x) = \frac{1}{2}(1 - x^2)$.

15. $l = \pi$, $f(x) = 0$, $u(0,t) = u(\pi,t) = 0$,

$$F(x,t) = \begin{cases} x, & 0 < x < \frac{\pi}{2} \\ \pi - x, & \frac{\pi}{2} \leq x < \pi. \end{cases}$$

16. $a = 1$, $l = \pi$, $F(x,t) = e^{-t}$, $u_x(0,t) = u_x(\pi,t) = 0$, $f(x) = \cos 2x$.

17. Solve the following heat equation with periodic boundary conditions.

$$u_t(x,t) = u_{xx}(x,t), \quad -\pi < x < \pi, \ t > 0,$$
$$u(-\pi, t) = u(\pi, t) \quad u_x(\pi, t) = u_x(\pi, t), \ t > 0,$$
$$u(x,0) = f(x) \ -\pi < x < \pi.$$

18. Solve the nonhomogeneous heat equation

$$u_t(x,t) = a^2 u_{xx}(x,t) + F(x,t), \quad 0 < x < l, \ t > 0$$

subject to the initial condition

$$u(x,0) = 0, \ 0 < x < l$$

and the boundary conditions

$$u(0,t) = u(l,t) = 0, \ t > 0.$$

19. The equation

$$u_t(x,t) - u_{xx}(x,t) + bu(x,t) = F(x,t), \quad 0 < x < l, \ b \in \mathbb{R}, \ t > 0$$

describes heat conduction in a rod taking into account the radioactive losses if $b > 0$. Verify that the equation reduces to

$$v_t(x,t) - v_{xx}(x,t) = e^{bt} F(x,t)$$

with the transformation $u(x,t) = e^{-bt} v(x,t)$.

20. Solve the heat equation

$$u_t(x,t) = u_{xx}(x,t), \ 0 < x < 1, \ t > 0$$

subject to the initial condition

$$u(x,0) = f(x), \ 0 < x < 1$$

and the boundary conditions

$$u(0,t) = u(1,t) + u_x(1,t), \quad t > 0.$$

In Exercises 21–23 solve the heat equation

$$u_t(x,t) = a^2 u_{xx}(x,t) - bu(x,t), \quad 0 < x < 1, \ t > 0$$

subject to the given initial condition and given boundary conditions.

21. $u(x,0) = f(x)$, $\quad u(0,t) = u(l,t) = 0$.

22. $u(x,0) = \sin \frac{\pi}{2l} x$, $\quad u(0,t) = u_x(l,t) = 0$.

23. $u(x,0) = f(x)$, $\quad u_x(0,t) = u_x(l,t) = 0$.

24. Solve the heat equation

$$u_t(x,t) = a^2 u_{xx}(x,t), \quad 0 < x < 1, \; t > 0$$

subject to the initial condition $u(x,0) = T$-constant and the boundary conditions

$$u(0,t) = 0, \quad u_x(l,t) = Ae^{-t}, \; t > 0.$$

Hint: Look for a solution $u(x,t)$ of the type

$$u(x,t) = f(x)e^{-t} + v(x,t),$$

where $v(x,t)$ satisfies a homogeneous heat equation and homogeneous boundary conditions.

6.3 The Heat Equation in Higher Dimensions.

In this section, we will study the two and three dimensional heat equation in rectangular and circular domains. We begin this section with the fundamental solution of the heat equation on the whole spaces $\mathbb{R}^2$ and $\mathbb{R}^3$.

6.3.1 Green Function of the Higher Dimensional Heat Equation.

Consider the two dimensional heat equation on the whole plane

(6.3.1)
$$\begin{cases} u_t(x,y,t) = c^2 \left(u_{xx}(x,y,t) + u_{yy}(x,y,t) \right), & (x,y) \in \mathbb{R}^2, \; t > 0 \\ u(x,y,0) = f(x,y), & (x,y) \in \mathbb{R}^2. \end{cases}$$

One method to find the *fundamental solution* or *Green function* of the two dimensional heat equation in (6.3.1) is to work similarly to the one dimensional case. The other method is outlined in Exercise 1 of this section.
We will find a solution $G(x,y,t)$ of the heat equation of the form

$$G(x,y,t) = \frac{1}{t}g(\zeta), \quad \text{where } \zeta = \sqrt{\frac{x^2 + y^2}{t}},$$

and g is a function to be determined.

Using the chain rule we find

$$G_t(x,y,t) = -\frac{1}{t^2}g(\zeta) - \frac{\sqrt{x^2+y^2}}{t^{\frac{5}{2}}},$$

and

$$G_{xx}(x,y,t) = \frac{1}{t^{\frac{3}{2}}}\left[\frac{1}{\sqrt{t}}\frac{x^2}{x^2+y^2}g''(\zeta) + \frac{y^2}{(x^2+y^2)\sqrt{x^2+y^2}}g'(\zeta)\right],$$

$$G_{yy}(x,y,t) = \frac{1}{t^{\frac{3}{2}}}\left[\frac{1}{\sqrt{t}}\frac{y^2}{x^2+y^2}g''(\zeta) + \frac{x^2}{(x^2+y^2)\sqrt{x^2+y^2}}g'(\zeta)\right].$$

If we substitute the above derivatives into the heat equation, after rearrangement we obtain

$$-g(\zeta)\frac{\sqrt{x^2+y^2}}{2\sqrt{t}}g'(\zeta) = c^2\left(g''(\zeta) + \frac{1}{\zeta}g'(\zeta)\right).$$

The last equation can be written in the form

$$c^2\frac{d}{d\zeta}\left(\zeta g'(\zeta)\right) + \frac{1}{2}\frac{d}{d\rho}\left(\zeta^2 g(\zeta)\right) = 0.$$

If we integrate the last equation, then we obtain

$$c^2 g'(v) + \frac{1}{2}g(\zeta) = 0.$$

The general solution of the last equation is

$$g(\zeta) = Ae^{-\frac{\zeta^2}{4c^2}},$$

and so

$$G(x,y,t) = Ae^{-\frac{x^2+y^2}{4c^2t}}.$$

Usually the constant A is chosen such that

$$\iint\limits_{\mathbb{R}^2} G(x,y,t)\,dx\,dy = 1.$$

Therefore,

(6.3.2) $$\qquad\qquad G(x,y,t) = \frac{1}{\sqrt{4\pi ct}}\, e^{-\frac{x^2+y^2}{4c^2t}}.$$

The function $G(x,y,t)$ given by (6.3.2) is called the *fundamental solution* or the *Green function* of the two dimensional heat equation on the plane $\mathbb{R}^2$. Some properties of the Green function $G(x,y,t)$ are given in Exercise 2 of this section.

The Green function is of fundamental importance because of the following result, stated without a proof.

Theorem 6.3.1. *If $f(x,y)$ is a bounded and continuous function on the whole plane $\mathbb{R}^2$, then*

$$(6.3.3) \qquad u(x,y,t) = \iint_{\mathbb{R}^2} G(x - \xi, y - \eta, t) f(\xi, \eta) \, dx \, dy$$

is the unique solution of problem (6.3.1).

The Green function $G(x, y, z, t)$ for the three dimensional heat equation

$$u_t = c^2 \left(u_{xx} + u_{yy} + u_{zz} \right), \qquad (x, y, z) \in \mathbb{R}^3, \ t > 0$$

is given by

$$G(x, y, z, t) = \frac{1}{(4\pi c t)^{\frac{3}{2}}} \, e^{-\frac{x^2 + y^2 + z^2}{4c^2 t}}.$$

6.3.2 The Heat Equation on a Rectangle.

Consider a very thin, metal rectangular plate which is heated. Let

$$R = \{(x, y) : 0 < x < a, \ 0 < y < b\}$$

be the rectangular plate with boundary ∂R. Let the boundary of the plate be held at zero temperature for all times and let the initial heat distribution of the plate be given by $f(x, y)$. The heat distribution $u(x, y, t)$ of the plate is described by the following initial boundary value problem.

$$(6.3.4) \qquad \begin{cases} u_t = c^2 \Delta_{x,y} u = c^2 \left(u_{xx} + u_{yy} \right), & (x, y) \in R, \ t > 0 \\ u(x, y, 0) = f(x, y), & (x, y) \in R \\ u(x, y, t) = 0, & (x, y) \in \partial R, \ t > 0 \end{cases}$$

If a solution $u(x, y, t)$ of the above problem (6.3.4) is of the form

$$u(x, y, t) = W(x, y)T(t), \quad 0 < x < a, \ 0 < y < b, \ t > 0,$$

then the heat equation becomes

$$W(x, y)T'(t) = c^2 \Delta W(x, y) T(t).$$

Working exactly as in Chapter 5 we will be faced with solving the Helmholtz eigenvalue problem

$$(6.3.5) \qquad \begin{cases} \Delta W(x, y) + \lambda W(x, y) = 0, & (x, y) \in R \\ W(x, y) = 0, & (x, y) \in \partial R \end{cases}$$

and the first order ordinary differential equation

(6.3.6) $T'(t) + c^2 \lambda T(t) = 0, \quad t > 0.$

The eigenvalue problem (6.3.5) was solved in Chapter 5, where we found that its eigenvalues λ_{mn} and corresponding eigenfunctions $W_{mn}(x, y)$ are given by

$$\lambda_{mn} = \frac{m^2\pi^2}{a^2} + \frac{n^2\pi^2}{b^2}, \quad W_{mn}(x, y) = \sin\left(\frac{m\pi x}{a}\right) \sin\left(\frac{n\pi y}{b}\right), \quad (x, y) \in R.$$

A general solution of Equation (6.3.6), corresponding to the above found λ_{mn}, is given by

$$T_{mn}(t) = a_{mn} e^{-c^2 \lambda_{mn} t}.$$

Therefore, the solution of problem (6.3.4) will be of the form

(6.3.7) $$u(x, y, t) = \sum_{m=1}^{\infty} \sum_{n=1}^{\infty} a_{mn} e^{-c^2 \lambda_{mn} t} \sin\left(\frac{m\pi x}{a}\right) \sin\left(\frac{n\pi y}{b}\right).$$

Using the initial condition of the function $u(x, y, t)$ and the orthogonality property of the the the eigenfunctions

$$W_{m,n}(x, y) \equiv \sin\left(\frac{m\pi x}{a}\right) \sin\left(\frac{n\pi y}{b}\right), \quad m, n = 1, 2, \ldots$$

on the rectangle $[0, a] \times [0, b]$, from (6.3.7) we find that the coefficients a_{mn} are given by

(6.3.8) $$a_{mn} = \frac{4}{ab} \int_0^a \int_0^b f(x, y) \sin\left(\frac{m\pi x}{a}\right) \sin\left(\frac{n\pi y}{b}\right) dy\, dx.$$

Example 6.3.1. Solve the following heat plate initial boundary value problem.

$$\begin{cases} u_t(x, y, t) = \dfrac{1}{\pi^2}\left(u_{xx}(x, y, t) + u_{yy}(x, y, t)\right), & 0 < x < 1,\ 0 < y < 1,\ t > 0 \\ u(x, y, 0) = \sin 3\pi x \sin \pi y, & 0 < x < 1,\ 0 < y < 1,\ t > 0 \\ u(0, y, t) = u(1, y, t) = u(x, 0, t) = u(x, 1, t) = 0, & 0 < x < 1,\ 0 < y < 1. \end{cases}$$

Solution. From the initial condition $f(x, y) = \sin 3\pi x \sin \pi y$ and Equation (6.3.8) for the coefficients a_{mn}, we find

$$a_{mn} = 4 \int_0^1 \int_0^1 \sin 3\pi x \sin \pi y \sin m\pi x \sin n\pi y\, dy\, dx = \begin{cases} 1, & m = 3,\ n = 1 \\ 0, & \text{otherwise.} \end{cases}$$

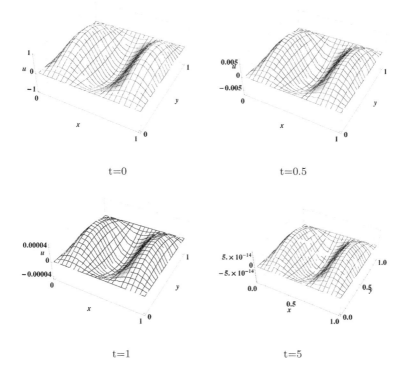

$t=0$ $t=0.5$

$t=1$ $t=5$

Figure 6.3.1

Therefore, the solution of this problem, using Equation (6.3.7), is given by

$$u(x, y, t) = e^{-10t} \sin 3\pi x \sin \pi y.$$

The plots of the heat distribution of the plate at several time instances are displayed in Figure 6.3.1.

Other types of the two dimensional heat equation on a rectangle (nonhomogeneous equation with homogeneous or nonhomogeneous Dirichlet or Neumann boundary conditions) can be solved exactly as in the one dimensional case. Therefore we will take only one example.

Example 6.3.2. Let R be the square

$$R = \{(x, y) : 0 < x < 1, \ 0 < y < 1\}$$

and ∂R its boundary. Solve the initial boundary value heat problem:

$$\begin{cases} u_t(x, y, t) = \dfrac{1}{\pi^2}\big(u_{xx}(x, y, t) + u_{yy}(x, y, t)\big) + xyt, & (x, y) \in R, \ t > 0 \\ u(x, y, 0) = \sin 3\pi x \sin \pi y, & (x, y) \in R, \ t > 0 \\ u(x, y, t) = 0, & (x, y) \in \partial R, \ t > 0. \end{cases}$$

Solution. The corresponding homogeneous wave equation is

$$\begin{cases} v_{tt}(x,y,t) = \dfrac{1}{\pi^2}\left(v_{xx}(x,y,t) + v_{yy}(x,y,t)\right), & (x,y) \in R, \ t > 0 \\ v(x,y,0) = \sin 3\pi x \, \sin \pi y, & (x,y) \in R, \ t > 0 \\ v(0x,y,t) = 0, & (x,y) \in \partial R, \ t > 0. \end{cases}$$

and it was solved in Example 6.3.1:

$$v(x,y,t) = e^{-10t} \sin 3\pi x \, \sin \pi y.$$

Now we solve the corresponding nonhomogeneous wave equation

$$\begin{cases} w_t(x,y,t) = \dfrac{1}{\pi^2}\left(w_{xx}(x,y,t) + w_{yy}(x,y,t)\right) + xyt, & (x,y) \in R, \ t > 0 \\ w(x,y,0) = \sin 3\pi x \, \sin \pi y, & (x,y) \in R, \ t > 0 \\ w(x,y,t) = 0, & (x,y) \in \partial R, \ t > 0. \end{cases}$$

For each fixed $t > 0$ expand the functions xyt and $w(x,y,t)$ in a double Fourier sine series:

$$xyt = 4t \sum_{m=1}^{\infty} \sum_{n=1}^{\infty} \frac{(-1)^m(-1)^n}{mn\pi^2} \sin m\pi x \, \sin n\pi y, \quad (x,y) \in R, \ t > 0$$

and

$$w(x,y,t) = \sum_{m=1}^{\infty} \sum_{n=1}^{\infty} w_{mn}(t) \sin m\pi x \, \sin n\pi y$$

where

$$w_{mn}(t) = 4 \int_0^1 \int_0^1 w(x,y,t) \sin m\pi x \, \sin n\pi y \, dy \, dx.$$

If we substitute the above expansions into the heat equation and compare the Fourier coefficients we obtain

$$w'_{mn}(t) + \left(m^2 + n^2\right) w_{mn}(t) = 4t \frac{(-1)^{m+n}}{mn\pi^2}.$$

The solution of the last equation, subject to the initial condition $w_{mn}(0) = 0$, is

$$w_{mn}(t) = \frac{4(-1)^m(-1)^n}{mn(m^2 + n^2)} e^{-(m^2+n^2)t}\left[1 - e^{m^2+n^2}\right],$$

and so

$$w(x,y,t) = \sum_{m=1}^{\infty} \sum_{n=1}^{\infty} \frac{4(-1)^m(-1)^n}{mn(m^2 + n^2)} e^{-(m^2+n^2)t}\left[1 - e^{m^2+n^2}\right] \sin m\pi x \, \sin n\pi y.$$

Therefore, the solution of the original problem is given by

$$u(x,y,t) = v(x,y,t) + w(x,y,t).$$

Remark. For the solution of the three dimensional heat equation on a rectangular solid by the separation of variables see Exercise 3 of this section.

6.3.3 The Heat Equation in Polar Coordinates.

In this part we will solve the problem of heating a circular plate.
Consider a circular plate D of radius a,

$$D = \{(x,y) : x^2 + y^2 < a^2\}$$

with boundary ∂D.

If $u(x,y,t)$ is the temperature of the plate at moment t and a point (x,y) of the plate, then the heat distribution of the plate is modeled by the heat initial boundary value problem

$$
\begin{cases}
u_t(x,y,t) = c^2\big(u_{xx}(x,y,t) + u_{yy}(x,y,t)\big), & (x,y) \in D,\ t > 0, \\
u(x,y,t) = 0, & (x,y) \in \partial D,\ t > 0, \\
u(x,y,0) = f(x,y), & (x,y) \in D.
\end{cases}
$$

Instead of working with cartesian coordinates when solving this problem, it is much more convenient to use polar coordinates.

Recall that the polar coordinates and the Laplacian in polar coordinates are given by

$$x = r\cos\varphi, \quad y = r\sin\varphi, \quad -\pi \le \varphi < \pi,\ 0 \le r < \infty$$

and

$$\Delta\, u(r,\varphi) = u_{rr} + \frac{1}{r}u_r + \frac{1}{r^2}u_{\varphi\varphi}.$$

Let us have a circular plate which occupies the disc of radius a and centered at the origin. If the temperature of the disc plate is denoted by $u(r,\varphi,t)$, then the heat initial boundary value problem in polar coordinates is given by

$$
(6.3.9)\quad
\begin{cases}
u_{tt} = c^2\big(u_{rr} + \dfrac{1}{r}u_r + \dfrac{1}{r^2}u_{\varphi\varphi}\big), & 0 < r \le a,\ 0 \le \varphi < 2\pi,\ t > 0, \\
u(a,\varphi,t) = 0, & 0 \le \varphi < 2\pi,\ t > 0, \\
u(r,\varphi,0) = f(r,\varphi), & 0 < r < a,\ 0 \le \varphi < 2\pi.
\end{cases}
$$

Now, we can separate out the variables. Let

$$u(r,\varphi,t) = \Phi(r,\varphi)T(t), \quad 0 < r \le a,\ -\pi \le \varphi < \pi.$$

Working exactly in the same way as in the oscillations of a circular membrane problem, discussed in Chapter 5, we will be faced with the eigenvalue problem

$$
(6.3.10)\quad
\begin{cases}
\Phi_{rr} + \dfrac{1}{r}\Phi_r + \dfrac{1}{r^2}\Phi_{\varphi\varphi} + \lambda^2\Phi(r,\varphi) = 0, & 0 < r \le a,\ 0 \le \varphi < 2\pi, \\
\Phi(a,\varphi) = 0,\quad |\lim_{r\to 0+}\Phi(r,\varphi)| < \infty, & 0 \le \varphi < 2\pi, \\
\Phi(r,\varphi) = \Phi(r,\varphi + 2\pi), & 0 < r \le a,\ 0 \le \varphi < 2\pi.
\end{cases}
$$

and the differential equation

(6.3.11) $T'(t) + \lambda^2 c^2 T(t) = 0, \quad t > 0.$

After separating the variables in the Helmholtz equation in problem (6.3.10), we find that its eigenvalues and corresponding eigenfunctions are given by

$$\lambda_{mn} = \frac{z_{mn}}{a}, \quad m = 0, 1, 2, \ldots; \ n = 1, 2, \ldots$$

and

$$\Phi_{mn}^{(c)}(r, \varphi) = J_m\Big(\frac{z_{mn}}{a}r\Big) \cos m\varphi, \quad \Phi_{mn}^{(s)}(r, \varphi) J_m\Big(\frac{z_{mn}}{a}r\Big) \sin m\varphi,$$

where z_{mn} is the nth zero of the Bessel function $J_m(\cdot)$ of order m.

For the above parameters λ_{mn}, the general solution of (6.3.11) is given by

$$T_{mn}(t) = A_{mn}e^{-\lambda_{mn}c^2 t},$$

and so, the solution $u = u(r, \varphi, t)$ of heat distribution problem (6.3.9) is given by

(6.3.12) $u = \displaystyle\sum_{m=0}^{\infty} \sum_{n=1}^{\infty} A_{mn}e^{-\lambda_{mn}c^2 t} J_m\Big(\frac{z_{mn}}{a}r\Big) \big[a_m \cos m\varphi + b_m \sin m\varphi\big].$

The coefficients in (6.3.12) are found by using the initial condition in (6.3.9):

$$f(r, \varphi) = u(r, \varphi, 0) = \sum_{m=0}^{\infty} \sum_{n=1}^{\infty} \big[a_m \cos m\varphi + b_m \sin m\varphi\big] A_{mn} J_m\Big(\frac{z_{mn}}{a}r\Big).$$

From the orthogonality property of the Bessel eigenfunctions $\{J_m(\cdot), \ m = 0, 1, \ldots\}$, as well as the orthogonality of $\{\sin m\varphi, \ \cos m\varphi, \ m = 0, 1, \ldots\}$ we obtain

(6.3.13)
$$\begin{cases} a_m A_{mn} = \dfrac{\int\limits_0^{2\pi}\int\limits_0^a f(r, \varphi) J_m\big(\frac{z_{mn}}{a}r\big) r \cos m\varphi\, dr\, d\varphi}{\int\limits_0^{2\pi}\int\limits_0^a J_m^2\big(\frac{z_{mn}}{a}r\big) r \cos^2 m\varphi\, dr\, d\varphi}, \\[3em] b_m A_{mn} = \dfrac{\int\limits_0^{2\pi}\int\limits_0^a f(r, \varphi) J_m\big(\frac{z_{mn}}{a}r\big) r \sin m\varphi\, dr\, d\varphi}{\int\limits_0^{2\pi}\int\limits_0^a J_m^2\big(\frac{z_{mn}}{a}r\big) r \sin^2 m\varphi\, dr\, d\varphi}. \end{cases}$$

Let us take an example.

Example 6.3.3. Solve the heat disc problem (6.3.9) if $a = c = 1$ and

$$f(r, \varphi) = (1 - r^2)r^2 \sin 2\varphi, \quad 0 \le r \le 1, \ 0 \le \varphi < 2\pi.$$

Display the solution $u(r, \varphi, t)$ of the temperature at several time moments t.

Solution. Since

$$\int_0^{2\pi} \sin 2\varphi \cos m\varphi \, d\varphi = 0, \quad m = 0, 1, 2, \ldots,$$

and

$$\int_0^{2\pi} \sin 2\varphi \sin m\varphi \, d\varphi = 0, \quad \text{for every } m \ne 2,$$

from (6.3.13) we have $a_m A_{mn} = 0$ for every $m = 0, 1, 2, \ldots$ and every $n = 1, 2, \ldots$; and $b_m A_{mn} = 0$ for every $m \ne 2$ and every $n = 1, 2, \ldots$.

For $m = 2$ and $n \in \mathbb{N}$, from (6.3.13) we have

(6.3.14)
$$b_2 A_{2n} = \frac{\int_0^{2\pi} \int_0^1 (1 - r^2)r^2 \sin 2\varphi \, J_2(z_{2n}r)r \sin 2\varphi \, dr \, d\varphi}{\int_0^{2\pi} \int_0^1 J_2^2(z_{2n}r)r \sin^2 2\varphi \, dr \, d\varphi}$$

$$= \frac{\int_0^1 (1 - r^2)r^3 J_2(z_{2n}r) \, dr}{\int_0^1 r J_2^2(z_{2n}r) \, dr}.$$

The numbers z_{2n} in (6.3.14) are the zeros of the Bessel function $J_2(x)$.

Taking $a = 1$, $p = 2$ and $\lambda = z_{2n}$ in Example 3.2.12 of Chapter 3, for the integral in the numerator in (6.3.14) we have

(6.3.15)
$$\int_0^1 (1 - r^2)r^3 J_2(z_{2n}r) \, dr = \frac{2}{z_{2n}^2} J_4(z_{2n}).$$

In Theorem 3.3.11 of Chapter 3 we proved the following formula. If λ_k are the zeros of the Bessel function $J_\mu(x)$, then

$$\int_0^1 x J_\mu^2(\lambda_k x) \, dx = \frac{1}{2} J_{\mu+1}^2(\lambda_k).$$

If $\mu = 2$ in this formula, then for the denominator in (6.3.15) we have

(6.3.16)
$$\int_0^1 r J_2^2(z_{2n}r)\, dr = \frac{1}{2} J_3^2(z_{2n}).$$

If we substitute (6.3.15) and (6.3.16) into (6.3.14) we obtain

$$b_2 A_{2n} = \frac{4 J_4(z_{2n})}{z_{2n}^2 J_3^2(z_{2n})}.$$

The right hand side of the last equation can be simplified using the following recurrence formula for the Bessel functions.

$$J_{\mu+1}(x) + J_{\mu-1}(x) = \frac{2\mu}{x} J_\mu(x).$$

(See Section 3.3. of Chapter 3.) Taking $\mu = 3$ and $x = z_{2n}$ in this formula, and using the fact that $J_2(z_{2n}) = 0$ it follows that $z_{2n} J_4(z_{2n}) = 6 J_3(z_{2n})$.
Thus,

$$b_2 A_{2n} = \frac{24}{z_{2n}^3 J_3(z_{2n})},$$

and so the solution $u(r, \varphi, t)$ of the problem is given by

$$u(r, \varphi, t) = 24 \left(\sum_{n=1}^{\infty} e^{-z_{2n}t} \frac{1}{z_{2n}^3 J_3(z_{2n})} J_2(z_{2n}r) \right) \sin 2\varphi.$$

See Figure 6.3.2 for the temperature distribution of the disc at the given instances.

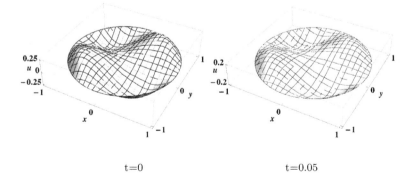

t=0   t=0.05

Figure 6.3.2

6.3.4 The Heat Equation in Cylindrical Coordinates.

In this section we will solve the problem of heat distribution in a cylinder. We will use the cylindrical coordinates

$$x = r \cos \varphi, \quad y = r\varphi, \ z = z$$

and the fact that the Laplacian Δ in the cylindrical coordinates is given by

$$\Delta = \frac{\partial^2}{\partial r^2} + \frac{1}{r}\frac{\partial}{\partial r} + \frac{1}{r^2}\frac{\partial^2}{\partial \varphi^2} + \frac{\partial^2}{\partial z^2}$$

(see Appendix E).

In the cylinder C, defined in cylindrical coordinates (r, φ, z) by

$$C = \{(r, \varphi, z) : 0 < r < a, \ 0 \le \varphi < 2\pi, \ 0 < z < l\}$$

consider the initial boundary value heat problem

(6.3.17)
$$\begin{cases} u_t(r, \varphi, z, t) = c^2 \Delta u(r, \varphi, z, t), \quad (r, \varphi, z) \in C, \ t > 0 \\ u(r, \varphi, z, 0) = f(r, \varphi, z), \quad (r, \varphi, z) \in C \\ \Big[u_r(r, \varphi, z, t) + k_1 u(r, \varphi, z, t)\Big]_{r=0} = 0, \\ \Big[u_z(r, \varphi, z, t) - k_2 u(r, \varphi, z, t)\Big]_{z=0} = 0, \\ \Big[u_z(r, \varphi, z, t) + k_3 u(r, \varphi, z, t)\Big]_{z=l} = 0. \end{cases}$$

If the function $u(r, \varphi, z, t)$ is of the form

$$u(r, \varphi, z, t) = R(r)\Phi(\varphi)Z(z)T(t),$$

then the variables will be separated and in view of the boundary conditions we will obtain the differential equation

(6.3.18)
$$T'(t) + \lambda T(t) = 0,$$

and the eigenvalue problems

(6.3.19)
$$\begin{cases} Z''(z) + (\lambda - \mu)Z(z) = 0, \\ Z'(0) - k_2(0)Z(0) = 0, \ Z'(l) + k_3(l)Z(l) = 0, \\ r^2 R''(r) + r R'(r) + (r^2\mu - \nu)R(r) = 0, \\ R(a) = 0, \ \big[R_r(r) + k_1 R(r)\big]_{r=0}, \ |\lim_{r \to 0^+} R(r)| < \infty, \\ \Phi''(\varphi) + \nu\Phi(\varphi), \quad \Phi(\varphi) = \Phi(\varphi + 2\pi), \quad 0 \le \varphi < 2\pi. \end{cases}$$

The eigenvalues ν and the corresponding eigenfunctions of the last eigen-value problem in (6.3.19) are given by

(6.3.20) $\nu = n, \quad \Phi_n(\varphi) = \{\sin n\varphi, \ \cos n\varphi \ \ n - 0, 1, 2, \dots\}.$

For the above values of ν, the eigenvalues of the second eigenvalue problem for the function $R(r)$ and the corresponding eigenfunctions are given by

(6.3.21) $\mu = z_{mn}, \quad R_{mn}(r) = J_n\left(\dfrac{z_{mn}}{a}r\right), \ \ m, n, = 0, 1, 2, \dots,$

where z_{mn} is the mth positive solution of the equation

(6.3.22) $z_{mn} J_n'(z_{mn}) + ak_1 J_n(z_{mn}).$

The eigenvalues $\lambda - \mu$ and the corresponding eigenfunctions of the first eigenvalue problem in (6.3.19) are given by

(6.3.23) $\lambda - \mu = \varsigma_j, \quad Z_j(z) = \sin(\varsigma_j z + \alpha_j), \ \ j = 0, 1, 2, \dots,$

where ς_j are the positive solutions of the equation

$$\cot(l\varsigma_j) = \frac{\varsigma_j^2 - k_2 k_3}{\varsigma_j(k_2 + k_3)},$$

and α_j are given by

$$\alpha_j = \arctan\left(\frac{\varsigma_j}{k_1}\right).$$

For all of these eigenvalues, a solution of the differential equation (6.3.18) is given by

(6.3.24) $T_{mnj}(t) = e^{-c^2\left(\frac{z_{mn}^2}{a^2} + \varsigma_j^2\right)t},$

and so from (6.3.20), (6.3.21), (6.3.22), (6.3.23) and (6.3.24) it follows that the solution $u = u(r, \varphi, z, t)$ of the heat problem for the cylinder is given by

$$u = \sum_{j=0}^{\infty} \sum_{m=0}^{\infty} \sum_{n=0}^{\infty} \left[A_{jmn} e^{-c^2\left(\frac{z_{mn}^2}{a^2} + \varsigma_j^2\right)t} J_n\left(\frac{z_{mn}}{a}r\right) \cos n\varphi \sin\left(\varsigma_j z + \alpha_k\right) \right.$$

$$\left. + B_{jmn} e^{-c^2\left(\frac{z_{mn}^2}{a^2} + \varsigma_j^2\right)t} J_n\left(\frac{z_{mn}}{a}r\right) \sin n\varphi \sin\left(\varsigma_j z + \alpha_k\right) \right],$$

where the coefficients A_{jmn} and B_{jmn} are determined using the initial con-dition $u(r, \varphi, 0, z) = f(r, \varphi, z)$ and the orthogonality property of the Bessel function $J_n(\cdot)$ and the sine and cosine functions:

$$A_{jmn} = 4 \frac{\int_0^a \int_0^{2\pi} \int_0^l r f(r, \varphi, z) J_n\left(\frac{z_{mn}}{a}r\right) \cos n\varphi \sin\left(\varsigma_j z + \alpha_j\right) dr \, d\varphi \, dz}{\pi a^2 \epsilon_n \left[1 + \dfrac{(h_2 h_3 + \varsigma_k^2)(h_2 + h_3)}{h_2^2 + \varsigma_k^2)(h_3^2 + \varsigma_k^2)}\right] J_n(z_{mn}^2) \left[1 + \dfrac{a^2 h_1^2 - n^2}{z_{mn}}\right]}$$

where

$$\epsilon_n = \begin{cases} 2, & n = 0, \\ 1, & n \neq 0. \end{cases}$$

Example 6.3.4. Consider an infinitely long circular cylinder in which the temperature $u(r, \varphi, z, t)$ is a function of the time t and radius r only, i.e., $u = u(r, t)$. Solve the heat problem

$$\begin{cases} u_t(r, t) = c^2 \left(\dfrac{\partial^2 u}{\partial r^2} + \dfrac{1}{r} \dfrac{\partial u}{\partial r} \right), & 0 < r < a,\ t > 0 \\ u(r, 0) = f(r), & 0 < r < a \\ |\lim_{r \to 0+} u(r, t)| < \infty, & u(a, t) = 0,\ t > 0. \end{cases}$$

Solution. We look for the solution $u(r, t)$ of the above problem in the form

$$u(r, t) = R(r)T(t).$$

Separating the variables and using the given boundary conditions we obtain

$$u(r, t) = \sum_{n=1}^{\infty} A_n J_0 \left(\frac{z_{n0}}{a} r \right) e^{-c z_{n0} t},$$

where z_{n0} is the nth positive zero of the Bessel function $J_0(\cdot)$.

The coefficients A_n are evaluated using the initial condition and the orthogonality property of the Bessel function:

$$\int_0^a r J_0(z_{m0} r) J_0(z_{n0} r)\, dr = 0, \quad \text{if } m \neq n$$

$$\int_0^a r J_0^2(z_{m0} r) = \frac{a}{2} J_1^2(z_{n0} r)\, dr = 0, \quad \text{if } m = n.$$

Using the above formulas we obtain

$$A_n = \frac{2}{a^2 J_1(z_{n0} a)} \int_0^a r f(r) J_0(z_{n0} r)\, dr.$$

6.3.5 The Heat Equation in Spherical Coordinates.

In this section we will solve the heat equation in a ball. The derivation of the solution is exactly the same as in the problem of vibrations of a ball in spherical coordinates, so we will omit many of the details.

Consider the initial boundary value heat problem

(6.3.25)
$$\begin{cases} u_t = \dfrac{1}{r^2} \dfrac{\partial}{\partial r} \left(r^2 \dfrac{\partial u}{\partial r} \right) + \dfrac{1}{r^2 \sin^2 \theta} \dfrac{\partial}{\partial \theta} \left(\sin \theta \dfrac{\partial u}{\partial \theta} \right) + \dfrac{1}{r^2 \sin^2 \theta} \dfrac{\partial^2 u}{\partial \theta^2} \\ u(a, \varphi, \theta, t) = 0, \quad 0 \leq \varphi < 2\pi,\ 0 \leq \theta < \pi, \\ u(r, \varphi, \theta, 0) = f(r, \varphi, \theta), \quad 0 \leq r < a,\ 0 \leq \varphi < 2\pi,\ 0 \leq \theta < \pi. \end{cases}$$

If we assume a solution of the above problem of the form

$$u(r, \varphi, \theta, t) = R(r) \, \Phi(\varphi) \, \Theta(\theta) \, T(t),$$

then working exactly as in the vibrations of the ball problem we need to solve the ordinary differential equation

$$T'(t) + \lambda c^2 T(t) = 0, \quad t > 0$$

and the three eigenvalue problems

$$\begin{cases} \dfrac{d}{dr}\left(r^2 \dfrac{dR}{dr}\right) + (\lambda^2 r^2 - \nu)R = 0, \quad 0 < r < a, \\[2mm] R(a) = 0, \quad |\lim_{r \to 0+} R(r)| < \infty, \\[2mm] \dfrac{d^2 \Phi}{d\theta^2} + \mu \Phi = 0, \quad 0 < \varphi < 2\pi, \\[2mm] \Phi(\varphi + 2\pi) = \Phi(\varphi), \quad 0 \le \varphi \le 2\pi, \\[2mm] \dfrac{d^2 \Theta}{d\theta^2} + \cot \varphi \, \dfrac{d\Theta}{d\theta} + \left(\nu - \dfrac{\mu}{\sin^2 \theta}\right)\Theta = 0, \\[2mm] |\lim_{\theta \to 0+} \Theta(\theta)| < \infty, \quad |\lim_{\theta \to \pi-} \Theta(\theta)| < \infty. \end{cases}$$

Using the results of the vibrations of the ball problem we obtain that the solution $u = u(r, \varphi, \theta)$ of the heat ball problem (6.3.25) is given by

(6.3.26)
$$u = \sum_{m=0}^{\infty} \sum_{n=0}^{\infty} \sum_{j=1}^{\infty} A_{mnj} e^{-c^2 \frac{z_{nj}}{a} t} J_{n+\frac{1}{2}}\left(\frac{z_{nj}}{a} r\right) \left[a_m \cos m\varphi \, P_n^{(m)}(\cos \theta) \right. $$
$$\left. + b_m \sin m\varphi \, P_n^{(m)}(\cos \theta) \right],$$

where z_{nj} is the jth positive zero of the Bessel function $J_{n+\frac{1}{2}}(x)$, $P_n^{(m)}(x)$ are the associated Legendre polynomials of order m (discussed in Chapter 3), given by

$$P_n^{(m)}(x) = (1 - x^2)^{\frac{m}{2}} \frac{d^m}{dx^m}\left(P_n(x)\right),$$

and $P_n(x)$ are the Legendre polynomials of order n.

The coefficients $A_{mnj} a_m$ and $A_{mnj} b_m$ in (6.3.26) are determined using the initial conditions given in (6.3.25) and the orthogonality properties of the eigenfunctions $R(r)$, $\Phi(\varphi)$, $\Theta(\theta)$.

Exercises for Section 6.3.

1. Let $v(x,t)$ and $w(y,t)$ be solutions of the one dimensional heat equations

$$v_t(x,t) = c^2 v_{xx}(x,t), \quad x, \, y \in \mathbb{R}, \; t > 0,$$
$$w_t(y,t) = c^2 w_{yy}(y,t), \quad x, \, y \in \mathbb{R}, \; t > 0,$$

respectively.

Show that $u(x,y,t) = v(x,t) \, w(y,t)$ is a solution of the two dimensional heat equation

$$u_t = c^2 (u_{xx} + u_{yy}), \quad (x,y) \in \mathbb{R}^2, \; t > 0$$

and deduce that the Green function $G(x,y,t)$ for the two dimensional heat equation is given by

$$G(x,y,t) = \frac{1}{\sqrt{4\pi ct}} \, e^{-\frac{x^2+y^2}{4c^2 t}}.$$

Generalize the method for the three dimensional heat equation.

2. Show that the Green function $G(x,y,t)$ satisfies the following:

   (a) $\int\limits_{-\infty}^{\infty} \int\limits_{-\infty}^{\infty} G(x,y,t) \, dy \, dx = 1$ for every $t > 0$.

   (b) $\lim\limits_{t \to 0+} G(x,y,t) = 0$ for every fixed $(x,y) \neq (0,0)$.

   (c) $G(x,y,t)$ is infinitely differentiable function on $\mathbb{R}^2$.

3. Let V be the three dimensional box defined by $V = \{(x,y,z) : 0 < x < a, \, 0 < y < b, \, 0 < z < c\}$ with boundary ∂V. Use the separation of variables method to find the solution of the following three dimensional heat initial boundary value problem.

$$\begin{cases} u_t = k^2 (u_{xx} + u_{yy} + u_{zz}), & (x,y,z) \in V, \; t > 0, \\ u(x,y,z,0) = f(x,y,z), & (x,y,z) \in V, \\ u(x,y,z,t) = 0, & (x,y,z) \in \partial V. \end{cases}$$

4. Let R be the rectangle defined by $R = \{(x,y) : 0 < x < a, \, 0 < y < b\}$ with boundary ∂R. Use the separation of variables method to find the general solution of the following two dimensional boundary value heat problem.

$$\begin{cases} u_t(x,y,t) = u_{xx}(x,y,t) + u_{yy}(x,y,t), & (x,y) \in R, \; t > 0, \\ u(0,y,t) = 0, & (x,y) \in \partial R, \; t > 0. \end{cases}$$

Let R be the rectangle defined by $R = \{(x,y) : 0 < x < a,\ 0 < y < b\}$ with boundary ∂R.

In Exercises 5–8 solve the following two dimensional initial boundary heat problem.

$$\begin{cases} u_t(x,y,t) = u_{xx}(x,y,t) + u_{yy}(x,y,t), & (x,y) \in R,\ t > 0, \\ u(x,y,t) = 0, & (x,y) \in \partial R,\ t > 0, \\ u(x,y,0) = f(x,y), & (x,y) \in R,\ t > 0 \end{cases}$$

if it is given that

5. $a = 1,\ b = 2,\ f(x,y) = 3\sin 6\pi x \sin 2\pi y + 7\sin \pi x \sin \frac{3\pi}{2}y$.

6. $a = b = \pi,\ f(x,y) = 1$.

7. $a = b = \pi,\ f(x,y) = \sin x \sin y + 3\sin 2x \sin y + 7\sin 3x \sin 2y$.

8. $a = b = \pi,\ f(x,y) = g(x)h(y),\ g(x) = h(x) = \begin{cases} x, & 0 \le x \le \frac{\pi}{2} \\ \pi - x, & \frac{\pi}{2} \le x \le \pi. \end{cases}$

For Exercises 9–11, let S be the square defined by $S = \{(x,y) : 0 < x < \pi,\ 0 < y < \pi\}$ whose boundary is ∂S.

9. Solve the following nonhomogeneous initial boundary value problem.

$$\begin{cases} u_t = u_{xx} + u_{yy} + \sin 2x \sin 3y, & (x,y) \in S,\ t > 0, \\ u(x,y,t) = 0, & (x,y) \in \partial S,\ t > 0, \\ u(x,y,0) = \sin 4x \sin 7y, & (x,y) \in S. \end{cases}$$

10. Use the separation of variables method to solve the following diffusion initial boundary value problem.

$$\begin{cases} u_t = u_{xx} + u_{yy} + 2k_1 u_x + 2k_2 u_y - k_3 u, & (x,y) \in S,\ t > 0, \\ u(x,y,t) = 0, & (x,y) \in \partial S,\ t > 0, \\ u(x,y,0) = f(x,y), & (x,y) \in S. \end{cases}$$

11. Solve the following nonhomogeneous problem.

$$\begin{cases} u_t = u_{xx}(x,y,t) + u_{yy} + F(x,y,t), & (x,y) \in S,\ t > 0, \\ u(x,y,t) = 0, & (x,y) \in \partial S,\ t > 0, \\ u(x,y,0) = & (x,y) \in S. \end{cases}$$

Hint: $F(x,y,t) = \sum_{m=1}^{\infty} \sum_{n=1}^{\infty} F_{mn}(t) \sin mx \sin ny$.

12. Find the solution $u = u(r, t)$ of the following initial boundary heat problem in polar coordinates.

$$\begin{cases} u_t = c^2 \left(u_{rr} + \dfrac{1}{r} u_r \right), & 0 < r < a, \ t > 0, \\ |\lim\limits_{r \to 0+} u(r, t)| < \infty, & u(a, t) = 0, \ t > 0, \\ u(r, 0) = T_0, & 0 < r < a. \end{cases}$$

13. Find the solution $u = u(r, \varphi, t)$ of the following initial boundary heat problem in polar coordinates.

$$\begin{cases} u_t = u_{rr} + \dfrac{1}{r} u_r + \dfrac{1}{r^2} u_{\varphi\varphi}, & 0 < r < 1, \ 0 < \varphi < 2\pi, \ t > 0, \\ |\lim\limits_{r \to 0+} u(r, \varphi, t)| < \infty, & u(1, \varphi, t) = 0, \ 0 < \varphi < 2\pi, \ t > 0, \\ u(r, \varphi, 0) = (1 - r^2) r \sin \varphi, & 0 < r < 1, \ 0 < \varphi < 2\pi. \end{cases}$$

14. Find the solution $u = u(r, \varphi, t)$ of the following initial boundary heat problem in polar coordinates.

$$\begin{cases} u_t = u_{rr} + \dfrac{1}{r} u_r + \dfrac{1}{r^2} u_{\varphi\varphi}, & 0 < r < 1, \ 0 < \varphi < 2\pi, \ t > 0, \\ |\lim\limits_{r \to 0+} u(r, \varphi, t)| < \infty, & u(1, \varphi, t) = \sin 3\varphi, \ 0 < \varphi < 2\pi, \ t > 0, \\ u(r, \varphi, 0) = 0, & 0 < r < 1, \ 0 < \varphi < 2\pi. \end{cases}$$

15. Find the solution $u = u(r, z, t)$ of the following initial boundary heat problem in cylindrical coordinates.

$$\begin{cases} u_t = u_{rr} + \dfrac{1}{r} u_r + u_{zz}, & 0 < r < 1, \ 0 < z < 1, \ t > 0, \\ u(r, 0, t) = u(r, 1, t) = 0, & 0 < r < 1, \ t > 0, \\ u(1, z, t) = 0, & 0 < z < 1, \ t > 0, \\ u(r, z, 0) = 1, & 0 < r < 1. \end{cases}$$

16. Find the solution $u = u(r, t)$ of the following initial boundary heat problem in polar coordinates.

$$\begin{cases} \dfrac{\partial u(r, t)}{\partial t} - \dfrac{c^2}{r^2} \dfrac{\partial}{\partial r} \left(r^2 \dfrac{\partial}{\partial r} \right), & 0 < r < a, \ t > 0, \\ |\lim\limits_{r \to 0+} u(r, t)| < \infty, & u(1, t) = 0, \ t > 0, \\ u(r, 0) = 1, & 0 < r < 1. \end{cases}$$

17. Find the solution $u = u(r, t)$ of the heat ball problem

$$\begin{cases} u_t = cx^2 \left(\dfrac{\partial^2 u}{\partial r^2} + \dfrac{2}{r} \dfrac{\partial u}{\partial r} \right), & 0 < r < 1, \ t > 0 \\[2mm] u(r, 0) = f(r) & 0 < r < 1, \\[2mm] |\lim\limits_{r \to 0+} u(r, t)| < \infty, & u(1, t) = 0, \ t > 0. \end{cases}$$

Hint: Introduce a new function $v(r, t)$ by $v(r, t) = ru(r, t)$.

18. Find the solution $u = u(r, \varphi, \theta, t) = u(r, \theta, t)$ (u is independent of φ) of the heat ball problem in spherical coordinates.

$$
\begin{cases}
u_t = \dfrac{\partial^2 u}{\partial r^2} + \dfrac{2}{r}\dfrac{\partial u}{\partial r} + \dfrac{1}{r^2}\dfrac{\partial^2 u}{\partial \theta^2} + \dfrac{\cot\theta}{r^2}\dfrac{\partial u}{\partial \theta}, \quad 0 < r < 1,\ 0 < \theta < \pi,\ t > 0 \\[2mm]
u(r, \theta, 0) = \begin{cases} 1, & 0 \le \theta < \frac{\pi}{2} \\ 0, & \frac{\pi}{2} \le \theta \le \pi, \end{cases} \quad 0 < r < 1, \\[2mm]
u(1, \theta, t) = 0, \quad 0 \le \theta \le \pi,\ t > 0.
\end{cases}
$$

6.4 Integral Transform Methods for the Heat Equation.

In this section we will apply the Laplace and Fourier transforms to solve the heat equation and similar partial differential equations. We begin with the Laplace transform method.

6.4.1 The Laplace Transform Method for the Heat Equation.

The the Laplace transform method for solving the heat equation is the same as the one used to solve the wave equation.

As usual, we use capital letters for the Laplace transform with respect to $t > 0$ for a given function $g(x, t)$. For example, we write

$$
\mathcal{L}\, u(x, t) = U(x, s), \quad \mathcal{L}\, y(x, t) = Y(x, s), \quad \mathcal{L}\, f(x, t) = F(x, s).
$$

Let us take several examples to illustrate the Laplace method.

Example 6.4.1. Solve the initial boundary value heat problem

$$
\begin{cases}
u_t(x, t) = u_{xx}(x, t) + a^2 u(x, t) + f(x), \quad 0 < x < \infty,\ t > 0, \\
u(0, t) = u_x(0, t) = 0, \quad t > 0.
\end{cases}
$$

Solution. If $U(s, t) = \mathcal{L}\left(u(x, t)\right)$ and $F(s) = \mathcal{L}\left(f(x)\right)$, then the heat equation is transformed into

$$
U_t(s, t) = s^2 U(s, t) - su(0, t) - u_x(0, t) + a^2 U(s, t) + F(s).
$$

In view of the boundary conditions for the function $u(x, t)$ the above equation becomes

$$
U_t(s, t) - (s^2 + a^2)U(x, t)+ = F(s).
$$

The general solution of the last equation (treating it as an ordinary differential equation with respect to t) is given by

$$U(s,t) = Ce^{(s^2+a^2)t} - \frac{F(s)}{s^2 + a^2}.$$

From the fact that every Laplace transform tends to zero as $s \to \infty$ it follows that $C = 0$. Thus,

$$U(s,t) = -\frac{F(s)}{s^2 + a^2},$$

and so by the convolution theorem for the Laplace transform we have

$$u(x,t) = -\frac{1}{a} \int_0^x f(x-y) \sin ay \, dy.$$

Example 6.4.2. Using the Laplace transform method solve the initial boundary value heat problem

$$\begin{cases} u_t(x,t) = u_{xx}(x,t), & 0 < x < 1, \ t > 0, \\ u(0,t) = 0, & u(1,t) = 1, \ t > 0, \\ u(x,0) = 0, & 0 < x < 1. \end{cases}$$

Solution. If $U = U(x,s) = \mathcal{L}\left(u(x,t)\right)$, then applying the Laplace transform to both sides of the heat equation and using the initial condition we have

$$\frac{d^2U}{dx^2} - sU = 0.$$

The general solution of the above ordinary differential equation is

$$U(x,s) = c_1 \cosh\left(\sqrt{s}x\right) + c_2 \sinh\left(\sqrt{s}x\right).$$

From the boundary conditions for the function $u(x,t)$ we have

$$U(0,s) = \mathcal{L}\left(u(0,t)\right) = 0, \quad U(1,s) = \mathcal{L}\left(u(1,t)\right) = \frac{1}{s}.$$

The above boundary conditions for $U(x,s)$ imply

$$U(x,s) = \frac{1}{s} \frac{\sinh\left(\sqrt{s}x\right)}{\sinh\left(\sqrt{s}\right)} = \frac{1}{s} \frac{e^{(x+1)\sqrt{s}} - e^{-(x+1)\sqrt{s}}}{1 - e^{-2\sqrt{s}}}.$$

Using the geometric series

$$\frac{1}{1 - e^{-2\sqrt{s}}} = \sum_{n=0}^{\infty} e^{-2n\sqrt{s}}$$

we obtain

$$U(x,s) = \sum_{n=0}^{\infty} \left[\frac{e^{-(2n+1-x)\sqrt{s}}}{s} - \frac{e^{-(2n+1+x)\sqrt{s}}}{s} \right].$$

Therefore (from Table A of Laplace transforms given in Appendix A) we have

$$u(x,t) = \sum_{n=0}^{\infty} \left[\mathcal{L}^{-1}\left(\frac{e^{-(2n+1-x)\sqrt{s}}}{s} \right) - \mathcal{L}^{-1}\left(\frac{e^{-(2n+1+x)\sqrt{s}}}{s} \right) \right]$$

$$= \sum_{n=0}^{\infty} \left[erf\left(\frac{2n+1-x}{2\sqrt{t}} \right) - erf\left(\frac{2n+1+x}{2\sqrt{t}} \right) \right],$$

where the *error function* $erf(\cdot)$ is defined by

$$erf(x) = \frac{2}{\sqrt{\pi}} \int_{x}^{\infty} e^{-t^2} \, dt.$$

Example 6.4.3. Using the Laplace transform method solve the boundary value problem

$$\begin{cases} u_t(x,t) = u_{xx}(x,t), & -1 < x < 1, \ t > 0, \\ u(x,0) = 1, & x > 0, \\ u_x(1,t) + u(1,t) = u_x(-1,t) + u(1,t) = 0, & t > 0. \end{cases}$$

Solution. If $U = U(x,s) = \mathcal{L}\left(u(x,t)\right)$, then from the initial condition we have

$$\frac{d^2 U}{dx^2} - sU = -1.$$

The solution of the last equation, in view of the boundary conditions

$$\left[\frac{dU}{dx} + U(x,s) \right]_{x=1} = 0, \quad \left[\frac{dU}{dx} + U(x,s) \right]_{x=-1} = 0,$$

is

(6.4.1) $$U(x,s) = \frac{1}{s} - \frac{\cosh\left(\sqrt{s}x\right)}{s\left[\sqrt{s}\sinh\left(\sqrt{s}\right) + \cosh\left(\sqrt{s}\right)\right]}.$$

The first function in (6.4.1) is easily invertible. To find the inverse Laplace transform of the second function is not as easy. Let

$$F(s) = \frac{\cosh\left(\sqrt{s}x\right)}{s\left[\sqrt{s}\sinh\left(\sqrt{s}\right) + \cosh\left(\sqrt{s}\right)\right]}.$$

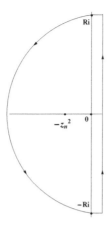

Figure 6.4.1

The singularities of this function are at $s = 0$ and all points s which are solutions of the equation

(6.4.2)
$$\sqrt{s} \sinh \sqrt{s} + \cosh \sqrt{s} = 0.$$

These singularities are simple poles (check !). Consider the contour integral

$$\int_{C_R} F(z)\, e^{tz}\, dz,$$

where the contour C_R is given in Figure 6.4.1.

Letting $R \to \infty$, from the formula for the inverse Laplace transform by residues (see Appendix F) we have

(6.4.3) $\mathcal{L}^{-1} F(s) = Res\left(F(z)e^{zt}, z = 0\right) + \displaystyle\sum_{n=1}^{\infty} Res\left(F(z)e^{zt}, z = -z_n^2\right),$

where $-z_n^2$ are the negative solutions of Equation (6.4.2).
   To calculate the above residues we use a result from Appendix F for evaluating residues. For the pole $z = 0$ we have

$$Res\{F(z)\, e^{zt}, z = 0\} = \lim_{z \to 0} z F(z)\, e^{zt}$$
$$= \lim_{z \to 0} z \frac{\cosh\left(\sqrt{z}\, x\right) e^{zt}}{\left[\sqrt{z}\, \sinh(\sqrt{z} + \cosh(\sqrt{z})\right]} = 1.$$

For the poles $z = -z_n^2$, using l'Hopital's rule and Euler's formula

$$e^{i\alpha} = \cos\alpha + i\sin\alpha$$

we have

$$Res\left\{F(z)e^{zt}, z = -z_n^2\right\} = \lim_{z \to -z_n^2} (z + z_n^2)F(z)e^{zt}$$

(6.4.4)
$$= \cosh\left(-z_n ix\right)e^{-z_n^2 t} \lim_{z \to -z_n^2} \frac{z + z_n^2}{\left[\sqrt{z}\sinh\left(\sqrt{z} + \cosh\left(\sqrt{z}\right)\right)\right]}$$

$$= -2\frac{\cos\left(z_n x\right)e^{-z_n^2 t}}{z_n\left(\sin\left(z_n\right) + z_n\cos\left(z_n\right)\right)}.$$

Therefore, from (6.4.1), (6.4.3) and (6.4.4) it follows that

$$u(x,t) = \mathcal{L}^{-1}\left(\frac{1}{s}\right) + \left[-1 - 2\sum_{n=1}^{\infty} 2\frac{\cos\left(z_n x\right)e^{-z_n^2 t}}{z_n\left(\sin\left(z_n\right) + z_n\cos\left(z_n\right)\right)}\right]$$

$$= 2\sum_{n=1}^{\infty} \frac{\cos\left(z_n x\right)e^{-z_n^2 t}}{z_n\left[2\sin\left(z_n\right) + z_n\cos\left(z_n\right)\right]}.$$

Example 6.4.4. Find the solution of the problem

(6.4.5)
$$\begin{cases} u_t(x,t) = a^2 u_{xx}(x,t), & 0 < x < l,\ t > 0, \\ \lim_{x \to 0+} u(x,t) = \delta(t), & u(l,t) = 0,\ t > 0, \\ u(x,0) = 0, & 0 < x < l, \end{cases}$$

where $\delta(t)$ is the Dirac impulse "function" concentrated at $t = 0$.

Solution. If $U = U(x,s) = \mathcal{L}\left(u(x,t)\right)$, then using the result

$$\int_{-\infty}^{\infty} f(x)\delta(x)\,dx = f(0),$$

for every continuous function f on $\mathbb{R}$ which vanishes outside an open interval, the initial boundary value problem (6.4.5) is reduced to the problem

$$\begin{cases} U_{xx}(x,s) - \dfrac{s}{a^2}U(x,s) = 0, & 0 < x < l,\ s > 0 \\ \lim_{x \to 0+} U(x,s) = 1, & U(l,s) = 0,\ s > 0. \end{cases}$$

The solution of the last problem is given by

$$U(x,s) = \frac{\sinh\left(\frac{l-x}{a}\sqrt{s}\right)}{\sinh\left(\frac{l}{a}\sqrt{s}\right)}.$$

To find the inverse Laplace of the above function we proceed as in Example 6.4.2. We write the function $U(x,s)$ in the form

$$U(x,s) = \frac{\sinh\left(\frac{l-x}{a}\sqrt{s}\right)}{\sinh\left(\frac{l}{a}\sqrt{s}\right)} = \frac{e^{-\frac{x}{a}\sqrt{s}} - e^{-\frac{2l-x}{a}\sqrt{s}}}{1 - e^{-\frac{2l}{a}\sqrt{s}}}$$

$$(6.4.6) \qquad = \left(e^{-\frac{x}{a}\sqrt{s}} - e^{-\frac{2l-x}{a}\sqrt{s}}\right)\sum_{n=0}^{\infty} e^{-\frac{2ln}{a}\sqrt{s}}$$

$$= \sum_{n=0}^{\infty} e^{-\frac{(2nl+x)}{a}\sqrt{s}} - \sum_{n=0}^{\infty} e^{-\frac{(2nl-x)}{a}\sqrt{s}}.$$

From Table A in Appendix A we have

$$\mathcal{L}^{-1}\left(e^{-y\sqrt{s}}\right) = g(y,t) = \frac{y}{2\sqrt{\pi}\,t^{\frac{3}{2}}}\,e^{-\frac{y^2}{4}t}.$$

Therefore, from (6.4.6) it follows that

$$u(x,t) = \sum_{n=0}^{\infty} g\left(\frac{2nl+x}{a}, t\right) - \sum_{n=0}^{\infty} g\left(\frac{2nl-x}{a}, t\right).$$

Since the function $g(y,t)$ is odd with respect to the variable y, from the last equation we obtain

$$u(x,t) = \frac{1}{2\sqrt{\pi}\,t^{\frac{3}{2}}}\sum_{n=-\infty}^{\infty} (2nl+x)\,e^{-\frac{(2nl+x)^2}{4a^2l}t}.$$

6.4.2 The Fourier Transform Method for the Heat Equation.

Fourier transforms, like the Laplace transform, can be used to solve the heat equation, and similar equations, in one and higher spatial dimensions.

The fundamentals of Fourier transforms were developed in Chapter 2. As usual, we use capital letters for the Fourier transform with respect to $x \in \mathbb{R}$ for a given function $g(x,t)$. For example, we write

$$\mathcal{F}u(x,t) = U(\omega,t), \quad \mathcal{F}y(x,t) = Y(\omega,t), \quad \mathcal{F}f(x,t) = F(\omega,t).$$

Let us take first an easy example.

Example 6.4.5. Using the Fourier transform solve the transport boundary value problem

$$\begin{cases} u_t(x,t) = c^2 u_{xx}(x,t), & -\infty < x < \infty, \ t > 0 \\ u(x,0) = f(x), & -\infty < x < \infty. \end{cases}$$

Solution. If $U = U(\omega, t) = \mathcal{F}\left(u(x, t)\right)$ and $F(\omega) = \mathcal{F}\left(f(x)\right)$, then the above problem is reduced to the problem

$$\frac{dU}{dt} = -c^2\omega^2\,\omega U, \quad U(\omega, 0) = F(\omega).$$

The solution of the last problem is $U(\omega, t) = F(\omega)e^{-c^2\omega^2 t}$.

Now, from the inverse Fourier transform formula it follows that

$$
\begin{aligned}
u(x, t) &= \frac{1}{2\pi}\int_{-\infty}^{\infty} U(\omega, t)e^{i\omega x}\, d\omega = \frac{1}{2\pi}\int_{-\infty}^{\infty} F(\omega)e^{-c^2\omega^2 t}e^{i\omega x}\, d\omega \\
&= \frac{1}{2\pi}\int_{-\infty}^{\infty}\left(\int_{-\infty}^{\infty} f(\xi)e^{-i\omega\xi}\, d\xi\right)e^{-c^2\omega^2 t}e^{i\omega x}\, d\omega \\
&= \frac{1}{2\pi}\int_{-\infty}^{\infty}\left(\int_{-\infty}^{\infty} e^{-c^2\omega^2 t}e^{i\omega(x-\xi)}\, d\omega\right)f(\xi)\, d\xi \\
&= \frac{1}{\pi}\int_{-\infty}^{\infty}\left(\int_{0}^{\infty} e^{-c^2\omega^2 t}\cos\left(\omega(x-\xi)\right)\, d\omega\right)f(\xi)\, d\xi \\
&= \frac{1}{2c\sqrt{\pi t}}\int_{-\infty}^{\infty} e^{-\frac{(x-\xi)^2}{4c^2 t}}\, f(\xi)\, d\xi.
\end{aligned}
$$

In the above we used the following result.

$$(6.4.7) \qquad \int_{0}^{\infty} e^{-c^2\omega^2 t}\cos\left(\omega\lambda\right)\, d\omega = \frac{1}{2c}\sqrt{\frac{\pi}{t}}.$$

This result can be easily derived by differentiating the function

$$g(\omega, \lambda) = \int_{0}^{\infty} e^{-c^2\omega^2 t}\cos\left(\omega\lambda\right)\, d\omega$$

with respect to λ and integrating by parts to obtain

$$\frac{\partial g(\omega, \lambda)}{\partial \lambda} = -\frac{\lambda}{2c^2 t}\, g(\omega, \lambda),$$

from which the result (6.4.7) follows.

Example 6.4.6. Using the Fourier cosine transform solve the initial boundary value problem

$$
\begin{cases}
u_t(x,t) = c^2 u_{xx}(x,t), & 0 < x < \infty, \; t > 0 \\
u_x(0,t) = f(t), & 0 < t < \infty, \\
u(x,0) = 0, & 0 < x < \infty \\
\lim_{x \to \infty} u(x,t) = \lim_{x \to \infty} u_x(x,t) = 0, & t > 0.
\end{cases}
$$

Solution. If $U = U(\omega,t) = \mathcal{F}_c\left(u(x,t)\right)$ and $F(\omega) = \mathcal{F}_c\left(f(x)\right)$, then taking the Fourier cosine transform from both sides of the heat equation and using the integration by parts formula (twice) and the boundary condition $u_x(0,t) = f(t)$ we obtain

$$
\frac{dU(\omega,t)}{dt} = c^2 \int_0^\infty u_{xx}(x,t) \, \cos\left(\omega\, x\right) dx = c^2 \left[u_x(x,t) \, \cos\left(\omega\, x\right) \right]_{x=0}^\infty
$$

$$
+ c^2 \omega \int_0^\infty u_x(x,t) \, \sin\left(\omega\, x\right) dx = -c^2 f(t) + c^2 \left[\omega\, u(x,t) \, \sin\left(\omega\, x\right) \right]_{x=0}^\infty
$$

$$
- c^2 \omega^2 \int_0^\infty u(x,t) \, \cos\left(\omega\, x\right) dx = -c^2 f(t) - c^2 \omega^2 \, U(\omega,t),
$$

i.e.,

$$
\frac{dU(\omega,t)}{dt} + c^2 \omega^2 \, U(\omega,t) = -c^2 f(t).
$$

The solution of the last differential equation, in view of the initial condition

$$
U(\omega,0) = \int_0^\infty u(x,0) \, \cos\left(\omega\, x\right) dx = 0,
$$

is given by

$$
U(\omega,t) = -c^2 \int_0^t f(\tau) \, e^{-c^2 \omega^2 (t-\tau)} \, d\tau.
$$

If we take the inverse Fourier cosine transform, then we obtain that the solution $u(x,t)$ of the given problem is

$$u(x,t) = \frac{2}{\pi} \int\limits_0^\infty U(\omega,t) \cos \omega x \, d\omega$$

$$= -\frac{2c^2}{\pi} \int\limits_0^\infty \left(\int\limits_0^t f(\tau) e^{-c^2 \omega^2 (t-\tau)} \, d\tau \right) \cos \omega x \, d\omega$$

$$= -\frac{2c^2}{\pi} \int\limits_0^t \left(\int\limits_0^\infty e^{-c^2 \omega^2 (t-\tau)} \cos \omega x \, d\omega \right) f(\tau) \, d\tau$$

$$= -\frac{c}{\sqrt{\pi}} \int\limits_0^t \frac{f(\tau)}{\sqrt{t-\tau}} e^{-\frac{x^2}{4c^2(t-\tau)}} \quad \text{(by result (6.4.7) in Example 6.4.5).}$$

Exercises for Section 6.4.

In Problems 1–11, use the Laplace transform to solve the indicated initial boundary value problem on $(0, \infty)$, subject to the given conditions.

1. $u_t(x,t) = u_{xx}(x,t)$, $0 < x < 1$, $t > 0$, subject to the following conditions: $u(x,0) = 0$, $u(0,t) = 1$, $u(1,t) = u_0$, u_0 is a constant.

2. $u_t(x,t) = u_{xx}(x,t)$, $0 < x < \infty$, $t > 0$, subject to the following conditions: $\lim\limits_{x\to\infty} u(x,t) = u_1$, $u(0,t) = u_0$, $u(x,0) = u_1$, u_0 and u_1 are constants.

3. $u_t(x,t) = u_{xx}(x,t)$, $0 < x < \infty$, $t > 0$, subject to the following conditions: $\lim\limits_{x\to\infty} u(x,t) = u_0$, $u_x(0,t) = u(0,t)$, $u(x,0) = u_0$, u_0 is a constant.

4. $u_t(x,t) = u_{xx}(x,t)$, $0 < x < \infty$, $t > 0$, subject to the following conditions: $\lim\limits_{x\to\infty} u(x,t) = u_0$, $u(0,t) = f(t)$, $u(x,0) = 0$, u_0 is a constant.

5. $u_t(x,t) = u_{xx}(x,t)$, $0 < x < \infty$, $t > 0$, subject to the following conditions: $\lim\limits_{x\to\infty} u(x,t) = 60$, $u(x,0) = 60$, $u(0,t) = 60 + 40\,\mathcal{U}_2(t)$, where $\mathcal{U}_2(t) = \begin{cases} 1, & 0 \le t \le 2 \\ 0, & 2 < t < \infty. \end{cases}$

6. $u_t(x,t) = u_{xx}(x,t)$, $-\infty < x < 1$, $t > 0$, subject to the following conditions: $\lim\limits_{x\to-\infty} u(x,t) = u_0$, $u_x(1,t) + u(1,t) = 100$, $u(x,0) = 0$, u_0 is a constant.

7. $u_t(x,t) = u_{xx}(x,t)$, $0 < x < 1$, $t > 0$, subject to the following conditions: $u(x,0) = u_0 + u_0 \sin x$, $u(0,t) = u(1,t) = u_0$, u_0 is a constant.

8. $u_t(x,t) = u_{xx}(x,t) - hu(x,t)$, $0 < x < \infty$, $t > 0$, subject to the following conditions: $\lim\limits_{x \to \infty} u(x,t) = 0$, $u(x,0) = 0$, $u(0,t) = u_0$, u_0 is a constant.

9. $u_t(x,t) = u_{xx}(x,t)$, $0 < x < \infty$, $t > 0$, subject to the following conditions: $\lim\limits_{x \to \infty} u(x,t) = 0$, $u(x,0) = 10e^{-x}$, $u(0,t) = 10$.

10. $u_t(x,t) = u_{xx}(x,t) + 1$, $0 < x < 1$, $t > 0$, subject to the following conditions: $u(0,t) = u(1,t) = 0$, $u(x,0) = 0$.

11. $u_t(x,t) = c^2\big(u_{xx}(x,t) + (1+k)u_x(x,t) + ku(x,t)\big)$, $0 < x < \infty$, $t > 0$, subject to the following conditions: $u(0,t) = u_0$, $\lim\limits_{x \to \infty} |u(x,t)| < \infty$, $u(x,0) = 0$, u_0 is a constant.

12. Find the solution $u(r,t)$ of the initial boundary value problem

$$u_t(r,t) = u_{rr}(r,t) + \frac{2}{r}u_r(r,t), \quad 0 < r < 1,\ t > 0$$
$$\lim\limits_{r \to 0+} |u(r,t)| < \infty, \quad u_r(1,t) = 1,\ t > 0$$
$$u(r,0) = 0, \quad 0 \le r < 1.$$

13. Solve the boundary value problem

$$u_t(x,t) = u_{xx}(x,t) + u(x,t) + A\cos x, \quad 0 < x < \infty,\ t > 0$$
$$u(0,t) = Be^{-3t}, \quad u_x(0,t) = 0,\ t > 0.$$

14. Solve the heat equation

$$u_t(x,t) = u_{xx}(x,t), \quad 0 < x < \infty,\ t > 0,$$

subject to the following conditions:

(a) $\lim\limits_{x \to 0+} u(x,t) = \delta(t)$, $\lim\limits_{x \to \infty} u(x,t) = 0$, $u(x,0) = 0$, $0 < x < \infty$, where $\delta(t)$ is the Dirac impulse function, concentrated at $t = 0$.
(b) $\lim\limits_{x \to 0+} u(x,t) = f(t)$, $\lim\limits_{x \to \infty} u(x,t) = 0$, $u(x,0) = 0$, $0 < x < \infty$.

In Problems 15–19, use the Fourier transform to solve the indicated initial boundary value problem on the indicated interval, subject to the given conditions.

15. $u_t(x,t) = c^2 u_{xx}(x,t)$, $-\infty < x < \infty$, $t > 0$ subject to the following condition: $u(x,0) = \mu(x)$.

16. $u_t(x,t) = c^2 u_{xx}(x,t) + f(x,t)$, $-\infty < x < \infty$, $t > 0$ subject to the following condition: $u(x,0) = 0$.

17. $u_t(x,t) = c^2 u_{xx}(x,t)$, $-\infty < x < \infty$, $t > 0$ subject to the condition

$$u(x,0) = \begin{cases} 1, & |x| < 1 \\ 0, & |x| > 1. \end{cases}$$

18. $u_t(x,t) = c^2 u_{xx}(x,t)$, $-\infty < x < \infty$, $t > 0$ subject to the condition $u(x,0) = e^{-|x|}$.

19. $u_t(x,t) = t2u_{xx}(x,t)$, $-\infty < x < \infty$, $t > 0$ subject to the condition $u(x,0) = f(x)$.

In Problems 20–25, use one of the Fourier transforms to solve the indicated heat equation, subject to the given initial and boundary conditions.

20. $u_t(x,t) = c^2 u_{xx}(x,t)$, $0 < x < \infty$, $t > 0$ subject to the conditions $u_x(0,t) = \mu(t)$, and $u(x,0) = 0$.

21. $u_t(x,t) = c^2 u_{xx}(x,t) + f(x,t)$, $0 < x < \infty$, $t > 0$ subject to the conditions $u(0,t) = u(x,0) = 0$.

22. $u_t(x,t) = u_{xx}(x,t)$, $0 < x < \infty$, $t > 0$ subject to the initial conditions

$$u(x,0) = \begin{cases} 1, & 0 < x < 1 \\ 0, & 1 \le x < \infty. \end{cases}$$

23. $u_t(x,t) = u_{xx}(x,t)$, $0 < x < \infty$, $t > 0$ subject to the initial conditions $u(x,0) = 0$, $u_t(x,0) = 1$ and the boundary condition $\lim_{x \to \infty} u(x,t) = 0$.

24. $u_t(x,t) = c^2 u_{xx}(x,t) + f(x,t)$, $0 < x < \infty$, $t > 0$, subject to the initial condition $u(x,0) = 0$, $0 < x < \infty$, and the boundary condition $u(0,t) = 0$.

25. $u_t(x,t) = c^2 u_{xx}(x,t) + f(x,t)$, $0 < x < \infty$, $t > 0$, subject to the initial condition $u_x(x,0) = 0$, $0 < x < \infty$, and the boundary condition $u(0,t) = 0$.

6.5 Projects Using Mathematica.

In this section we will see how Mathematica can be used to solve several problems involving the heat equation. In particular, we will develop several Mathematica notebooks which automate the computation of the solution of this equation. For a brief overview of the computer software Mathematica consult Appendix H.

Project 6.5.1. Solve the following heat distribution problem for the unit disc.

$$\begin{cases} u_t(r,\varphi,t) = u_{rr} + \dfrac{1}{r}u_r + \dfrac{1}{r^2}u_{\varphi\varphi}, & 0 < r \le 1,\ 0 \le \varphi < 2\pi,\ t > 0, \\ |\lim_{r \to 0+} u(r,\varphi,t)| < \infty, & u(1,\varphi,t) = 0,\ 0 \le \varphi < 2\pi,\ t > 0, \\ u(r,\varphi,0) = f(r,\varphi) = (1 - r^2)r\sin\varphi, & 0 < r < 1,\ 0 \le \varphi < 2\pi. \end{cases}$$

All calculation should be done using Mathematica. Display the plot of the function $u(r,\varphi,t)$ at several time instances t and find the hottest places on the disc plate for the specified instances.

Solution. The solution of the heat problem, as we derived in Section 6.3, is given by

$$u(r,\varphi,t) = \sum_{m=0}^{\infty} \sum_{n=0}^{\infty} J_m(\lambda_{mn}\,r)e^{-\lambda_{mn}^2\,t^2}\left[A_{mn}\cos m\varphi + B_{mn}\sin m\varphi\right],$$

where

$$A_{mn} = \frac{1}{\pi J_{m+1}^2(z_{mn})}\int_0^1\int_0^{2\pi} f(r,\varphi)\cos m\varphi\, J_m(z_{mn}\,r)\,r\,d\varphi\,dr,$$

$$B_{mn} = \frac{1}{\pi J_{m+1}^2(z_{mn})}\int_0^1\int_0^{2\pi} f(r,\varphi)\sin m\varphi\, J_m(z_{mn}\,r)\,r\,d\varphi\,dr.$$

Define the nth positive zero of the Bessel function $J_m(\cdot)$:

$In[1] := z[m\_, n\_]{:=}N[\text{BesselJZero}[m, n, 0]];$

Now define the eigenfunctions of the Helmholtz equation:

$In[2]{:=}\lambda[m\_, n\_] := z[m, n];$

Next, define the eigenfunctions:

$In[3]{:=}\text{EFc}[r\_, \varphi\_, m\_, n\_]{:=}\text{BesselJ}[m, \lambda[m, n]\, r]\ \text{Cos}[m\,\varphi];$

$In[4]{:=}\text{EFs}[r\_, \varphi\_, m\_, n\_]{:=}\text{BesselJ}[m, \lambda[m, n]\, r]\ \text{Sin}[m\,\varphi];$

Define the initial condition:

$In[5] := f[r\_, \varphi\_] := (1 - r^2)\, r\, \mathrm{Sin}[\varphi];$

Define the coefficients A_{mn} and B_{mn}:

$In[6] := A[m\_, n\_] := 1/(\mathrm{BesselJ}[m + 1, \lambda[m, n]^2)$
$\mathrm{Integrate}\big[f[r, \varphi]\,\mathrm{EFc}[r, \varphi, m, n]\, r, \{r, 0, 1\}, \{\varphi, 0, 2\,\mathrm{Pi}\}\big];$

$In[7] := B[m\_, n\_] := 1/(\mathrm{BesselJ}[m + 1, \lambda[m, n])^2$
$\mathrm{Integrate}\big[f[r, \varphi]\,\mathrm{EFs}[r, \varphi, m, n]\, r, \{r, 0, 1\}, \{\varphi, 0, 2\,\mathrm{Pi}\}\big];$

Define the kth partial sum which will be the approximation of the solution:

$In[8] := u[r\_, \varphi\_, t\_, k\_] := \mathrm{Sum}\Big[\big(A[m, n]\,\mathrm{EFc}[r, \varphi, m, n]$

$+ A[m, n]\,\mathrm{EFs}[r, \varphi, m, n]\big)\,\mathrm{Exp}[-\lambda[m, n])^2\, t],\ \{m, 0, k\},\ \{n, 0k\}\Big];$

To plot the solution $u(r, \varphi, t)$ at $t = 0$ (taking $k = 5$) we use

$In[9] := \mathrm{ParametricPlot3D}\Big[\{r\mathrm{Cos}[\varphi], r\,\mathrm{Sin}[\varphi], u[r, \varphi, 0, 5]\}, \{\varphi, 0, 2\,\mathrm{Pi}\},$

$\{r, 0, 1\}, \mathrm{Ticks} -> \{\{-1, 0, 1\}, \{-1, 0, 1\}, \{-.26, 0, 0.26\}\}, \mathrm{RegionFunction}$
$-> \mathrm{Function}[\{r, \varphi, u\}, r <= 1], \mathrm{BoxRatios} -> \mathrm{Automatic}, \mathrm{AxesLabel}$
$-> \{x, y, u\}\Big];$

Let us find the points on the disc with maximal temperature $u(r, \varphi)$ at the specified time instances t.

At the initial moment $t = 0$ we have that

$In[10] := \mathrm{FindMaximum}\big[\{u[r, \varphi, 0], 0 <= r < 1\,\&\&\,0 <= \varphi < 2\,\mathrm{Pi}\}, \{r, \varphi\}\big]$

$Out[10] = \{0.387802, \{r- > 0.605459, \varphi- > 1.5708\}\}$

At the moment $t = 0.2$ we have that

$In[11] := \mathrm{FindMaximum}\big[\{u[r, \varphi, 0.2], 0 <= r < 1\,\&\&\, <= \varphi < 2\,\mathrm{Pi}\}, \{r, \varphi\}\big]$

$Out[11] = \{0.297189, \{r- > 0.534556, \varphi- > 1.5708\}\}$

At the moment $t = 0.4$ we have that

$In[12] := \mathrm{FindMaximum}\big[\{u[r, \varphi, 0.4], 0 <= r < 1\,\&\&\,0 <= \varphi < 2\,\mathrm{Pi}\}, \{r, \varphi\}\big]$

$Out[12] = \{0.224781, \{r- > 0.507249, \varphi- > 1.5708\}\}$

At the moment $t = 0.6$ we have that

$In[13] := \mathrm{FindMaximum}\big[\{u[r, \varphi, 0.6], 0 <= r < 1\,\&\&\, <= \varphi < 2\,\mathrm{Pi}\}, \{r, \varphi\}\big]$

$Out[13] = \{0.168849, \{r- > 0.493909, \varphi- > 1.5708\}\}$

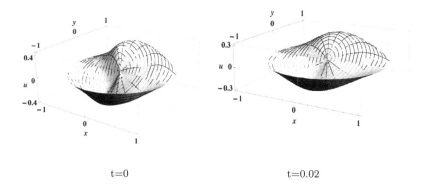

t=0   t=0.02

Figure 6.5.1

The plots of $u(r, \varphi, t)$ at the time instances $t = 0$ and $t = 0.02$ are displayed in Figure 6.5.1.

Project 6.5.2. Use the Laplace transform to solve the heat boundary value problem

$$\begin{cases} u_t(x,t) = u_{xx}(x,t), & 0 < x < \infty,\ t > 0, \\ u(x,0) = 1, & 0 < x < \infty, \\ u(0,t) = 10, & t > 0. \end{cases}$$

Solution. First we define the heat expression:

$In[1] := heat = D[u[x,t], \{t,1\}] - D[u[x,t], \{x,2\}];$

$Out[1] = u^{(0,1)}[x,t] - u^{(2,0)}[x,t];$

Next, take the Laplace transform:

$In[2] := LaplaceTransform[heat, t, s];$

$Out[2] = s\, LaplaceTransform[u[x,t], t, s]$
$- LaplaceTransform[[u^{(2,0)}[x,t], t, s] - u[x,0];$

Define:

$In[3] := U[x\_, s\_] := LaplaceTransform[u[x,t], t, s];$

Define the initial condition:

$In[4] := f[x\_] := 1;$

Using the initial condition define the expression:

$In[5] := eq = s\, U[x,s] - D[U[x,s], \{x,2\}] - f[x];$

Next find the general solution of

$$\frac{d^2 U(x,s)}{dx^2} - sU(x,s) - f(x) = 0$$

$In[6] := gensol = DSolve[eq == 0, U[x, s], x]$

$Out[6] := \{\{U[x, s]- > \frac{1}{s} + e^{\sqrt{s}\,x}C[1] + e^{-\sqrt{s}\,x}C[2]\}\}$

$In[7] := Sol = U[x, s]/.gensol[[1]];$

Define the boundary condition:

$In[8] := b[t\_] = 10;$

From the boundary conditions and the bounded property of the Laplace transform as $s \to \infty$ we find the constants $C[1]$ and $C[2]$:

$In[9] := BC = LaplaceTransform[b[t], t, s];$

$Out[9] = \frac{10}{s}$

$In[10] := BC = (Sol/.\{x \to 0, C[1]- > 0\})$

$Out[10] = \frac{9}{s} = C[2]$

$In[11] := FE=Sol/.\{C[2]- > \frac{9}{s}, C[1] \to 0\}$

$Out[11] = \frac{1}{s} + 9\,\frac{e^{-\sqrt{s}\,x}}{s};$

Find the inverse Laplace of the last expression:

$In[12] := InverseLaplace\left[\frac{1}{s} + 9\,\frac{e^{-\sqrt{s}\,x}}{s}, s, t\right]$

$Out[12] = 1 + 9\frac{\sqrt{\frac{t}{x^2}}\,x\,Erfc\left[\frac{x}{\sqrt{t}}\right]}{\sqrt{t}}.$

Plots of $u[x, t]$ at several time instances t are displayed in Figure 6.5.2.

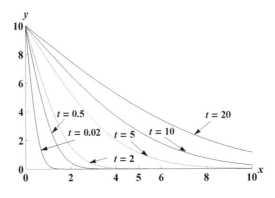

Figure 6.5.2

Project 6.5.3. Use the Fourier transform to solve the heat boundary value problem

$$
\begin{cases}
u_t(x,t) = \dfrac{1}{4} u_{xx}(x,t), & -\infty < x < \infty, \ t > 0, \\[2mm]
u(x,0) = f(x) = \begin{cases} 10, & -1 < x < 1 \\ 0, & \text{otherwise,} \end{cases} & -\infty < x < \infty.
\end{cases}
$$

Do all calculations using Mathematica.

Solution. First we define the heat expression:

$In[1] := heat = D[u[x,t],\{t,1\}] - 1/4\,D[u[x,t],\{x,2\}];$

$Out[1] = u^{(0,1)}[x,t] - \frac{1}{4}u^{(2,0)}[x,t];$

Next, take the Fourier transform:

$In[2] := FourierTransform[heat, x, z];$

$Out[2] =$FourierTransform$[u^{(0,1)}[x,t], x, z] - \frac{1}{4}$ FourierTransform$[u[x,t], x, z];$

Further define:

$In[3] := U[z\_,t\_] :=$FourierTransform$[u[x,t], x, z];$

Define the initial condition:

$In[4] := f[x\_] :=$Piecewise$[\{\{20, -1 < x < 1\}, \{0, x \geq 1\}, \{0, x \leq -1\}\}];$

$In[5] := eq = D[U[z,t],\{t,1\}] - z^2/4U[z,t];$

Next find the general solution of

$$
\frac{dU(z,t)}{dt} + \frac{z^2}{4}U(z,t) = 0
$$

$In[6] := gensol = DSolve[eq == 0, U[z,t], t]$

$Out[6] := \{\{U[z,t] - > e^{-\frac{tz^2}{4}}C[1]\}\}$

$In[7] := Sol = U[z,t]/.gensol[[1]];$

From the initial condition we find the constant $C[1]$:

$In[8] :=$IC$=$FourierTransform$[f[x], x, z];$

$Out[8] = \dfrac{20\,\sqrt{\frac{2}{\pi}}\,Sin[z]}{z}$

$In[9] := IC=(Sol/.\{z \to 0,\})$

$Out[9] = \dfrac{20\,\sqrt{\frac{2}{\pi}}\,Sin[z]}{z} = C[1]$

$In[10] :=$Final$=$Sol$/.\{C[1]- > \dfrac{20\,\sqrt{\frac{2}{\pi}}\,Sin[z]}{z}\}$

$$Out[10] = \frac{20\,e^{-\frac{tz^2}{4}}\sqrt{\frac{2}{\pi}}\,Sin[z]}{z};$$

To find the inverse Fourier transform we use the convolution theorem.
First define:

$$In[11] := g[x\_,t\_] := \text{InverseFourierTransform}\left[\frac{20\sqrt{\frac{2}{\pi}}\,Sin[z]}{z}, z, x\right];$$

$$In[12] := h[x\_,t\_] := \text{InverseFourierTransform}\left[e^{-\frac{tz^2}{4}}, z, x\right];$$

The solution $u(x,t)$ is obtained by

$$In[13] := u[x\_,t\_] := \frac{1}{\sqrt{2\pi}}\,\text{Convolve}\left[h[y,t],g[y,t],y,x\right];$$

$$In[14] := \text{FullSimplify}\,[\%]$$

$$Out[14] = 10\left(-Erf\left[\frac{-1+x}{\sqrt{t}}\right] + Erf\left[\frac{1+x}{\sqrt{t}}\right]\right)$$

CHAPTER 7

LAPLACE AND POISSON EQUATIONS

The purpose of this chapter is to study the two dimensional equation

$$\Delta u(x,y) \equiv u_{xx}(x,y) + u_{yy}(x,y) = -f(x,y), \ (x,y) \in \Omega \subseteq \mathbb{R}^2$$

and its higher dimensional version

$$\Delta u \equiv u_{xx} + u_{yy} + u_{zz} = -f(x,y,z), \ (x,y,z) \in \Omega \subseteq \mathbb{R}^3.$$

When $f \equiv 0$ in the domain Ω the homogeneous equation is called the *Laplace equation*. The nonhomogeneous equation usually is called the *Poisson equation*.

The Laplace equation is one of the most important equations in mathematics, physics and engineering. This equation is important in many applications and describes many physical phenomena, e.g., gravitational and electromagnetic potentials, and fluid flows. The Laplace equation can also be interpreted as a two or three dimensional heat equation when the temperature does not change in time, in which case the corresponding solution is called the *steady-state solution*. The Laplace equation plays a big role in several mathematical disciplines such as complex analysis, harmonic analysis and Fourier analysis. The Laplace equation is an important representative of the very large class of elliptic partial differential equations.

The first section of this chapter is an introduction, in which we state and prove several important properties of the Laplace equation, such as the maximum principle and the uniqueness property. The fundamental solution of the Laplace equation is discussed in this section. In the next several sections we will apply the Fourier method of separation of variables for constructing the solution of the Laplace equation in rectangular, polar and spherical coordinates. In the last section of this chapter we will apply the Fourier and Hankel transforms to solve the Laplace equation.

7.1 The Fundamental Solution of the Laplace Equation.

Let Ω be a domain in the plane $\mathbb{R}^2$ or in the space $\mathbb{R}^3$. As usual, we denote the boundary of Ω by $\partial\Omega$. The following two boundary value problems will be investigated throughout this chapter:

The first problem, called the *Dirichlet boundary value problem*, is to find a function $u(\mathbf{x})$, $\mathbf{x} \in \Omega$ with the properties

(D)
$$\begin{cases} \Delta u(\mathbf{x}) = -F(\mathbf{x}), & \mathbf{x} \in \Omega \\ u(\mathbf{z}) = f(\mathbf{z}), & \mathbf{z} \in \partial\Omega. \end{cases}$$

where F and f are given functions.

The second problem, called the *Neumann boundary value problem*, is to find a function $u(\mathbf{x})$, $\mathbf{x} \in \Omega$ with the properties

$$
\text{(N)} \qquad
\begin{cases}
\Delta u(\mathbf{x}) = -F(\mathbf{x}), & \mathbf{x} \in \Omega, \\
u_{\mathbf{n}}(\mathbf{z}) = g(\mathbf{z}), & \mathbf{z} \in \partial\Omega,
\end{cases}
$$

where F and g are given functions, and $u_{\mathbf{n}}$ is the derivative of the function u in the direction of the outward unit normal vector $\mathbf{n}$ to the boundary $\partial\Omega$.

The following theorem, the *Maximum Principle* for the Laplace equation is true for any of the above two problems in any dimension; we will prove it for the two dimensional homogeneous Dirichlet boundary value problem. The proof for the three dimensional Laplace equation is the same.

Theorem 7.1.1. *Let $\Omega \subset \mathbb{R}^2$ be a bounded domain with boundary $\partial\Omega$ and let $\overline{\Omega} = \Omega \cup \partial\Omega$. Suppose that $u(x, y)$ is a nonconstant, continuous function on $\overline{\Omega}$ and that $u(x, y)$ satisfies the Laplace equation*

$$
\text{(7.1.1)} \qquad u_{xx}(x, y) + u_{yy}(x, y) = 0, \quad (x, y) \in \Omega.
$$

Then, $u(x, y)$ achieves its largest, and also its smallest value on the boundary $\partial\Omega$.

Proof. We will prove the first assertion. The second assertion is obtained from the first assertion, applied to the function $-u(x, y)$.

For any $\epsilon > 0$ consider the function

$$
w(x, y) = u(x, y) + \epsilon(x^2 + y^2).
$$

Then, from (7.1.1) we obtain

$$
\text{(7.1.2)} \qquad w_{xx}(x, y) + w_{yy}(x, y) = 4\epsilon \quad (x, y) \in \Omega.
$$

Since the function $w(x, y)$ is continuous on the closed bounded region $\overline{\Omega}$, w achieves its maximum value at some point $(x_0, y_0) \in \overline{\Omega}$. If we have $(x_0, y_0) \in \Omega$, then from calculus of functions of several variables it follows that

$$
w_x(x_0, y_0) = w_y(x_0, y_0) = 0, \quad w_{xx}(x_0, y_0) \le 0, \quad w_{yy}(x_0, y_0) \le 0,
$$

which is a contradiction to (7.1.2). Therefore, $(x_0, y_0) \in \partial\Omega$.

Let

$$
A = \max\{u(x, y) : (x, y) \in \partial\Omega\}, \quad B = \max\{x^2 + y^2, \ (x, y) \in \partial\Omega\}.
$$

Then, from $u(x, y) \le w(x, y)$ on $\overline{\Omega}$ and $\partial\Omega \subseteq \overline{\Omega}$ it follows that

$$
\begin{aligned}
\max_{(x,y)\in\overline{\Omega}} u(x, y) &\le \max_{(x,y)\in\overline{\Omega}} w(x, y) = \max_{(x,y)\in\partial\Omega} w(x, y) \\
&= \max_{(x,y)\in\partial\Omega} u(x, y) + \epsilon \max_{(x,y)\in\partial\Omega} (x^2 + y^2) = A + \epsilon B.
\end{aligned}
$$

Thus,

$$\max_{(x,y)\in\overline{\Omega}} u(x,y) \leq A + \epsilon B,$$

and since $\epsilon > 0$ was arbitrary we have

$$\max_{(x,y)\in\overline{\Omega}} u(x,y) = A. \quad\blacksquare$$

With this theorem, it is very easy to establish the following uniqueness result.

Corollary 7.1.1. *If $\Omega \subset \mathbb{R}^2$ is a bounded domain, then the Dirichlet problem (D) has a unique solution.*

Proof. If u_1 and u_2 are two solutions of the Dirichlet problem (D), then the function $u = u_1 - u_2$ is a solution of the problem

$$\begin{cases} \Delta u(x,y) = 0, & (x,y) in\ \Omega, \\ u(x,y) = 0, & (x,y) \in \partial\Omega. \end{cases}$$

By Theorem 7.1.1, the function u achieves its maximum and minimum values on the boundary $\partial\Omega$. Since these boundary values are zero, u must be zero in Ω. $\blacksquare$

Remark. A function u with continuous second order partial derivatives in Ω that satisfies the Laplace equation

$$\Delta u(\mathbf{x}) = 0, \quad \mathbf{x} \in \Omega$$

is called *harmonic* in Ω.

The following are some properties of harmonic functions which can be easily verified. See Exercise 1 of this section.

(a) A finite linear combination of harmonic functions is also a harmonic function.

(b) The real and imaginary parts of any analytic function in a domain $\Omega \subseteq \mathbb{R}^2$ are harmonic functions in Ω.

(c) If $f(z)$ is an analytic function in a domain $\Omega \subseteq \mathbb{R}^2$, then $|f(z)|$ is a harmonic function in that part of Ω where $f(z) \neq 0$.

(d) Any harmonic function in a domain Ω is infinitely differentiable in that domain.

Example 7.1.1. Show that the following functions are harmonic in the indicated domain Ω.

(a) $u(x, y) = x^2 - y^2$, $\Omega = \mathbb{R}^2$.

(b) $u(x, y) = e^x \sin y$, $\Omega = \mathbb{R}^2$.

(c) $u(x, y) = \ln(x^2 + y^2)$, $\Omega = \mathbb{R}^2 \setminus \{(0, 0)\}$.

(d) $u(x, y, z) = \dfrac{1}{\sqrt{x^2 + y^2 + z^2}}$, $\Omega = \mathbb{R}^3 \setminus \{(0, 0, 0)\}$.

Solution. (a) $u(x, y)$ is the real part of the analytic function z^2, $z = x + iy$ in the whole plane.

(b) $u(x, y)$ is the real part of the analytic function e^z, $z = x + iy$ in the whole plane.

(c) $u(x, y)$ is the real part of the analytic function $\ln z$, $z = x + iy$, $\mathbb{R}^2 \setminus \{(0, 0)\}$.

(d) We write the function $u(x, y, z)$ in the form

$$u = \frac{1}{r}, \quad r = \sqrt{x^2 + y^2 + z^2}.$$

Using the chain rule we obtain

$$u_x = -\frac{x}{r^3}, \quad u_y = -\frac{y}{r^3}, \quad u_z = -\frac{z}{r^3}$$

$$u_{xx} = -\frac{r^3 - 3x^2 r}{r^6}, \quad u_{yy} = -\frac{r^3 - 3y^2 r}{r^6}, \quad u_{zz} = -\frac{r^3 - 3z^2 r}{r^6}.$$

Therefore,

$$\Delta u(x, y, z) = -\frac{r^3 - 3x^2 r}{r^6} - \frac{r^3 - 3x^2 r}{r^6} - \frac{r^3 - 3z^2 r}{r^6}$$

$$= -\frac{3r^3 - 3(x^2 + y^2 + z^2)r}{r^6} = -\frac{3r^3 - 3r^3}{r^6} = 0.$$

The following *mean value property* is important in the study of the Laplace equation. We will prove it only for the two dimensional case since the proof for higher dimensions is very similar.

Theorem 7.1.2. Mean Value Property. *If $u(x, y)$ is a harmonic function in a domain $\Omega \subseteq \mathbb{R}^2$, then for any point $(a, b) \in \Omega$*

$$u(a, b) = \frac{1}{2\pi r} \int_{\partial D_r(z)} u(x, y)\, ds = \frac{1}{r^2 \pi} \int_{D_r(z)} u(x, y)\, dx\, dy,$$

where $D_r = D_r(z)$ *is any open disc with its center at the point* $z = (a, b)$ *and radius* r *such that* $\overline{D_r(a, b)} \subseteq \Omega$.

Proof. Let $D_\epsilon(z)$ be the open disc with its center at the point $z = (a, b)$ and radius $\epsilon < r$ and let $A_\epsilon = D_r(z) \setminus D_\epsilon(z)$. Let us consider the harmonic function

$$v(x, y) = \frac{1}{\sqrt{(x - a)^2 + (y - b)^2}}$$

in the annulus A_ϵ. See Figure 7.1.1.

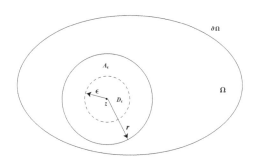

Figure 7.1.1

We will apply Green's formula (see Appendix E):

$$\int_{\partial A_\epsilon} (v u_{\mathbf{n}} - u\, v_{\mathbf{n}})\, ds = \iint_{A_\epsilon} (u\Delta v - v\Delta u)\, dx\, dy.$$

Since u and v are harmonic in A_ϵ from the above formula we have

$$\int_{\partial A_\epsilon} u v_{\mathbf{n}}\, ds = \int_{\partial A_\epsilon} v u_{\mathbf{n}}\, ds = \frac{1}{r} \int_{\partial A_\epsilon} u_{\mathbf{n}}\, ds.$$

Applying Green's formula again, now for u and $v = 1$ in the annulus A_ϵ, we have

$$\int_{\partial A_\epsilon} u_{\mathbf{n}}\, ds.$$

Thus,

$$\int_{\partial A_\epsilon} u v_{\mathbf{n}}\, ds = 0.$$

Passing to polar coordinates, from the above we have

$$\frac{1}{\epsilon} \int_0^{2\pi} u(a + \epsilon \cos \varphi, b + \epsilon \sin \varphi)\, d\varphi = \frac{1}{r} \int_0^{2\pi} u(r\, \varphi, r\, \varphi)\, d\varphi.$$

Taking lim as $\epsilon \to 0$ from both sides of the above equation and using the fact that u is a continuous function it follows that

$$2\pi u(a, b) = \frac{1}{r} \int\limits_{\partial D_r} u \, ds,$$

which is the first part of the assertions. The second assertion is left as an exercise. ∎

The converse of the mean value property for harmonic functions is also true. Even though the property is true for any dimensional domains we will state and prove it for the three dimensional case.

Theorem 7.1.3. *If u is a twice continuously differentiable function on a domain $\Omega \subseteq \mathbb{R}^3$ and satisfies the mean value property*

$$u(\mathbf{x}) = \frac{1}{\frac{4}{3}r^3\pi} \int\limits_{B(\mathbf{x},r)} u(\mathbf{y}) \, d\mathbf{y} = \frac{1}{4r^2\pi} \int\limits_{\partial B(\mathbf{x},r)} u(\mathbf{y}) \, dS(\mathbf{y}),$$

for every ball $B(\mathbf{x}, r)$ with the property $\overline{B(\mathbf{x}, r)} \subset \Omega$, then u is harmonic in Ω.

Proof. Suppose contrarily that u is not harmonic in Ω. Then $\Delta u(\mathbf{x}_0) \neq 0$ for some $\mathbf{x}_0 \in \Omega$. We may assume that $\Delta u(\mathbf{x}_0) > 0$. Since u is a twice continuously differentiable function in Ω there exists an $r_0 > 0$ such that $\Delta u(\mathbf{x}) > 0$ for every $\mathbf{x} \in B(\mathbf{x}_0, r)$ and every $r < r_0$. Now, define the function

$$f(r) = \frac{1}{\frac{4}{3}r^3\pi} \int\limits_{B(\mathbf{x}_0,r)} u(\mathbf{y}) \, d\mathbf{y} = \frac{1}{\frac{4}{3}\pi} \int\limits_{B(\mathbf{0},1)} u(\mathbf{x}_0 + r\mathbf{z}) \, d\mathbf{z}, \quad r < r_0.$$

Because u satisfies the mean value property we have $f(r) = u(\mathbf{x}_0)$. Thus, $f'(r) = 0$. On the other hand, from the Gauss–Ostrogradski formula we have

$$f'(r) = \frac{1}{\frac{4}{3}\pi} \int\limits_{B(\mathbf{0},1)} grad\left(u(\mathbf{x}_0 + r\mathbf{z})\right) \cdot \mathbf{z} \, d\mathbf{z}$$

$$= \frac{1}{\frac{4}{3}\pi} \int\limits_{\partial B(\mathbf{0},1)} \Delta u(\mathbf{x}_0 + r\mathbf{z})) \, d\mathbf{z} > 0,$$

which is a contradiction. ∎

Now, we will discuss the fundamental solution of the Laplace equation. Consider the Laplace equation in the whole plane without the origin.

(7.1.3) $u_{xx}(x, y) + u_{yy}(x, y) = 0, \quad (x, y) \in \mathbb{R}^2 \setminus \{(0,0)\}.$

We will find a solution $u(x, y)$ of (7.1.3) of the form

$$u(x, y) = f(r), \quad r = \sqrt{x^2 + y^2},$$

where $f(r)$ is a twice differentiable function which will be determined.
Using the chain rule we have

$$u_x(x, y) = \frac{x}{r} f'(r), \quad u_y(x, y) = \frac{y}{r} f'(r), \quad (x, y) \neq (0, 0)$$

$$u_{xx}(x, y) = \frac{1}{r} f'(r) - \frac{x^2}{r^3} f'(r) + \frac{x^2}{r^2} f''(r),$$

$$u_{yy}(x, y) = \frac{1}{r} f'(r) - \frac{y^2}{r^3} f'(r) + \frac{y^2}{r^2} f''(r), \quad (x, y) \neq (0, 0),$$

and so the Laplace equation becomes

$$\frac{1}{r} f'(r) + f''(r), \quad r \neq 0.$$

The general solution of the last equation is given by

$$f(r) = C_1 \ln |r| + C_2, \quad r \neq 0,$$

where C_1 and C_2 are arbitrary constants. For a particular choice of these
constants, the function

(7.1.4) $$\Phi(x, y) = -\frac{1}{2\pi} \ln(x^2 + y^2)$$

is called the *fundamental solution* or *Green function* of the two dimensional
Laplace equation (7.1.3) in $\mathbb{R}^2 \setminus \{(0, 0)\}$.

For the three dimensional Laplace equation in the whole space without the
origin

(7.1.5) $$\Delta u(x, y, z) = 0, \quad (x, y, z) \in \mathbb{R}^3 \setminus \{(0, 0, 0)\},$$

working similarly as in the two dimensional Laplace equation we obtain that
the fundamental solution of (7.1.5) is given by

(7.1.6) $$\Phi(x, y, z) = \frac{1}{4\pi} \frac{1}{\sqrt{x^2 + y^2 + z^2}}.$$

The significance of the fundamental solution of the Laplace equation is
because of the following two integral representation theorems.

Theorem 7.1.4. *If F is a twice continuously differentiable function on $\mathbb{R}^n$ $(n = 2, 3)$ which vanishes outside a closed and bounded subset of $\mathbb{R}^n$, then*

$$u(\mathbf{x}) = \int_{\mathbb{R}^n} \Phi(\mathbf{x}, \mathbf{y}) \, F(\mathbf{y}) \, dy$$

is the unique solution of the Poisson equation

$$\Delta u(\mathbf{x}) = -F(\mathbf{x}), \quad \mathbf{x} \in \mathbb{R}^n,$$

where $\Phi(\cdot, \cdot)$ is the fundamental solution of the Laplace equation on $\mathbb{R}^n \setminus \{\mathbf{0}\}$, given by

$$\Phi(\mathbf{x}, \mathbf{y}) = \begin{cases} \frac{1}{2\pi} \ln \frac{1}{|\mathbf{x}-\mathbf{y}|}, & \mathbf{x} = (x, y) \ \text{if} \ n = 2 \\[2mm] \frac{1}{4\pi} \frac{1}{|\mathbf{x}-\mathbf{y}|}, & \mathbf{x} = (x, y, z) \ \text{if} \ n = 3. \end{cases}$$

Theorem 7.1.5. *Let Ω be a domain in $\mathbb{R}^n$, $n = 2, 3$, with a smooth boundary $\partial\Omega$. If $u(\mathbf{x})$ is a continuously differentiable function on $\overline{\Omega} = \Omega \cup \partial\Omega$ and harmonic in Ω, then*

$$u(\mathbf{x}) = \int_{\partial\Omega} \left[\Phi(\mathbf{x}, \mathbf{y}) \, u_{\mathbf{n}}(\mathbf{y}) - u(\mathbf{y}) \Phi_{\mathbf{n}(\mathbf{y}}(\mathbf{x}, \mathbf{y}) \right] dS(\mathbf{y}), \quad \mathbf{x} \in \Omega,$$

where $\Phi(\cdot, \cdot)$ is the fundamental solution of the Laplace equation on $\mathbb{R}^n \setminus \{\mathbf{0}\}$, given by

$$\Phi(\mathbf{x}, \mathbf{y}) = \begin{cases} \frac{1}{2\pi} \ln \frac{1}{|\mathbf{x}-\mathbf{y}|}, & \mathbf{x} = (x, y) \ \text{if} \ n = 2 \\[2mm] \frac{1}{4\pi} \frac{1}{|\mathbf{x}-\mathbf{y}|}, & \mathbf{x} = (x, y, z) \ \text{if} \ n = 3. \end{cases}$$

and $u_{\mathbf{n}}(\mathbf{y})$ and $\Phi_{\mathbf{n}}(\mathbf{x}, \mathbf{y})$ are the derivatives of u and Φ, respectively, in the direction of the unit outward vector $\mathbf{n}$ to the boundary $\partial\Omega$ at the boundary point $\mathbf{y} \in \partial\Omega$.

Proof. First, from the Gauss–Ostrogradski formula (see Appendix E) applied to two harmonic functions u and v in the domain Ω we obtain

$$(7.1.7) \qquad \int_{\partial\Omega} \left[v(\mathbf{y}) \, u_{\mathbf{n}(\mathbf{y})}(\mathbf{y}) - u(\mathbf{y}) v_{\mathbf{n}(\mathbf{y})}(\mathbf{y}) \right] dS(\mathbf{y}) = 0.$$

For a fixed point $\mathbf{x} \in \Omega$ and sufficiently small $\epsilon > 0$ consider the closed ball $\overline{B_\epsilon(\mathbf{x})} = \{\mathbf{y} \in \mathbb{R}^n : |\mathbf{y} - \mathbf{x}| \le \epsilon\} \subset \Omega$. Let $\Omega_\epsilon = \Omega \setminus \overline{B_\epsilon(\mathbf{x})}$ (remove the closed ball $\overline{B_\epsilon(\mathbf{x})}$ from Ω). Since $\Phi(\mathbf{x}, \mathbf{y})$ is harmonic in Ω_ϵ we can apply

Equation (7.1.7) to the functions $u(\mathbf{x})$ and $v(x) = \Phi(\mathbf{x}, \mathbf{y})$ in the domain Ω_ϵ with boundary $\partial\Omega_\epsilon = \partial\Omega \cup \{\mathbf{y} : |\,\mathbf{y} - \mathbf{x}\,| = \epsilon\}$ and obtain

(7.1.8)
$$\int_{\partial\Omega} \left[\Phi(\mathbf{x}, \mathbf{y})\, u_{\mathbf{n}}(\mathbf{y}) - u(\mathbf{y})\Phi_{\mathbf{n}(\mathbf{y})}(\mathbf{x}, \mathbf{y})\right] dS(\mathbf{y})$$
$$= \int_{\{\mathbf{y}:\, |\mathbf{y}-\mathbf{x}|=\epsilon\}} \left[\Phi(\mathbf{x}, \mathbf{y})\, u_{\mathbf{n}}(\mathbf{y}) - u(\mathbf{y})\Phi_{\mathbf{n}(\mathbf{y})}(\mathbf{x}, \mathbf{y})\right] dS(\mathbf{y}).$$

On the sphere $\{\mathbf{y} : |\,\mathbf{y} - \mathbf{x}\,| = \epsilon\}$ we have

$$\Phi(\mathbf{x}, \mathbf{y}) = \begin{cases} -\frac{1}{2\pi} \ln \epsilon, & n = 2 \\ \frac{1}{4\pi\epsilon}, & n = 3, \end{cases}$$

and thus,

$$\Phi_{\mathbf{n}(\mathbf{y})}(\mathbf{x}, \mathbf{y}) = \begin{cases} -\frac{1}{2\pi\epsilon}, & n = 2 \\ -\frac{1}{4\pi\epsilon^2}, & n = 3. \end{cases}$$

From the above fact and the continuity of the function $u(\mathbf{x})$ in Ω it follows that

$$\lim_{\epsilon \to 0} \int_{\{\mathbf{y}:\, |\mathbf{y}-\mathbf{x}|=\epsilon\}} \left[u(\mathbf{y}) - u(\mathbf{x})\right]\Phi_{\mathbf{n}(\mathbf{y})}(\mathbf{x}, \mathbf{y})\, dS(\mathbf{y}) = 0$$

and therefore, the right hand side of (7.1.8) approaches 0 as $\epsilon \to 0$, which concludes the proof. ∎

Now, we will introduce the *Green function* for a domain.

Let Ω be a domain with smooth boundary $\partial\Omega$. The *Green function* $G(\mathbf{x}, \mathbf{y})$ for the domain Ω is defined by

$$G(\mathbf{x}, \mathbf{y}) = \Phi(\mathbf{x}, \mathbf{y}) + h(\mathbf{x}, \mathbf{y}), \quad \mathbf{x}, \mathbf{y} \in \Omega,$$

where $h(\mathbf{x}, \mathbf{y})$ is a harmonic function on Ω in each variable $\mathbf{x}, \mathbf{y} \in \Omega$ separately and it is such that the function $G(\cdot, \cdot)$ satisfies the boundary condition

$$G(\mathbf{x}, \mathbf{y}) = 0$$

if at least one of the points $\mathbf{x}$ and $\mathbf{y}$ is on the boundary $\partial\Omega$.

Using the Divergence Theorem of Gauss–Ostrogradski, just as in Theorem 7.1.5, it can be shown that the solutions of the Dirichlet and Neumann boundary value problems (D) and (N), respectively, are given by the following two theorems.

Theorem 7.1.6. *Let* Ω *be a domain in* $\mathbb{R}^2$ *or* $\mathbb{R}^3$ *with a smooth boundary* $\partial\Omega$. *If* $F(\mathbf{x})$ *is a continuous function with continuous first order partial derivatives on* Ω *and* $f(\mathbf{x})$ *is a continuous function on* $\overline{\Omega}$, *then*

$$u(\mathbf{x}) = \int_\Omega G(\mathbf{x}, \mathbf{y})\, F(\mathbf{y})\, d\mathbf{y} - \int_{\partial\Omega} f(\mathbf{z})\, G_{\mathbf{n}}(\mathbf{x}, \mathbf{z})\, dS(\mathbf{z}), \quad \mathbf{x} \in \Omega$$

is the unique solution of the Dirichlet boundary value problem (D)

$$\begin{cases} \Delta\, u(\mathbf{x}) = -F(\mathbf{x}), & \mathbf{x} \in \Omega, \\ u(\mathbf{z}) = f(\mathbf{z}), & \mathbf{z} \in \partial\Omega. \end{cases}$$

Theorem 7.1.7. *Let* Ω *be a domain in* $\mathbb{R}^2$ *or* $\mathbb{R}^3$ *with a smooth boundary* $\partial\Omega$. *If* $F(\mathbf{x})$ *is a continuous function with continuous first order partial derivatives on* Ω *and* $g(\mathbf{x})$ *is a continuous function on* $\overline{\Omega}$, *then*

$$u(\mathbf{x}) = \int_\Omega G(\mathbf{x}, \mathbf{y})\, F(\mathbf{y})\, d\mathbf{y}\, d\mathbf{y} - \int_{\partial\Omega} g(\mathbf{z})\, G(\mathbf{x}, \mathbf{z})\, dS(\mathbf{z}) + C, \quad \mathbf{x} \in \Omega$$

is a solution of the Neumann boundary value problem (N)

$$\begin{cases} \Delta\, u(\mathbf{x}) = -F(\mathbf{x}), & \mathbf{x} \in \Omega, \\ u_{\mathbf{n}}(\mathbf{x}) = g(\mathbf{x}), & \mathbf{x} \in \partial\Omega, \end{cases}$$

where C *is any numerical constant, provided that the following condition is satisfied*

$$\int_\Omega F(\mathbf{y})\, d\mathbf{y} = \int_{\partial\Omega} g(\mathbf{z})\, dS(\mathbf{z}).$$

For some basic properties of the Green function see Exercise 2 of this section.

Now, let us take a few examples.

Example 7.1.2. Verify that the expression

$$G(\mathbf{x}, \mathbf{y}) = \Phi(\mathbf{x}, \mathbf{y}) - \Phi\left(|\,\mathbf{x}\,|\,\mathbf{y}\,, \frac{\mathbf{x}}{|\,\mathbf{x}\,|}\right)$$

is actually the Green function for the unit ball $|\,\mathbf{x}\,| < 1$ in $\mathbb{R}^n$ for $n = 2, 3$, where $\Phi(\cdot, \cdot)$ is the fundamental solution of the Laplace equation.

Solution. Let us consider the function $h(\mathbf{x}, \mathbf{y})$, defined by

$$h(\mathbf{x}, \mathbf{y}) = \Phi\left(|\,\mathbf{x}\,|\,\mathbf{y}, \frac{\mathbf{x}}{|\,\mathbf{x}\,|}\right).$$

It is straightforward to verify that for every $\mathbf{x}$ and $\mathbf{y}$ the following identities hold.

$$(7.1.9) \quad \begin{cases} \left| |\mathbf{x}| \mathbf{y} - \dfrac{\mathbf{x}}{|\mathbf{x}|} \right| = |\mathbf{y}| \left| \mathbf{x} - \dfrac{\mathbf{y}}{|\mathbf{y}|^2} \right|, \\[2mm] \left| |\mathbf{x}| \mathbf{y} - \dfrac{\mathbf{x}}{|\mathbf{x}|} \right| = |\mathbf{x}| \left| \mathbf{y} - \dfrac{\mathbf{x}}{|\mathbf{x}|^2} \right|, \end{cases}$$

and

$$(7.1.10) \quad \left| |\mathbf{x}| \mathbf{y} - \dfrac{\mathbf{x}}{|\mathbf{x}|} \right| = \left| |\mathbf{y}| \mathbf{x} - \dfrac{\mathbf{y}}{|\mathbf{y}|} \right|.$$

From the first identity in (7.1.9) it follows that the function $h(\mathbf{x}, \mathbf{y})$ is harmonic with respect to $\mathbf{x}$ in the unit ball, while from the second identity in (7.1.9) we have $h(\mathbf{x}, \mathbf{y})$ is harmonic with respect to $\mathbf{y}$ in the unit ball. From the definition of $h(\mathbf{x}, \mathbf{y})$ and identity (7.1.10) we have

$$h(\mathbf{x}, \mathbf{y}) = \Phi(\mathbf{x}, \mathbf{y}), \quad \text{when either } |\mathbf{x}| = 1 \text{ or } |\mathbf{y}| = 1.$$

Therefore, the above defined function $G(\cdot, \cdot)$ satisfies all the properties for a function to be a Green function.

If we take $n = 2$, for example, then from the definition of the fundamental solution of the Laplace equation in $\mathbb{R}^2 \setminus \{\mathbf{0}\}$ and the above discussion, it follows that the Green function in the unit disc of the complex plane $\mathbb{R}^2$ is given by

$$G(\mathbf{x}, \mathbf{y}) = -\frac{1}{2\pi} \ln \frac{|\mathbf{x} - \mathbf{y}|}{\left| |\mathbf{x}| \mathbf{y} - \frac{\mathbf{x}}{|\mathbf{x}|} \right|}, \quad |\mathbf{x}| < 1, \ |\mathbf{y}| < 1.$$

Example 7.1.3. Find the Green function for the upper half plane.

Solution. For a point $\mathbf{x} = (x, y)$, define its reflection by $\mathbf{x}^* = (x, -y)$. If

$$U^+ = \{\mathbf{x} = (x, y) \in \mathbb{R}^2 : y > 0\}$$

is the upper half plane, then define the function $g(\mathbf{x}, \mathbf{y})$ on U^+ by

$$g(\mathbf{x}, \mathbf{y}) = \Phi(\mathbf{x}^*, \mathbf{y}), \quad \mathbf{x}, \mathbf{y} \in U^+,$$

where $\Phi(\cdot, \cdot)$ is the fundamental solution of the Laplace equation in $\mathbb{R}^2 \setminus \{\mathbf{0}\}$. Using the obvious identity

$$|\mathbf{x} - \mathbf{y}^*| = |\mathbf{y} - \mathbf{x}^*|, \quad \mathbf{x}, \mathbf{y} \in \mathbb{R}^2$$

and the facts that $\mathbf{x}^* \notin U^+$ if $\mathbf{x} \in U^+$, and the same for $\mathbf{y}$, it follows that

$$g(\mathbf{x}, \mathbf{y}) = g(\mathbf{y}, \mathbf{x}), \quad \mathbf{x}, \mathbf{y} \in U^+$$

and also, $g(\mathbf{x}, \mathbf{y})$ is harmonic with respect to $\mathbf{x}$ in U^+ for every fixed $\mathbf{y} \in U^+$ and $g(\mathbf{x}, \mathbf{y})$ is harmonic with respect to $\mathbf{y}$ in U^+ for every fixed $\mathbf{x} \in U^+$.

Further, from the definitions of the functions $g(\mathbf{x}, \mathbf{y})$ and $\Phi(\mathbf{x}, \mathbf{y})$ it follows that $g(\mathbf{x}, \mathbf{y}) = \Phi(\mathbf{x}, \mathbf{y})$ for every $\mathbf{x} \in \partial U^+$ or every $\mathbf{y} \in \partial U^+$.

Therefore,

$$G(\mathbf{x}, \mathbf{y}) = \Phi(\mathbf{x}, \mathbf{y}) - \Phi(\mathbf{x}^*, \mathbf{y}), \quad \mathbf{x}, \mathbf{y} \in U^+$$

is the Green function for the domain U^+.   ∎.

In the next several examples we apply Theorem 7.1.4 to harmonic functions in domains in $\mathbb{R}^2$ or $\mathbb{R}^3$, giving important integral representations of the harmonic functions in the domains.

Example 7.1.4. If u and v are two harmonic functions in a domain Ω in $\mathbb{R}^2$ or $\mathbb{R}^3$ with a smooth boundary $\partial\Omega$ and continuously differentiable on $\overline{\Omega} = \Omega \cup \partial\Omega$, then

$$\int_{\partial\Omega} u(\mathbf{z}) v_{\mathbf{n}}(\mathbf{z}\, ds(\mathbf{z}) = \int_{\partial\Omega} v(\mathbf{z}) u_{\mathbf{n}}(\mathbf{z}\, ds(\mathbf{z}),$$

where $u_{\mathbf{n}}$ and $v_{\mathbf{n}}$ are the derivatives of the functions u and v, respectively, in the direction of the outward unit normal vector $\mathbf{n}$ to the boundary $\partial\Omega$.

Proof. For $n = 3$ apply the Gauss–Ostrogradski formula (see Appendix E)

$$\iiint_{\Omega} div\big(\mathbf{F}(\mathbf{x})\big)\, d\mathbf{x} = \iint_{\partial\Omega} \big(\mathbf{F}(\mathbf{z})\big) \cdot \mathbf{n}(\mathbf{z})\, d\mathbf{z}$$

to $\mathbf{F}(\mathbf{x}) = u(\mathbf{x})\, grad\big(v(\mathbf{x})\big) - v(\mathbf{x})\, grad\big(u(\mathbf{x})\big)$ and use the fact that u and v are harmonic in Ω to obtain the result.   ∎

Example 7.1.5. If u is a harmonic function in a domain Ω in $\mathbb{R}^2$ or $\mathbb{R}^3$ with a smooth boundary $\partial\Omega$ and continuously differentiable on $\overline{\Omega} = \Omega \cup \partial\Omega$, then

$$u(\mathbf{x}) = -\int_{\partial\Omega} G_{\mathbf{n}}(\mathbf{x}, \mathbf{z})) f(\mathbf{z})\, ds(\mathbf{z}), \quad \mathbf{x} \in \Omega$$

is the solution of the Dirichlet boundary value problem

$$\begin{cases} \Delta u(\mathbf{x}) = 0, & \mathbf{x} \in \Omega, \\ u(\mathbf{z}) = f(\mathbf{z}), & \mathbf{z} \in \partial\Omega. \end{cases}$$

Proof. Apply Theorem 7.1.6.   ∎

Exercises for Section 7.1.

1. Show that the following are true.

   (a) Any finite linear combination of harmonic functions is a harmonic function.

   (b) The real and imaginary parts of any analytic function in a domain $\Omega \subseteq \mathbb{R}^2$ are harmonic functions in Ω.

   (c) If $f(z)$ is an analytic function in a domain Ω, then $\mid f(z) \mid$ is a harmonic function in the set of points $z \in \Omega$ where $f(z) \neq 0$.

   (d) Any harmonic function in a domain Ω is infinitely differentiable in that domain.

2. Show that the Green function G for any domain Ω in $\mathbb{R}^2$ or $\mathbb{R}^3$ is symmetric:
$$G(\mathbf{x}, \mathbf{y}) = G(\mathbf{y}, \mathbf{x}), \quad \mathbf{x}, \mathbf{y} \in \Omega.$$

   Hint: Use the Gauss–Ostrogradski formula.

3. Find the maximum value of the harmonic function $u(x, y) = x^2 - y^2$ in the closed unit disc $x^2 + y^2 \leq 1$.

4. Find the harmonic function $u(x, y)$ in the upper half plane $\{(x, y) : y > 0\}$, subject to the boundary condition
$$u(x, 0) = \frac{x}{x^2 + 1}.$$

5. Find the harmonic function $u(x, y, z)$, $(x, y, z) \in \mathbb{R}^3$, in the lower half space $\{(x, y, z) : z < 0\}$, subject to the boundary condition
$$u(x, y, 0) = \frac{1}{(1 + x^2 + y^2)^{\frac{3}{2}}}.$$

6. Find the Green function of the right half plane $RP = \{(x, y) : x > 0\}$ and solve the Poisson equation
$$\begin{cases} \Delta\, u(\mathbf{x}) = f(\mathbf{x}), & \mathbf{x} \in RP \\ u(\mathbf{x}) = h(\mathbf{x}, & \mathbf{x} \in \partial RP. \end{cases}$$

7. Find the Green function of the right half plane $UP = \{(x, y) : y > 0\}$ subject to the boundary condition
$$G_\mathbf{n}(\mathbf{x}, \mathbf{y}) = 0, \quad (\mathbf{x}, \mathbf{y}) \in \partial UP$$

and solve the Poisson equation

$$\begin{cases} \Delta u(\mathbf{x}) = f(\mathbf{x}), & \mathbf{x} \in UP \\ u_n(\mathbf{x}) = h(\mathbf{x}, & \mathbf{x} \in \partial UP. \end{cases}$$

8. Using the Green function for the unit ball in $\mathbb{R}^n$, $n = 2, 3$, derive the Poisson formula

$$u(\mathbf{x}) = \frac{1}{\omega_n} \int\limits_{|\mathbf{y}|=1} \frac{1 - |\mathbf{x}|^2}{|\mathbf{y} - \mathbf{x}|^2} f(\mathbf{y}) \, dS(\mathbf{y}),$$

where ω_n is the area of the unit sphere in $\mathbb{R}^n$ which gives the solution of the Poisson boundary value problem

$$\begin{cases} \Delta u(\mathbf{x}) = -f(\mathbf{x}), & |\mathbf{x}| < 1 \\ u(\mathbf{x}) = g(\mathbf{x}), & |\mathbf{x}| = 1. \end{cases}$$

9. Using the Green function for the disc $D_R = \{\mathbf{x} = (x, y) : x^2 + y^2 < R^2\}$, solve the following Poisson equation with Neumann boundary values

$$\begin{cases} \Delta u(\mathbf{x}) = -f(\mathbf{x}), & \mathbf{x} \in D_R \\ u_n(\mathbf{x}) = g(\mathbf{x}), & |\mathbf{x}| = R, \end{cases}$$

provided that the function g satisfies the boundary condition

$$\int\limits_{\partial B_R} g(\mathbf{y}) \, dS(\mathbf{y}).$$

10. In the domain $\{(x, y) : x^2 + y^2 \le 1, y \ge 0\}$ solve the boundary value problem

$$\begin{cases} \Delta u(x, y) = 0, & x^2 + y^2 < 1, \ y > 0 \\ u(x, y) = \begin{cases} 1, & x^2 + y^2 = 1 \\ 0, & y = 0. \end{cases} \end{cases}$$

Hint: Use the Green function.

7.2 Laplace and Poisson Equations on Rectangular Domains.

In this section we will discuss the *Separation of Variables Method* for solving the Laplace and Poisson equations. We will consider the Laplace and Poisson equations with homogeneous and nonhomogeneous Dirichlet and Neumann boundary conditions on rectangular domains. We already discussed this method in the previous two chapters when we solved the wave and heat equations. The idea and procedure of the separation of variables method is the same and therefore we will omit many details in our discussions.

1^0. *Laplace Equation on a Rectangle.* First let us have Dirichlet boundary conditions.

Consider the following Laplace boundary value problem with Dirichlet boundary values.

(U) $$\begin{cases} u_{xx}(x,y) + u_{yy}(x,y) = 0, & 0 < x < a, \ 0 < y < b \\ u(x,0) = f_1(x), & u(x,b) = f_2(x), \ 0 < x < a \\ u(0,y) = g_1(y), & u(a,y) = g_2(y), \ 0 < y < b. \end{cases}$$

We split the above problem (U) into the two problems

(V) $$\begin{cases} v_{xx}(x,y) + v_{yy}(x,y) = 0, & 0 < x < a, \ 0 < y < b \\ v(x,0) = f_1(x), & v(x,b) = f_2(x), \ 0 < x < a \\ v(0,y) = v(a,y) = 0, & 0 < y < b, \end{cases}$$

and

(W) $$\begin{cases} w_{xx}(x,y) + w_{yy}(x,y) = 0, & 0 < x < a, \ 0 < y < b \\ w(x,0) = w(x,b) = 0, & 0 < x < a \\ w(0,y) = g_1(y), & w(a,y) = g_2(y), \ 0 < y < b \end{cases}$$

with homogeneous Dirichlet boundary values on two parallel sides. If v and w are the solutions of problems (V) and (W), respectively, then $u = v + w$ is the solution of problem (W).

From symmetrical reasons, it is enough to solve either problem (V) or (W). Let us consider problem (V).

If we assume that the solution of (V) is of the form $v(x,y) = X(x)Y(y)$, then after substituting the derivatives into the Laplace equation and using the homogeneous boundary conditions in (V), we have the eigenvalue problem

(7.2.1) $$X''(x) + \lambda^2 X(x) = 0, \quad X(0) = X(a) = 0$$

and the differential equation

(7.2.2) $$Y''(y) - \lambda^2 Y(y) = 0.$$

The eigenvalues and corresponding eigenfunctions of (7.2.1) are

$$\lambda_n = \frac{n\pi}{a}, \quad X_n(x) = \sin\frac{n\pi}{a}x, \ n = 1, 2, \ldots$$

For these eigenvalues λ_n, the general solution of (7.2.2) is given by

$$Y_n(y) = A_n \cosh \lambda_n y + B_n \sinh \lambda_n y.$$

Therefore,

$$v(x, y) = \sum_{n=1}^{\infty} \left[A_n \cosh \lambda_n y + B_n \sinh \lambda_n y\right] \sin \lambda_n x.$$

The above coefficients A_n and B_n are determined from the boundary conditions in problem (V):

$$f_1(x) = v(x, 0) = \sum_{n=1}^{\infty} A_n \sin \lambda_n x,$$

$$f_2(x) = v(x, b) = \sum_{n=1}^{\infty} \left[A_n \cosh \lambda_n b + B_n \sinh \lambda_n b\right] \sin \lambda_n x.$$

The above two series are Fourier sine series and therefore,

(7.2.3)
$$\begin{cases} A_n = \dfrac{2}{a} \displaystyle\int_0^a f_1(x) \sin \lambda_n x \, dx, \\[4mm] A_n \cosh \lambda_n b + B_n \sinh \lambda_n b = \dfrac{2}{b} \displaystyle\int_0^a f_2(x) \sin \lambda_n x \, dx, \end{cases}$$

from which A_n and B_n are determined.

Let us take an example.

Example 7.2.1. Find the solution $u(x, y)$ of the Laplace equation in the rectangle $0 < x < \pi$, $0 < y < 1$, subject to the boundary conditions

$$\begin{cases} u(x, 0) = f(x), & u(x, \pi) = f(x), \ 0 < x < \pi \\ u(0, y) = f(y), & u(\pi, y) = f(y), \ 0 < y < \pi, \end{cases}$$

where

$$f(x) = \begin{cases} x, & 0 < x < \frac{\pi}{2} \\ \pi - x, & \frac{\pi}{2} < x < \pi. \end{cases}$$

Display the plot of the solution $u(x, y)$.

Solution. Consider the following two problems.

$$\begin{cases} v_{xx}(x,y) + v_{yy}(x,y) = 0, & 0 < x < \pi, \ 0 < y < \pi \\ v(x,0) = f(x), & v(x,b) = f(x), \ 0 < x < a \\ v(0,y) = v(\pi,y) = 0, & 0 < y < \pi, \end{cases}$$

and

$$\begin{cases} w_{xx}(x,y) + w_{yy}(x,y) = 0, & 0 < x < \pi, \ 0 < y < \pi \\ w(x,0) = w(x,\pi) = 0, & 0 < x < \pi \\ w(0,y) = f(y), & w(\pi,y) = f(y), \ 0 < y < \pi. \end{cases}$$

It is enough to solve only the first problem for the function v. The eigenvalues λ_n for this problem are $\lambda_n = n$ and so from (7.2.3) we have

$$A_n = \frac{2}{\pi} \int_0^\pi f(x) \sin nx \, dx,$$

$$A_n \cosh n\pi + B_n \sinh n\pi = \frac{2}{\pi} \int_0^\pi f(x) \sin nx \, dx.$$

If we solve the integral for the given function $f(x)$ we obtain

$$A_n = \frac{4}{\pi} \frac{\sin \frac{n\pi}{2}}{n^2}, \quad A_n \cosh n\pi + B_n \sinh n\pi = \frac{4}{\pi} \frac{\sin \frac{n\pi}{2}}{n^2},$$

from which we obtain that the solution $v(x,y)$ is given by

$$v(x,y) = \frac{4}{\pi} \sum_{n=1}^\infty \frac{\sin \frac{n\pi}{2}}{n^2 \sinh n\pi} \left[\sinh(ny) + \sinh(n(\pi - y)) \right] \sin nx.$$

Because of symmetry reasons, we can interchange the roles of x and y and obtain the solution of problem for the function w:

$$w(x,y) = \frac{4}{\pi} \sum_{n=1}^\infty \frac{\sin \frac{n\pi}{2}}{n^2 \sinh n\pi} \left[\sinh nx + \sinh n(\pi - x) \right] \sin ny.$$

Therefore, the solution of the original problem is given by

$$u(x,y) = \frac{4}{\pi} \sum_{n=1}^\infty \frac{\sin \frac{n\pi}{2}}{n^2 \sinh n\pi} \left\{ \left[\sinh ny + \sinh n(\pi - y) \right] \sin nx \right.$$

$$\left. + \left[\sinh nx + \sinh n(\pi - x) \right] \sin ny \right\}.$$

The plot of the solution $u(x,y)$ is displayed in Figure 7.2.1.

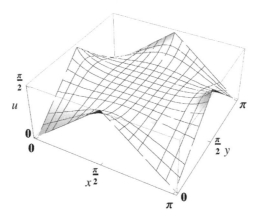

Figure 7.2.1

Remark. The solution of the Laplace equation on a rectangle, or any planar or space domain with piecewise smooth boundary, subject to Neumann boundary conditions does not always exist, and even if exists it is not unique (recall Theorem 7.1.7 of the previous section).

Example 7.2.2. Consider the problem

$$\begin{cases} u_{xx}(x,y) + u_{yy}(x,y) = 0, \quad 0 < x < \pi, \ 0 < y < \pi \\ u_x(0,y) = u_x(\pi,y) = 0, \quad 0 < y < \pi \\ u_y(x,0) = 1, \quad u_y(x,\pi) = K, \ 0 < x < \pi, \end{cases}$$

where K is a constant. Find a solution of this problem.

Solution. Looking for a solution of the form $u(x,y) = X(x)Y(y)$ we find that

$$X''(x) + \lambda X(x) = 0, \quad X'(0) = X'(\pi) = 0$$

and

$$Y''(y) + \lambda Y(y) = 0.$$

Solving the eigenvalue problem we find that eigenvalues and corresponding eigenfunctions are

$$\lambda_0 = 0, \ X_0(x) = 1; \quad \lambda_n = n^2, \ X_n(x) = \cos nx.$$

The corresponding solutions $Y(y)$ are given by

$$Y_0(y) = A_0 + B_0 y, \quad Y_n(y) = A_n \cosh ny + B_n \sinh ny.$$

Therefore,

$$u(x,y) = A_0 + B_0 y + \sum_{n=1}^{\infty} \left[A_n \cosh ny + B_n \sinh ny \right] \cos nx.$$

From the boundary condition $u_y(x, 0) = 1$ it follows that

$$B_0 + \sum_{n=1}^{\infty} nB_n \cos nx = 1, \quad 0 < x < \pi,$$

from which we find (using the Fourier cosine series)

$$B_0 = 1, \quad B_n = 0.$$

From the other boundary condition $u_y(x, \pi) = K$ it follows that

$$B_0 + \left[nA_n \sinh n\pi + nB_n \cosh n\pi \right] \cos nx = K, \quad n = 1, 2, \dots.$$

The left hand side of the last equation is a Fourier cosine series and so, from the uniqueness theorem for Fourier cosine expansion, it follows that

$$B_0 = K, \quad nA_n \sinh n\pi + nB_n \cosh n\pi = 0, \ n = 1, 2, \dots.$$

Thus, we must have

$$K = B_0 = 1, \quad A_n = B_n = 0, \ n = 1, 2, \dots.$$

Therefore, a solution of the given problem exists only if $K = 1$, in which case we have the following infinitely many solutions

$$u(x, u) = A_0 + y,$$

where A_0 is any constant.

The next example is about the Laplace equation on a rectangle with mixed boundary conditions.

Example 7.2.3. Using the separation of variables method, solve the problem

(7.2.4) $\qquad u_{xx}(x, y) + u_{yy}(x, y) = 0, \ 0 < x < a, \ 0 < y < b$

(7.2.5) $\qquad u(0, y) = 0, \ u(a, y) = f(y), \ 0 < y < b$

(7.2.6) $\qquad u_y(x, 0) = 0, \ u_y(x, b) = 0, \ 0 < x < a.$

Solution. Taking $u(x, y) = X(x)Y(y)$ and separating the variables we have

$$\frac{Y''(y)}{Y(y)} = -\frac{X''(x)}{X(x)} = -\lambda,$$

and using the homogeneous boundary conditions (7.2.5) and (7.2.6) we have

(7.2.7) $\qquad Y''(y) + \lambda Y(y) = 0, \qquad Y(0) = Y'(b) = 0,$

(7.2.8) $\qquad X''(x) - \lambda X(x) = 0, \qquad X(0) = 0.$

If we examine all possible cases for λ: $\lambda > 0$, $\lambda < 0$ and $\lambda = 0$ in the Sturm–Liouville problem (7.2.7), we find that the eigenvalues and corresponding eigenfunctions of (7.2.7) are

$$\lambda_0 = 0; \ Y_0(y) = 1, \quad \lambda_n = \frac{n^2\pi^2}{b^2}, \quad Y_n(y) = \cos\frac{n\pi}{b}y, \ n = 1, 2, \dots.$$

For the found λ_n, the solutions of (7.2.8) are

$$X_0(x) = A_0 x, \quad X_n(x) = A_n \sinh\left(\frac{n\pi}{b}x\right).$$

Therefore,

$$u(x, y) = A_0 x + \sum_{n=1}^{\infty} A_n \sinh\left(\frac{n\pi}{b}x\right) \cos\frac{n\pi}{b}y.$$

From the boundary condition

$$f(y) = u(a, y) = A_0 a + \sum_{n=1}^{\infty} A_n \sinh\frac{n\pi}{b}a \cos\frac{n\pi}{b}y$$

and the orthogonality of the functions $\{\cos\frac{n\pi}{b}y : n = 0, 1, \dots\}$ on the interval $[0, b]$ we find

$$A_0 = \frac{1}{ab}\int_0^b f(y)\, dy, \quad A_n = \frac{1}{b\sinh\frac{n\pi}{b}a}\int_0^b f(y)\cos\frac{n\pi}{b}y\, dy.$$

2^0. *Poisson Equation on a Rectangle.* Now, we consider the following Poisson boundary value problem with general Dirichlet boundary values.

(P)
$$\begin{cases} u_{xx}(x, y) + u_{yy}(x, y) = F(x, y), & 0 < x < a, \ 0 < y < b \\ u(x, 0) = f_1(x), \quad u(x, b) = f_2(x), \ 0 < x < a \\ u(0, y) = g_1(y), \quad u(a, y) = g_2(y), \ 0 < y < b. \end{cases}$$

We cannot directly separate the variables, but as with the Laplace boundary value problem with general Dirichlet boundary conditions, we can split problem (P) into the following two problems.

(PH)
$$\begin{cases} v_{xx}(x, y) + v_{yy}(x, y) = F(x, y), & 0 < x < a, \ 0 < y < b \\ v(x, 0) = 0, \quad v(x, b) = 0, \ 0 < x < a \\ v(0, y) = v(a, y) = 0, \quad 0 < y < b, \end{cases}$$

and

$$(L) \quad \begin{cases} w_{xx}(x,y) + w_{yy}(x,y) = 0, & 0 < x < a, \ 0 < y < b \\ w(x,0) = f_1(x), \quad w(x,b) = f_2(x), \ 0 < x < a \\ w(0,y) = g_1(y), \quad w(a,y) = g_2(y), \ 0 < y < b. \end{cases}$$

If $v(x,y)$ is the solution of problem (PH) and $w(x,y)$ is the solution of problem (L), then the solution $u(x,y)$ of problem (P) will be $u(x,y) = v(x,y) + w(x,y)$.

Problem (L), being a Laplace Dirichlet boundary value problem, has been solved in *Case* 1^0. For problem (PH), notice first that the boundary value on every side of the rectangle is homogeneous. This suggests that $u(x,y)$ may be of the form

$$v(x,y) = \sum_{m=1}^{\infty} \sum_{n=1}^{\infty} F_{mn} \sin \frac{m\pi}{a} x \sin \frac{n\pi}{b} y,$$

where F_{mn} are coefficients that will be determined. If we substitute $v(x,y)$ into the Poisson equation in problem (PH) (we assume that we can differentiate term by term the double Fourier series for $v(x,y)$) we obtain

$$F(x,y) = -\frac{\pi^2}{a^2 b^2} \sum_{m=1}^{\infty} \sum_{n=1}^{\infty} F_{mn} \left(b^2 m^2 + a^2 n^2 \right) \sin \frac{m\pi}{a} x \sin \frac{n\pi}{b} y.$$

Recognizing the above equation as the double Fourier sine series of $F(x,y)$ we have

$$F_{mn} = -\frac{\pi^2}{a^2 b^2 \left(b^2 m^2 + a^2 n^2 \right)} \int_0^a \int_0^b F(x,y) \sin \frac{m\pi}{a} x \sin \frac{n\pi}{b} y \, dy \, dx.$$

Example 7.2.4. Solve the problem

$$\begin{cases} u_{xx}(x,y) + u_{yy}(x,y) = xy, & 0 < x < \pi, \ 0 < y < \pi \\ u(x,0) = 0, \quad u(x,\pi) = 0, \ 0 < x < \pi \\ u(0,y) = u(\pi,y) = 0, & 0 < y < \pi. \end{cases}$$

Solution. In this case we have $a = b = \pi$ and $F(x,y) = xy$. Therefore,

$$F_{mn} = -\frac{1}{\pi^2 \left(m^2 + n^2 \right)} \int_0^\pi \int_0^\pi xy \sin mx \sin ny \, dy \, dx$$

$$= -\frac{1}{\pi^2 \left(m^2 + n^2 \right)} \left(\int_0^\pi x \sin mx \, dx \right) \left(\int_0^\pi x \sin nx \, dx \right)$$

$$= -\frac{1}{\pi^2 \left(m^2 + n^2 \right)} \frac{(-1)^{m-1}}{m} \frac{(-1)^{n-1}}{n}.$$

The solution of the problem is given by

$$u(x, y) = -\frac{1}{\pi^2} \sum_{m=1}^{\infty} \sum_{n=1}^{\infty} \frac{(-1)^{m+n}}{mn(m^2 + n^2)} \sin mx \sin ny.$$

3^0. *Laplace Equation on an Infinite Strip.* Consider the boundary value problem

$$\begin{cases} u_{xx}(x, y) + u_{yy}(x, y) = 0, \quad 0 < x < \infty, \ 0 < y < b \\ u(x, 0) = f_1(x), \quad u(x, b) = f_2(x), \ 0 < x < \infty \\ u(0, y) = g(y), \quad | \lim_{x \to \infty} u(x, y) \,|< \infty, \ 0 < y < b. \end{cases}$$

The solution of this problem is $u(x, y) = v(x, y) + w(x, y)$, where $v(x, y)$ is the solution of the problem

(*)
$$\begin{cases} v_{xx}(x, y) + v_{yy}(x, y) = 0, \quad 0 < x < \infty, \ 0 < y < b \\ v(x, 0) = 0, \quad v(x, b) = 0, \ 0 < x < \infty \\ v(0, y) = g(y), \quad 0 < y < b \\ | \lim_{x \to \infty} v(x, y) \,|< \infty, \quad 0 < y < b, \end{cases}$$

and $w(x, y)$ is the solution of the problem

(**)
$$\begin{cases} w_{xx}(x, y) + w_{yy}(x, y) = 0, \quad 0 < x < \infty, \ 0 < y < b \\ w(x, 0) = f_1(x), \quad w(x, b) = f_2(x), \ 0 < x < \infty \\ w(0, y) = 0, \quad 0 < y < b \\ | \lim_{x \to \infty} w(x, y) \,|< \infty, \quad 0 < y < b. \end{cases}$$

Problem (*) can be solved by the separation of variables method. Indeed, if $v(x, y) = X(x)Y(y)$, then after separating the variables in the Laplace equation and using the boundary conditions at $y = 0$ and $y = b$ we obtain the eigenvalue problem

(EY) $Y''(y) + \lambda Y(y) = 0, \quad Y(0) = Y(b) = 0$

and the differential equation

(EX) $X''(x) - \lambda X(x) = 0, \quad x > 0, \ | \lim_{x \to \infty} X(x) \,|< \infty.$

The eigenvalues and corresponding eigenfunctions of problem (EY) are given by

$$\lambda_n = \frac{n^2 \pi^2}{b^2}, \quad Y_n(y) = \sin \sqrt{\lambda_n}\, y, \ n = 1, 2, \ldots.$$

The solution of the differential equation (EX) corresponding to the above eigenvalues, in view of the fact that $X(x)$ remains bounded as $x \to \infty$, is given by

$$X_n(x) = e^{-\sqrt{\lambda_n}\, x}.$$

Therefore, the solution $v(x, y)$ of problem $(*)$ is given by

$$v(x, y) = \sum_{n=1}^{\infty} A_n\, e^{-\sqrt{\lambda_n}\, x} \sin \frac{n\pi}{b} y,$$

where the coefficients A_n are determined from the boundary condition

$$g(y) = v(0, y) = \sum_{n=1}^{\infty} A_n \sin \frac{n\pi}{b} y,$$

and the orthogonality of the eigenfunctions $\sin \frac{n\pi}{b} y$ on the interval $[0, b]$:

$$A_n = \frac{2}{b} \int_0^b g(y) \sin \frac{n\pi}{b} y\, dy.$$

Now, let us turn our attention to problem $(**)$. If $w(x, y) = X(x)Y(y)$ is the solution of problem $(**)$, then, after separating the variables, from the Laplace equation we obtain the problem

$$X''(x) + \mu^2 X(x) = 0, \quad X(0) = 0, \quad |\lim_{x \to \infty} X(x)| < \infty$$

and the equation

$$Y''(y) - \mu^2 Y(y) = 0, \quad 0 < y < b,$$

where μ is any nonzero parameter.

A solution of the first problem is given by

$$X_\mu(x) = \sin \mu x,$$

and the general solution of the second equation can be written in the special form

$$Y(y) = \frac{A(\mu)}{\sinh \mu b} \sinh \mu y + \frac{B(\mu)}{\sinh \mu b} \sinh \mu(b - y).$$

Therefore,

$$w(x, y) = \sin \mu x \left[\frac{A(\mu)}{\sinh \mu b} \sinh \mu y + \frac{B(\mu)}{\sinh \mu b} \sinh \mu(b - y) \right]$$

is a solution of the Laplace equation for any $\mu > 0$, which satisfies all conditions in problem $(**)$ except the boundary conditions at $y = 0$ and $y = b$. Since the Laplace equation is homogeneous it follows that

$$w(x,y) = \int\limits_0^\infty \sin \mu x \left[\frac{A(\mu)}{\sinh \mu b} \sinh \mu y + \frac{B(\mu)}{\sinh \mu b} \sinh \mu(b - y) \right] d\mu$$

will also be a solution of the Laplace equation. Now, it remains to find $A(\mu)$ and $B(\mu)$ such that the above function will satisfy the boundary conditions at $y = 0$ and $y = b$:

$$f_1(x) = w(x,0) = \int\limits_{=0}^\infty B(\mu) \sin \mu x \, d\mu,$$

$$f_2(x) = w(x,b) = \int\limits_0^\infty A(\mu) \sin \mu x \, d\mu.$$

We recognize the last equations as Fourier sine transforms and so $A(\mu)$ and $B(\mu)$ can be found by taking the inverse Fourier transforms. Therefore, we have solved problem $(**)$ and, consequently, the original problem of the Laplace equation on the semi-infinite strip with the prescribed general Dirichlet condition is completely solved.

We will return to this problem again in the last section of this chapter, where we will discuss the integral transforms for solving the Laplace and Poisson equations.

Now, we will take an example of the Laplace equation on an unbounded rectangular domain, and the use of the Fourier transform is not necessary.

Example 7.2.5. Solve the problem

$$\begin{cases} u_{xx}(x,y) + u_{yy}(x,y) = 0, & 0 < x < \infty, \ 0 < y < \pi, \\ u(x,0) = 0, & u(x,\pi) = 0, \ 0 < x < \infty, \\ u(0,y) = y(\pi - y), & 0 < y < \pi, \\ \lim\limits_{x\to\infty} u(x,y) = 0, & 0 < y < \pi. \end{cases}$$

Solution. Let $u(x,y) = X(x)Y(y)$. After separating the variables and using the boundary conditions at $y = 0$ and $y = \pi$, we obtain

$$Y''(y) + \lambda Y(y) = 0, \quad Y(0) = Y(\pi) = 0$$

and the differential equation

$$X''(x) - \lambda X(x) = 0, \quad x > 0, \ \lim\limits_{x\to\infty} X(x) = 0.$$

If we solve the eigenvalue problem for $Y(y)$ we obtain

$$\lambda_n = n^2, \quad Y_n(y) = \sin ny.$$

For these λ_n the general solution of the differential equation for $X(x)$ is given by

$$X_n(x) = A_n e^{-\sqrt{\lambda_n}\, x} + B_n e^{\sqrt{\lambda_n}\, x}.$$

To ensure that $u(x, y)$ is bounded as $x \to \infty$, we need to take $B_n = 0$. Therefore,

$$X_n(x) = A_n e^{-nx},$$

and so,

$$u(x, y) = \sum_{n=1}^{\infty} A_n e^{-nx} \sin ny, \quad 0 < x < \infty,\ 0 < y < \pi.$$

The coefficients A_n are determined from the boundary condition

$$y(\pi - y) = u(0, y) = \sum_{n=1}^{\infty} A_n \sin ny, \quad 0 < y < \pi$$

and the orthogonality of the system $\{\sin ny : n = 1, 2, \ldots\}$ on $[0, \pi]$:

$$A_n = \frac{2}{\pi} \int_0^{\pi} y(\pi - y) \sin ny\, dy = \frac{2(1 - \cos n\pi)}{n^3} = \begin{cases} 0, & n \text{ even} \\ \frac{4}{n^3}, & n \text{ odd.} \end{cases}$$

Therefore, the solution $u(x, y)$ of the problem is given by

$$u(x, y) = \frac{8}{\pi} \sum_{n=1}^{\infty} \frac{1}{(2n - 1)^3} e^{-(2n-1)x} \sin(2n - 1)y.$$

The following example is from the field of electrostatics.

Example 7.2.6. Let V be the potential of an electrostatic field in the domain $S = \{(x, y) : 0 < x < a,\ 0 < y < \infty\}$. For given $0 < h < a$, define the function $c = c(x)$ by

$$c(x) = \begin{cases} c_1, & 0 < x < h,\ 0 < y < \infty, \\ c_2, & h < x < a,\ 0 < y < \infty, \end{cases}$$

where c_1 and c_2 are given constants. Consider the problem

(7.2.9) $\qquad \mathrm{div}\big(c\, \mathrm{grad}\, u(x, y)\big) = 0, \quad (x, y) \in S$

(7.2.10) $\qquad \lim_{y \to \infty} u(x, y) = 0, \quad 0 < x < a.$

We need to find the solution (potential) $u(x, y)$ of the above problem of the form

$$u(x, y) = \begin{cases} v(x, y), & 0 < x < h, \ 0 < y < \infty \\ w(x, y), & h < x < a, \ 0 < y < \infty, \end{cases}$$

where the functions $u(x, y)$ and $w(x, y)$ satisfy the boundary conditions

(7.2.11)
$$\begin{cases} v(0, y) = 0, & v(a, y) = 0, \ 0 < y < \infty \\ w(x, 0) = V, & 0 < x < a \\ \lim\limits_{y \to \infty} v(x, y) = \lim\limits_{y \to \infty} w(x, y) = 0, & 0 < x < a. \end{cases}$$

Solution. Recall that for a vector field $\mathbf{F}(x, y) = \big(f(x, y), g(x, y)\big)$ and a scalar function $u(x, y)$, *grad* and *div* are defined by

$$\operatorname{div} \mathbf{F}(x, y) = f_x(x, y) + g_y(x, y), \quad \operatorname{grad} u(x, y) = \big(u_x(x, y), u_y(x, y)\big).$$

Now, since $c(x)$ is piecewise constant and $u(x, y)$ is a continuous function, problem (7.2.9), (7.2.10), in view of the boundary conditions (7.2.11), is reduced to the problems

$$\begin{cases} \Delta v(x, y) = 0, & \Delta w(x, y) = 0, \ 0 < x < a, \ 0 < y < \infty \\ v(h, y) = w(h, y), & c_1 v_y(h, y) = c_2 w_y(h, y), \ 0 < y < \infty \\ \lim\limits_{y \to \infty} v(x, y) = \lim\limits_{y \to \infty} w(x, y) = 0, & 0 < x < a. \end{cases}$$

We try to separate the variables. If $u(x, y) = X(x)Y(y)$, then from (7.2.9) after separating the variables we obtain that

(7.2.12)
$$\begin{cases} \big(cX'(x)\big)' + c\lambda X(x) = 0, \\ X(0) = X(a) = 0 \end{cases}$$

and

(7.2.13)
$$Y''(y) - \lambda Y(y) = 0, \quad \lim\limits_{y \to \infty} Y(y) = 0.$$

If we write the function $X(x)$ in the form

$$X(x) = \begin{cases} X_1(x), & 0 < x < h \\ X2(x), & h < x < a, \end{cases}$$

then, from (7.2.12) and the continuity conditions, we have the eigenvalue problems

$$\begin{cases} X_1''(x) + \lambda X_1(x) = 0, & X_1(0) = 0 \\ X_2''(x) + \lambda X_2(z) = 0, & X_2(a) = 0 \end{cases}$$

and the continuity conditions

$$X_1(h) = X_2(h), \quad c_1 X_1'(h) = c_2 X_2'(h).$$

The solution of these problems, in view of the continuity conditions, is

$$X_{1n}(x) = \frac{\sin \sqrt{\lambda_n}\, x}{\sin \sqrt{\lambda_n}}, \quad X_{2n}(x) = \frac{\sin \sqrt{\lambda_n}\,(a - x)}{\sin \sqrt{\lambda_n}\,(a - h)},$$

where λ_n are the positive solutions of the equation

(7.2.14) $\qquad\qquad c_1 \cot \sqrt{\lambda}\, h + \epsilon_2 \cot \sqrt{\lambda}\,(a - h).$

For the above found λ_n, the general solution of (7.2.13) is given by

$$Y_n(y) = A_n\, e^{-\sqrt{\lambda_n}\, y},$$

and thus, the general solution $u(x, y)$ of the original problem has the form

(7.2.15) $\qquad u(x, y) = \sum_{n=1}^{\infty} A_n\, e^{-\sqrt{\lambda_n}\, y}\, X_n(x), \quad 0 < x < a,\ 0 < y < \infty.$

The solution of the problem now is obtained if the coefficients A_n are chosen so that the above representation of $u(x, y)$ satisfies the initial condition

$$V = u(x, 0) = \sum_{n=1}^{\infty} A_n\, X_n(x), \quad 0 < x < a.$$

If we use the orthogonality condition

$$\int_0^b c\, X_n(x) X_m(x)\, dx = 0, \quad \text{if } m \neq n,$$

then we find

$$A_n = \frac{V}{\| X_n \|^2} \int_0^b X_n^2(x)\, dx,$$

where, based on Equation (7.2.14), we have

$$\| X_n \|^2 = \int_0^a c(x) X_n^2(x)\, dx = c_1 \int_0^h X_{1n}^2(x)\, dx + c_2 \int_h^a X_{2n}^2(x)\, dx$$

$$= \frac{c_1\, h}{2 \sin^2\left(\sqrt{\lambda_n}\, h\right)} + \frac{c_2\,(a - h)}{2 \sin^2 \sqrt{\lambda_n}\,(a - h)},$$

i.e.,

$$(7.2.16) \qquad \| X_n \|^2 = \frac{c_1 h}{2 \sin^2 \sqrt{\lambda_n} \, h} + \frac{c_2 (a - h)}{2 \sin^2 \sqrt{\lambda_n} \, (a - h)}.$$

Therefore, the solution of the original problem is given by (7.1.15), where A_n are determined by (7.2.16), λ_n are the positive solutions of Equation (7.2.14) and $X_n(x)$ are given by

$$X_n(x) = \begin{cases} X_{1n}(x), & 0 < x < h \\ X_{2n}(x), & h < x < a \end{cases} = \begin{cases} \dfrac{\sin \sqrt{\lambda_n} \, x}{\sin \sqrt{\lambda_n}}, & 0 < x < h \\[2mm] \dfrac{\sin \sqrt{\lambda_n} \, (a - x)}{\sin \sqrt{\lambda_n} \, (a - h)}, & h < x < a. \end{cases}$$

Exercises for Section 7.2

1. Using the separation of variables method solve the problem

$$u_{xx}(x, y) + u_{yy}(x, y) = 0, \quad 0 < x < a, \ 0 < y < b$$
$$u(x, 0) = f_1(x), \quad u(x, b) = f_2(x), \ 0 < x < a$$
$$u(0, y) = g_1(y), \quad u(a, y) = g_2(y), \ 0 < y < b.$$

for the given data.

(a) $a = b = 1$, $f_1(x) = 100$, $f_2(x) = 200$, $g_1(y) = g_2(y) = 0$.

(b) $a = b = \pi$, $f_1(x) = 0$, $f_2(x) = 1$, $g_1(y) = g_2(y) = 1$.

(c) $a = 1$, $b = 2$, $f_1(x) = 0$, $f_2(x) = x$, $g_1(y) = g_2(y) = 0$.

(d) $a = 2$, $b = 1$, $f_1(x) = 100$, $f_2(x) = 0$, $g_1(y) = 0$, $g_2(y) = 100y(1 - y)$.

(e) $a = b = 1$, $f_1(x) = 7 \sin 7\pi x$, $f_2(x) = \sin \pi x$, $g_1(y) = \sin 3\pi y$, $g_2(y) = \sin 6\pi y$.

(f) $a = b = 1$, $f_1(x) = 0$, $f_2(x) = 100$, $g_1(y) = 0$, $g_2(y) = 0$.

(g) $u(x, 0) = f_1(x) = 0$, $u(x, b) = f_2(x) = f(x)$, $u(0, y) = g_1(y) = 0$, $u(a, y) = g_2(y) = 0$.

In problems 2–9 solve the Laplace equation in $[0, a] \times [0, b]$ subject to the given boundary conditions.

2. $u(x,0) = f(x)$, $u(x,b) = 0$, $0 < x < a$; $u(0,y) = 0$, $u(a,y) = 0$, $0 < y < b$.

3. $u_y(x,0) = 0$, $u_y(x,1) = 0$, $0 < x < a$; $u(0,y) = 0$, $u(1,y) = 1 - y$, $0 < y < b$.

4. $u_x(0,y) = u(0,y)$, $u(0,y) = 1$, $0 < y < b$; $u(x,0) = 0$, $u(x,\pi) = 0$, $0 < x < a$.

5. $u_x(0,y) = u(0,y)$, $u_x(a,y) = 0$, $0 < y < b$; $u(x,0) = 0$, $u(x,b) = f(x)$, $0 < x < a$.

6. $u_x(0,y) = 0$, $u(a,y) = 1$, $0 < y < b$; $u(x,0) = 1$, $u(x,b) = 1$, $0 < x < a$.

7. $u(0,y) = 1$, $u(a,y) = 1$, $0 < y < b$; $u_y(x,0) = 0$, $u(x,b) = 0$, $0 < x < a$.

8. $u(x,0) = 1$, $u(x,1) = 4$, $0 < x < \pi$; $u(0,y) = 0$, $u(\pi,y) = 0$.

9. $u(x,0) = 0$, $u_y(x,1) = 0$, $0 < x < 1$; $u(0,y) = 0$, $u(1,y) = 0$, $0 < y < 1$.

10. Determine conditions on the function $f(x)$ on the segment $[0,1]$ which guarantee a solution of the Neumann problem
$$u_{xx}(x,y) + u_{yy}(x,y) = 0, \quad 0 < x < 1,\ 0 < y < 1$$
$$u_y(x,0) = 0, \quad u_y(x,1) = 0,\ 0 < x < 1$$
$$u_x(0,y) = f(y), \quad u_x(1,y) = 0,\ 0 < y < 1.$$

In Problems 11–14 solve the Laplace equation $\Delta u(x,y) = 0$ on the indicated semi-infinite strip with the given boundary conditions.

11. $0 < x < \pi$, $0 < y < \infty$; $u(0,y) = u(\pi,y) = 0$; $u(x,0) = f(x)$; $|\lim_{y \to \infty} u(x,y)| < \infty$.

12. $0 < x < \frac{\pi}{2}$, $0 < y < \infty$; $u(0,y) = u(\frac{\pi}{2},y) = e^{-y}$; $u(x,0) = 1$, $|\lim_{y \to \infty} u(x,y)| < \infty$.

13. $0 < x < \infty$, $0 < y < b$; $u(x,0) = u_y(x,b) = 0$; $u(0,y) = f(y)$;
$| \lim_{x \to \infty} u(x,y) | < \infty$.

14. $0 < x < \infty$, $0 < y < b$; $u_y(x,0) = u_y(x,b) + hu(x,b) = 0$;
$u(0,y) = f(y)$; $| \lim_{x \to \infty} u(x,y) | = 0$, $0 < y < b$.

In Problems 15–20 solve the Poisson equation $\Delta u(x,y) = -F(x,y)$ on the indicated rectangle $0 < x < a$ $0 < y < b$, with the given boundary values and given function $F(x,y)$.

15. $a = b = 1$, $F(x,y) = -1$, $u(x,y) = 0$ on all sides of the rectangle.

16. $a = b = 1$, $F(x,y) = -x$, $u(x,y) = 0$ on all sides of the rectangle.

17. $a = b = 1$, $F(x,y) = -\sin \pi x$, $u(x,0) = u(0,y) = u(1,y) = 0$, $u(x,1) = x$.

18. $a = b = 1$, $F(x,y) = -xy$, $u(x,0) = u(0,y) = u(1,y) = 0$, $u(x,1) = x$.

19. $a = b = \pi$, $F(x,y) = e^{2y} \sin x$, $u(x,0) = u(0,y) = u(\pi,y) = 0$ $u(x,\pi) = f(x)$.

20. Let $C = \{(x,y,z) : 0 < x < 1,\ 0 < y < 1,\ 0 < z < 1\}$ be the unit cube. Show that the solution of the Laplace equation

$$u_{xx}(x,y,z) + u_{yy}(x,y,z) + u_{zz}(x,y,z) = 0, \quad (x,y,z) \in C,$$

subject to the Dirichlet boundary conditions $u(1,y,z) = g(y,z)$, $0 < y < 1$ $0 < z < 1$; and $u(x,y,z) = 0$ on all other sides of the cube is given by

$$u(x,y,z) = \sum_{m=1}^{\infty} \sum_{m=1}^{\infty} A_{mn} \sinh \lambda_{mn} x \sin m\pi y \sin n\pi z,$$

where

$$A_{mn} = \frac{4}{\sinh \lambda_{mn}} \int_0^1 \int_0^1 g(y,z) \sin m\pi y \sin n\pi z \, dz \, dy$$

and

$$\lambda_{mn} = \pi \sqrt{m^2 + n^2}.$$

Hint: Take $u(x,y,z) = X(x)Y(y)Z(z)$ and separate the variables.

7.3 Laplace and Poisson Equations on Circular Domains.

In this section we will study the two and three dimensional the Laplace equation in circular domains. We begin by considering the Dirichlet problem of the Laplace equation on a disc.

7.3.1 Laplace Equation in Polar Coordinates.

We will consider and solve the following boundary value problem: Find a twice differentiable function $u(r, \varphi)$ in the disc

$$D = \{(r, \varphi) : r < a\}$$

that is continuous on the closed disc $\overline{D} = D \cup \partial D$, such that

(7.3.1) $\Delta u(r, \varphi) = 0, \quad 0 < r < a, \ 0 \le \varphi < 2\pi$

(7.3.2) $u(a, \varphi) = f(\varphi), \quad 0 \le \varphi < 2\pi,$

where $f(\varphi)$ is a prescribed function on the circle $C = \partial D$. We will assume that the given function f is continuous and differentiable on the circle C, even though the results which will be proved are valid for a more general class of functions f.

As usual, when working in a disc, we use the polar coordinates (r, φ) in which, as we have seen in the previous chapters, the Laplacian is given by

(7.3.3) $\Delta u(r, \varphi) = \dfrac{1}{r} \dfrac{\partial}{\partial r} \left(r \dfrac{\partial u}{\partial r} \right) + \dfrac{1}{r^2} \dfrac{\partial^2 u}{\partial \varphi^2} \equiv \dfrac{\partial^2 u}{\partial r^2} + \dfrac{1}{r} \dfrac{\partial u}{\partial r} + \dfrac{1}{r^2} \dfrac{\partial^2 u}{\partial \varphi^2}.$

We will find the solution of the problem (7.3.1), (7.3.2) of the form

$$u(r, \varphi) = R(r)\Phi(\varphi).$$

If we substitute $u(r, \varphi)$ into Equation (7.3.1), after separating the variables, we obtain

$$\frac{r(rR''(r) + R'(r))}{R(r)} = -\frac{\Phi''(\varphi)}{\Phi(\varphi)} = \lambda,$$

where λ is a constant. Therefore, we obtain the two differential equations

(7.3.4) $\Phi''(\varphi) + \lambda\Phi(\varphi) = 0, \quad 0 \le \varphi < 2\pi,$

and

(7.3.5) $r^2 R''(r) + rR'(r) - \lambda R(r) = 0, \quad 0 < r \le a.$

Since $u(r, \varphi)$ is a 2π-periodic function of φ we have

$$\Phi(\varphi) = \Phi(\varphi + 2\pi)$$

for every φ. Using this condition (as we did in the section on the wave equation on circular domains), we have

$$\sqrt{\lambda} = n, \quad n = 0, 1, 2, \ldots.$$

Consequently, the general solution of Equation (7.3.4) is given by

$$\Phi_n(\varphi) = A_n \cos n\varphi + B_n \sin n\varphi.$$

For these $\lambda_n = n^2$, the solution of Equation (7.3.5), as a Cauchy–Euler equation (see Appendix D), is

$$R(r) = Ar^n + Br^{-n}.$$

Since $u(r, \varphi)$ is a continuous function in the whole disc D, the function $R(r)$ cannot be unbounded for $r = 0$. Thus, $B = 0$, and so

$$R(r) = r^n, \quad n = 0, 1, 2, \ldots.$$

Therefore,

$$u(r, \varphi) = \sum_{n=0}^{\infty} r^n \left(A_n \cos n\varphi + B_n \sin n\varphi \right), \quad 0 \le r \le a,\ 0 \le \varphi < 2\pi$$

are particular solutions of our problem, provided that the above series converge.

We find the coefficients A_n and B_n so that the above $u(r, \varphi)$ satisfies the boundary condition

$$f(\varphi) = u(a, \varphi) = \sum_{n=0}^{\infty} a^n \left(A_n \cos n\varphi + B_n \sin n\varphi \right), \quad 0 \le \varphi < 2\pi.$$

Recognizing that the above right hand side is the Fourier series of $f(\varphi)$, we have

$$A_n = \frac{1}{a^n \pi} \int_0^{2\pi} f(\psi) \cos n\psi \, d\psi,$$

$$B_n = \frac{1}{a^n \pi} \int_0^{2\pi} f(\psi) \sin n\psi \, d\psi; \quad n = 0, 1, 2, \ldots.$$

Therefore, the solution $u = u(r, \varphi)$ of our Dirichlet problem for the Laplace equation on the disc is given by

$$(7.3.6) \quad u = \frac{a_0}{2} + \sum_{n=1}^{\infty} \left(\frac{r}{a}\right)^n \left(a_n \cos n\varphi + b_n \sin n\varphi \right), \quad 0 \le r \le a,\ 0 \le \varphi < 2\pi$$

where

$$(7.3.7) \qquad a_n = \frac{1}{\pi} \int_0^{2\pi} f(\psi) \cos n\psi \, d\varphi, \quad b_n = \frac{1}{\pi} \int_0^{\pi} f(\psi) \sin n\psi \, d\psi.$$

Using the assumption that $f(\varphi)$ is a continuous and differentiable function on the circle C and using the Weierstrass test for uniform convergence, it can be easily shown that the series in (7.3.6) converges, it can be differentiated term by term and it is a continuous function on the circle C. By construction, every term of the above series satisfies the Laplace equation and therefore the solution of the original problem is given by (7.3.6).

Now, using (7.3.6) and (7.3.7) we will derive the very important *Poisson integral formula*, which gives the solution of the Dirichlet problem for the Laplace equation on the disc D by means of an integral (*Poisson integral*).

If we substitute the coefficients a_n and b_n, given by (7.3.7), into (7.3.6) and interchange the order of summation and integration we obtain

$$u(r, \varphi) = \frac{1}{\pi} \int_0^{2\pi} \left[\frac{1}{2} + \sum_{n=1}^{\infty} \left(\frac{r}{a}\right)^n \left(\cos n\psi \cos n\varphi + \sin \psi \sin n\varphi \right) \right] f(\psi) \, d\psi$$

$$= \frac{1}{\pi} \int_0^{2\pi} \left[\frac{1}{2} + \sum_{n=1}^{\infty} \left(\frac{r}{a}\right)^n \cos n(\varphi - \psi) \right] f(\psi) \, d\psi.$$

To simplify our further calculations we let $z = \frac{r}{a}$. Notice that $|z| < 1$. Using Euler's formula and the geometric sum formula

$$\cos \alpha = \frac{e^{i\alpha} + e^{-i\alpha}}{2}, \quad \sum_{n=1}^{\infty} w^n = \frac{w}{1 - w}, \quad |w| < 1,$$

we obtain

$$u(r, \varphi) = \frac{1}{\pi} \int_0^{2\pi} \left[\frac{1}{2} + \sum_{n=1}^{\infty} \left(\frac{r}{a}\right)^n \cos n(\varphi - \psi) \right] f(\psi) \, d\psi$$

$$= \frac{1}{2\pi} \int_0^{2\pi} \left[1 + \sum_{n=1}^{\infty} \left(z e^{(\varphi - \psi)ni} \right)^n + \sum_{n=1}^{\infty} \left(z e^{-(\varphi - \psi)ni} \right)^n \right] f(\psi) \, d\psi$$

$$= \frac{1}{2\pi} \int_0^{2\pi} \left[1 + \frac{z e^{(\varphi - \psi)ni}}{1 - z e^{(\varphi - \psi)ni}} + \frac{z e^{-(\varphi - \psi)ni}}{1 - z e^{-(\varphi - \psi)ni}} \right] f(\psi) \, d\psi$$

$$= \frac{1}{2\pi} \int_0^{2\pi} \frac{1 - z^2}{1 - 2z \cos (\varphi - \psi) + z^2} f(\psi) \, d\psi$$

$$= \frac{1}{2\pi} \int_0^{2\pi} \frac{a^2 - r^2}{r^2 - 2ar \cos (\varphi - \psi) + a^2} f(\psi) \, d\psi.$$

Hence,

$$(7.3.8) \qquad u(r, \varphi) = \frac{1}{2\pi} \int_0^{2\pi} \frac{a^2 - r^2}{r^2 - 2ar \cos(\varphi - \psi) + a^2} f(\psi) \, d\psi.$$

Equation (7.3.8) is called the *Poisson formula* for the harmonic function $u(r, \varphi)$ for the disc D with boundary values $f(\varphi)$.

If we introduce the function

$$(7.3.9) \qquad\qquad P(r, \varphi) = \frac{a^2 - r^2}{r^2 - 2ar \cos \varphi + a^2},$$

known as the *Poisson kernel*, then the Poisson integral formula (7.3.8) can be written in the form

$$(7.3.10) \qquad\qquad u(r, \varphi) = \frac{1}{2\pi} \int_0^{2\pi} P(r, \varphi - \psi) f(\psi) \, d\pi.$$

Example 7.3.1. Let $D = \{(x, y) : x^2 + y^2 < 1\}$ be the unit disc. Find a harmonic function $u(x, y)$ on D which is continuous on the closed disc $\overline{D}$ and which on the boundary ∂D has value $f(x, y) = x^2 - 3xy - 2y^2$.

Solution. We pass to polar coordinates. From (7.3.6) it follows that the required function $u(r, \varphi)$ is given by

$$u(r, \varphi) = \frac{a_0}{2} + \sum_{n=1}^{\infty} r^n \big(a_n \cos n\varphi + b_n \sin n\varphi\big), \quad 0 \leq r \leq 1, \, 0 \leq \varphi < 2\pi,$$

where a_n and b_n are the Fourier coefficients of the function $f(x, y)$, which will be found from the boundary condition

$$\cos^2 \varphi - 3 \cos \varphi \sin \varphi - 2 \sin^2 \varphi = \frac{a_0}{2} + \sum_{n=1}^{\infty} \big(a_n \cos n\varphi + b_n \sin n\varphi\big).$$

From

$$\cos^2 \varphi - 3 \cos \varphi \sin \varphi - 2 \sin^2 \varphi = \frac{1 + \cos 2\varphi}{2} - \frac{3}{2} \sin 2\varphi - \big(1 - \cos \varphi\big),$$

by comparison of the coefficients we obtain

$$a_0 = -1, \quad a_2 = \frac{3}{2}, \, b_2 = -\frac{3}{2}$$

and the other coefficients a_n and b_n are all zero. Therefore, the required function is

$$u(r, \varphi) = -\frac{1}{2} - \frac{3}{2}r^2 \cos 2\varphi + \frac{3}{2}r^2 \sin 2\varphi.$$

We can rewrite the function $u(r, \varphi)$ as

$$u(r, \varphi) = -\frac{1}{2} - \frac{3}{2}r^2(\cos^2 \varphi - \sin^2 \varphi) + 3r^2 \sin \varphi \cos \varphi,$$

from which we find that the solution u in Cartesian coordinates is given by

$$u(x, y) = -\frac{1}{2} - \frac{3}{2}(x^2 - y^2) + 3xy.$$

Example 7.3.2. Let A be the annulus defined by

$$A = \{(r, \varphi) : a < r < b, \ 0 \le \varphi < 2\pi\}.$$

Find the solution $u(r, \varphi)$ of the problem

$$\begin{cases} \Delta\, u(r, \varphi) = 0, & (r, \varphi) \in A \\ u(a, \varphi) = f(\varphi), & 0 \le \varphi < 2\pi \\ u(b, \varphi) = g(\varphi), & 0 \le \varphi < 2\pi. \end{cases}$$

Solution. If we take $u(r, \varphi) = R(r)\Phi(\varphi)$ and work similarly as in the Laplace equation on the disc we will obtain Equation (7.3.5) for $R(r)$, $a < r < b$ and Equation (7.3.4) for $\Phi(\varphi)$. Since Φ is 2π-periodic we obtain that the general solution of (7.3.4) is given by

$$\Phi_n(\varphi) = A_n \cos n\varphi + B_n \sin n\varphi, \quad n = 0, 1, 2, \ldots$$

For $n = 1, 2, \ldots$ we have two independent solutions of the Euler–Cauchy equation (7.3.5):

$$R(r) = r^n, \quad R(r) = r^{-n},$$

while for $n = 0$ the two independent solutions of (7.3.6) are

$$R(r) = \ln r, \quad R(r) = 1.$$

Therefore, the solution of the Laplace equation on the annulus A will be

$$u(r, \varphi) = a_0 \ln r + b_0 + \sum_{n=1}^{\infty} \left[(a_n r^n + b_n r^{-n}) \cos n\varphi + (c_n r^n + d_n r^{-n}) \sin n\varphi \right],$$

where the coefficients, a_n, b_n, c_n and d_n are determined if we compare the equations obtained from the boundary conditions at $r = a$ and $r = b$:

$$a_0 \ln a + b_0 + \sum_{n=1}^{\infty} \left[(a_n a^n + b_n a^{-n}) \cos n\varphi + (c_n a^n + d_n a^{-n}) \sin n\varphi = f(\varphi) \right]$$

$$a_0 \ln b + b_0 + \sum_{n=1}^{\infty} \left[(a_n b^n + b_n b^{-n}) \cos n\varphi + (c_n b^n + d_n b^{-n}) \sin n\varphi = g(\varphi) \right],$$

to the Fourier expansions of $f(\varphi)$ and $f(\varphi)$:

$$f(\varphi) = \frac{a_0}{2} + \sum_{n=1}^{\infty} (a_n \cos n\varphi + b_n \sin n\varphi)$$

$$g(\varphi) = \frac{c_0}{2} + \sum_{n=1}^{\infty} (c_n \cos n\varphi + d_n \sin n\varphi).$$

Example 7.3.3. Find the harmonic function $u(r, \varphi)$ on the unit disc such that
$$u(1, \varphi) = \begin{cases} 1, & 0 < \varphi < \pi \\ 0, & \pi < \varphi < 2\pi. \end{cases}$$

Solution. The Fourier coefficients a_n and b_n of the function f are

$$a_0 = \frac{1}{\pi} \int_0^{2\pi} f(\psi) \, d\psi = 1, \quad a_n = \frac{1}{\pi} \int_0^{2\pi} f(\psi) \cos n\psi \, d\psi = \frac{1}{\pi} \int_0^{\pi} \cos n\psi \, d\psi = 0,$$

$$b_n = \frac{1}{\pi} \int_0^{2\pi} f(\psi) \sin n\psi \, d\psi = \frac{1}{\pi} \int_0^{\pi} \sin n\psi \, d\psi = \frac{1}{\pi} \frac{1 - (-1)^n}{n}.$$

Therefore, by (7.3.6) the function $u(r, \varphi)$ is given by

$$u(r, \varphi) = \frac{a_0}{2} + \sum_{n=1}^{\infty} (r^n (a_n \cos n\varphi + b_n \sin n\varphi)$$

$$= \frac{1}{2} + \frac{2}{\pi} \sum_{n=1}^{\infty} \frac{r^{2n-1}}{2n - 1} \sin (2n - 1)\varphi.$$

Using the Poisson integral formula (7.3.8), the function $u(r, \varphi)$ can be represented in the form

$$u(r, \varphi) = \frac{1}{2\pi} \int_0^{2\pi} \frac{1 - r^2}{r^2 - 2r \cos (\varphi - \psi) + 1} f(\psi) \, d\psi$$

$$= \frac{1 - r^2}{2\pi} \int_0^{\pi} \frac{\sin \psi}{r^2 - 2r \cos (\varphi - \psi) + 1} \, d\psi.$$

Example 7.3.4. Solve the following Laplace equation boundary value problem on the semi-disc.

$$\begin{cases} \Delta u(r, \varphi) = 0, \quad 0 < r < 1 \, 0 < \varphi < \pi \\ u(1, \varphi) = c, \quad 0 < \varphi < \pi \\ u(r, 0) = u(r, \pi) = 0, \quad 0 < r < \pi \\ |\lim_{r \to 0+} u(r, \varphi)| < \infty, \quad 0 < \varphi < \pi. \end{cases}$$

Solution. If we let $u(r, \varphi) = R(r)\Phi(\varphi)$ and substitute in the Laplace equation (after separating the variables) we obtain

$$\Phi''(\varphi) + \lambda \Phi(\varphi) = 0, \quad \varphi)(0) = \Phi(\pi) = 0,$$
$$r^2 R''(r) + r R'(r) - \lambda R(r) = 0, \quad 0 < r < 1, \ |\lim_{r \to 0+} R(r)| < \infty.$$

Solving the above eigenvalue problem for $\Phi(\varphi)$ we obtain that

$$\lambda_n = n^2, \quad \Phi_n(\varphi) = \sin n\varphi.$$

For these λ_n, in view of the fact that $R(r)$ is bounded in a neighborhood of $r = 0$, a particular solution of the Cauchy–Euler differential equation for $R(r)$ is

$$R(r) = r^n, \quad 0 \leq r \leq 1.$$

Therefore, the general solution of the Laplace equation on the semi-disc is given by

$$u(r, \varphi) = \sum_{n=1}^{\infty} a_n r^n \sin n\varphi.$$

We find the coefficients a_n from the boundary condition

$$c = u(1, \varphi) = \sum_{n=1}^{\infty} a_n \sin n\varphi.$$

From the orthogonality property of the functions $\sin n\varphi$ on $[0, \pi]$ it follows that

$$a_n = \frac{2}{\pi} c \int_0^{\pi} \sin n\varphi \, d\varphi = \frac{2}{\pi} c \frac{1 - (-1)^n}{n}.$$

Therefore, the solution of the original boundary value problem is given by

$$u(r, \varphi) = \frac{4c}{\pi} \sum_{n=1}^{\infty} \frac{r^{2n-1}}{2n - 1} \sin (2n - 1)\varphi.$$

7.3.2. Poisson Equation in Polar Coordinates.

We consider the following Poisson boundary value problem on the unit disc.

$$\begin{cases} \Delta u(r, \varphi) = -F(r, \varphi), & 0 < r < 1, \ 0 \le \varphi < 2\pi \\ u(1, \varphi) = f(\varphi), & 0 \le \varphi < 2\pi. \end{cases}$$

Let the solution $u(r, \varphi)$ of the problem be of the form

$$u(r, \varphi) = v(r, \varphi) + w(r, \varphi),$$

where $v(r, \varphi)$ and $w(r, \varphi)$ are the solutions of the following problems, respectively,

(L)
$$\begin{cases} \Delta v(r, \varphi) = 0, & 0 < r < 1, \ 0 \le \varphi < 2\pi \\ u(a, \varphi) = f(\varphi), & 0 \le \varphi < 2\pi \end{cases}$$

and

(P)
$$\begin{cases} \Delta w(r, \varphi) = -F(r, \varphi), & 0 < r < 1, \ 0 \le \varphi < 2\pi \\ w(1, \varphi) = 0, & 0 \le \varphi < 2\pi. \end{cases}$$

The first problem (L) has been already solved in part 7.3.1. Therefore, it is enough to solve the second problem (P). We will look for the solution of problem (P) in the form

$$w(r, \varphi) = \sum_{m=0}^{\infty} \sum_{n=0}^{\infty} W_{mn}(r, \varphi),$$

where $W_{mn}(r, \varphi)$ are the eigenfunctions of the Helmholtz eigenvalue problem

(H)
$$\begin{cases} \Delta w(r, \varphi) + \lambda^2 \, w(r, \varphi) = 0, & 0 \le r < 1, \ 0 \le \varphi < 2\pi \\ w(1, \varphi) = 0, & 0 \le \varphi < 2\pi. \end{cases}$$

We solved problem (H) in the vibration of a circular drum in Chapter 5, where we found that its eigenvalues λ_{mn}^2 are obtained by solving the equation

$$J_m(\lambda_{mn}) = 0$$

and the corresponding eigenfunctions $W_{mn}(r, \varphi)$ are given by

$$W_{mn}(r, \varphi) = J_m(\lambda_{mn} \, r) \, \cos m\varphi, \qquad W_{mn}(r, \varphi) = J_m(\lambda_{mn} \, r) \, \cos m\varphi.$$

Thus,

$$(7.3.11) \qquad w(r, \varphi) = \sum_{m=0}^{\infty} \sum_{n=0}^{\infty} J_m(\lambda_{mn} \, r) \big[a_{mn} \, \cos m\varphi + b_{mn} \, \sin m\varphi\big].$$

Now, from
$$\Delta W_{mn} + \lambda_{mn}^2 W_{mn} = 0, \quad m, n = 0, 1, 2, \ldots$$

we have

(7.3.12) $\Delta w(r, \varphi) = \sum_{m=0}^{\infty} \sum_{n=0}^{\infty} \Delta W_{mn}(r, \varphi) = -\sum_{m=0}^{\infty} \sum_{n=0}^{\infty} \lambda_{mn}^2 W_{mn}(r, \varphi).$

Therefore, from (7.3.11), (7.3.12) and from the Poisson equation in (P) we have

$$\sum_{m=0}^{\infty} \sum_{n=0}^{\infty} \lambda_{mn}^2 J_m(\lambda_{mn} r) \left[a_{mn} \cos m\varphi + b_{mn} \sin m\varphi \right] = F(r, \varphi).$$

From the last expansion and from the orthogonality of the eigenfunction with respect to the weight r we find

(7.3.13)
$$\begin{cases} A_{mn} = \dfrac{\epsilon_m}{\pi J_{m+1}^2(\lambda_{mn})} \displaystyle\int_0^1 \int_0^{2\pi} f(\varphi) J_m(\lambda_{mn}r) \, r \cos m\varphi \, d\varphi \, dr, \\[2em] B_{mn} = \dfrac{2}{\pi J_{m+1}^2(\lambda_{mn})} \displaystyle\int_0^1 \int_0^{2\pi} f(\varphi) J_m(\lambda_{mn} r) \, r \sin m\varphi \, d\varphi \, dr, \end{cases}$$

where
$$\epsilon_m = \begin{cases} 1, & m = 0, \\ 2, & m > 0. \end{cases}$$

Example 7.3.5. Solve the boundary value problem for the Poisson equation.

$$\begin{cases} \Delta u(r, \varphi) = -2 - r^3 \cos 3\varphi, & 0 \le r < 1, \ 0 \le \varphi < 2\pi \\ u(1, \varphi) = 0, & 0 \le \varphi < 2\pi. \end{cases}$$

Solution. From Equation (7.3.13) we have $B_{mn} = 0$ for every m and n and $A_{mn} = 0$ for every $m \ne 3$ and $m \ne 0$ and every n. For $m = 0$ from (7.3.13) we have

$$A_{0n} = \frac{1}{\pi \lambda_{0n}^2 J_1^2(\lambda_{0n})} \int_0^1 \int_0^{2\pi} (2 + r^3 \cos 3\varphi) J_0(\lambda_{0n}r) \, r \, d\varphi \, dr$$

$$= \frac{4}{J_1^2(\lambda_{0n})} \int_0^1 J_0(\lambda_{0n}r) \, r \, dr$$

$$= \frac{4}{\lambda_{0n}^2 J_1^2(\lambda_{0n})} \frac{1}{\lambda_{0n}} J_1(\lambda_{0n}) = \frac{4}{\lambda_{0n}^3 J_1(\lambda_{0n})}.$$

For $m = 3$ we have

$$A_{3n} = \frac{2}{\pi \lambda_{3n}^2 \, J_4^2(\lambda_{3n})} \int_0^1 \int_0^{2\pi} (2 + r^3 \cos 3\varphi) \cos 3\varphi \, J_3(\lambda_{3n} r) \, r \, d\varphi \, dr$$

$$= \frac{4}{\lambda_{3n}^2 \, J_4^2(\lambda_{3n})} \int_0^1 r^4 J_3(\lambda_{3n} r) \, dr$$

$$= \frac{4}{\lambda_{3n}^2 \, J_4^2(\lambda_{3n})} \frac{1}{\lambda_{3n}} J_4(\lambda_{3n}) = \frac{4}{\lambda_{3n}^3 \, J_4(\lambda_{0n})}.$$

In the above calculations we have used the following fact from the Bessel functions (see Chapter 3):

$$\int_0^1 r^{n+1} J_n(\lambda r) \, dr = \frac{1}{\lambda} J_{n+1}(\lambda).$$

Therefore, the solution of the original problem is given by

$$u(r, \varphi) = 4 \sum_{n-1}^{\infty} \left[\frac{1}{\lambda_{0n}^3 \, J_1(\lambda_{0n})} J_0(\lambda_{0n} r) + \frac{1}{\lambda_{3n}^3 \, J_4(\lambda_{3n})} J_3(\lambda_{3n} r) \cos 3\varphi \right].$$

7.3.3 Laplace Equation in Cylindrical Coordinates.

In this part we will solve the Laplace equation in cylindrical coordinates $x = r \cos \varphi$, $y = r \sin \varphi$, $z = z$. Recall that the Laplacian in these coordinates is given by

$$\Delta u(r, \varphi, z) \equiv u_{rr} + \frac{1}{r} u_r + \frac{1}{r^2} u_{\varphi\varphi} + u_{zz}.$$

Let C be the cylinder given by

$$C = \{(r, \varphi, z) : 0 < r < a, \, 0 < \varphi < 2\pi, \, 0 < z < b\}.$$

We will find a unique, twice differentiable function $u = u(r, \varphi, z)$ on C which satisfies the boundary value problem

$$(7.3.14) \qquad u_{rr} + \frac{1}{r} u_r + \frac{1}{r^2} u_{\varphi\varphi} + u_{zz} = 0, \quad (r, \varphi, z) \in C,$$

$$(7.3.15) \qquad u(a, \varphi, z) = 0, \quad 0 < \varphi < 2\pi, \, 0 < z < b,$$

$$(7.3.16) \qquad u(r, \varphi, 0) = f(r, \varphi), \, 0 < r < a, \, 0 < \varphi < 2\pi,$$

$$(7.3.17) \qquad u(r, \varphi, b) = g(r, \varphi), \, 0 < r < a, \, 0 < \varphi < 2\pi.$$

Naturally, the function $u(r, \varphi, z)$ is 2π-periodic with respect to φ:

$$u(r, \varphi, z) = u(r, \varphi + 2\pi, z), \quad 0 < r < a, \, 0 < \varphi < 2\pi, \, 0 < z < b$$

and bounded at $r = 0$:

$$| \lim_{r \to 0+} u(r, \varphi, z) | < \infty, \quad 0 < \varphi < 2\pi, \ 0 < z < b.$$

If we search for the solution of this problem of the form

$$u(r, \varphi, z) = R(r)\Phi(\varphi)Z(z),$$

then substituting it into Equation (7.3.14), after separating the variables and using the boundary conditions (7.3.15), (7.3.16) and (7.3.17), we will obtain the problems

(7.3.18) $\qquad Z''(z) - \lambda Z(z) = 0, \quad 0 < z < b,$

(7.3.19) $\qquad \Phi''(\varphi) + \mu^2 \Phi(\varphi) = 0, \ \Phi(\varphi) = \Phi(\varphi + 2\pi),$

(7.3.20) $\qquad r^2 R'' + R' + (\lambda r^2 - \mu^2)R = 0, \ |\lim_{r \to 0} R(r)| < \infty$

Because $\Phi(\varphi)$ is a parodic function we have $\mu = \mu_n = n, \ n = 0, 1, 2, \ldots$, and so two independent solutions of Equation (7.3.19) are given by

$$\Phi_n(\varphi) = \cos n\varphi, \quad \Phi_n(\varphi) = \sin n\varphi.$$

For these found $\mu_n = n$, from the bounded condition for the function $R(r)$ at $r = 0$, we find a solution of the Bessel equation in (7.3.20) of order n, given by

$$R_{mn}(r) = J_n\left(\sqrt{\lambda_{mn}}\, a\, r\right)$$

where λ_{mn} are obtained by solving the equation

$$J_n\left(\sqrt{\lambda_{mn}}\, a\right) = 0,$$

found from the boundary condition at $r = a$.

Therefore, the general solution $u = u(r, \varphi, z)$ of our original problem is of the form

$$u = \sum_{m=0}^{\infty} \sum_{n=0}^{\infty} \left[A_{mn} J_n\left(\sqrt{\lambda_{mn}}\, ar\right) \cos n\varphi + B_{mn} J_n\left(\sqrt{\lambda_{mn}}\, ar\right) \sin n\varphi \right] Z_{mn}(z),$$

where $Z_{mn}(z)$ are solutions of (7.3.18).

As usual, we split the original problem into two boundary value problems:

$$u(r, \varphi, z) = u_1(r, \varphi, z) + u_2(r, \varphi, z),$$

where the harmonic functions $u_1(r, \varphi, z)$ and $u_2(r, \varphi, z)$ satisfy the conditions

$$\begin{cases} u_1(a, \varphi, z) = 0, & u_1(r, \varphi, 0) = f(r, \varphi), \ u_1(r, \varphi, b) = 0 \\ u_2(a\varphi, z) = 0, & u_2(r, \varphi, 0) = 0, \ u_2(r, \varphi, b) = g(r, \varphi). \end{cases}$$

It is enough to consider only $u_1(r, \varphi, z)$. In this case we have

$$Z_{mn}(z) = \sinh \sqrt{\lambda_{mn}} \, (b - z).$$

Therefore the solution $u = u(r, \varphi, z)$ of the given original problem is

$$u = \sum_{m=0}^{\infty} \sum_{n=0}^{\infty} \left[A_{mn} \cos n\varphi + B_{mn} \sin n\varphi \right] J_n\left(\sqrt{\lambda_{mn}} \, ar \right) \frac{\sinh \sqrt{\lambda_{mn}}(b - z)}{\sinh \sqrt{\lambda_{mn}} b}$$

$$+ \sum_{m=0}^{\infty} \sum_{n=0}^{\infty} \left[C_{mn} \cos n\varphi + D_{mn} \sin n\varphi \right] J_n\left(\sqrt{\lambda_{mn}} r \right) \frac{\sinh \sqrt{\lambda_{mn}} \, z}{\sinh \sqrt{\lambda_{mn}} b},$$

where the coefficients A_{mn}, B_{mn}, C_{mn} and D_{mn} are found from the boundary functions $f(r, \varphi)$ and $g(r, \varphi)$ and the orthogonality of the Bessel functions $J_n(\cdot)$ with respect to the weight function r:

$$A_{mn} = \frac{2}{a^2 \pi \epsilon_n \left(J_n'(\sqrt{\lambda_{mn}} \, a)^2 \right)} \int_0^{2\pi} \int_0^a f(r, \varphi) \cos n\varphi \, J_n\left(\sqrt{\lambda_{mn}} \, r \right) r \, dr \, d\varphi$$

$$B_{mn} = \frac{2}{a^2 \pi \epsilon_n \left(J_n'(\sqrt{\lambda_{mn}} \, a)^2 \right)} \int_0^{2\pi} \int_0^a f(r, \varphi) \sin n\varphi \, J_n\left(\sqrt{\lambda_{mn}} \, r \right) r \, dr \, d\varphi$$

$$C_{mn} = \frac{2}{a^2 \pi \epsilon_n \left(J_n'(\sqrt{\lambda_{mn}} \, a)^2 \right)} \int_0^{2\pi} \int_0^a g(r, \varphi) \cos n\varphi \, J_n\left(\sqrt{\lambda_{mn}} \, r \right) r \, dr \, d\varphi$$

$$D_{mn} = \frac{2}{a^2 \pi \epsilon_n \left(J_n'(\sqrt{\lambda_{mn}} \, a)^2 \right)} \int_0^{2\pi} \int_0^a f(r, \varphi) \sin n\varphi \, J_n\left(\sqrt{\lambda_{mn}} \, r \right) r \, dr \, d\varphi,$$

and

$$\epsilon_n = \begin{cases} 2, & n = 0 \\ 1, & n \neq 0. \end{cases}$$

7.3.4 Laplace Equation in Spherical Coordinates.

Let us consider the following problem: Find a function u, harmonic inside a three dimensional ball, continuous on the closure of the ball and assumes on the sphere a given continuous function f. For convenience only, we will take the ball to be the unit ball, centered at the origin. As in many cases when working with a three dimensional ball, we use spherical coordinates.

As usual, by (r, θ, φ) we denote the spherical coordinates. Let $B = \{(r, \theta, \varphi) : 0 \leq r < 1\}$ be the unit ball with boundary S. We will find a function $u(r, \theta, \varphi)$ which satisfies the boundary value problem

$$(7.3.21) \qquad \qquad \Delta u(r, \theta, \varphi) = 0, \quad (r, \theta, \varphi) \in B,$$

$$(7.3.22) \qquad \qquad u(r, \theta, \varphi) = f(\theta, \varphi), \quad (r, \theta, \varphi) \in S,$$

where $f(\theta, \varphi)$ is a given function. Let us recall from the previous chapters that the Laplacian $\Delta u(r, \theta, \varphi)$ in spherical coordinates is given by

$$(7.3.23) \quad \Delta u \equiv \frac{1}{r^2} \frac{\partial}{\partial r} \left(r^2 \frac{\partial u}{\partial r} \right) + \frac{1}{r^2 \sin \theta} \frac{\partial}{\partial \theta} \left(\sin \theta \frac{\partial u}{\partial \theta} \right) + \frac{1}{r^2 \sin^2 \theta} \frac{\partial^2 u}{\partial \varphi^2}.$$

As many times before, we look for the solution of this problem of the form

$$u(r, \theta, \varphi) = R(r) Y(\theta, \varphi).$$

If we now substitute $u(r, \theta, \varphi)$ into (7.3.21), using (7.3.23), after separating the variables, we obtain the Euler–Cauchy equation

$$(7.3.24) \quad r^2 R''(r) + 2r R(r) - \lambda R(r) = 0, \quad 0 < r \le 1, \quad | \lim_{r \to 0+} R(r) | < \infty,$$

and the Helmholtz equation

$$(7.3.25) \qquad \frac{1}{\sin \theta} \frac{\partial}{\partial \theta} \left(\sin \theta \frac{\partial Y}{\partial \theta} \right) + \frac{1}{\sin^2 \theta} \frac{\partial^2 Y}{\partial \varphi^2} + \lambda Y = 0.$$

The function $Y(\theta, \varphi)$ also satisfies the periodic and boundary conditions

$$(7.3.26) \qquad \begin{cases} Y(\theta, \varphi + 2\pi) = Y(\theta, \varphi), & 0 < \theta, \ 0 < \varphi < 2\pi \\ | \lim_{\theta \to 0+} Y(\theta, \varphi) | < \infty, & | \lim_{\theta \to \pi-} Y(\theta, \varphi) | < \infty. \end{cases}$$

The bounded solutions $Y(\theta, \varphi)$ of the eigenvalue problem (7.3.25), (7.3.26) are called *spherical functions*. An excellent analysis of these important functions is given in the book by N. Asmar [13]. To solve this eigenvalue problem we use again the separation of variables method. If $Y(\theta, \varphi)$ is of the form

$$Y(\theta, \varphi) = P(\theta) \Phi(\varphi),$$

then we obtain the two problems

$$(7.3.27) \qquad \Phi''(\varphi) + \mu \Phi(\varphi) = 0, \quad \Phi(\varphi + 2\pi) = \Phi(\varphi),$$

and

$$(7.3.28) \qquad \begin{cases} \sin^2 \theta \left[\frac{1}{\sin \theta} \frac{d}{d\theta} \left(\sin \theta \frac{dP}{d\theta} \right) + \lambda \right] = \mu, \\ | \lim_{\theta \to 0+} P(\theta) | < \infty, \ | \lim_{\theta \to \pi-} P(\theta) | < \infty. \end{cases}$$

The eigenvalues of (7.3.24) are $\mu_m = m^2$, $m = 0, 1, 2 \dots$; the corresponding eigenfunctions are

$$\{ 1, \ \cos \varphi, \ \sin \varphi, \dots, \cos m\varphi, \ \sin m\varphi, \dots, \}.$$

If we now introduce in problem (7.3.28) the new variable x by $x = \cos \theta$ and write $X(x) = P(\theta)$, then for the found $\mu_m = m^2$, problem (7.3.28) becomes

(7.3.29)
$$\begin{cases} \dfrac{d}{dx}\left((1-x^2)\dfrac{dX(x)}{dx}\right) + \left(\lambda - \dfrac{m^2}{1-x^2}\right)X(x) = 0, \quad -1 < x < 1 \\[2mm] |\lim_{x \to -1+} X(x)| < \infty, \quad |\lim_{x \to 1-} X(x)| < \infty. \end{cases}$$

As we saw in the section on vibrations of the ball in Chapter 5, as well in the section on heat distribution in the ball in Chapter 6, problem (7.3.29) has a bounded solution only if $\lambda = n(n+1)$ for some nonnegative integer n and in this case, the bounded solutions $X_{mn}(x)$ are given by

$$X_{mn}(x) = P_n^{(m)}(x), \quad -1 < x < 1,$$

where $P_n^{(m)}(x)$ are the associated Legendre polynomials. Therefore, the eigenfunctions of problem (7.3.28) are

$$\{P_n(\cos \theta),\ P_n^{(m)}(\cos \theta) \cos m\theta,\ P_n^{(m)}(\cos \theta) \sin m\theta : n \geq 0,\ m \in \mathbb{N}\},$$

where $P_n(\cdot)$ is the Legendre polynomial of order n and $P_n^{(m)}(\cdot)$ is the associated Legendre polynomial.

Now, let us solve Equation (7.3.24) for the found $\lambda = n(n+1)$. Two independent solutions of this equation are

$$R(r) = r^n, \quad R(r) = r^{-(n+1)}, \ 0 < r \leq 1.$$

Since $R(r)$ is bounded,

$$R(r) = r^n$$

is the only one (up to a multiplicative constant) of problem (7.3.24).

Now, consider the series

$$\sum_{n=0}^{\infty} \sum_{m=0}^{n} \left[A_{mn} \cos m\varphi + B_{mn} \sin m\varphi\right] P_n^{(m)}(\cos \theta).$$

Since the above series is uniformly convergent for each fixed $0 \leq r < 1$ it can be differentiated twice term by term and since each member of the series is a harmonic function on the unit ball the series is a harmonic function on the unit ball. Therefore, the general solution of the original problem is given by

(7.3.30) $\quad u(r, \theta, \varphi) = \displaystyle\sum_{n=0}^{\infty} \sum_{m=0}^{n} \left[A_{mn} \cos m\varphi + B_{mn} \sin m\varphi\right] r^n\, P_n^{(m)}(\cos \theta).$

From the boundary condition

$$f(\theta, \varphi) = u(1, \theta, \varphi) = \sum_{m=0}^{\infty} \sum_{n=0}^{\infty} \left[A_{mn} \cos m\varphi + B_{mn} \sin m\varphi \right] P_n^{(m)}(\cos \theta),$$

using the orthogonality property of the associated Legendre polynomials, discussed in Section 3.3.2 of Chapter 3, we obtain that the coefficients A_{mn} and B_{mn} in (3.3.30) are determined by the formulas

$$(7.3.31) \quad \begin{cases} A_{mn} = \dfrac{1}{N_{mn}} \displaystyle\int_0^{2\pi} \int_0^{\pi} f(\theta, \varphi) \, P_n^{(m)}(\cos \theta) \, \cos m\varphi \, \sin \theta \, d\theta \, d\varphi \\[2mm] B_{mn} = \dfrac{1}{N_{mn}} \displaystyle\int_0^{2\pi} \int_0^{\pi} f(\theta, \varphi) \, P_n^{(m)}(\cos \theta) \, \sin m\varphi \, \sin \theta \, d\theta \, d\varphi \\[2mm] N_{mn} = \dfrac{2\pi c_n}{2n+1} \dfrac{(m+n)!}{(n-m)!}, \end{cases}$$

where

$$c_n = \begin{cases} 2, & m = 0 \\ 1, & m > 0. \end{cases}$$

Therefore, the solution of our original problem (7.3.19) is given by (7.3.27), where A_{mn} and B_{mn} are given by (7.3.31).

Example 7.3.6. Find the harmonic function $u(r, \theta)$ in the unit ball

$$\{(r, \theta, \varphi) : \leq r < 1\}$$

such that

$$u(1, \theta) = \sin^4 \theta, \quad 0 \leq \theta < \pi.$$

Solution. Notice that we assumed that the function $u(r, \theta)$ is independent on the variable φ. Therefore, the general solution of the Laplace equation in the ball is given by

$$u(r, \theta) = \sum_{n=0}^{\infty} a_n r^n P_n(\cos \theta),$$

where $P_n(\cdot)$ are the Legendre polynomials of order n. To find the coefficients a_n we use the boundary condition at $r = 1$:

$$(7.3.32) \qquad \sin^4 \theta = u(1, \theta) = \sum_{n=0}^{\infty} a_n P_n(\cos \theta).$$

Using the orthogonality property of the Legendre polynomials, from the above expansion we have

$$a_n \int_0^\pi P_n^2(\cos\theta)\sin\theta\,d\theta = \int_0^\pi \sin^4\theta\,P_n(\cos\theta)\sin\theta\,d\theta.$$

From Chapter 3 we know that

$$\int_0^\pi P_n^2(\cos\theta)\sin\theta\,d\theta = \frac{2}{2n+1},$$

and so

$$a_n = \frac{2n+1}{2}\int_0^\pi \sin^4\theta\,P_n(\cos\theta)\sin\theta\,d\theta.$$

It is not straightforward to find the above integral. In this situation it is much easier to use only (7.3.32) if we introduce the variable x by $x = \cos\theta$. Then from (7.3.32) we have

$$(7.3.33)\quad 1-2x^2+x^4 = a_0 P_0(x)+a_1 P_1(x)+a_2 P_2(x)+a_3 P_3(x)+a_4 P_4(x)+\ldots.$$

If we recall from Chapter 3 that the first 5 Legendre polynomials are

$$P_0(x) = 1, \quad P_1(x) = x, \quad P_2(x) = \frac{1}{2}(3x^2-1),$$

$$P_3(x) = \frac{1}{2}(5x^3-3x), \quad P_4(x) = \frac{1}{8}(35x^4-30x^2+3),$$

then from (7.3.33), by comparison we obtain

$$a_0 = \frac{22}{35}, \quad a_1 = a_3 = 0, \quad a_2 = -\frac{2}{7}, \quad a_4 = \frac{8}{35}, \quad \text{and } a_n = 0 \ \text{ for } \ n \geq 5.$$

Therefore, the solution $u(r,\theta)$ of the Poisson equation is given by

$$u(r,\theta) = \frac{22}{35} - \frac{2}{7}r^2 P_2(\cos\theta) + \frac{8}{35}r^4 P_4(\cos\theta),$$

from which we obtain

$$u(r,\theta) = \frac{22}{35} - \frac{1}{7}r^2\left(3\cos^2\theta - 1\right) + \frac{1}{35}r^4\left(35\cos^4\theta - 30\cos^2\theta + 3\right).$$

The Poisson equation on the unit ball can be solved in a similar way as in the unit disc. For the solution of the Poisson boundary value problem see Exercise 19 of this section.

Exercises for Section 7.3.

1. Use the separation of variables method to solve the Neumann problem for the Laplace equation in a disc, i.e., solve the boundary value problem

$$\Delta u(r, \varphi) = 0, \quad 0 < r < a, \ 0 \le \varphi < 2\pi.$$
$$u_{\mathbf{n}}(a, \varphi) = f(\varphi), \quad 0 \le \varphi < 2\pi,$$

where $\mathbf{n}$ is the outward unit normal vector to the circle, provided that

$$\int_0^{2\pi} f(\varphi)\, d\varphi = 0.$$

2. Use the separation of variables method to solve the Neumann problem for the Laplace equation outside a disc, i.e., solve the boundary value problem

$$\Delta u(r, \varphi) = 0, \quad a < r < \infty, \ 0 \le \varphi < 2\pi.$$
$$u_{\mathbf{n}}(a, \varphi) = f(\varphi), \quad 0 \le \varphi < 2\pi,$$
$$|\lim_{r \to \infty} u(r, \varphi)| < \infty, \quad 0 \le \varphi < 2\pi,$$

where $\mathbf{n}$ is the inward unit normal vector to the circle, provided that

$$\int_0^{2\pi} f(\varphi)\, d\varphi = 0.$$

In Exercises 3–7 find the harmonic function $u(r, \varphi)$ in the unit disc which on the boundary assumes the given function $f(\varphi)$.

3. $f(\varphi) = A \sin \varphi.$

4. $f(\varphi) = A \sin^3 \varphi + B.$

5. $f(\varphi) = \begin{cases} A \sin \varphi, & 0 \le \varphi < \pi, \\ \frac{A}{3} \sin^3 \varphi, & \pi < \varphi < 2\pi. \end{cases}$

6. $f(\varphi) = \begin{cases} 1, & 0 \le \varphi < \pi, \\ 0, & \pi < \varphi < 2\pi. \end{cases}$

7. $f(\varphi) = \frac{1}{2}(\pi - \varphi).$

8. Solve the Dirichlet problem for the Laplace equation in an annulus, i.e., solve the problem

$$\begin{cases} \Delta u(r, \varphi) = 0, & a < r < b, \ 0 < \varphi < 2\pi \\ u(a, \varphi) = f(\varphi), & u(b, \varphi) = g(\varphi), \ 0 \le \varphi < 2\pi. \end{cases}$$

What is the solution if $b = 1$, $f(\varphi) = 0$ and $g(\varphi) = 1 + 2\sin \varphi$?

In Exercises 9–10 solve the Laplace equation

$$\Delta u(r, \varphi) = 0, \quad (r, \varphi) \in \Omega$$

on the indicated domain Ω and the indicated boundary conditions.

9. $\Omega = \{(r, \varphi) : 0 < r < a, \ 0 < \varphi < \frac{\pi}{2}\}$, $u_\varphi(r, 0) = u_\varphi(r, \frac{\pi}{2}) = 0$,
$$u(a, \varphi) = \begin{cases} 1, & 0 < \varphi < \frac{\pi}{4} \\ 0, & \frac{\pi}{4} < \varphi < \frac{\pi}{2}. \end{cases}$$

10. $\Omega = \{(r, \varphi) : 0 < r < 2, \ 0 < \varphi < \pi\}$, $u_\varphi(r, 0) = u_\varphi(r, \pi) = 0$,
$$u(2, \varphi) = \begin{cases} c, & 0 < \varphi < \frac{\pi}{2} \\ 0, & \frac{\pi}{2} < \varphi < \pi. \end{cases}$$

In Exercises 11–13 solve the Poisson boundary value problem

$$\begin{cases} \Delta u(r, \varphi) = F(r, \varphi), & 0 < r < 1, \ 0 < \varphi < 2\pi \\ u(1, \varphi) = f(\varphi), & 0 < \varphi < 2\pi. \end{cases}$$

11. $F(r, \varphi) = 1, \ f(\varphi) = 0.$

12. $F(r, \varphi) = 2 + r^3 \cos 3\varphi, \ f(\varphi) = 0.$

13. $F(r, \varphi) = 1, \ f(\varphi) = \sin 2\varphi.$

In Exercises 14–15 solve the Laplace boundary value problem on a cylinder, i.e., find the function $u(r, \varphi, z)$ which satisfies the problem

$$\begin{cases} \Delta u(r, \varphi, z) = 0, & 0 < r < 1, \ 0 < \varphi < 2\pi, \ 0 < z < 2 \\ u(1, \varphi, z) = f(\varphi, z), & u(r, \varphi, 0) = g(r, \varphi). \end{cases}$$

14. $f(\varphi, z) = 0, \ g(r, \varphi) = 1, \ h(r, \varphi) = 0.$

15. $f(\varphi, z) = 0, \ g(r, \varphi) = G(r), \ h(r, \varphi) = H(r).$

In Exercises 16–19 find the function $u(r, \theta, \varphi) = u(r, \theta)$ which satisfies the problem

$$\begin{cases} \Delta u \equiv \dfrac{1}{r^2} \dfrac{\partial}{\partial r}\left(r^2 \dfrac{\partial u}{\partial r}\right) + \dfrac{1}{r^2 \sin \theta} \dfrac{\partial}{\partial \theta}\left(\sin \theta \dfrac{\partial u}{\partial \theta}\right) = 0, & 0 < r < 1, \ 0 < \theta < \pi \\ \\ u(1, \theta, \varphi) = f(\theta), & 0 < \theta < \pi. \end{cases}$$

16. $f(\theta) = \begin{cases} 4, & 0 < \theta < \frac{\pi}{2} \\ 0, & \frac{\pi}{2} < \theta < \pi. \end{cases}$

17. $f(\theta) = 1 - \cos\theta, \ 0 < \theta < \pi.$

18. $f(\theta) = 2 + \cos^2\theta, \ 0 < \theta < \pi.$

19. $f(\theta) = \begin{cases} 4\cos\theta, & 0 < \theta < \frac{\pi}{2} \\ 0, & \frac{\pi}{2} < \theta < \pi. \end{cases}$

20. Solve the Laplace equation outside the unit ball with given Dirichlet boundary values, i.e., solve the problem

$$\begin{cases} \Delta u(r,\theta,\varphi) = 0, & r > 1, \ 0 < \varphi < 2\pi, \ 0 < \theta < \pi \\ u(1,\theta,\varphi) = f(\theta,\varphi), & 0 < \theta < \pi, \ 0 < \varphi < 2\pi \\ |\lim_{r\to+\infty} u(r,\theta,\varphi)| < \infty, & 0 < \theta < \pi, \ 0 < \varphi < 2\pi. \end{cases}$$

21. Solve the Laplace equation inside the unit ball with given Newmann boundary values, i.e., solve the problem

$$\begin{cases} \Delta u(r,\theta,\varphi) = 0, & 0 < r < 1, \ 0 < \varphi < 2\pi, \ 0 < \theta < \pi \\ u_{\mathbf{n}}(1,\theta,\varphi) = f(\theta,\varphi), & 0 < \theta < \pi, \ 0 < \varphi < 2\pi \\ |\lim_{r\to 0+} u(r,\theta,\varphi)| < \infty, & 0 < \theta < \pi, \ 0 < \varphi < 2\pi, \end{cases}$$

where $\mathbf{n}$ is the outward unit normal vector to the sphere and $f(\theta,\varphi)$ is a given function which satisfies the condition

$$\int_0^{2\pi} \int_0^{\pi} f(\theta,\varphi)\sin\theta\, d\theta\, d\varphi = 0.$$

22. Show that the solution $u(r,\theta,\varphi)$ of the Laplace equation between two concentric balls with given Dirichlet boundary values

$$\begin{cases} \Delta u(r,\theta,\varphi) = 0, & r_1 < r < r_2, \ 0 < \varphi < 2\pi, \ 0 < \theta < \pi \\ u(r_2,\theta,\varphi) = f(\theta,\varphi), & 0 < \theta < \pi, \ 0 < \varphi < 2\pi \\ u(r_1,\theta,\varphi) = 0, & 0 < \theta < \pi, \ 0 < \varphi < 2\pi \\ |\lim_{r\to+\infty} u(r,\theta,\varphi)| < \infty, & 0 < \theta < \pi, \ 0 < \varphi < 2\pi \end{cases}$$

is given by

$$u(r,\theta,\varphi) = \sum_{n=0}^{\infty} a_n \left[\left(\frac{r}{r_2}\right)^n - \left(\frac{r_2}{r}\right)^{n+1} \right] P_n(\cos\theta),$$

where the coefficients a_n are determined by

$$a_n = \frac{2n+1}{2} \frac{1}{\left(\frac{r}{r_1}\right)^n - \left(\frac{r_2}{r_1}\right)^{n+1}} \int_0^{\pi} f(\theta) \, P_n(\cos\theta) \, d\theta.$$

Hint: Use the solution of the Laplace equation inside the ball and the solution of the Laplace equation outside a ball. (see Exercise 20 of this section.)

7.4 Integral Transform Methods for the Laplace Equation.

In this section we will use the Fourier and Hankel transform to solve the Laplace equation on unbounded domains.

7.4.1 The Fourier Transform Method for the Laplace Equation.

Let us apply the Fourier transform to several examples of elliptic boundary value problems.

Example 7.4.1. Use the Fourier transform to solve the following boundary value problem on the right half plane.

$$\begin{cases} u_{xx}(x,y) + u_{yy}(x,y) = 0, & x > 0, \ -\infty < y < \infty \\ u(0,y) = f(y), & -\infty < y < \infty \\ \lim_{x \to +\infty} u(x,y) = 0 & -\infty < y < \infty. \end{cases}$$

Solution. Let

$$U(x,\omega) = \mathcal{F}\left(u(x,y)\right) = \int_{-\infty}^{\infty} u(x,y) \, e^{-i\omega y} \, dy$$

and

$$U(0,\omega) = \mathcal{F}\left(f\right) = F(\omega).$$

If we apply the Fourier transform to the Laplace equation, then in view of the boundary condition for the function $u(x,y)$ at $x = 0$, we obtain the ordinary differential equation

$$\frac{d^2 U(x,\omega)}{dx^2} - \omega^2 U(x,\omega) = 0, \quad x \geq 0.$$

The general solution of the above equation is given by

$$U(x,\omega) = C_1(\omega)e^{|\omega| x} + C_2(\omega)e^{-|\omega| x}.$$

In view of the condition $\lim\limits_{x \to +\infty} U(x, \omega) = 0$ we have $C_1(\omega) = 0$, while from $U(0, \omega) = F(\omega)$ it follows that $C_2(\omega) = F(\omega)$. Therefore,

$$U(x, \omega) = F(\omega) e^{-|\omega| x}.$$

Now, using the fact that

$$\left(\mathcal{F}^{-1}\left(e^{-|\omega| x}\right) \right)(y) = \frac{1}{\pi} \frac{x}{x^2 + y^2},$$

(see the Table B of Appendix B), from the convolution theorem for the Fourier transform we have

$$(7.4.1) \qquad u(x, y) = \frac{1}{\pi} \left(f * \frac{x}{x^2 + y^2} \right)(x) = \frac{1}{\pi} \int\limits_{-\infty}^{\infty} f(t) \frac{x}{(x^2 + (y - t)^2} \, dt.$$

The function

$$P_x(y) = \frac{1}{\pi} \frac{x}{x^2 + (y - t)^2}$$

is called the *Poisson kernel* for the Laplace equation on the right half plane $\{(x, y) : x > 0\}$, and Equation (7.4.1) is known as the *Poisson integral formula* for harmonic functions on the right half plane.

Example 7.4.2. Solve the following Dirichlet boundary value problem in the given semi-infinite strip

$$\begin{cases} u_{xx}(x, y) + u_{yy}(x, y) = 0, & 0 < x < \infty, \ 0 < y < b \\ u(x, 0) = f(x), \quad u(x, b) = 0, \ 0 < x < \infty \\ u(0, y) = 0, \quad \lim\limits_{x \to +\infty} u(x, y) = 0, \ 0 < y < b, \end{cases}$$

and taking $b \to \infty$ solve the boundary problem for the Laplace equation in the first quadrant $\{(x, y) : 0 < x < \infty, \ 0 < y < \infty\}$.

Solution. Since in the boundary conditions of the problem are not derivatives we use the Fourier sine transform . Let

$$U(\omega, y) = \mathcal{F}_s\left(u(x, y)\right) = \int\limits_{0}^{\infty} u(x, y) \sin(\omega x) \, dx,$$

and

$$U(\omega, 0) = \mathcal{F}_s(f) = F(\omega).$$

If we apply the Fourier sine transform to the Laplace equation, then in view of the boundary condition for the function $u(x, y)$ at $x = 0$, we obtain the ordinary differential equation

$$\frac{d^2 U(\omega, y)}{dy^2} - \omega^2 U(\omega, y) = 0, \quad 0 \le y < \infty.$$

The general solution of the above equation is given by

$$U(\omega, y) = C_1(\omega) \sinh(\omega y) + C_2(\omega) \cosh(\omega y).$$

In view of the boundary conditions $U(0, \omega) = F(\omega)$ and $U(\omega, b) = 0$ we have

$$C_2(\omega) = F(\omega),$$
$$C_1(\omega) \sinh(\omega b) + C_2(\omega) \cosh(\omega b) = 0.$$

Solving the above system for $C_1(\omega)$ and $C_2(\omega)$ we obtain

$$U(\omega, y) = F(\omega) \frac{\sinh(\omega(b - y))}{\sinh(\omega b)}.$$

The inverse Fourier sine transform implies that the solution of the original problem is given by

$$u(x, y) = \frac{2}{\pi} \int_0^\infty U(\omega, y) \sin(\omega y) \, d\omega$$

(7.4.2)
$$= \frac{2}{\pi} \int_0^\infty U(\omega, y) \sin(\omega y) F(\omega) \frac{\sinh(\omega(b - y))}{\sinh(\omega b)} \, d\omega$$

$$= \frac{2}{\pi} \int_0^\infty \int_0^\infty f(t) \sin(\omega t) \sin(\omega y) \frac{\sinh(\omega(b - y))}{\sinh(\omega b)} \, dt \, d\omega.$$

Now in (7.4.2) take $b \to \infty$. Assuming that we can pass the limit inside the integrals, from the following limit (obtained by l'Hospital's rule)

$$\lim_{b \to \infty} \frac{\sinh(\omega(b - y))}{\sinh(\omega b)} = e^{-\omega y}$$

it follows that

$$u(x, y) = \frac{2}{\pi} \int_0^\infty f(t) \left(\int_0^\infty \sin(\omega t) \sin(\omega y) e^{\omega y} \, d\omega \right) dt.$$

If for the inside integral we use the formula

$$\sin(\omega t) \sin(\omega y) = \frac{1}{2} [\cos(\omega(t - y)) - \cos(\omega(t + y))]$$

and the integration by parts formula, then we obtain

$$\int_0^\infty \sin(\omega t) \sin(\omega y) e^{\omega y} \, d\omega = \frac{1}{2} \left\{ \int_0^\infty \left[\cos(\omega(t - y)) \cos(\omega(t + y)) \right] e^{\omega y} \, d\omega \right\}$$

$$= \frac{1}{2} \left[\frac{y}{((x - t)^2 + y^2)} - \frac{y}{((x + t)^2 + y^2)} \right].$$

Therefore, the solution of the given problem is

$$u(x,y) = \frac{y}{\pi} \int\limits_0^\infty \left[\frac{y}{((x-t)^2 + y^2)} - \frac{y}{((x+t)^2 + y^2)} \right] f(t)\, dt.$$

Example 7.4.3. Solve the boundary value problem

$$\begin{cases} u_{xx}(x,y) + u_{yy}(x,y) = 0, & 0 < x < a,\ 0 < y < \infty \\ u(0,y) = 0, & u(a,y) = g(y),\ 0 < x < \infty \\ u_y(x,0) = 0, & 0 < x < a. \end{cases}$$

Solution. Since the boundary condition involves a derivative with respect to y of the required function we will use the Fourier cosine transform (with respect to y). Let

$$U(x,\omega) = \mathcal{F}_c\left(u(x,y)\right) = \int\limits_0^\infty u(x,y) \cos(\omega y)\, dx,$$

and

$$U(0,\omega) = \mathcal{F}_c\left(f\right) = F(\omega).$$

If we apply the Fourier cosine transform to the Laplace equation, then in view of the boundary condition for the function $u(x,y)$ at $y = 0$, we obtain the ordinary differential equation

$$\frac{d^2 U(x,\omega)}{dx^2} - \omega^2 U(x,\omega) - u_y(x,0) = 0, \quad 0 \le x < \pi.$$

i.e.,

$$\frac{d^2 U(x,\omega)}{dx^2} - \omega^2 U(x,\omega), \quad 0 \le x < \pi.$$

The general solution of the above equation is given by

$$U(x,\omega) = C_1(\omega) \sinh\left(|\omega| x\right) + C_2(\omega) \cosh\left(\omega x\right).$$

From the boundary condition $u(0,y) = 0$ it follows that $U(0,\omega) = 0$, and therefore, $C_2(\omega) = 0$. From the other boundary condition $u(a,y) = f(y)$ we have

$$C_1(\omega) \sinh(\omega a) = F(\omega),$$

and so

$$U(x,\omega) = F(\omega) \frac{\sinh(\omega x)}{\sinh(\omega a)}.$$

From the inversion formula for the Fourier cosine transform we obtain

$$u(x, y) = \frac{2}{\pi} \int_0^\infty U(x, \omega) \cos (\omega y) \, d\omega$$

$$= \frac{2}{\pi} \int_0^\infty \int_0^\infty f(t) \cos (\omega t) \cos (\omega y) \frac{\sinh (\omega x)}{\sinh (\omega a)} \, dt \, d\omega,$$

and after interchanging the order of integration it follows that

$$u(x, y) = \frac{2}{\pi} \int_0^\infty \left(\int_0^\infty \cos (\omega t) \cos (\omega y) \frac{\sinh (\omega x)}{\sinh (\omega a)} \, d\omega \right) f(t) \, dt.$$

Now, we will introduce the *two dimensional Fourier transform* which can be applied to solve the Laplace equation on unbounded domains.

The Fourier transform of an integrable function $f(\mathbf{x})$, $\mathbf{x} = (x, y) \in \mathbb{R}^2$ is defined by

$$\hat{f}(\mathbf{w}) = \mathcal{F} (f(\mathbf{x})) = \iint_{\mathbb{R}^2} f(\mathbf{x}) \, e^{-i \mathbf{w} \cdot \mathbf{x}} \, d\mathbf{x},$$

where for $\mathbf{x} = (x, y)$ and $\mathbf{w} = (\xi, \eta)$, $\mathbf{w} \cdot \mathbf{x}$ is the usual dot product defined by

$$\mathbf{w} \cdot \mathbf{x} = \xi x + \eta y.$$

The higher dimensional Fourier transform is defined similarly.

The basic properties of the two dimensional Fourier transform are just like those of the one dimensional Fourier transform and they are summarized in the following theorem.

Theorem 7.4.1. *Suppose that $f(\mathbf{x})$, $\mathbf{x} = (x, y)$, is integrable on $\mathbb{R}^2$. Then*

(a) *For any fixed $\mathbf{a} \in \mathbb{R}^2$,*

$$\mathcal{F} (f(\mathbf{x} - \mathbf{a})) = e^{-i \mathbf{w} \cdot \mathbf{a}} \mathcal{F} (f(\mathbf{x})), \quad \mathcal{F} (e^{i \mathbf{a} \cdot \mathbf{x}} f(\mathbf{x})) = \mathcal{F} (\mathbf{w} - \mathbf{a}).$$

(b) *If the partial derivatives $f_x(\mathbf{x})$ and $f_y(\mathbf{y})$ exist and are integrable on $\mathbb{R}^2$, then*

$$\mathcal{F} (f_x) = ix \, \mathcal{F} (f), \quad \mathcal{F} (f_y) = iy \, \mathcal{F} (f).$$

(c) *If g and the convolution $f * g$ are integrable on $\mathbb{R}^2$, then*

$$\mathcal{F} (f * g) = (\mathcal{F} (f)) (g))$$

(*The convolution $f * g$ of the functions f and g is defined by*

$$(f * g)(\mathbf{x}) = \iint\limits_{\mathbb{R}^2} f(\mathbf{y}) g(\mathbf{x} - \mathbf{y}) \, d\mathbf{y}.)$$

(d) *If f is integrable and continuous on $\mathbb{R}^2$ and $F(\omega) = \mathcal{F}(f)$ is also integrable on $\mathbb{R}^2$, then*

$$f(\mathbf{x}) = \frac{1}{4\pi^2} \iint\limits_{\mathbb{R}^2} e^{i\mathbf{w} \cdot \mathbf{x}} F(\mathbf{w}) \, d\mathbf{w}.$$

Example 7.4.4. Solve the three dimensional boundary value problem

$$\begin{cases} u_{xx}(x, y, z) + u_{yy}(x, y, z) + u_{zz}(x, y, z) = 0, & x, \ y \in \mathbb{R}, \ 0 < z < \infty \\ u(x, y, 0) = f(x, y), & -\infty < x < \infty, \ -\infty < y < \infty \\ \lim\limits_{|\mathbf{x}| \to 0} u(x, y, z) = 0, & |\mathbf{x}| = \sqrt{x^2 + y^2 + z^2}. \end{cases}$$

Solution. We apply the two dimensional Fourier transform with respect to x and y. Let $U(\xi, \eta, z) = \mathcal{F}(u)$ and $F(\xi, \eta) = \mathcal{F}(f)$. If we use part (b) of Theorem 7.4.1, then the three dimensional Laplace equation and the boundary conditions become

$$\begin{cases} U_{zz}(\xi, \eta, z) - (\xi^2 + \eta^2)U(\xi, \eta, z) = 0, \\ U(\xi, \eta, 0) = F(\xi, \eta), & \lim\limits_{\xi^2 + \eta^2 + z^2 \to 0} U(\xi, \eta, z) = 0. \end{cases}$$

The solution of the above differential equation, in view of the boundary conditions, is given by

$$U(\xi, \eta, z) = F(\xi, \eta) \, e^{-\sqrt{\xi^2 + \eta^2} \, z}.$$

Using the convolution property (c) of Theorem 7.4.1 we obtain

$$u(x, y, z) = (f * g)(x, y, z),$$

where

$$g(x, y, z) = \mathcal{F}^{-1}\left(e^{-\sqrt{\xi^2 + \eta^2} \, z}\right).$$

To find the function $g(x, y, z)$ is not a trivial matter. Using the inversion Fourier transform formula, given in part (d) in the above theorem, we have

$$g(x, y, z) = \frac{1}{4\pi^2} \int_{-\infty}^{\infty} \int_{-\infty}^{\infty} e^{(x\xi + y\eta)i} e^{-\sqrt{\xi^2 + \eta^2} \, z} \, d\xi \, d\eta.$$

It can be shown that

$$g(x, y, z) = \frac{1}{2\pi} \frac{z}{(x^2 + y^2 + z^2)^{\frac{3}{2}}}.$$

(See Exercise 2, page 247, in the book by G. B. Folland, [6] for details.)
Therefore,

$$u(x, y, z) = \frac{1}{2\pi} \int_{-\infty}^{\infty} \int_{-\infty}^{\infty} f(\xi, \eta) \, g(x - \xi, y - \eta) \, d\xi \, d\eta$$

$$= \frac{z}{2\pi} \int_{-\infty}^{\infty} \int_{-\infty}^{\infty} \frac{f(\xi, \eta)}{\left[(x - \xi)^2 + (y - \eta)^2 + z^2\right]^{\frac{3}{2}}} \, d\xi \, d\eta.$$

7.4.2 The Hankel Transform Method.

In this section we will introduce the *Hankel transform* and we will apply it to solve the Laplace equation on some unbounded domains.

Let $f(r)$ be a function defined on $[0, \infty)$. The *Hankel transform* of order n, n nonnegative integer, of the function f, denoted by $F_n(= \mathcal{H}_n(f)$, is defined by

$$\mathcal{H}_n(f)(\omega) = \int_0^{\infty} r f(r) J_n(\omega r) \, dr,$$

where $J_n(\cdot)$ is the Bessel function of the first kind of order n, provided that the improper integral exists.

For the Hankel transform we have the inversion formula

$$f(r) = \mathcal{H}_n^{-1} bigl(F_n(\omega)) = \int_0^{\infty} \omega F_n(\omega) J_n(\omega r) \, d\omega.$$

The Hankel transform can be defined of any order $\mu \geq -\frac{1}{2}$, but for our purposes we will need the Hankel transform only of nonnegative integer order n.

Important cases for our applications in solving partial differential equations are $n = 0$ and $n = 1$ and the most important properties (for our applications) of the Hankel transform are the following.

Property 1. *Let $f(r)$ be a function defined on $r \geq 0$. If $f(r)$ and $f'(r)$ are bounded at the origin $r = 0$ and satisfy the boundary conditions*

$$\begin{cases} \lim_{r \to \infty} \sqrt{r} \, f(r) = \lim_{r \to \infty} \sqrt{r} \, f(r) = \lim_{r \to \infty} \sqrt{r} \, f'(r) = 0, \\ |\lim_{r \to 0+} r f'(r) J_n(r)| < \infty, \quad |\lim_{r \to 0} r f(r) J_n'(r)| < \infty, \end{cases}$$

then

$$(7.4.3) \qquad \int_0^\infty r J_n(\omega r) \left[f''(r) + \frac{1}{r} f'(r) - \frac{n^2}{r^2} f(r) \right] dr = -\omega^2 F_n(\omega),$$

where $F_n(\cdot)$ is the Hankel transform of f of order n.

Proof. We integrate by parts.

$$\int_0^\infty r J_n(\omega r) \left[f''(r) + \frac{1}{r} f'(r) - \frac{n^2}{r^2} f(r) \right] dr$$

$$= \int_0^\infty r \left[\frac{d}{dr} \left(r \frac{df(r)}{dr} \right) - \frac{n^2}{r^2} f(r) \right] J_n(\omega r) \, dr$$

$$= \left[r f'(r) J_n(\omega r) - \omega r J_n'(\omega r) \right]_{r=0}^{r=\infty} - \omega^2 \int_0^\infty r J_n(\omega r) \, d\omega = -\omega^2 F_n(\omega).$$

Notice that in the above we have used the facts that the Bessel function $J_n(\omega r)$ is bounded at zero and that $\lim\limits_{r \to \infty} \sqrt{r}\, J_n(\omega r) = 0.$ ∎

Property 2. *For the Bessel function of order 0 the following is true:*

$$(7.4.4) \qquad \int_0^\infty \frac{r J_0(\omega r)}{\sqrt{r^2 + z^2}} \, dr = \frac{1}{\omega} e^{-\omega z}.$$

This property can be justified very easy using the definition of the Hankel transform.

Property 3. *For the Bessel function of order 0 the following is true:*

$$(7.4.5.) \qquad \int_0^\infty J_0(\omega r) e^{-z\omega} \, d\omega = \frac{1}{\sqrt{r^2 + z^2}}.$$

The property can be verified by using the series expansion of the Bessel functions $J_o(\omega r)$ (from Chapter 3) and integrating term by term the series expansion.

For some additional properties of the Hankel transform see Exercise 11 of this section.

Example 7.4.5. Find the steady temperature function $u(r, \varphi, z) = u(r, z)$ in the semi-infinite cylinder $\{r, \varphi, z) : 0 \le r < \infty, \ 0 \le \varphi < 2\pi, \ 0 \le z < \infty\}$ in the following two cases:

(a) The temperature on the boundary $z = 0$ is equal to $f(r)$.

(b) The temperature on the boundary $z = 0$ is equal to T for $r < a$, and it is equal to 0 for $r > a$.

Solution. For part (a) we need to solve the boundary value problem

(7.4.6) $$\Delta\, u(r, z) = \frac{1}{r}\frac{\partial}{\partial r}\left(r\frac{\partial u}{\partial r}\right) + \frac{\partial^2 u}{\partial z^2} = 0, \ r > 0, \ z > 0,$$

(7.4.7) $$u(r, 0) = f(r), \ \lim_{z \to \infty} u(r, z) = 0, \ r > 0,$$

(7.4.8) $$\lim_{r \to \infty} u(r, z) = \lim_{r \to \infty} u_r(r, z) = 0, \ z > 0.$$

Let

$$U(\omega, z) = \mathcal{H}\left(u(r, z)\right) = \int_0^\infty r J_0(\omega r) u(r, z)\, dr,$$

$$F(\omega) = \mathcal{H}\left(f(r)\right) = \int_0^\infty r J_0(\omega r) f(r)\, dr$$

be the Hankel transforms of order 0 of $u(r, z)$ and $f(r)$, respectively.

If we multiply both sides of (7.4.6) by $r J_0(\omega r)$, then using the integration by parts formula and the boundary conditions (7.4.8) we have

$$\int_0^\infty r J_0(\omega r)\frac{\partial^2 u}{\partial z^2}\, dr = -\int_0^\infty J_0(\omega r)\frac{\partial}{\partial r}\left(r\frac{\partial u}{\partial r}\right) dr$$

$$= -\left[r J_0(\omega r)\frac{\partial u}{\partial r}\right]_{r=0}^{r=\infty} + \omega\int_0^\infty r J_0'(\omega r)\frac{\partial u}{\partial r}\, dr = \omega\int_0^\infty r J_0'(\omega r)\frac{\partial u}{\partial r}\, dr$$

$$= \omega\left[r u(r, z) J_0'(\omega r)\right]_{r=0}^{r=\infty} - \omega\int_0^\infty u(r, z)\frac{\partial}{\partial r}\left(r J_0'(\omega r)\right) dr$$

$$= -\omega\int_0^\infty u(r, z)\frac{\partial}{\partial r}\left(r J_0'(\omega r)\right) dr$$

$$= -\omega\int_0^\infty u(r, z) J_0'(\omega r)\, dr - \omega^2\int_0^\infty r u(r, z) J_0''(\omega r)\, dr.$$

Thus,

$$(7.4.9) \quad \int_0^\infty r J_0(\omega r) \frac{\partial^2 u}{\partial z^2}\, dr = -\omega \int_0^\infty u(r,z) J_0'(\omega r) dr - \omega^2 \int_0^\infty r u(r,z) J_0''(\omega r) dr.$$

Since $J_0(\cdot)$ is the Bessel function of order zero we have

$$J_0''(\omega\, r) + \frac{1}{r} J_0'(\omega\, r) + \omega^2 J_0(\omega\, r) = 0.$$

If we find $J_0'(\omega r)$ from the last equation and substitute in (7.4.9) we obtain

$$\int_0^\infty r J_0(\omega\, r) \frac{\partial^2 u}{\partial z^2}\, dr = \omega^2 \int_0^\infty r J_0(\omega\, r)\, dr,$$

which implies the boundary value problem

$$\begin{cases} \dfrac{d^2 U(r,z)}{dz^2} - \omega^2 U(r,z) = 0, \\[2mm] U(\omega,0) = F(\omega), \quad \lim_{z\to\infty} U(\omega,z) = 0, \ \ 0 < z < \infty. \end{cases}$$

The solution of the last boundary value problem is given by

$$U(r,z) = F(\omega) e^{-\omega z}.$$

Taking the inverse Hankel transform we obtain that the solution $u = u(r,z)$ of the original boundary value problem is given by

$$u = \int_0^\infty \omega J_0(\omega r) F(\omega) e^{-\omega z}\, d\omega = \int_0^\infty \left(\int_0^\infty \sigma J_0(\omega\, \sigma) f(\sigma)\, d\sigma \right) \omega J_0(\omega r) e^{-\omega z}\, d\omega.$$

The solution of (b) follows from (a) by substituting the given function f:

$$u(r,z) = \int_0^\infty \left(\int_0^\infty \sigma J_0(\omega\, \sigma) f(\sigma)\, d\sigma \right) \omega J_0(\omega\, r) e^{-\omega z}\, d\omega$$

$$= T \int_0^\infty \left(\int_0^a \sigma J_0(\omega\, \sigma) f(\sigma)\, d\sigma \right) \omega J_0(\omega\, r) e^{-\omega z}\, d\omega$$

$$= T \int_0^\infty \left(\int_0^{a\omega} \frac{\rho}{\omega^2} J_0(\rho)\, d\rho \right) \omega J_0(\omega\, r) e^{-\omega z}\, d\omega$$

$$= T \int_0^\infty \left(\int_0^{a\omega} \rho J_0(\rho)\, d\rho \right) \frac{1}{\omega} J_0(\omega\, r) e^{-\omega z}\, d\omega.$$

If we use the following identity for the Bessel function

$$\int_0^c \rho J_0(\rho)\, d\rho = c J_1(c),$$

(see the section for the Bessel functions in Chapter 3), then the above solution $u(r, z)$ can be written as

$$u(r, z) = aT \int_0^\infty J_0(\omega r)\, J_1(a\omega)\, e^{-\omega z}\, d\omega.$$

Example 7.4.6. Find the solution $u = u(r, \varphi, z) = u(r, z)$ of the boundary value problem

(a) $$\Delta u = \frac{\partial^2 u}{\partial r^2} + \frac{1}{r}\frac{\partial}{\partial r} + \frac{\partial^2 u}{\partial z^2} = 0, \quad 0 < r < \infty, \ z \in \mathbb{R},$$

(b) $$\lim_{r \to \infty} u(r, z) = \lim_{|z| \to \infty} u(r, z) = 0,$$

(c) $$\lim_{r \to 0} r^2 u(r, z) = \lim_{r \to 0} r u_r(r, z) = f(z).$$

Solution. Let

$$U(\omega, z) = \mathcal{H}\left(u(r, z)\right) = \int_0^\infty r J_0(\omega r) u(r, z)\, dr,$$

$$F(\omega) = \mathcal{H}\left(f(r)\right) = \int_0^\infty r J_0(\omega r) f(r)\, dr$$

be the Hankel transforms of order 0 of $u(r, z)$ and $f(r)$, respectively. If we take $n = 0$ in Property 1, then we obtain

$$\frac{d^2 U(\omega, z)}{dz^2} - \omega^2 U(\omega, z) + \left[r u_r(r, z) J_0(\omega r) - \omega r J_0'(\omega r) \right]_{r=0}^{r=\infty} = 0$$

From the given boundary conditions (b) at $r = \infty$ and the boundary conditions (c), in view of the following properties for the Bessel functions

$$J_0(0) = 1, \quad J_0'(r) = -J_1(r), \quad \lim_{r \to 0+} \frac{1}{r} J_0(\omega r) = \frac{\omega}{2},$$

it follows that

$$\frac{d^2 U(\omega, z)}{dz^2} - \omega^2 U(\omega, z) = f(z), \quad -\infty < z < \infty.$$

Using $U(\omega, z) \to 0$ as $|z| \to \infty$ we obtain that the solution of the above differential equation is given by

$$U(\omega, z) = -\frac{1}{2} \int_{-\infty}^{\infty} e^{-|\omega - \tau|} f(\tau) \, d\tau.$$

If we take the inverse Hankel transform of the above $U(\omega, z)$ and use (7.4.5) (after changing the integration order) we obtain that the solution $u(x, z)$ of our original problem is given by

$$u(x, z) = \int_{0}^{\infty} \omega \, J_0(\omega r) \, U(\omega, z) \, d\omega = -\frac{1}{2} \int_{-\infty}^{\infty} \left(\int_{0}^{\infty} e^{-|\omega - \tau|} J_0(\omega r) \, d\omega \right) f(\tau) \, d\tau$$

$$= -\frac{1}{2} \int_{-\infty}^{\infty} \frac{f(\tau)}{\sqrt{r^2 + (z - \tau)^2}} \, d\tau.$$

Exercises for Section 7.4.

In Problems 1–9, use one of the Fourier transforms to solve the Laplace equation

$$u_{xx}(x, y) + u_{yy}(x, y) = 0, \quad (x, y) \in D$$

on the indicated domain D, subject to the given boundary value conditions.

1. $D = \{(x, y) : 0 < x < \pi, \, 0 < y < \infty\}$; $u(0, y) = 0$, $u(\pi, y) = e^{-y}$, for $y > 0$ and $u_y(x, 0) = 0$ for $0 < x < \pi$.

2. $D = \{(x, y) : 0 < x < \infty, \, 0 < y < 2\}$; $u(0, y) = 0$, for $0 < y < 2$ and $u(x, 0) = f(x)$, $u(x, 2) = 0$ for $0 < x < \infty$.

3. $D = \{(x, y) : 0 < x < \infty, \, 0 < y < \infty\}$; $u(0, y) = e^{-y}$, for $0 < y < \infty$ and $u(x, 0) = e^{-x}$ for $0 < x < \infty$.

4. $D = \{(x, y) : -\infty < x < \infty, \, 0 < y < \infty\}$; $u(x, 0) = f(x)$ for $-\infty < x < \infty$, $\displaystyle\lim_{|x|, \, y \to \infty} u(x, y) = 0$.

5. $D = \{(x, y) : -\infty < x < \infty, \, 0 < y < \infty\}$; $u(x, 0) = \begin{cases} 1, & |x| < 1 \\ 0, & |x| > 1. \end{cases}$

6. $D = \{(x, y) : -\infty < x < \infty, \, 0 < y < \infty\}$; $u(x, 0) = \frac{1}{4 + x^2}$.

7. $D = \{(x, y) : -\infty < x < \infty, \, 0 < y < \infty\}$; $u(x, 0) = \cos x$.

8. $D = \{(x, y) : 0 < x < 1, \ 0 < y < \infty\}; \ u(0, y) = 0, \ u(1, y) = e^{-y}$, for $y > 0$ and $u(x, 0) = 0$ for $0 < x < 1$.

9. $D = \{(x, y) : 0 < x < \infty, \ 0 < y < \infty\}; \ u(x, 0) = 0$ for $x > 0$ and

$$u(0, y) = \begin{cases} 1, & 0 < y < 1 \\ 0, & y > 1. \end{cases}$$

10. Using the two dimensional Fourier transform show that the solution $u(x, y)$ of the equation

$$u_{xx}(x, y) + u_{yy}(x, y) - u(x, y) = -f(x, y), \ x, y \in \mathbb{R},$$

where f is a square integrable function on $\mathbb{R}^2$, is given by

$$u(x, y) = \frac{1}{4\pi} \int_0^\infty \left(\int_{-\infty}^\infty \int_{-\infty}^\infty e^{-\frac{(x-\xi)^2 + (y-\eta)^2}{4t}} \, d\xi \, d\eta \right) \frac{e^{-t}}{t} f(t) \, dt.$$

Hint: Use the result

$$\mathcal{F}^{-1}\left(\frac{1}{1 + \xi^2 + \eta^2}\right)(x, y) = \frac{1}{2} \int_0^\infty \frac{e^{-t}}{t} e^{-\frac{x^2 + y^2}{4t}} \, dt.$$

11. Using properties of the Bessel functions prove the following identities for the Hankel transform.

(a) $\mathcal{H}_0\left(\frac{1}{x}\right)(\omega) = \frac{1}{\omega}$.

(b) $\mathcal{H}_0\left(e^{-ax}\right)(\omega) = \frac{a}{\sqrt{(a^2 + \omega^2)^3}}$.

(c) $\mathcal{H}_0\left(\frac{1}{x} e^{-ax}\right)(\omega) = \mathcal{L}\left(J_0(ax)\right)(\omega) = \frac{1}{\sqrt{a^2 + \omega^2}}$.

(d) $\mathcal{H}_0\left(e^{-a^2 x^2}\right)(\omega) = \frac{1}{2a^2} e^{-\frac{\omega^2}{4a^2}}$.

(e) $\mathcal{H}_n\left(f(ax)\right)(\omega) = \frac{1}{a} \mathcal{H}\left(f(x)\right)\left(\frac{\omega}{a}\right)$.

(f) $\mathcal{H}_n\left(f(ax)\right)(\omega) = \frac{1}{a} \mathcal{H}\left(f(x)\right)\left(\frac{\omega}{a}\right)$.

In Problems 12–13, use the Hankel transform to solve the partial differential equation

$$\frac{1}{r} \frac{\partial}{\partial r}\left(r \frac{\partial u}{\partial r}\right) + \frac{\partial^2 u}{\partial z^2} = 0, \quad 0 < r < \infty, \ 0 < z < \infty,$$

subject to the given boundary conditions.

12. $\lim\limits_{z,\to\infty} u(r,z) = 0,\ \lim\limits_{r,\to\infty} u(r,z) = 0,\quad u(r,0) = \begin{cases} 1, & 0 \le r < a \\ \\ 0, & 1 < r < \infty. \end{cases}$

13. $\lim\limits_{z,\to\infty} u(r,z) = 0,\ \lim\limits_{r,\to\infty} u(r,z) = 0,\quad u(r,0) = \dfrac{1}{\sqrt{a^2 + r^2}}.$

7.5 Projects Using Mathematica.

In this section we will see how Mathematica can be used to solve several problems involving the Laplace and Poisson equations. In particular, we will develop several Mathematica notebooks which automate the computations of the solutions of these equations.

Project 7.5.1. Use the separation of variables method to solve the Laplace equation on the given square.

$$\begin{cases} u_{xx}(x,y) + u_{yy}(x,y) = 0, & 0 < x < \pi,\ 0 < y < \pi, \\ u(x,0) = f(x) = \begin{cases} x, & 0 < x < \frac{\pi}{2} \\ \pi - x, & \frac{\pi}{2} < x < \pi, \end{cases} \\ u(0,y) = f(y) = 0, & 0 < y < \pi. \end{cases}$$

All calculation should be done using Mathematica. Display the plot of $u(x,y)$.

Solution. In Example 7.2.1 we solved this problem and we found out that its solution $u(x,y)$ is given by

$$u(x,y) = \frac{4}{\pi} \sum_{n=1}^{\infty} \frac{\sin \frac{n\pi}{2}}{n^2 \sinh n\pi} \left[\sinh(ny) + \sinh\left(n(\pi - y)\right)\right] \sin nx.$$

We split the problem into two problems

$$\begin{cases} v_{xx}(x,y) + v_{yy}(x,y) = 0, & 0 < x < \pi,\ 0 < y < \pi, \\ v(x,0) = f(x) = \begin{cases} x, & 0 < x < \frac{\pi}{2} \\ 1 - x, & \frac{\pi}{2} < x < \pi, \end{cases} \\ v(0,y) = f(y) = 0, & 0 < y < \pi. \end{cases}$$

and

$$\begin{cases} w_{xx}(x,y) + w_{yy}(x,y) = 0, & 0 < x < \pi,\ 0 < y < \pi, \\ w(x,0) = f(x) = \begin{cases} x, & 0 < x < \frac{\pi}{2} \\ 1 - x, & \frac{\pi}{2} < x < \pi, \end{cases} \\ w(0,y) = f(y) = 0, & 0 < y < \pi. \end{cases}$$

The solution of the original problem is $u(x, y) = v(x, y) + w(x, y)$. Because of symmetry we need to solve only the problem for the function $w(x, y)$.

Let us use Mathematica to "derive" this result. Let $w[X, Y] = X[x] Y[y]$. After separating the variables we obtain the equations

$$\begin{cases} X''(x) + \lambda^2 X(x) = 0, & X(0) = 0, \ X(\pi) = 0 \\ Y''(y) - \lambda^2 X(x) = 0. \end{cases}$$

First define the boundary function $f(x)$:

$In[1] := f[x\_] := \text{Piecewise}[\{\{x, 0 < x < \frac{\pi}{2}\}, \{\pi - x, x > 0\}\}];$

Solve the eigenvalue problem for $X(x)$:

$In[2] := \text{DSolve}[X''[x] + \lambda^2 * X[x] == 0, X[0] == 0\}, X, x];$

$Out[2] = \{\{X- > \text{Function}[\{x\}, C[2] \ \text{Sin}[x\lambda]]\}\}$

Use the boundary condition $X(\pi) = 0$:

$In[3] := \text{Reduce}[\text{Sin}[\lambda * Pi] == 0, \lambda, \text{Integers}];$

$Out[3] = C[1] \in \text{Integers} \&\&\lambda == C[1];$

Define the eigenvalues $\lambda_n = n$:

$In[4] := \lambda[n\_] := n;$

Now solve the equation for $Y(y)$:

$In[5] := \text{DSolve}[Y''[x] - \lambda * Y[y] == 0, Y, y];$

$Out[5] = \{\{Y- > \text{Function}[\{y\}, e^{ny}C[1] + e^{-ny}C[2]]\}\}$

Define the eigenfunctions $X_n(x)$ and the functions $Y_n(x)$ by

$In[6] := X[x\_, n\_] := \text{Sin}[n * x]; \text{medskip}$
$In[7] := Y[y\_, n\_] := A[n] * e^{-n*y} + B[n] * e^{n*y};$

Now find $A[n]$, $B[n]$ from the conditions $w(x, 0) = w(x, \pi) = f(x)$:

$In[8] := a[n\_] = \text{FullSimplify}[\text{Integrate}[f[x] * \text{Sin}[\lambda[n] * x], \{x, 0, Pi\}],$

$\text{Assumptions} - > \{\text{Element}[n, \text{Integers}]\}]$

$Out[8] = \frac{2Sin\left[\frac{n\pi}{2}\right]}{n^2}$

$In[9] := b[n\_] = \text{FullSimplify}[\text{Integrate}[(Sin[\lambda[n] * x])^2, \{x, 0, Pi\}]$

$Out[9] = \frac{\pi}{2}$

$In[10] := \text{Solve}[(A[n]+B[n]) * b[n] == a[n] \&\& (A[n] * e^{-n*yPi} + B[n] * e^{n*Pi}) * b[n] == a[n], \{A[n], B[n]\}, \text{Reals}];$

$In[11] := \text{FullSimplify}[\%]$

$Out[11] = \left\{\left\{A[n] \to \frac{4e^{n\pi}Sin\left[\frac{n\pi}{2}\right]}{(1+e^{n\pi})n^2\pi}, B[n] \to \frac{4Sin\left[\frac{n\pi}{2}\right]}{(1+e^{n\pi})n^2\pi}\right\}\right\}$

Now, define the Nth partial sum of the solution $w(x, y)$:

$In[12] := w[x_-, y_-, N_-] := \text{Sum } [X[x, n] * Y[y, n], \{n, 1, N\}];$

The partial sum of the solution $v(x, y)$ is given by

$In[13] := v[x_-, y_-, N_-] := \text{Sum } [X[y, n] * Y[x, n], \{n, 1, N\}];$

Thus, the partial sums of the solution of the original problem are given by

$In[14] := u[x_-, y_-, N_-] := v[x, y, N] + w[x, y, N];$

The plot of $u[x, y, 100]$ is displayed in Figure 7.5.1.

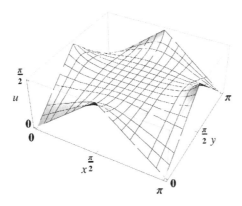

Figure 7.5.1

Project 7.5.2. Plot the solution of the boundary value problem

$$\begin{cases} u_{xx}(x, y) + u_{yy}(x, t) = 0, & 0 < x < \pi, \ 0 < y < \pi, \\ u(x, 0) = 0, & 0 < x < \pi, \\ u(0, y) = 0, & 0 < y < \pi. \end{cases}$$

Solution. In Example 7.2.4. we found that the solution of this problem is given by

$$u(x, y) = \frac{1}{\pi^2} \sum_{m=1}^{\infty} \sum_{n=1}^{\infty} \frac{(-1)^{m+n}}{mn(m^2 + n^2)} \sin mx \sin nx.$$

Define the Nth double partial sum:

$In[1] := u[x_-, y_-, N_-] := \frac{1}{\pi^2} \text{Sum } \left[\frac{(-1)^{m+n}}{m * n(m^2 + n^2)} \right.$

$\left. \sin m * x \sin n * x, \{m, 1, N\}, \{n, 1, N\} \right];$

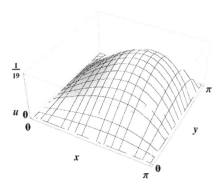

Figure 7.5.2

The plot of $u[x, y, 100]$ is displayed in Figure 7.5.2.

Project 7.5.3. Use the Fourier transform to solve the boundary value problem

$$\begin{cases} u_{xx}(x, y) + u_{yy}(x, y) = 0, & -\infty < x < \infty, \ 0 < y < \infty \\ u(x, 0) = f(x), & -\infty < x < \infty \\ \lim_{y \to \infty} u(x, y) = 0, & -\infty < x < \infty, \\ \lim_{|x| \to \infty} u(x, y) = 0, & 0 < y < \infty. \end{cases}$$

Plot the solution $u(x, y)$ of the problem for the boundary value functions

(a) $\quad f(x) = \begin{cases} \frac{1}{2}, & 0 < x < \frac{1}{2} \\ 0, & x \le 0 \ \text{or} \ x \ge \frac{1}{2}. \end{cases}$

(b) $\quad f(x) = x^2 e^{-x^4}, \quad -\infty < x < \infty.$

In each case plot the level sets of the solution.

Solution. (a) Define the boundary function $f(x)$:

$In[1] := f[x\_] := \text{Piecewise} \ [\{\{0, x \le 0\}, \{\frac{1}{2}, 0 < x \le \frac{1}{2}\}, \{0, x > \frac{1}{2}\}\}];$

Next, define the Laplace operator $L u = u_{xx} + u_{yy}$:

$In[2] := Lapu = D[u[x, y], x, x] + D[u[x, y], y, y];$

Find the Fourier transform of the Laplace operator with respect to x:

$In[3] :=$FourierTransform $[D[u[x, y], x, x], x, \omega,$ FourierParameters $\to \{1, -1\}]+$ FourierTransform $[D[u[x, y], y, y], x, \omega,$ FourierParameters $\to \{1, -1\}]$

$Out[3] = -\omega^2$ FourierTransform $[u[x,y],x,\omega]]$
$+$ FourierTransform $u^{(2,0)}[[x,y],y,\omega]$,FourierParameters $\to \{1,-1\}];$

Remark. In Mathematica, by default the Fourier transform of a function f and the inverse Fourier transform of a function F are defined by

$$\mathcal{F}(f)(\omega) = \frac{1}{\sqrt{2\pi}} \int_{\infty}^{\infty} f(x)e^{i\omega x}\, dx,$$

$$\mathcal{F}^{-1}(F)(x) = \frac{1}{\sqrt{2\pi}} \int_{\infty}^{\infty} F(\omega)e^{-i\omega x}\, d\omega.$$

$In[] := F[\omega] =$ FourierTransform $f[[x],x,\omega];$

$In[] :=$ InverseFourierTransform $F[\omega],\omega,x];$

Define the differential expression:

$In[4] :=$ de $= -\omega^2 * U[\omega],y]+$ D $[U[\omega],y],\{y,2\}]$

$Out[4] = -\omega^2 U[\omega],y] + U^{(0,2)}[\omega],y]$

Next solve the differential equation:

$In[5] :=$ ftu=DSolve $[$ de $== 0, U[\omega,y],y]$

$Out[5] = \{\{U[\omega,y] \to e^{y|\omega|}C[1] + e^{-y|\omega|}C[2]\}\}$

Write the solution in the standard form.

$In[6] :=$ Sol $= U[\omega,y]/.$ ftu $[[1]]$

$Out[6] = e^{y|\omega|}C[1 + e^{-y|\omega|}C[2]$

Use the boundary conditions to obtain:

$In[7] := U[\omega,y] = F[\omega]\ e^{-y\ Abs[\omega]};$

Find the inverse Fourier transform of $e^{-y|\omega|}$:

$In[8] := p[x_-] =$ InverseFourierTransform $[Exp[-y\ Abs[\omega]],\omega,x,$
FourierParameters $\to \{1,-1\}]$

$Out[8] = \frac{y}{\pi(x^2+y^2)}$

Use the convolution theorem to find the solution $u(x,y)$:

$In[9] := u[x_-,y_-] =$ Convolve $[f[t],p[t],t,x]$

$Out[9] = \frac{ArcCot\left[\frac{2y}{1-2x}\right]+ArcTan\left[\frac{x}{y}\right]}{\pi}$

Alternatively, we integrate numerically:

$In[10] := \text{Clear } [u];$

$In[11] := u[x\_, y\_] = \text{If } [y > 0, \text{NIntegrate } [\frac{1}{\pi} \frac{y \, f[t]}{(x-t)^2 + y^2}, \{t, \infty, \infty\}], f[x]]];$

$In[12] := \text{Plot } 3D[u[x, y], \{x, -4, 4\}, \{y, 0.005, 4\}, \text{PlotRange} \to \text{All}]$

$In[13] := \text{ContourPlot } [u[x, y], \{x, -4, 4\}, \{y, 0.005, 4\},$
$\text{FrameLabel} \to \{"x", "y"\}, \text{ContourLabels} \to \text{True}]$

The plots of the solutions $u(x, y)$ and their level sets for (a) and (b) are displayed in Figures 7.5.3 and Figure 7.5.4, respectively.

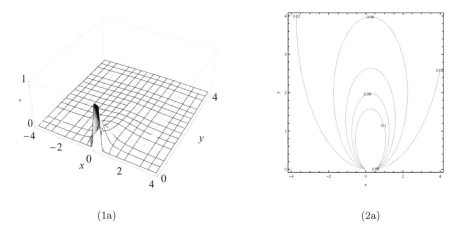

(1a)   (2a)

Figure 7.5.3

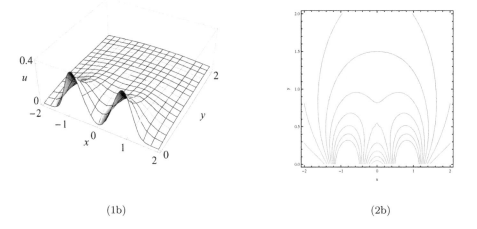

(1b)   (2b)

Figure 7.5.4

CHAPTER 8

FINITE DIFFERENCE NUMERICAL METHODS

Despite the methods developed in the previous chapters for solving the wave, heat and Laplace equations, very often these analytical methods are not practical, and we are forced to find approximate solutions of these equations.

In this chapter we will discuss the *finite difference* numerical methods for solving partial differential equations.

The first two sections are devoted to some mathematical preliminaries required for developing these numerical methods. Some basic facts of linear algebra, numerical iterative methods for linear systems and finite differences are described in these sections.

All the numerical results of this chapter and various graphics were produced using the mathematical software Mathematica.

8.1 Basics of Linear Algebra and Iterative Methods.

Because of their importance in the study of systems of linear equations, we briefly review some facts from linear algebra. More detailed properties of matrices discussed in this section can be found in many linear algebra textbooks, such as the book by J. W. Demmel [9].

A matrix A of type $m \times n$ is a table of the form

$$
(8.1.1) \qquad A = \begin{pmatrix} a_{11} & a_{12} & \cdots & a_{1n} \\ a_{21} & a_{22} & \cdots & a_{2n} \\ \vdots & \vdots & \cdots & \vdots \\ a_{m1} & a_{m2} & \cdots & a_{mn} \end{pmatrix}
$$

with m rows and n columns. The matrix A will be denoted by $A = \left(a_{ij}\right)$.

The *transpose* matrix of the $m \times n$ matrix $A = \left(a_{ij}\right)$ is the $n \times m$ matrix $A^T = \left(a_{ji}\right)$.

An $n \times n$ matrix A is called *symmetric* if $A^T = A$.

An $n \times 1$ matrix $\mathbf{a} = \begin{pmatrix} a_1 \\ a_2 \\ \vdots \\ a_n \end{pmatrix}$ usually is called a *column vector*.

If $A = \left(a_{ij}\right)$ and $B = \left(b_{ij}\right)$ are two $m \times n$ matrices and c is a scalar, then cA and $A + B$ are defined by

$$
cA = \left(ca_{ij}\right), \quad A + B = \left(a_{ij} + b_{ij}\right).
$$

A set $S = \{\mathbf{a_1}, \mathbf{a_2}, \ldots, \mathbf{a_k}\}$ of finitely many $n \times 1$ vectors is said to be *linearly independent* if none of the vectors of S can be expressed as a linear combination of the other vectors in S.

If $A = (a_{ij})$ is an $m \times n$ matrix and $B = (b_{ij})$ is an $n \times p$ matrix, then their product AB is the $m \times p$ matrix $C = (c_{ij})$ defined by

$$c_{ij} = \sum_{k=1}^{n} a_{ik} b_{kj}, \quad i = 1, 2, \ldots, \; j = 1, 2, \ldots p.$$

In general, matrix multiplication is not a commutative operation.

Example 8.1.1. Compute the products AB and BA for the matrices

$$A = \begin{pmatrix} 0 & 4 & 5 \\ 1 & 2 & 3 \end{pmatrix}, \quad B = \begin{pmatrix} 2 & 4 \\ 2 & 2 \\ 3 & 2 \end{pmatrix}.$$

Solution. Because A is a 2×3 matrix and B is a 3×2 matrix, AB is the 2×2 matrix and BA is the 3×3 matrix

$$AB = \begin{pmatrix} 23 & 18 \\ 15 & 14 \end{pmatrix}, \quad BA = \begin{pmatrix} 4 & 16 & 22 \\ 2 & 10 & 16 \\ 2 & 16 & 22 \end{pmatrix}.$$

and so the products are not equal.

Two vectors $\mathbf{x} = \begin{pmatrix} x_1 \\ x_2 \\ \vdots \\ x_n \end{pmatrix}$ and $\mathbf{y} = \begin{pmatrix} y_1 \\ y_2 \\ \vdots \\ y_n \end{pmatrix}$ are said to be *orthogonal* if

$$\mathbf{x}^T \mathbf{y} = \sum_{k=1}^{n} x_k y_k = 0.$$

An $n \times n$ matrix is called a *square matrix*.

The $n \times n$ matrix

$$I_n = \begin{pmatrix} 1 & 0 & 0 & \cdots & 0 \\ 0 & 1 & 0 & \cdots & 0 \\ \vdots & \vdots & \vdots & \cdots & \vdots \\ 0 & 0 & 0 & \cdots & 1 \end{pmatrix}$$

is called the $n \times n$ *identity matrix*.

For any square $n \times n$ matrix A we have that $AI_n = I_n A = A$.

A square matrix is called *diagonal* if the only nonzero entries are on the main diagonal.

The *determinant* of a square $n \times n$ matrix A, denoted by $det(A)$ or $|A|$ can be defined inductively:

If $A = (a_{11})$ is an 1×1 matrix, then $det(A) = a_{11}$. If

$$A = \begin{pmatrix} a_{11} & a_{12} \\ a_{21} & a_{22} \end{pmatrix}$$

is a 2×2 matrix, then

$$det(A) = \begin{vmatrix} a_{11} & a_{12} \\ a_{21} & a_{22} \end{vmatrix} = a_{11}a_{22} - a_{12}a_{21}.$$

For an $n \times n$ matrix A given by (8.1.1) we proceed as follows: Let A_{ij} be the $(n-1) \times (n-1)$ matrix obtained by deleting the ith row and the jth column of A. Then we define

$$\det(A) = \begin{vmatrix} a_{11} & a_{12} & \cdots & a_{1n} \\ a_{21} & a_{22} & \cdots & a_{2n} \\ \vdots & \vdots & \cdots & \vdots \\ a_{n1} & a_{n2} & \cdots & a_{nn} \end{vmatrix} = \sum_{j=1}^{n} (-1)^{i+j} a_{ij} \det(A_{ij}).$$

A square $n \times n$ matrix A is said to be *invertible* or *nonsingular* if there exists an $n \times n$ matrix B such that

$$AB = BA = I_n.$$

In this case, the matrix B is called the *inverse* of A and it is denoted by A^{-1}.

A matrix A is invertible if and only if $\det(A) \neq 0$.

A nonzero $n \times 1$ column vector $\mathbf{x}$ is called an *eigenvector* of an $n \times n$ matrix A if there is a number (real or complex) λ such that

$$A\mathbf{x} = \lambda\mathbf{x}.$$

The number λ is called an *eigenvalue* of the matrix A with corresponding eigenvector $\mathbf{x}$.

The eigenvalues λ of a matrix A can be be found by solving the nth degree polynomial equation, called the *characteristic equation* of A,

$$\det(\lambda I_n - A) = 0.$$

Example 8.1.2. Find the eigenvalues and corresponding eigenvectors of the matrix

$$A = \begin{pmatrix} 4 & -6 \\ 3 & -7 \end{pmatrix}.$$

Solution. The eigenvalues of A can be found by solving the characteristic equation

$$\det(\lambda I_2 - A) = \begin{vmatrix} \lambda - 4 & 6 \\ -3 & \lambda + 7 \end{vmatrix} = \lambda^2 + 3\lambda - 10 = 0.$$

The solutions of the last equation are $\lambda = 2$ and $\lambda = -5$.

If $\mathbf{x} = \begin{pmatrix} x_1 \\ x_2 \end{pmatrix}$ is an eigenvector of A corresponding to the eigenvalue $\lambda = 2$, then

$$\begin{pmatrix} 4 & -6 \\ 3 & -7 \end{pmatrix} \begin{pmatrix} x_1 \\ x_2 \end{pmatrix} = 2 \begin{pmatrix} x_1 \\ x_2 \end{pmatrix}.$$

From the last equation we obtain the linear system

$$2x_1 - 6x_2 = 0$$
$$3x_1 - 9x_2 = 0.$$

This system has infinitely many solutions given by

$$\begin{pmatrix} x_1 \\ x_2 \end{pmatrix} = x_2 \begin{pmatrix} 3 \\ 1 \end{pmatrix}.$$

Taking $x_2 \neq 0$ we obtain the eigenvectors of A corresponding to $\lambda = 2$.
Similarly, the eigenvectors of A which correspond to $\lambda = -5$ are given by

$$\begin{pmatrix} x_1 \\ x_2 \end{pmatrix} = t \begin{pmatrix} 2 \\ 3 \end{pmatrix}, \quad t \neq 0.$$

For real symmetric matrices we have the following result.

Theorem 8.1.1. *If A is a square and real symmetric matrix, then*

1. *The eigenvalues of A are real numbers.*

2. *The eigenvectors corresponding to distinct eigenvalues are orthogonal to each other.*

3. *The matrix A is diagonalizable, i.e.,*

$$A = UDU^T$$

   where D is the diagonal matrix with the eigenvalues of the matrix A along the main diagonal, and U is the orthogonal matrix ($U^TU = I$) whose columns are the corresponding eigenvectors.

A real $n \times n$ matrix A is called *positive definite* if

$$\mathbf{x}^T A \mathbf{x} \geq 0$$

for every nonzero $n \times 1$ vector $\mathbf{x}$.

All eigenvalues of a real positive definite matrix are nonnegative numbers.

The notions of a *norm* of vectors and a norm of square matrices are important in the investigation of iterative methods for solving linear systems of algebraic equations.

A norm on the space $\mathbb{R}^n$ of all $n \times 1$ vectors, denoted by $\| \ \|$, is a real valued function on $\mathbb{R}^n$ which satisfies the following properties.

(1) $\|\mathbf{x}\| > 0$ if $\mathbf{x} \neq \mathbf{0}$ and $\|\mathbf{x}\| = 0$ only if $\mathbf{x} = \mathbf{0}$.

(2) $\|\alpha \mathbf{x}\| = |\alpha| \|\mathbf{x}\|$ for every $\alpha \in \mathbb{R}$ and every $\mathbf{x} \in \mathbb{R}^n$.

(3) $\|\mathbf{x} + \mathbf{y}\| \leq \|\mathbf{x}\| + \|\mathbf{y}\|$ for all $\mathbf{x}, \mathbf{y} \in \mathbb{R}^n$.

For a column vector $\mathbf{x} = \left(x_1, x_2, \ldots, x_n\right)^T \in \mathbb{R}^n$, the following three norms are used most often.

$$\|\mathbf{x}\|_\infty = \max_{1 \leq k \leq n} |x_k|, \quad \|\mathbf{x}\|_1 = \sum_{k=1}^{n} |x_k|, \quad \|\mathbf{x}\|_2 = \sqrt{\sum_{k=1}^{n} |x_k|^2}.$$

A *matrix norm* $\| \ \|$ is a real valued function on the space of all $n \times n$ matrices with the following properties.

(1) $\|A\| \geq 0$ and $\|A\| = 0$ only if $A = \mathbf{0}$ matrix.

(2) $\|\alpha A\| = |\alpha| \|A\|$ for every matrix A and every scalar α.

(3) $\|A + B\| \leq \|A\| + \|B\|$ for every $n \times n$ matrix A and B.

(4) $\|A B\| \leq \|A\| \|B\|$ for every $n \times n$ matrix A and B.

The norm of an $n \times n$ matrix A, induced by a vector norm $\| \cdot \|$ in $\mathbb{R}^n$, is defined by

$$\|A\| = \max\{\|A\mathbf{x}\| : \|\mathbf{x}\| = 1\}.$$

If $A = \left(a_{ij}\right)$ is an $n \times n$ real matrix, then the matrix norms of A, induced by the above three vector norms are given by

$$\|A\|_\infty = \max_{1 \leq i \leq n} \sum_{j=1}^{n} |a_{ij}|, \quad \|A\|_1 = \max_{1 \leq j \leq n} \sum_{i=1}^{n} |a_{ij}|, \quad \|A\|_2 = \sqrt{\lambda_{max}},$$

where λ_{max} is the largest eigenvalue of the symmetric and positive definite matrix $A^T A$.

If A is a real symmetric and positive definite matrix, then

$$\|A\|_2 = |\lambda_{max}|,$$

where λ_{max} is the eigenvalue of A of largest absolute value.

The following simple lemma will be needed.

Lemma 8.1.1. *If $\|\cdot\|$ is any matrix norm on the space of all $n \times n$ matrices, then*

$$\|A\| \geq |\lambda|,$$

where λ is any eigenvalue of the matrix A.

Proof. If λ is an eigenvalue of the matrix A, then there exists a nonzero column vector $\mathbf{x}$ such that $A\mathbf{x} = \lambda \mathbf{x}$. Therefore,

$$|\lambda| \, \|\mathbf{x}\| = \|\lambda \, x\| = \|A \, \mathbf{x}\| \leq \|A\| \, \|\mathbf{x}\|. \qquad \blacksquare$$

Most of the numerical methods for solving partial differential equations lead to systems of linear algebraic equations, very often of very large order. Therefore, we will review some facts about iterative methods for solving linear systems.

Associated with a linear system of n unknowns $x_1, x_2, \ldots, x_n$

$$(8.1.2) \qquad \begin{cases} a_{11}x_1 + a_{12}x_2 + \ldots + a_{1n}x_n = b_1 \\ a_{21}x_1 + a_{22}x_2 + \ldots + a_{2n}x_n = b_2 \\ \quad \vdots \\ a_{n1}x_1 + a_{12}x_2 + \ldots + a_{nn}x_n = b_n \end{cases}$$

are the following square matrix A and column vectors $\mathbf{b}$ and $\mathbf{x}$.

$$A = \begin{pmatrix} a_{11} & a_{12} & \cdots & a_{1n} \\ a_{21} & a_{22} & \cdots & a_{2n} \\ \vdots & \vdots & \cdots & \vdots \\ a_{n1} & a_{n2} & \cdots & a_{nn} \end{pmatrix} \quad \text{and} \quad \mathbf{b} = \begin{pmatrix} b_1 \\ b_2 \\ \vdots \\ b_n \end{pmatrix}, \quad \mathbf{x} = \begin{pmatrix} x_1 \\ x_2 \\ \vdots \\ x_n \end{pmatrix}.$$

Using these matrices, the linear system (8.1.2) can be written in the matrix form

$$(8.1.3) \qquad\qquad\qquad A\mathbf{x} = \mathbf{b}.$$

In our further discussion, the matrix A will be invertible, so the linear system (8.1.3) has a unique solution.

In general, there are two classes of methods for solving linear systems. One of them is *Direct Elimination Methods* and the other is *Iterative Methods*. The direct methods are based on the well-known Gauss elimination technique, which consists of applying row operations to reduce the given system of equations to an equivalent system which is easier to solve.

For very large linear systems, the direct methods are not very practical, and so iterative methods are almost always used. Therefore, our main focus in this section will be on the iterative methods for solving large systems of linear equations.

All of the iterative methods for the numerical solution of any system of equations (not only linear) are based on the following very important result from mathematical analysis, known as the *Banach Fixed Point Theorem* or *Contraction Mapping Theorem*.

Theorem 8.1.2. Contraction Mapping Theorem. *Let K be a closed and bounded subset of R^n, equipped with a norm $\| \ \|$. Suppose that $f : K \to K$ is a function such that there is a positive constant $q < 1$ with the property*

$$\|f(x) - f(\mathbf{y})\| \le q \|\mathbf{x} - \mathbf{y}\|$$

for all $\mathbf{x}$, $\mathbf{y} \in K$. Then, there is a unique point $\mathbf{x}^\star \in K$ such that $\mathbf{x}^\star = f(\mathbf{x}^\star)$. Moreover, the sequence $\{\mathbf{x}^{(k)} : k = 0, 1, 2, \ldots\}$, defined recursively by

$$\mathbf{x}^{(k+1)} = f(\mathbf{x}^{(k)}), \quad k = 0, 1, \ldots,$$

converges to $\mathbf{x}^\star$ for any initial point $\mathbf{x}^{(0)} \in C$.

The following inequalities hold and describe the rate of convergence of the sequence $\{\mathbf{x}^{(k)} : k = 0, 1, 2, \ldots\}$:

$$\|\mathbf{x}^{(k)} - \mathbf{x}^\star\| \le \frac{q}{1-q} \|\mathbf{x}^{(k)} - \mathbf{x}^{(k-1)}\|, \quad k = 1, 2, \ldots,$$

or equivalently

$$\|\mathbf{x}^{(k)} - \mathbf{x}^\star\| \le \frac{q^k}{1-q} \|\mathbf{x}^{(1)} - \mathbf{x}^{(0)}\|, \quad k = 1, 2, \ldots.$$

The proof of this theorem is far beyond the scope of this book, and the interested student is referred to the book by M. Rosenlicht [11], or the book by M. Reed and B. Simon [15].

Now we are ready to describe some iterative methods for solving a linear system.

Usually, a given system of linear equations

$$A\mathbf{x} = \mathbf{b}$$

is transformed to an equivalent linear system

$$\mathbf{x} = \mathbf{x} - C(A\mathbf{x} - \mathbf{b}) = (I_n - CA)\mathbf{x} + C\mathbf{b} = B\mathbf{x} + C\mathbf{b},$$

where C is some matrix (usually such that C is invertible). Then the $(k+1)^{th}$ approximation $\mathbf{x}^{(k+1)}$ of the exact solution $\mathbf{x}$ is determined recursively by

(8.1.4) $$\mathbf{x}^{(k+1)} = (I_n - CA)\mathbf{x}^{(k)} + C\mathbf{b}, \ \ k = 0, 1, \ldots.$$

For convergence of this iterative process we have the following theorem.

Theorem 8.1.3. *The iterative process*

$$\mathbf{x}^{(k+1)} = B\mathbf{x}^{(k)} + C\mathbf{b}$$

converges for any initial point $\mathbf{x}^{(0)}$ *if and only if the spectral radius*

$$\rho(B) := max\{|\lambda_i| < 1 : 1 \le i \le n\} < 1,$$

where $\lambda_1, \ldots, \lambda_n$ *are the eigenvalues of the matrix* B.

In practice, to find the eigenvalues of matrices is not a simple matter. Therefore, other sufficient conditions for convergence of the iterative process are used.

For a different choice of the iteration matrix C we have different iterative methods.

The Jacobi Iteration. Let us assume that the diagonal entries a_{ii} of the matrix $A = (a_{ij})$ are all nonzero, otherwise we can exchange some rows of the nonsingular matrix A, if necessary. Next, we split the matrix A into its lower diagonal part L, its diagonal part D and its upper diagonal part U:

$$L = \begin{pmatrix} 0 & 0 & 0 & \cdots & 0 & 0 \\ a_{21} & 0 & 0 & \cdots & 0 & 0 \\ a_{31} & a_{32} & 0 & \cdots & 0 & 0 \\ \vdots & \vdots & \vdots & \cdots & \vdots & \vdots \\ a_{n-1,1} & a_{n-1,2} & a_{n-1,3} & \cdots & 0 & 0 \\ a_{n1} & a_{n2} & a_{n3} & \cdots & a_{n,n-1} & 0 \end{pmatrix},$$

$$D = \begin{pmatrix} a_{11} & 0 & 0 & \cdots & 0 & 0 \\ 0 & a_{22} & 0 & \cdots & 0 & 0 \\ \vdots & \vdots & \vdots & \cdots & \vdots & \vdots \\ 0 & 0 & 0 & \cdots & a_{n-1,n} & 0 \\ 0 & 0 & 0 & \cdots & 0 & a_{n,n} \end{pmatrix}$$

and

$$U = \begin{pmatrix} 0 & a_{12} & a_{13} & \cdots & a_{1n-1} & a_{1n} \\ 0 & 0 & a_{23} & \cdots & a_{2,n-1} & a_{2n} \\ 0 & 0 & 0 & \cdots & a_{3,n-1} & a_{3n} \\ \vdots & \vdots & \vdots & \cdots & \vdots & \vdots \\ 0 & 0 & 0 & \cdots & 0 & a_{n-1,n} \\ 0 & 0 & 0 & \cdots & 0 & 0 \end{pmatrix}.$$

If we take $B = D$, then the matrix C in this case is given by

$$C = I_n - B^{-1}A = -D^{-1}(L + U),$$

with entries

$$c_{ij} = \begin{cases} -\dfrac{a_{ij}}{a_{ii}}, & \text{if } i \neq j \\ 0, & \text{if } i = j. \end{cases}$$

and so the iterative process is given by

$$(8.1.5) \qquad \mathbf{x}^{(k+1)} = -D^{-1}(L + U)\mathbf{x}^{(k)} + D^{-1}\mathbf{b}, \ k = 0, 1, 2, \ldots.$$

The iterative process (8.1.5) in coordinate form is given by

$$(8.1.6) \qquad x_i^{(k+1)} = \frac{1}{a_{ii}}\left(b_i - \sum_{\substack{j=1 \\ j \neq i}}^{n} a_{ij}x_i^{(k)}\right), \quad i = 1, \ldots, n; \ k = 1, 2, \ldots.$$

The iterative sequence given by (8.1.5) or (8.1.6) is called the *Jacobi Iteration*.

The Gauss–Seidel Iteration. The Gauss–Seidel iterative method is an improvement of the Jacobi method.

If already obtained iterations $x_j^{(k+1)}$, $1 \leq j \leq i-1$ from the Jacobi method are used in evaluating $x_j^{(k+1)}$, $i \leq j \leq n$, then we obtain the *Gauss–Seidel iterative* method.

$$(8.1.7) \ x_i^{(k+1)} = \frac{1}{a_{ii}}\left(b_i - \sum_{j=1}^{i-1} a_{ij}x_i^{(k+1)} - \sum_{j=i+1}^{n} a_{ij}x_i^{(k)}\right), \quad 1 \leq j \leq n, \ k \in \mathbb{N},$$

or in matrix form

$$(8.1.8) \qquad \mathbf{x}^{(k+1)} = D^{-1}\left(\mathbf{b} - \mathbf{x}^{(k+1)} - U\mathbf{x}^{(k)}\right), \ k \in \mathbb{N}.$$

An $n \times n$ matrix $A = (a_{ij})$ is called a *strongly diagonally dominant matrix* if

$$\sum_{\substack{j=1 \\ j \neq i}}^{n} |a_{ij}| < |a_{ii}|, \quad 1 \leq i \leq n.$$

For convergence of the Jacobi and Gauss–Seidel iterative methods we have the following theorem.

Theorem 8.1.4. *Let* $A = (a_{ij})$ *be an* $n \times n$ *nonsingular, strongly diagonally dominant matrix, then the Jacobi iteration* (8.1.5) *and Gauss–Seidel iteration* (8.1.8) *converge for every initial point* $\mathbf{x}^{(0)}$.

Proof. Let λ be any eigenvalue of the matrix C used in the iterative processes (8.1.5) or (8.1.8). If we use the ∞ matrix norm of C, then by Lemma 8.1.1. and the strictly dominance of the matrix C it follows that

$$|\lambda| \leq \|C\|_\infty = \max_{1 \leq i \leq n} \sum_{j=1}^{n} |c_{ij}| = \max_{1 \leq i \leq n} \sum_{\substack{j=1 \\ j \neq i}}^{n} \frac{|a_{ij}|}{|a_{ii}|}$$

$$= \max_{1 \leq i \leq n} \frac{1}{|a_{ii}|} \sum_{\substack{j=1 \\ j \neq i}}^{n} |a_{ij}| < \max_{1 \leq i \leq n} \frac{1}{|a_{ii}|} |a_{ii}| = 1.$$

The conclusion now follows from Theorem 8.1.2.

   There is more general result for convergence of the Jacobi and Gauss–Seidel iterations which we state without a proof.

Theorem 8.1.5. *If the matrix* A *is strictly positive definite, then the Jacobi and Gauss–Seidel iterative processes are convergent.*

Example 8.1.3. Solve the linear system

$$\begin{pmatrix} 9 & 2 & -1 & 3 \\ 2 & 8 & -3 & 1 \\ -1 & -3 & 10 & -2 \\ 3 & 1 & -2 & 8 \end{pmatrix} \begin{pmatrix} x_1 \\ x_2 \\ x_3 \\ x_4 \end{pmatrix} = \begin{pmatrix} 5 \\ -2 \\ 6 \\ -4 \end{pmatrix}$$

by the Jacobi and Gauss–Seidel method using 10 iterations.

Solution. First, we check that the matrix of the system is a diagonally dominant matrix. This can be done using Mathematica:

$In[1] := A = \{\{9, 2, -1, 3\}, \{2, 8, -3, 1\}, \{-1, -3, 10, -2\}, \{3, 1, -2, 8\}\}$
$Out[2] = \{\{9, 2, -1, 3\}, \{2, 8, -3, 1\}, \{-1, -3, 10, -2\}, \{3, 1, -2, 8\}\}$
$In[3] := Table[-Sum[Abs[A[[i, j]]], \{j, 1, i - 1\}] - Sum[Abs[A[[i, j]]],$
$\{j, i + 1, 4\}] + Abs[A[[i, i]]], \{i, 1, 4\}]$
$Out[4] = \{3, 2, 4, 2\}$
Since A is diagonally dominant, the Jacobi and Gauss–Seidell iterations, given by

$$\begin{cases} 9x_1^{(k+1)} = 5 - 2x_2^{(k)} + x_3^{(k)} - 3x_4^{(k)} \\ 8x_2^{(k+1)} = -2 - 2x_1^{(k)} + 3x_3^{(k)} - x_4^{(k)} \\ 10x_3^{(k+1)} = 6 + x_1^{(k)} + 3x_2(k) + 2x_4^{(k)} \\ 8x_4^{(k+1)} = -4 - 3x_2^{(k)} - x_2^{(k)} + 2x_3^{(k)} \end{cases}$$

and

$$\begin{cases} 9x_1^{(k+1)} = 5 - 2x_2^{(k)} + x_3^{(k)} - 3x_4^{(k)} \\ 8x_2^{(k+1)} = -2 - 2x_1^{(k)} + 3x_3^{(k)} - x_4^{(k)} \\ 10x_3^{(k+1)} = 6 + x_1^{(k)} + 3x_2(k) + 2x_4^{(k)} \\ 8x_4^{(k+1)} = -4 - 3x_2^{(k)} - x_2^{(k)} + 2x_3^{(k)}, \end{cases}$$

respectively, will converge for any initial point $\left(x_1^{(0)}, x_2^{(0)}, x_3^{(0)}, x_4^{(0)}\right)$. Taking the initial point $(0, -1, 1, 1)$ we have the results displayed in Table 8.1.1 (Jacobi) and Table 8.1.2 (Gauss–Seidel).

Table 8.1.1 (Jacobi)

$k \backslash x_k$	x_1	x_2	x_3	x_4
1	1.22222	0.25	0.1	−0.12
2	0.552778	−0.502431	0.772222	−0.964583
3	1.07454	0.0219618	0.311632	−0.451432
4	0.735778	−0.345343	0.623756	−0.827789
5	0.977534	−0.0965626	0.404417	−0.57681
6	0.814219	−0.270626	0.553423	−0.753401
7	0.92832	−0.151846	0.449554	−0.633148
8	0.850299	−0.234354	0.520648	−0.716751
9	0.904401	−0.177738	0.471374	−0.659406
10	0.86723	−0.216909	0.505238	−0.69909

Table 8.1.2 (Gauss–Seidel)

$k \backslash x_k$	x_1	x_2	x_3	x_4
1	0.555556	−0.138889	0.813889	−0.4875
2	0.839352	−0.0936921	0.558328	−0.663464
3	0.859567	−0.172586	0.501488	−0.675392
4	0.87476	−0.196208	0.493535	−0.680125
5	0.880703	−0.200084	0.49202	0.682248
6	0.882104	−0.200737	0.49154	−0.682812
7	0.882383	−0.200917	0.491401	−0.682929
8	0.882447	−0.20097	0.491368	−0.682954
9	0.882463	−0.200984	0.49136	−0.682961
10	0.882468	−0.200987	0.491359	−0.682962

The above tables were generated by the following Mathematica programs.

$Jac[A0_-, B0_-, X0_-, max_-] :=$
$Module[\{A = N[A0], B = N[B0], i, j, k = 0, n = Length[X0], X = X0,$
$Xin = X0\},$
$While[k < max,$

$For[i = 1, i <= n, i + +,$

$$X_{[[i]]} = \frac{1}{A_{[[i,i]]}} \left(B_{[[i]]} + A_{[[i,i]]} \star Xin_{[[i]]} - \sum_{j=1}^{n} A_{[[i,j]]} \star Xin_{[[j]]} \right)];$$

$Xin = X;$

$k = k + 1;];$

$Return[X];];$

$A = \{\{9, 2, -1, 3\}, \{2, 8, -3, 1\}, \{-1, -3, 10, -2\}, \{3, 1, -2, 8\}\};$

$B = \{5, -2, 6, -4\};$

$X = \{0, -1, 1, -1\};$

$X = Jac[A, B, X, 10]$

$GS[A0\_, B0\_, X0\_, max\_] :=$

$Module[\{A = N[A0], B = N[B0], i, j, k = 0, n = Length[X0], X = X0\},$

$While[k < max,$

$For[i = 1, i <= n, i + +,$

$$X_{[[i]]} = \frac{1}{A_{[[i,i]]}} \left(B_{[[i]]} + A_{[[i,i]]} \star X[[i]] - \sum_{j=1}^{n} A_{[[i,j]]} \star X_{[[j]]} \right)];$$

$k = k + 1;];$

$Return[X];];$

$X = \{0, -1, 1, -1\};$

$X = GS[A, B, X, 10]$

If we want the error to be incorporated in the program, which will allow us to stop the iteration when some required accuracy of the iteration sequence has been achieved, then we can use the extended Jacobi module and the extended Gauss–Seidel module. For tracking the error we can use, for example, the norm

$$\|x_i^{(k+1)} - x_i^{(k)}\|_\infty = \max_{1 \le i \le n} |x_i^{(k+1)} - x_i^{(k)}| \le \epsilon.$$

$generate[A\_List, b\_List] := Module[\{B, c, n\},$

$flag = True;$

$n = Length[A];$

$Do[If[A[[i, i]] == 0, flag = False], \{i, 1, n\}];$

$If[flag, \{B = Table[0, \{i, 1, n\}, \{j, 1, n\}];$

$c = Table[0, \{i, 1, n\}];$

$Do[$

$\{Do[If[i \ne j, B[[i, j]] = -A[[i, j]]/A[[i, i]]], \{j, 1, n\}];$

$c[[i]] = b[[i]]/A[[i, i]]\}, \{i, 1, n\}]; \};$

$If[!flag, Print["Anerrorhasoccuredintheirconstruction."]];$

$If[flag, \{B, c\}]$

$];$

$A = \{\{9, 2, -1, 3\}, \{2, 8, -3, 1\}, \{-1, -3, 10, -2\}, \{3, 1, -2, 8\}\};$

$b = \{5, -2, 6, -4\};$

$\{B, c\} = generate[A, b];$

```
ExtJac[A_List, b_List, x0_List, e_] := Module[{},
norm1[xx_List, yy_List] :=
Max[Sum[Abs[xx[[i]] − yy[[i]]], {i, 1, Length[xx]}]];
norm[xx_List] := Max[Sum[Abs[xx[[i]]], {i, 1, Length[xx]}]];
{B, c} = generate[A, b];
z = Table[0, {i, 1, Length[x0]}];
x[0] = x0;
G[x_List] := B.x + c;
x[k_ /; k > 0] := x[k] = G[x[k − 1]];
Do[
If[norm1[x[k], x[k − 1]] <= e, {z = x[k], savek = k,
Break[]}], {k, 1, 100}];
];

generate[A_List, b_List] := Module[{B, c, n},
flag = True;
n = Length[A];
Do[If[A[[i, i]] == 0, flag = False], {i, 1, n}];
If[flag, {B = Table[0, {i, 1, n}, {j, 1, n}];
c = Table[0, {i, 1, n}];
Do[
{Do[If[i ≠ j, B[[i, j]] = −A[[i, j]]/A[[i, i]]], {j, 1, n}];
c[[i]] = b[[i]]/A[[i, i]]}, {i, 1, n}]; }];
If[flag, {B, c}]
];

A = {{9, 2, −1, 3}, {2, 8, −3, 1}, {−1, −3, 10, −2}, {3, 1, −2, 8}};
b = {5, −2, 6, −4};
{B, c} = generate[A, b];
ExtGS[A_List, b_List, x0_List, e_] := Module[{},
norm1[xx_List, yy_List] :=
Max[Sum[Abs[xx[[i]] − yy[[i]]], {i, 1, Length[xx]}]];
norm[xx_List] := Max[Sum[Abs[xx[[i]]], {i, 1, Length[xx]}]];
{B, c} = generate[A, b];
z = Table[0, {i, 1, Length[x0]}];
x[0] = x0;
G[x_List] := B.x + c;
x[k_ /; k > 0] := x[k] = G[x[k − 1]];
Do[
If[norm1[x[k], x[k − 1]] <= e, {z = x[k], savek = k,
Break[]}], {k, 1, 100}]];
```

Taking $\epsilon = 10^{-8}$ and
$A = \{\{9, 2, -1, 3\}, \{2, 8, -3, 1\}, \{-1, -3, 10, -2\}, \{3, 1, -2, 8\}\};$

$b = \{5, -2, 6, -4\};$

$ExtpGS[A, b, \{0, 0, 0, 0\}, 0.00000001]$

The result is given in Table 8.1.3 (Extended Jacobi).

Table 8.1.3 Extended Jacobi

k	x_1	x_2	x_3	x_4
23	0.882469	−0.200988	0.491358	−0.682963

For Gauss–Seidel we have

$ExtGauss - Seidel[A, b, \{0, 0, 0, 0\}, 0.00000001]$

The result is given in Table 8.1.4 (Extended Gauss–Seidel).

Table 8.1.4 Extended Gauss–Seidel

k	x_1	x_2	x_3	x_4
14	0.882469	−0.200988	0.491358	−0.682963

If A is not a diagonally dominant matrix, then the Jacobi iteration process for the system $A\mathbf{x} = \mathbf{b}$ may not converge.

Example 8.1.4. Apply the Jacobi iterative method to the linear system

$$\begin{pmatrix} 1 & 2 & -1 & 3 \\ 2 & 2 & -3 & 1 \\ -1 & -3 & 1 & -2 \\ 3 & 1 & -2 & 6 \end{pmatrix} \begin{pmatrix} x_1 \\ x_2 \\ x_3 \\ x_4 \end{pmatrix} = \begin{pmatrix} 1 \\ -2 \\ 3 \\ 4 \end{pmatrix}.$$

Solution. Taking the initial approximation $(0, 0, 0, 0)$ in the Jacobi iterative method, the Jacobi module described above will generate the following approximations.

Table 8.1.5

k	x_1	x_2	x_3	x_4
1	1	−1.	−1.	0.666667
2	0.	−3.83333	−0.333333	0.
3	8.33333	−1.5	−8.5	1.19444
4	−8.08333	−22.6806	4.44444	−6.08333
5	69.0556	16.7917	−60.9583	9.96991
6	−123.45	−166.478	102.491	−56.9792
7	607.384	304.677	−505.927	124.302
8	−1487.19	−1429.43	1275.81	−522.447
9	5703.01	3661.13	−4727.57	1407.77
10	−16272.1	−13499.2	13873.9	−5036.88

This particular system has the exact solution

$$\left(\frac{35}{43}, -\frac{17}{43}, \frac{51}{43}, \frac{31}{43}\right) = \left(0.813953, -0.395349, 1.18605, 0.72093\right).$$

But from Table 8.1.5 we can see that the iterations given by the Jacobi method are getting worse instead of better. Therefore we can conclude that this iterative process diverges.

The next example shows that the Gauss–Seidel iterative method is divergent while the Jacobi iterative process is convergent.

Example 8.1.5. Discuss the convergence of the Jacobi and Gauss–Seidel iterative methods for the system

$$A \cdot \mathbf{x} = \mathbf{b},$$

where A is the matrix given by

$$A = \begin{pmatrix} 1 & 0 & 1 \\ -1 & 1 & 0 \\ 1 & 2 & -3 \end{pmatrix}.$$

Solution. It is obvious that the matrix A is not strictly diagonally dominant. For the Jacobi iterative method, the matrix $B = B_J$ in (8.1.7) is given by

$$B_J = D^{-1} \cdot (L + U) = \begin{pmatrix} 1 & 0 & 0 \\ 0 & 1 & 0 \\ 0 & 0 & -3 \end{pmatrix}^{-1} \cdot \left[\begin{pmatrix} 0 & 0 & 0 \\ -1 & 0 & 0 \\ 1 & 2 & 0 \end{pmatrix} + \begin{pmatrix} 0 & 0 & 1 \\ 0 & 0 & 0 \\ 0 & 0 & 0 \end{pmatrix} \right]$$

$$= \begin{pmatrix} 0 & 0 & 1 \\ -1 & 0 & 0 \\ -\frac{1}{3} & -\frac{2}{3} & 0 \end{pmatrix}.$$

Using Mathematica, in order to simplify the calculations, we find that the characteristic polynomial $p_B(\lambda)$ is given by

$$p_{B_J}(\lambda) = -\lambda^3 - \frac{1}{3}\lambda + \frac{2}{3}.$$

Again using Mathematica, we find that the eigenvalues of B_J (the roots of $p_{B_J}(\lambda)$) are

$$\lambda_1 = 0.747415, \quad \lambda_2 = -0.373708 + 0.867355\,i, \quad \lambda_3 = -0.373708 - 0.867355\,i.$$

Their moduli are

$$|\lambda_1| = 0.747415, \quad |\lambda_2| = 0.944438, \quad |\lambda_3| = 0.944438.$$

Therefore, the spectral radius $\rho(B_J) = 0.944438 < 1$ and so by Theorem 8.1.3 the Jacobi iterative method converges.

For the Gauss–Seidel iterative method, the matrix $B = B_{GS}$ in (8.1.8) is given by

$$B_{GS} = (L+D)^{-1} \cdot U = \left[\begin{pmatrix} 0 & 0 & 0 \\ -1 & 0 & 0 \\ 1 & 2 & 0 \end{pmatrix} + \begin{pmatrix} 1 & 0 & 0 \\ 0 & 1 & 0 \\ 0 & 0 & -3 \end{pmatrix} \right]^{-1} \cdot \begin{pmatrix} 0 & 0 & 1 \\ 0 & 0 & 0 \\ 0 & 0 & 0 \end{pmatrix}$$

$$= \begin{pmatrix} 0 & 0 & 1 \\ 0 & 0 & 1 \\ 0 & 0 & 1 \end{pmatrix}.$$

Using Mathematica (to simplify the calculations) we find that the characteristic polynomial $p_{B_{GS}}(\lambda)$ is given by

$$p_{B_{GS}}(\lambda) = -\lambda^3 + \lambda^2.$$

We find that the eigenvalues of B_{GS} (the roots of $p_{B_{GS}}(\lambda)$) are

$$\lambda_1 = 0, \quad \lambda_2 = 0, \quad \lambda_3 = 1.$$

Their moduli are

$$|\lambda_1| = 0, \quad |\lambda_2| = 0, \quad |\lambda_3| = 1.$$

Therefore, the spectral radius $\rho(B_{GS}) = 1$ and so by Theorem 8.1.3 the Gauss–Seidel iterative method diverges.

The strictly diagonal dominance is not a necessary condition for convergence of the Jacobi iterative process. There are linear systems for which their matrices are not strictly diagonally dominant, but nevertheless, the Jacobi iterative method produces convergent iterations. See Exercise 7 of this section.

The Conjugate Gradient Iteration. The Conjugate Gradient Method is the most popular iterative method for solving large linear systems. It is based on solving a minimization problem of a specific functional.

Let

$$(8.1.9) \qquad\qquad A\mathbf{x} = \mathbf{b}$$

be a given linear system, where A is a symmetric and positive definite matrix. For an arbitrary vector column $\mathbf{x} \in \mathbb{R}^n$ consider the real valued function

$$(8.1.10) \qquad\qquad f(\mathbf{x}) = \mathbf{x}^T A\mathbf{x} - 2\mathbf{b}^T\mathbf{x}.$$

We will show that the solution $\mathbf{x}$ of the system (8.1.9) minimizes the function $f(\mathbf{x})$ defined in (8.1.10).

Let $\widetilde{\mathbf{x}}$ be a solution of (8.1.9), i.e., let $A\widetilde{\mathbf{x}} = \mathbf{b}$. Since A is symmetric we have $\left(A\mathbf{x}\right)^T\mathbf{x} = \mathbf{x}^T\left(A\mathbf{x}\right)$ and since A is positive definite we have $\left(A\mathbf{x}\right)^T\mathbf{x} \geq 0$ for every vector $\mathbf{x}$. Therefore,

$$\begin{aligned} f(\mathbf{x}) - f(\widetilde{\mathbf{x}}) &= \mathbf{x}^T A\mathbf{x} - 2\mathbf{b}^T\mathbf{x} - \widetilde{\mathbf{x}}^T A\widetilde{\mathbf{x}} + 2\mathbf{b}^T\widetilde{\mathbf{x}} \\ &= \mathbf{x}^T A\mathbf{x} - 2\widetilde{\mathbf{x}}^T A\mathbf{x} + 2\widetilde{\mathbf{x}}^T A\widetilde{\mathbf{x}} \\ &= \left(\mathbf{x} - \widetilde{\mathbf{x}}\right)^T A\left(\mathbf{x} - \widetilde{\mathbf{x}}\right) \geq 0. \end{aligned}$$

Equality in the last inequality holds only if $\mathbf{x} = \widetilde{\mathbf{x}}$, and so the minimum of the function (8.1.10) is achieved only at the solution of the system (8.1.9).

There are different methods to find the minimizer of the function (8.1.10). One of the most popular methods is the *steepest descent method*. With this method, from a given approximation $\mathbf{x}^{(k)}$, the new approximation $\mathbf{x}^{(k+1)}$ is chosen, such that it assures the largest decrease of the function (8.1.10).

The new approximation $\mathbf{x}^{(k+1)}$ is obtained from the old approximation $\mathbf{x}^{(k)}$ by the formula

$$\mathbf{x}^{(k+1)} = \mathbf{x}^{(k)} + h\,\mathbf{c},$$

where the vector $\mathbf{c}$ is in the direction of the largest decrease of $f(\mathbf{x})$ (in the direction of the gradient $grad\, f(\mathbf{x})$), and the coefficient h is chosen such that it insures the largest decrease of f in that direction. The direction $\mathbf{c}$ of the largest decrease of $f(\mathbf{x}^{(k)} + h\,\mathbf{c})$ is such that the derivative of $f(\mathbf{x}^{(k)} + h\,\mathbf{c})$ with respect to h when $h = 0$ has largest modulus. From

$$\begin{aligned} f(\mathbf{x}^{(k)} + h\,\mathbf{c}) &= \left(A\mathbf{x}^{(k)} + hA\mathbf{c}\right)^T\left(\mathbf{x}^{(k)} + h\,\mathbf{c}\right) - 2\left(\mathbf{x}^{(k)} + h\,\mathbf{c}\right)^T\mathbf{b} \\ &= h^2\mathbf{c}^T A\mathbf{c} + 2\mathbf{c}^T\left(A\mathbf{x}^{(k)} - \mathbf{b}\right), \end{aligned}$$

i.e.,

$$(8.1.11) \qquad f(\mathbf{x}^{(k)} + h\,\mathbf{c}) - h^2\mathbf{c}^T A\mathbf{c} + 2\mathbf{c}^T\left(A\mathbf{x}^{(k)} - \mathbf{b}\right)$$

we have

$$\left| \frac{d}{dh} f(\mathbf{x}^{(k)} + h\,\mathbf{c}) \right|_{h=0} = 2\left| \mathbf{c}^T \mathbf{r}^{(k)} \right|,$$

where

$$\mathbf{r}^{(k)} = A\mathbf{x}^{(k)} - \mathbf{b}.$$

So the direction $\mathbf{c}$ of the steepest descent of f is in the direction of the vector

$$\mathbf{r}^{(k)} = A \cdot \mathbf{x}^{(k)} - \mathbf{b}.$$

From (8.1.11) it follows that

$$\frac{d}{dh} f(\mathbf{x}^{(k)} + h\,\mathbf{c}) = 2h\left(A\mathbf{r}^{(k)}\right)^T + \left(\mathbf{r}^{(k)}\right)^T \mathbf{r}^{(k)} = 0,$$

and so,

$$h = -\frac{\left(\mathbf{r}^{(k)}\right)^T \mathbf{r}^{(k)}}{\left(A\mathbf{r}^{(k)}\right)^T \mathbf{r}^{(k)}}.$$

Therefore, the new approximation $\mathbf{x}^{(k+1)}$ is given by the formula

$$(8.1.12) \qquad \mathbf{x}^{(k+1)} = \mathbf{x}^{(k)} - \frac{\left(\mathbf{r}^{(k)}\right)^T \mathbf{r}^{(k)}}{\left(A\mathbf{r}^{(k)}\right)^T \mathbf{r}^{(k)}} \mathbf{r}^{(k)}.$$

The approximation given by (8.1.12) is called the *steepest descent* iterative method.

It can be shown that the convergence rate of the steepest descent iterative method is of geometric order. More precisely, the following is true.

$$\|\mathbf{x}^{(k+1)} - \tilde{\mathbf{x}}\|_2 \le \left(\frac{\lambda_{max} - \lambda_{min}}{\lambda_{max} + \lambda_{min}} \right)^{k+1} \|\mathbf{x}^{(0)} - \tilde{\mathbf{x}}\|_2,$$

where λ_{max} is the largest and λ_{min} is the smallest eigenvalue of A.

The *conjugate gradient method* is a modification of the steepest gradient method. The algorithm for the conjugate gradient method can be summarized as follows:

$\mathbf{x}^{(0)}$, initial approximation

$$\mathbf{v}^{(0)} = \mathbf{r}^{(0)} = \mathbf{b} - A\mathbf{x}^{(0)}$$

$$h_k = \frac{\left(\mathbf{v}^{(k)}\right)^T \mathbf{r}^{(k)}}{\left(\mathbf{v}^{(k)}\right)^T A\mathbf{v}^{(k)}}$$

$$\mathbf{x}^{(k+1)} = \mathbf{x}^{(k)} + h_k \mathbf{v}^{(k)}$$

$$\mathbf{r}^{(k+1)} = \mathbf{r}^{(k)} - h_k A\mathbf{v}^{(k)}$$

$$t_k = -\frac{\left(\mathbf{v}^{(k)}\right)^T A\mathbf{r}^{(k+1)}}{\left(\mathbf{v}^{(k)}\right)^T A\mathbf{v}^{(k)}}$$

$$\mathbf{v}^{(k+1)} = \mathbf{r}^{(k+1)} + t_k \mathbf{v}^{(k)}$$

Example 8.1.6. Solve the linear system

$$\begin{pmatrix} 9 & 2 & -1 & 3 \\ 2 & 8 & -3 & 1 \\ -1 & -3 & 10 & -2 \\ 3 & 1 & -2 & 8 \end{pmatrix} \begin{pmatrix} x_1 \\ x_2 \\ x_3 \\ x_4 \end{pmatrix} = \begin{pmatrix} 5 \\ -2 \\ 6 \\ -4 \end{pmatrix}$$

by the conjugate gradient method using 7 iterations and display the differences between any two successive approximations.

Solution. First we check (with Mathematica) that A is a strictly positive definite matrix:

$In[1] := A = \{\{9,2,-1,3\},\{2,8,-3,1\},\{-1,-3,10,-2\},\{3,1,-2,8\}\};$
$In[2] := PositiveDefiniteMatrixQ[A]$
$Out[3] := True;$
$In[4] := r[j\_] := r[j] = b - A.Transpose[x[j]];$
$In[5] := v[1] := r[0];$
$In[6] := t[j\_] := t[j] = (Transpose[v[j]].r[j-1])$
$/(Transpose[v[j]].(A.v[j]));$
$In[7] := x[j\_] := x[/] = x[j-1] + t[j][[1,1]] * (Transpose[v[j]])[[1]];$
$In[8] := s[j\_] := s[j] = -(Transpose[v[j]].(A.r[j]))$
$/(Transpose[v[j]].(A.v[j]));$
$In[9] := v[j\_] := v[j] = r[k-1] + (s[j-1][[1,1]]) * v[j-1];$
$In[10] := d[\_] := d[j] = Sum[([x[j][[i]] - x[j-1][[i]])^2, \{i,1,Length[x[j]]\};$
$In[11] := b = Transpose[\{\{5,-2,6,-4\}\}];$
$In[12] := x[0] = \{1,1,1,1\};$
$In[13] := Table[x[j],\{j,0,6\}];$
$In[14] := A = \{\{9,2,-1,3\},\{2,8,-3,1\},\{-1,-3,10,-2\},\{3,1,-2,8\}\};$
$In[15] := b = Transpose[\{\{5.,-2.,6.,-4.\}\}];$

The results are given in Table 8.1.6.

Table 8.1.6 Conjugate Gradient

$k\backslash x_k$	x_1	x_2	x_3	x_4	d_k
0	1.	1.	1.	1.	
1	0.355752	0.19469	1.16106	-0.127434	2.36063
2	0.686608	-0.0650599	0.429966	-0.645698	0.980036
3	881608	-0.203023	0.490418	-0.681147	0.0619698
4	0.882469	-0.200988	0.491358	-0.682963	$9.06913 * 10^{-6}$
5	0.882469	-0.200988	0.491358	-0.682963	$3.08149 * 10^{-32}$

Compare the obtained results with the exact solution:
$In[16] := EXACT = Inverse[A].b$
$Out[16] := \{\frac{1787}{2025}, -\frac{407}{2025}, \frac{199}{405}, -\frac{461}{675}\}$
$In[17] := N[\%]$
$Out[17] := \{0.882469, -0.200988, 0.491358, -0.682963\}$

The following Mathematica program solves a system of linear equations by the conjugate gradient iterative method using prescribed error to stop the iterations.

$In[1] := ConjGrad[A\_List, b\_List, x0\_List, d\_] := Module[\{\};$
$norm[xx\_List, yy\_List] :=$
$Max[Sum[Abs[xx[[i]] - yy[[i]]], \{i, 1, Length[xx]\}]];$
$norm[xx\_List] := Max[Sum[Abs[xx[[i]]], \{i, 1, Length[xx]\}]];$
$z = Table[0, \{i, 1, Length[x0]\}];$
$x[0] = x0;$
$r[0] = b - A.x[0];$
$v[1] = r[0];$
$r[k\_/; k > 0] := r[k] = r[k - 1] - t[k] * A.v[k];$
$t[k\_/; k > 0] := t[k] = Dot[r[k - 1], r[k - 1]]/Dot[v[k], A.v[k]];$
$s[k\_/; k > 0] := s[k] = Dot[r[k], r[k]]/Dot[r[k - 1], r[k - 1]];$
$v[k\_/; k > 1] := v[k] = r[k - 1] + s[k - 1] * v[k - 1];$
$x[k\_/; k > 0] := x[k] = x[k - 1] + t[k] * v[k];$
$Do[$
$If[norm[x[k], x[k - 1]] <= d, \{appr = x[k], savek = k,$
$Break[ ]\}], \{k, 1, 100\}];$
$];$

If we execute the module by

$In[2] := A = \{\{9, 2, -1, 3\}, \{2, 8, -3, 1\}, \{-1, -3, 10, -2\}, \{3, 1, -2, 8\}\};$
$In[3] := b = \{5, -2, 6, -4\};$
$In[4] := ConjGradG[A, b, 1, 0.4, 1, -1, 0.00000001];$

taking $d = 10^{-8}$ we obtain the result

$Out[4] := Approx = \{0.882469, -0.200988, 0.491358, -0.682963\}.$

Exercises for Section 8.1.

1. Find the eigenvalues and the corresponding eigenvectors of the matrix

$$A = \begin{pmatrix} -1 & 0 & 3 & 0 \\ 0 & 9 & 0 & 0 \\ 3 & 0 & 1 & 0 \\ 0 & 0 & 0 & 5 \end{pmatrix}.$$

2. The *Euclidean (Frobenious) norm* $\|A\|_E$ of an $n \times n$ matrix $A = (a_{ij})$ is defined by

$$\|A\|_E = \sqrt{\sum_{i,j} a_{ij}^2}.$$

For the matrix

$$A = \begin{pmatrix} 1 & 0 & 1 \\ 2 & 3 & 0 \\ 2 & 1 & 4 \end{pmatrix}$$

compute the norms $\|A\|_E$, $\|A\|_1$, $\|A\|_2$, $\|A\|_\infty$ and verify the inequalities

$$\|A\|_2 \le \|A\|_E \le \sqrt{n}\|A\|_2, \quad \|A\|_2 \le \sqrt{\|A\|_1 \|A\|_\infty}.$$

3. The condition number of an invertible matrix A is defined by

$$\kappa(A) = \|A\| \, \|A^{-1}\|.$$

In the case of a symmetric matrix A, the condition number of the matrix is given by

$$\kappa(A) = \|A\|_2 \, \|A^{-1}\|_2.$$

Compute the condition number for the matrix

$$A = \begin{pmatrix} 1 & 0.99999 \\ 0.99999 & 1 \end{pmatrix}.$$

A "large" condition number of a matrix strongly affects results of any procedure that involves the matrix. Verify this by solving the systems

$$\begin{pmatrix} 1 & 0.99999 \\ 0.99999 & 1 \end{pmatrix} \begin{pmatrix} x_1 \\ x_2 \end{pmatrix} = \begin{pmatrix} 2.99999 \\ 2.99998 \end{pmatrix}$$

and

$$\begin{pmatrix} 0.99999 & 0.99999 \\ 0.99999 & 1 \end{pmatrix} \begin{pmatrix} y_1 \\ y_2 \end{pmatrix} = \begin{pmatrix} 2.99999 \\ 2.99998 \end{pmatrix}.$$

4. Show that the matrix

$$A = \begin{pmatrix} 2 & -1 & 0 \\ -1 & 2 & -1 \\ 0 & -1 & 2 \end{pmatrix}$$

is positive definite.

5. Solve the system

$$\begin{pmatrix} 4 & 1 & -1 & 1 \\ 1 & 4 & -1 & -1 \\ -1 & -1 & 5 & 1 \\ 1 & -1 & 1 & 3 \end{pmatrix} \begin{pmatrix} x_1 \\ x_2 \\ x_3 \\ x_4 \end{pmatrix} = \begin{pmatrix} -2 \\ -1 \\ 0 \\ 1 \end{pmatrix}$$

by

(a) the Jacobi iterative method
(b) the Gauss–Seidel iterative method
using the condition

$$\|\mathbf{x}^{(k+1)} - \mathbf{x}^{(k)}\|_2 \leq \epsilon$$

to stop the iteration when $\epsilon = 10^{-6}$.

6. Solve the system

$$\begin{pmatrix} 10 & -1 & -1 & -1 \\ -1 & 10 & -1 & -1 \\ -1 & -1 & 10 & -1 \\ -1 & -1 & -1 & 10 \end{pmatrix} \begin{pmatrix} x_1 \\ x_2 \\ x_3 \\ x_4 \end{pmatrix} = \begin{pmatrix} 34 \\ 23 \\ 12 \\ 1 \end{pmatrix}$$

by the conjugate gradient method taking only 2 iterations.

7. Show that the matrix

$$A = \begin{pmatrix} 4 & 2 & -2 \\ 1 & -3 & -1 \\ 3 & -1 & 4 \end{pmatrix}$$

is not strictly diagonally dominant but yet the Jacobi and Gauss–Seidel iterative methods for the system

$$A \cdot \mathbf{x} = \mathbf{b}$$

are convergent.

8.2 Finite Differences.

In cases in which no analytical expressions for the solutions of partial differential equations can be given, then numerical methods for the solutions are used. One among many numerical methods is *the finite difference method*. With this method, a particular differential equation is replaced with a *difference equation*, i.e., a system of linear equations which can be solved by many numerical methods. This method is very often used because of its simplicity and its easy computer implementation.

The numerical methods which will be discussed in the next several sections are developed by approximating derivatives.

Let us begin with a real-valued function f of a single variable x. We assume that f is sufficiently smooth; in most cases we will assume that f

has derivative of any order, i.e., $f \in C^\infty(\mathbb{R})$. First, let us recall the definition of the first derivative:

$$f'(x) = \lim_{h \to 0} \frac{f(x+h) - f(x)}{h}.$$

This means that

$$\frac{f(x+h) - f(x)}{h}$$

can be a good candidate for an approximation of $f'(x)$ for sufficiently small h. But, it is not obvious how good this approximation might be. The answer lies in the well-known *Taylor's Formula*.

Taylor's Formula. *Let a function* f *have* $n+1$ *continuous derivatives in the interval* $(x - l, x + l)$. *Then for any* h *such that* $x + h \in (x - l, x + l)$ *there is a number* $0 < \theta < 1$ *such that*

$$f(x+h) = f(x) + \frac{f'(x)}{1!}h + \frac{f''(x)}{2!}h^2 + \ldots + \frac{f^{(n)}(x)}{n!}h^n + \frac{f^{(n+1)}(x+\theta h)}{(n+1)!}h^{n+1}.$$

If the derivative $f^{(n+1)}$ is bounded, then we write

$$(8.2.1) \quad f(x+h) = f(x) + \frac{f'(x)}{1!}h + \frac{f''(x)}{2!}h^2 + \ldots + \frac{f^{(n)}(x)}{n!}h^n + O(h^{n+1}).$$

The above notation O (read "big-oh") has the following meaning:

If $f(x)$ and $g(x)$ are two functions such that there exist constants $M > 0$ and $\delta > 0$ with the property

$$|f(x)| \le M|g(x)| \quad \text{for} \ |x| < \delta,$$

then we write

$$f(x) = O(g(x)) \quad \text{as} \ x \to 0.$$

If we take $n = 1$ in (8.2.1), then we obtain

$$f(x+h) = f(x) + f'(x)h + O(h^2)$$

which can be written as

$$(8.2.2) \qquad\qquad f'(x) = \frac{f(x+h) - f(x)}{h} + O(h).$$

From the last expression we have one approximation of $f'(x)$:

$$(8.2.3) \qquad f'(x) \approx \frac{f(x+h) - f(x)}{h}, \quad \textit{forward difference approximation.}$$

Replacing h by $-h$ in (8.2.2) we obtain

$$f'(x) = \frac{f(x) - f(x-h)}{h} + O(h).$$

From the last expression we have another approximation of $f'(x)$:

(8.2.4) $f'(x) \approx \dfrac{f(x) - f(x-h)}{h}$,   *backward difference approximation.*

Replacing h by $-h$ in (8.2.1) we have

(8.2.5) $f(x-h) = f(x) - \dfrac{f'(x)}{1!}h + \ldots + (-1)^n \dfrac{f^{(n)}(x)}{n!} h^n + O(h^{n+1}).$

If we take $n = 1$ in (8.2.5) and (8.2.1) and subtract (8.2.5) from (8.2.1), then

$$f'(x) = \frac{f(x+h) + f(x-h)}{2h} + O(h^2).$$

Therefore, we have another approximation of the first derivative

(8.2.6) $f'(x) \approx \dfrac{f(x+h) + f(x-h)}{2h}$,   *central difference approximation.*

The *truncation errors* for the above approximations are given by

$$E_{for}(h) \equiv \frac{f(x+h) - f(x)}{h} - f'(x) = O(h)$$

$$E_{back}(h) \equiv \frac{f(x) - f(x-h)}{h} - f'(x) = O(h)$$

$$E_{cent}(h) \equiv \frac{f(x+h) + f(x-h)}{2h} - f'(x) = O(h^2).$$

Example 8.2.1. Let $f(x) = \ln x$. Approximate $f'(2)$ taking $h = 0.1$ and $h = 0.001$ in

(a) the forward difference approximation

(b) the backward difference approximation

(c) the central difference approximation

and compare the obtained results with the exact value.

Solution. From calculus we know that $f'(x) = \dfrac{1}{x}$ and so the exact value of $f'(2)$ is 0.5.

(a) For $h = 0.1$ we have

$$f'(2) \approx \frac{f(2 + 0.1) - f(2)}{0.1} = \frac{\ln(2.1) - \ln(2)}{0.1} \approx 0.4879.$$

For $h = 0.001$ we have

$$f'(2) \approx \frac{f(2 + 0.001) - f(2)}{0.001} = \frac{\ln(2.001) - \ln(2)}{0.001} \approx 0.49988.$$

(b) For $h = 0.1$ we have

$$f'(2) \approx \frac{f(2) - f(2 - 0.1)}{0.1} = \frac{\ln(2) - \ln(1.9)}{0.1} \approx 0.5129.$$

For $h = 0.001$ we have

$$f'(2) \approx \frac{f(2) - f(2 - 0.001)}{0.001} = \frac{\ln(2) - \ln(1.999)}{0.001} \approx 0.50018.$$

(c) For $h = 0.1$ we have

$$f'(2) \approx \frac{f(2 + 0.1) - f(2 - 0.1)}{0.1} = \frac{\ln(2.1) - \ln(1.9)}{2(0.1)} \approx 0.500417.$$

For $h = 0.001$ we have

$$f'(2) \approx \frac{f(2 + 0.001) - f(2 - 0.001)}{2(0.001)} = \frac{\ln(2.001) - \ln(1.999)}{0.1} \approx 0.5000.$$

If we use the Taylor series (8.2.1) and (8.2.5), then we obtain the following approximations for the second derivative:

$$(8.2.7) \quad f''(x) \approx \frac{f(x + 2h) - 2f(x + h) + f(x)}{h^2}, \quad \textit{forward approximation.}$$

$$(8.2.8) \quad f''(x) \approx \frac{f(x) - 2f(x - h) + f(x - 2h)}{h^2}, \quad \textit{backward approximation.}$$

$$(8.2.9) \quad f''(x) \approx \frac{f(x + h) - 2f(x) + f(x - h)}{h^2}, \quad \textit{central approximation.}$$

The approximations (8.2.7) and (8.2.8) for the second derivative are accurate of first order, and the central difference approximation (8.2.9) for the second derivative is accurate of second order.

Example 8.2.2. Approximate $f''(2)$ for the function $f(x) = \ln x$ taking $h = 0.1$ with the central difference approximation.

Solution. We find that $f''(x) = -1/x^2$ and so $f''(2) = -0.25$. With the central difference approximation for the second derivative we have that

$$f''(x) \approx \frac{f(x+h) - 2f(x) + f(x-h)}{h^2} = \frac{\ln 2.1 - 2\ln 2 + \ln 1.9}{0.1^2} \approx -0.25031.$$

When solving a problem described by an ordinary differential equation, usually we partition a part of the line $\mathbb{R}$ into many equal intervals, defined by the points

$$x_i = ih, \quad i = 0, \pm 1, \pm 2, \ldots,$$

and then apply some finite difference approximation scheme to the derivatives involved in the differential equation.

Let us apply the finite differences to a particular boundary value problem.

Example 8.2.3. Solve the boundary value problem

$$(8.2.10) \qquad \begin{cases} y''(x) = -y(x), & 0 \le x \le \dfrac{\pi}{2} \\ y(0) = 0, & y\left(\dfrac{\pi}{2}\right) = 1 \end{cases}$$

by the central difference approximation.

Solution. The exact solution of the given boundary value problem is given by $y(x) = \sin(3x)$. Divide the interval $[0, \pi/2]$ into n equal subintervals with the points

$$x_j = jh, \ h = \frac{\pi}{2n}, \ j = 0, 1, 2, \ldots, n.$$

If y_j is an approximation of $y(x_j)$ $(y_j \approx y(x_j))$, then using the central difference approximations for the second derivative, Equation (8.2.10) can be approximated by

$$\frac{y(x_{j-1}) - 2y(x_j) + y(x_{j+1})}{h^2} = -y_j, \ j = 1, 2, \ldots, n-1.$$

From the last equations we obtain $n-1$ linear equations

$$(8.2.11) \qquad y_{j-1} - 2y_j + y_{j+1} = -h^2 y_j, \ j = 1, 2, \ldots, n-1,$$

and together with the boundary conditions $y_0 = y(0) = 1$ and $y_n = y(\pi/2) = 0$ can be written in the matrix form

$$A \cdot \mathbf{y} = \mathbf{b},$$

where

$$A = \begin{pmatrix} -2+9h^2 & 1 & 0 & \cdots & 0 & 0 & 0 \\ 1 & -2+9h^2 & 1 & \cdots & 0 & 0 & 0 \\ 0 & 1 & -2+9h^2 & \cdots & 0 & 0 & 0 \\ \vdots & & & & & & \\ \vdots & & & & & & \\ 0 & 0 & 0 & \cdots & 1 & -2+9h^2 & 1 \\ 0 & 0 & 0 & \cdots & 0 & 1 & -2+9h^2 \end{pmatrix},$$

$$\mathbf{y} = \begin{pmatrix} y_1 \\ y_2 \\ y_3 \\ y_4 \\ \vdots \\ \vdots \\ y_{n-2} \\ y_{n-1} \end{pmatrix}, \quad \mathbf{b} = \begin{pmatrix} 0 \\ 0 \\ 0 \\ 0 \\ \vdots \\ \vdots \\ 0 \\ -1 \end{pmatrix}.$$

Now let us take $n = 15$. Using Mathematica we have

$In[1] := n = 15;$

$In[2] := h = Pi/(2*n);$

$In[3]:=yApprox = Inverse[A] \cdot b;$

$In[4]:=N[\%]$

$Out[4]:=\{\{0.104576\}, \{0.208005\}, \{0.309154\}, \{0.406912\}, \{0.500208\},$
$\{0.588018\}, \{0.66938\}, \{0.743401\}, \{0.809271\}, \{0.866265\},$
$\{0.91376\}, \{0.951234\}, \{0.978277\}, \{0.994592\}\}$

Compare this approximative solution with the exact solution.

$In[5]:=Table[Sin[j*h], \{j, 1, 15\}];$

$In[6]:=yExact = N[\%]$

$Out[6]:=\{0.104528, 0.207912, 0.309017, 0.406737, 0.5, 0.587785,$
$0.669131, 0.743145, 0.809017, 0.866025, 0.913545,$
$0.951057, 0.978148, 0.994522, 1.\}$

The approximations of the derivatives of a function of a single variable can be extended to functions of several variables. For example, if $f(x, y)$ is a given function of two variables $(x, y) \in D \subseteq \mathbb{R}^2$, for the first partial derivatives we have

$$f_x(x,y) = \frac{f(x+h,y) - f(x,y)}{h} + O(h), \quad \textit{forward difference},$$

$$f_x(x,y) = \frac{f(x,y) - f(x-h,y)}{h} + O(h), \quad \textit{backward difference},$$

$$f_x(x,y) = \frac{f(x+h,y) - f(x-h,y)}{2h} + O(h^2), \quad \textit{central difference},$$

$$f_y(x,y) = \frac{f(x,y+k) - f(x,y)}{k} + O(k), \quad \textit{forward difference},$$

$$f_y(x,y) = \frac{f(x,y) - f(x,y-k)}{k} + O(k), \quad \textit{backward difference},$$

$$f_y(x,y) = \frac{f(x,y+k) - f(x,y-k)}{2k} + O(k^2), \quad \textit{central difference}.$$

For the partial derivatives of second order we have

$$f_{xx}(x,y) = \frac{f(x+2h,y) - 2f(x+h,y) + f(x,y)}{h^2} + O(h^2), \quad \textit{forward},$$

$$f_{xx}(x,y) = \frac{f(x,y) - 2f(x-h,y) + f(x-2h,y)}{h^2} + O(h^2), \quad \textit{backward},$$

$$f_{xx}(x,y) = \frac{f(x+h,y) - 2f(x,y) + f(x-h,y)}{h^2} + O(h^4), \quad \textit{central}.$$

For the second partial derivatives with respect to y, as well as for the mixed partial derivatives, we have similar expressions.

In order to approximate the partial derivatives on a relatively large set of discrete of points in $\mathbb{R}^2$ we proceed in the following way. We partition a part of the plane $\mathbb{R}^2$ into equal rectangles by a grid of lines parallel to the coordinate axes Ox and Oy, defined by the points

$$x_i = ih, \ i = 0, \pm 1, \pm 2, \ldots; \ y_j = jk, \ j = 0, \pm 1, \pm 2, \ldots \text{ (see Figure 8.2.1)}.$$

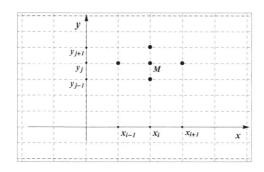

Figure 8.2.1

The following notation will be used.

$$f(M) = f(x_i, y_j) = f_{i,j}.$$

With this notation we have the following approximations for the first order partial derivatives.

$$f_x(x_i, y_j) \approx \frac{f_{i+1,j} - f_{i,j}}{h}, \qquad f_x(x_i, y_j) \approx \frac{f_{i,j} - f_{i-1,j}}{h}$$

$$f_y(x_i, y_j) \approx \frac{f_{i,j+1} - f_{i,j}}{k}, \qquad f_y(x_i, y_j) \approx \frac{f_{i,j} - f_{i,j-1}}{k},$$

$$f_x(x_i, y_j) \approx \frac{f_{i+1,j} - 2f_{i-1,j}}{2h}, \quad f_y(x_i, y_j) \approx \frac{f_{i,j+1} - 2f_{i,j-1}}{2k}.$$

For the second order partial derivatives we have the following approximations.

$$f_{xx}(x_i, y_j) \approx \frac{f_{i+2,j} - 2f_{i+1,j} + f_{i,j}}{h^2}, \quad f_{xx}(x_i, y_j) \approx \frac{f_{i,j} - 2f_{i-1,j} + fx_{i-2,j}}{h^2}$$

$$f_{yy}(x_i, y_j) \approx \frac{f_{i,j+2} - 2f_{i,j+1} + f_{i,j}}{k^2}, \quad f_{yy}(x_i, y_j) \approx \frac{f_{i,j} - 2f_{i,j-1} + f_{i,j-2}}{k^2},$$

$$f_{xx}(x_i, y_j) \approx \frac{f_{i+1,j} - 2f_{i,j} + f_{i-1,j}}{h^2}, \quad f_{yy}(x_i, y_j) \approx \frac{f_{i,j+1} - 2f_{i,j} + f_{i,j-1}}{k^2}$$

$$f_{xy}(x_i, y_j) \approx \frac{f_{i+1,j+1} - f_{i+1,j-1} - f_{i-1,j+1} + f_{i-1,j-1}}{4hk}.$$

Example 8.2.4. Using one of the above approximation formulas find an approximation of $f_{xx}(0.05, 0.05)$ for the function

$$f(x, y) = e^{-\pi^2 y} \sin \pi x.$$

Solution. If we take $h = k = 0.05$, then by the approximation formula

$$f_{xx}(x_i, y_j) \approx \frac{f_{i+1,j} - 2f_{i,j} + f_{i-1,j}}{h^2}$$

we have

$$f_{xx}(0.05, 0.05) \approx \frac{f_{2,1} - 2f_{1,1} + f_{0,1}}{h^2},$$

where $f_{2,1} = f(0.01, 0.05)$, $f_{1,1} = f(0.05, 0.5)$ and $f_{0,1} = f(0, 0.5)$. Calculating these f_{ij} for the given function we obtain that

$$f_{xx}(0.05, 0.05) \approx -68.7319.$$

Among several important characteristics of a numerical difference iteration applied to the solution of ordinary or partial differential equations (*convergence, consistency and stability*), we will consider only the notions of convergence and stability. Convergence means that the finite difference solution

approaches the true (exact) solution of the ordinary or partial differential
equation as the grid size approaches zero. A finite difference scheme is called
stable if the error caused by a small change (due to truncation, round-off) in
the numerical solution remains bounded. By a theorem which is beyond the
scope of this textbook, the stability of a finite difference scheme is a necessary
and sufficient condition for the convergence of the finite difference scheme.

Exercises for Section 8.2.

1. For each of the following functions estimate the first derivative of the
   function at $x = \pi/4$ taking $h = 0.1$ in the forward, backward and
   central finite difference approximation. Compare your obtained results with the exact values.

   (a) $f(x) = \sin x$.

   (b) $f(x) = \tan x$.

2. Use the table to estimate the first derivative at each mesh point. Estimate the second derivative at $x = 0.6$.

x	0.5	0.6	0.7
$f(x)$	0.47943	.56464	.64422

3. Let $f(x)$ have the first 4 continuous derivatives. Show that the following higher degree approximation for the first derivative $f'(x)$ holds.

   $$f'(x) = \frac{1}{12h}\left[f(x-2h) - 8f(x-h) + 8f(x+h) - f(x+2h)\right]$$
   $$+ O(h^4).$$

4. Using the central finite difference approximation with $h = 0.001$, estimate the second derivative of the function $f(x) = e^{x^2}$. Compare your
   result with the "exact" value.

5. Partition $[0,1]$ into 10 equal subintervals to approximate the solution
   of the following problem.

   $$y'' + xy = 0, \quad 0 < x < 1$$
   $$y(0) = 0, \quad y(1) = 1.$$

===

8.3 Finite Difference Methods for Laplace and Poisson Equations.

In this section we will apply the finite difference method to the Laplace and Poisson equations on domains with different shapes, subject to Dirichlet and Neumann conditions.

1^0. *Rectangular Domain. Dirichlet Boundary Condition.*

Let us consider the Poisson equation, subject to the Dirichlet boundary conditions on the two dimensional rectangular domain D

(8.3.1)
$$\begin{cases} \Delta u(x,y) = F(x,y), \ (x,y) \in D = \{a < x < b, \ c < y < d\} \\ u(x,y) = B(x,y), \ (x,y) \in S = \partial D \ \text{— the boundary of D.} \end{cases}$$

We partition the rectangle D into equal rectangles by a grid of lines parallel to the coordinate axes Ox and Oy, defined by the points

$$x_i = a + ih, \ i = 0,1,2,\ldots,m; \ \ y_j = c + jk, \ \ j = 0,1,2,\ldots,n,$$

where

$$h = \frac{b-a}{m} \quad k = \frac{d-c}{n}.$$

(See Figure 8.3.1.)

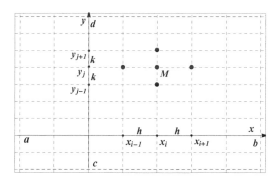

Figure 8.3.1

Then, using the central finite difference approximations for the second partial derivatives, an approximation to the Poisson/Laplace equation (8.3.1) at each interior point (node) (i,j), $1 \le i \le m-1; \ 1 \le j \le n-1$ is given by

(8.3.2)
$$\frac{u_{i-1,j} - 2u_{i,j} + u_{i+1,j}}{h^2} + \frac{u_{i,j-1} - 2u_{i,j} + u_{i,j+1}}{k^2} = F_{i,j},$$

where $F_{i,j} = F(x_i, y_j), \ 1 \le i \le m-1; \ 1 \le j \le n-1.$

At the boundary points (nodes) $(0, j)$, (m, j), $j = 0, 1, \ldots, n$ and $(i, 0)$, (i, n), $i = 0, 1, \ldots, m$ we have

$$(8.3.3) \qquad\qquad u_{i,j} = B(x_i, y_j) \equiv B_{i,j}.$$

In applications, usually we take $h = k$, and then Equations (8.3.2) become

$$(8.3.4) \qquad u_{i-1,j} - 4u_{i,j} + u_{i+1,j} + u_{i,j-1} + u_{i,j+1} = h^2 F_{i,j},$$

for $1 \leq i \leq m - 1$; $1 \leq j \leq n - 1$.

In order to write the matrix form of the linear system (8.3.2) for the $(m - 1)(n - 1)$ unknowns

$$u_{1,1}, u_{1,2}, \ldots, u_{1,n-1}, u_{2,1}, u_{2,2}, \ldots, u_{2,n-1}, \ldots, u_{m-1,1}, u_{m-1,2}, \ldots, u_{m-1,n-1},$$

we label the nodes in the rectangle from left to right and from top to bottom. With this enumeration, the matrix $(m - 1)(n - 1) \times (m - 1)(n - 1)$ matrix A of the system has the block form

$$A = \begin{pmatrix} U & -I & \mathbf{O} & \mathbf{O} & \cdots & \mathbf{O} & \mathbf{O} \\ -I & U & -I & \mathbf{O} & \cdots & \mathbf{O} & \mathbf{O} \\ \mathbf{O} & -I & U & -I & \cdots & \mathbf{O} & \mathbf{O} \\ & & \vdots & & & & \\ & & \vdots & & & & \\ \mathbf{O} & \mathbf{O} & \mathbf{O} & \mathbf{O} & \cdots & -I & U \end{pmatrix},$$

where I is the identity matrix of order $n - 1$, $\mathbf{O}$ is the zero matrix of order $n - 1$, and the $(n - 1) \times (n - 1)$ matrix U is given by

$$U = \begin{pmatrix} 4 & -1 & 0 & 0 & \cdots & 0 & 0 \\ 0 & 4 & -1 & 0 & \cdots & 0 & 0 \\ 0 & 0 & 4 & -1 & \cdots & 0 & 0 \\ & & \vdots & & & & \\ & & \vdots & & & & \\ 0 & 0 & 0 & 0 & \cdots & -1 & 4 \end{pmatrix}.$$

Even though the matrix A is a nice sparse matrix (formed by many identical matrix blocks that have many, many zeros) it is quite large even for relatively small m and n, and therefore some iterative method is required to solve the system.

We illustrate the finite difference method for the Poisson equation with the following example.

Example 8.3.1. Using a finite difference method solve the boundary value problem

$$\begin{cases} \Delta u\,(x,y) = -2, & (x,y) \in D, \\ u(x,y) = 0, & (x,y) \in \partial D, \end{cases}$$

where D is the square $D = \{(x,y) : -1 < x < 1, \ -1 < y < 1\}$, whose boundary is denoted by ∂D.

Solution. Taking a uniform grid with increment $h = \dfrac{2}{n}$ in both coordinate axes, from (8.3.3) it follows that

$$(8.3.5) \qquad u_{i-1,j} - 4u_{i,j} + u_{i+1,j} + u_{i,j-1} + u_{i,j+1} = -2h^2,$$

with boundary values

$$(8.3.6) \qquad u_{0,j} = u_{n,j} = 0, \ j = 0, 1, \ldots, n; \quad u_{i,0} = u_{i,n} = 0, \ 0 \le i \le n.$$

We label the nodes from left to right and from top to bottom. So, we start with the top left corner of the rectangle. With this labeling, taking $n = 5$, i.e., $h = 0.4$ in (8.3.5) and using the boundary conditions (8.3.6) we obtain the linear system

$$A \cdot \mathbf{u} = \mathbf{b},$$

where the 16×16 matrix A is given by

$$A = \begin{pmatrix}
4 & 0 & 0 & 0 & -1 & 0 & 0 & 0 & 0 & 0 & 0 & 0 & 0 \\
0 & 4 & -1 & 0 & 0 & -1 & 0 & 0 & 0 & 0 & 0 & 0 & 0 \\
0 & -1 & 4 & -1 & 0 & 0 & -1 & 0 & 0 & 0 & 0 & 0 & 0 \\
0 & 0 & -1 & 4 & -1 & 0 & 0 & -1 & 0 & 0 & 0 & 0 & 0 \\
-1 & 0 & 0 & -1 & 4 & 0 & 0 & 0 & -1 & 0 & 0 & 0 & 0 \\
0 & -1 & 0 & 0 & 0 & 4 & -1 & 0 & 0 & -1 & 0 & 0 & 0 \\
0 & 0 & -1 & 0 & 0 & -1 & 4 & -1 & 0 & 0 & -1 & 0 & 0 \\
0 & 0 & 0 & -1 & 0 & 0 & -1 & 4 & -1 & 0 & 0 & -1 & 0 \\
0 & 0 & 0 & 0 & -1 & 0 & 0 & -1 & 4 & 0 & 0 & 0 & -1 \\
0 & 0 & 0 & 0 & 0 & -1 & 0 & 0 & 0 & 4 & -1 & 0 & 0 \\
0 & 0 & 0 & 0 & 0 & 0 & -1 & 0 & 0 & -1 & 4 & -1 & 0 \\
0 & 0 & 0 & 0 & 0 & 0 & 0 & -1 & 0 & 0 & -1 & 4 & -1 \\
0 & 0 & 0 & 0 & 0 & 0 & 0 & 0 & -1 & 0 & 0 & -1 & 4
\end{pmatrix}$$

and

$$\mathbf{u} = \begin{pmatrix} u_{1,1} \\ \vdots \\ u_{1,4} \\ u_{2,1} \\ \vdots \\ u_{2,4} \\ u_{3,1} \\ \vdots \\ u_{3,4} \\ u_{4,1} \\ \vdots \\ u_{4,4} \end{pmatrix}, \quad \mathbf{b} = \begin{pmatrix} 0.32 \\ \vdots \\ 0.32 \\ \vdots \\ 0.32 \\ \vdots \\ 0.32 \\ \vdots \\ 0.32 \\ \vdots \\ 0.32 \end{pmatrix}_{16 \times 1}.$$

If we solve this system directly (using Mathematica) we obtain the following approximation of the Poisson equation.

.266667, .373333, .373333, .266667, .373333, .533333, .533333, .373333, .373333, .533333, .533333, .373333, .266667, .373333, .373333, .266667.

Compare this approximate solution with the exact solution

.277176, .385245, .385245, .277176, .385245, .549954, .549954, .385245, .385245, .549954, .549954, .385245, .277176, .385245, .385245, .277176.

The exact solution of the given Poisson equation, obtained in Section 7.2, is given by

$$u(x, y) = 1 - y^2 - \frac{32}{\pi^3} \sum_{k=1}^{\infty} \frac{(-1)^k \cosh\left((2k+1)\frac{\pi x}{2}\right) \cos\left((2k+1)\frac{\pi y}{2}\right)}{(2k+1)^3 \cosh\left((2k+1)\frac{\pi}{2}\right)}.$$

2^0. *Rectangular Domain. Neumann Boundary Condition.*
    Now we will approximate the solution of the Poisson equation on the unit square, subject to mixed boundary conditions on the boundary of the square. Let D be the unit square

$$D = \{(x, y) : 0 < x < 1, 0 < y < 1\}.$$

Consider the boundary value problem

$$\begin{cases} \Delta u(x, y) = F(x, y), & (x, y) \in D, \\ u(x, 0) = f_1(x), & u(x, 1) = f_2(x), \quad 0 < x < 1 \\ u_x(0, y) = g_1(y), & u_x(1, y) = g_2(y), \quad 0 < y < 1. \end{cases}$$

We make an $n \times n$ grid with grid size $h = \dfrac{1}{n}$. Let us denote a node (x_i, y_j) by (i, j).

At all interior nodes (i, j), for which $i = 2, 3, \ldots, n-2$, $j = 1, 2, \ldots, n-1$, we apply the approximation scheme (8.3.4):

$$u_{i-1,j} - 4u_{i,j} + u_{i+1,j} + u_{i,j-1} + u_{i,j+1} = h^2 F_{i,j}.$$

For the interior nodes $(1, j)$, and $(n-1, j)$, $j = 1, 2, \ldots, n-1$ the situation is slightly more complicated since we do not know the values of the function $u(x, y)$ at the boundary nodes $(i, 0)$, $i = 1, 2, \ldots, n-1$ and (n, j), $j = 1, 2, \ldots, n-1$. In order to approximate the values of the function $u(x, y)$ at these nodes we extend the grid with additional points (so called *fictitious points*) $(-1, j)$ and $(n+1, j)$ $j = 1, 2, \ldots, n-1$ to the left and right of the boundary points $x = 0$ and $x = 1$, respectively. (See Figure 8.3.2.)

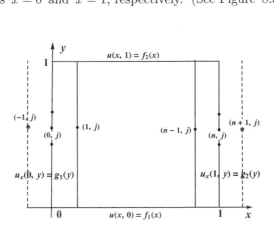

Figure 8.3.2

If we apply the central finite difference approximation for $u_x(0, y_j)$ and $u_x(1, y_j)$ we have

$$\begin{cases} u_x(0, y_j) \approx \dfrac{u(0 + h, y_j) - u(0 - h, y_j)}{2h}, \\[2mm] u_x(1, y_j) \approx \dfrac{u(1 + h, y_j) - u(1 - h, y_j)}{2h}. \end{cases}$$

Therefore,

(8.3.7) $u_{1,j} - u_{-1,j} = 2h f_1(y_j), \quad u_{n+1\,j} - u_{n-1,j} = 2h f_2(y_j).$

We will eliminate $u_{-1,j}$ and $u_{n+1,j}$ if we use the Poisson equation at the nodes $(0, j)$ and (n, j):

(8.3.8) $\begin{cases} u_{-1,j} + u_{1,j} + u_{0,j-1} + u_{0,j+1} - 4u_{0,j} = F_{0,j}, \\ u_{n+1,j} + u_{n-1,j} + u_{n,j-1} + u_{n,j+1} - 4u_{n,j} = F_{n,j}. \end{cases}$

From (8.3.7) and (8.3.8) it follows that

(8.3.9)
$$\begin{cases} 4u_{0,j} = 2u_{1,j} + u_{0,j+1} + u_{0,j-1} - 2h\,g_1(y_j) - F_{0,j}, \\ 4u_{n,j} = 2u_{n-1,j} + u_{n,j+1} + u_{n,j-1} + 2h\,g_2(y_j) - F_{n,j}. \end{cases}$$

We solve the system given by (8.3.4) and (8.3.9) for the n^2-1 unknowns $u_{i,j}$, $i = 2, \ldots n-2$, $j = 1, \ldots n-1$ and $u_{0,j}$, $u_{1,j}$, $u_{n-1,j}$, $u_{n,j}$, $j = 1, 2, \ldots, n-1$.

We illustrate the above discussion by an example for the Laplace equation.

Example 8.3.2. Find an approximation of the boundary value problem

$$\begin{cases} u_{xx} + u_{yy} = 0, \quad 0 < x < 1, \ 0 < y < 1, \\ u(x,0) = 0, \quad u(x,1) = 1 + x^2, \ 0 < x < 1, \\ u_x(0,y) = 1, \quad u_x(1,y) = 0, \ 0 < y < 1 \end{cases}$$

on a 5×5 uniform grid.

Solution. For this example we have

$$F(x,y) = 0, \quad f_1(x) = 0, \ f_2(x) = 1 + x^2,$$
$$g_1(y) = 1, \quad g_2(y) = 0,$$

and $n = 5$, $h = 0.2$.

Therefore, the system (8.3.4) and (8.3.9) is given by

$$\begin{cases} 4u_{i,j} - u_{i-1,j} - u_{i+1,j} - u_{i,j-1} - u_{i,j+1} = 0, \ i = 1,2,3,4 \ j = 1,2,3,4 \\ u_{i,0} = 0, \ i = 0,1,2,3,4,5 \\ u_{i,5} = 1 + (ih)^2, \ i = 0,1,2,3,4,5 \\ 4u_{0,j} - 2u_{1,j} - u_{0,j+1} - u_{0,j-1} = -2h, \ j = 1,2,3,4, \\ 4u_{5,j} - 2u_{4,j1} - u_{5,j+1} - u_{5,j-1} = 2, \ j = 1,2,3,4. \end{cases}$$

If we enumerate the nodes (i,j) from bottom to top and left to right, then we obtain that the system to be solved can be written in the matrix form

(8.3.10) $A \cdot \mathbf{u} = \mathbf{b}$,

where the 24×24 matrix has the block form

$$A = \begin{pmatrix} T & -2I & O & O & O & O \\ -I & T & -I & O & O & O \\ O & -I & T & -I & O & O \\ O & O & -I & T & -I & O \\ O & O & O & -I & T & -I \\ O & O & O & O & -2I & T \end{pmatrix},$$

where I is the identity 5×5 matrix, $\mathbf{O}$ is the zero 5×5 matrix, and the 5×5 matrix T is given by

$$
T = \begin{pmatrix}
4 & -1 & 0 & 0 & 0 \\
-1 & 4 & -1 & 0 & 0 \\
0 & -1 & 4 & -1 & 0 \\
0 & 0 & -1 & 4 & -1 \\
0 & 0 & 0 & -1 & 4
\end{pmatrix}.
$$

The matrix $\mathbf{u}$ of the 24 unknowns is given by

$$
\mathbf{u} = \begin{pmatrix} u_{0,1} & u_{0,2} & \cdots & u_{0,4} & \cdots & \cdots & u_{4,1} & u_{4,2} & \cdots & u_{4,4} & u_{5,1} & u_{5,2} & \cdots & u_{5,4} \end{pmatrix}^T
$$

and the 24×1 matrix $\mathbf{b}$ is given by

$$
\mathbf{b} = \begin{pmatrix} -2h & -2h & -2h & 1-2h & 0 & 0 & 0 & 1+h^2 & 0 & 0 & 0 & 1+4h^2 & \dots 0 & 0 & 1+25h^2 \end{pmatrix}^T.
$$

The above system was solved by the Gauss–Seidel iterative method (taking 10 iterations) and the results are given in Table 8.3.1.

Table 8.3.1

$j\backslash i$	0	1	2	3	4	5
1	−0.06946	0.049656	0.112271	0.169276	0.202971	0.206501
2	0.022854	0.155810	0.230152	0.361864	0.436107	0.420063
3	0.179795	0.320579	0.290663	0.611922	0.4361072	0.808039
4	0.478023	0.656047	0.785585	1.035632	1.182048	1.293034

3^0. *Nonrectangular Domain. Dirichlet Boundary Condition.*
  Now we will approximate the Poisson equation on a nonrectangular domain D with a curved boundary $\Gamma = \partial D$, subject to Dirichlet boundary conditions. We consider only Dirichlet boundary conditions. For problems with Neumann boundary conditions the interested reader is referred to the book [16] by G. Evans, J. Blackledge and P. Yardley.

  Consider the boundary value problem

$$
\begin{cases}
\Delta u(x,y) = F(x,y), & (x,y) \in D, \\
u(x,y) = f(x,y), & (x,y) \in \Gamma.
\end{cases}
$$

  We cover the domain D with a uniform grid of squares whose sides are parallel to the coordinate axes. Let h be the grid size. To apply a finite difference method, the given domain D is replaced by the set of those squares which completely lie in the domain D. (See Figure 8.3.3.)

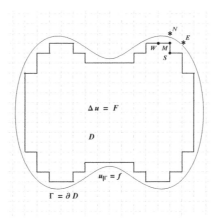

Figure 8.3.3

If an interior node of the squares is such that its four neighboring nodes either lie entirely in the domain D or none of them is outside the domain, then we use Equations (8.3.4) to approximate the Poisson equation. Special consideration should be taken for those nodes which are close to the boundary Γ for which at least one of their neighboring nodes is outside the domain. The node $M(x_i, y_j)$ in Figure 8.3.4 is an example of such a node. Its neighboring nodes E and N are outside the region D. To approximate the second partial derivatives in the Poisson equation at the node M we proceed as follows. In order to approximate u_{xx} we denote by $A(x, y_j)$ the point on the segment ME which is exactly on the boundary Γ and let $\Delta x = \overline{MA}$ (see Figure 8.3.4).

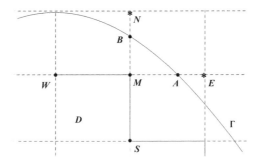

Figure 8.3.4

By Taylor's formula for the function $u(x, y)$ at the points A and W we have

$$u(W) = u(M) - u_x(M)h + \frac{1}{2}u_{xx}(M)h^2 + \dots$$

and

$$u(A) = u(M) + u_x(M)\Delta x + \frac{1}{2}u_{xx}(M)(\Delta x)^2 + \ldots.$$

From the last two equations we obtain the following approximation for u_{xx} at the node (i, j).

(8.3.11) $$u_{xx}(M) \approx 2\left[\frac{u(A)}{\Delta x(\Delta x + h)} + \frac{u(W)}{h(\Delta x + h)} - \frac{u(M)}{h\Delta x}\right].$$

Working similarly with the point B (see Figure 8.3.4) we have an approximation for u_{yy} at the node (i, j).

(8.3.12) $$u_{yy}(M) \approx 2\left[\frac{u(B)}{\Delta y(\Delta y + h)} + \frac{u(W)}{h(\Delta y + h)} - \frac{u(M)}{h\Delta y}\right],$$

where $\Delta y = \overline{MB}$.

From (8.3.11) and (8.3.12) we have the following approximation of the Poison equation at the node M:

(8.3.13)
$$\frac{u(A)}{\Delta x(\Delta x + h)} + \frac{u(B)}{\Delta y(\Delta y + h)} + \frac{u(W)}{h(\Delta x + h)}$$
$$+ \frac{u(S)}{h(\Delta y + h)} - \frac{\Delta x + \Delta y}{h\Delta x\Delta y}u(M) \approx \frac{1}{2}F(M).$$

In (8.3.13), $u(M)$, $u(W)$ and $u(S)$ are unknown and should be determined.

Example 8.3.3. Taking a uniform grid of step size $h = \frac{1}{4}$ solve the Laplace equation

$$u_{xx} + u_{yy} = 0, \quad x > 0, \ y > 0, \ \frac{x^2}{4} + y^2 < 1,$$

subject to the boundary conditions

$$\begin{cases} u(0, y) = y^2, & 0 \leq y \leq 1, \\ u(x, 0) = \dfrac{x^2}{4}, & 0 \leq x \leq 2, \\ u(x, y) = 1, & \dfrac{x^2}{4} + y^2 = 1, \ y > 0. \end{cases}$$

Solution. The uniform grid and the domain are displayed in Figure 8.3.5.

For the nodes $(6, 1)$ and (i, j), $i = 1, 2, \ldots, 5$, $j = 1, 2$ we use (8.3.4):

(8.3.14) $$4u_{i,j} - u_{i,j+1} - u_{i,j-1} - u_{i-1,j} - u_{i+1,j} = 0.$$

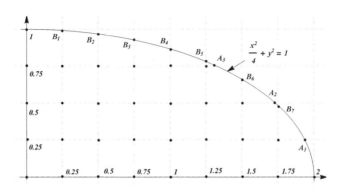

Figure 8.3.5

For the nodes $(5,3)$, $(6,2)$ and $(7,1)$ we use approximations $(8.3.13)$:

$(8.3.15)$
$$
\begin{cases}
\dfrac{u(A_3)}{\Delta x_3(\Delta x_3 + h)} + \dfrac{u(B_5)}{\Delta y_5(\Delta y_5 + h)} + \dfrac{u_{4,3}}{h(\Delta x_3 + h)} \\[2mm]
+ \dfrac{u_{5,2}}{h(\Delta y_5 + h)} - \dfrac{\Delta x_3 + \Delta y_5}{h\Delta x_3 \Delta y_5} u_{5,3} = 0 \\[3mm]
\dfrac{u(A_2)}{\Delta x_2(\Delta x_2 + h)} + \dfrac{u(B_6)}{\Delta y_6(\Delta y_6 + h)} + \dfrac{u_{5,2}}{h(\Delta x_2 + h)} \\[2mm]
+ \dfrac{u_{6,1}}{h(\Delta y_6 + h)} - \dfrac{\Delta x_2 + \Delta y_6}{h\Delta x_2 \Delta y_6} u_{6,2} = 0 \\[3mm]
\dfrac{u(A_1)}{\Delta x_1(\Delta x_1 + h)} + \dfrac{u(B_7)}{\Delta y_7(\Delta y_7 + h)} + \dfrac{u_{6,1}}{h(\Delta x_1 + h)} \\[2mm]
+ \dfrac{u_{7,0}}{h(\Delta y_7 + h)} - \dfrac{\Delta x_1 + \Delta y_7}{h\Delta x_1 \Delta y_7} u_{7,1} = 0.
\end{cases}
$$

Using the second order Taylor approximation for $u_{i-1,3}$, $i = 1,2,3,4$ we obtain the following 4 equations.

$(8.3.16)$
$$
u_{i-1,3} + u_{i+1,3} + \left[hu(B_i) + u_{i,2}\Delta y_i \right] \frac{2h}{(h + \Delta y_i)\Delta y_i}
$$
$$
- 2\frac{h + \Delta y_i}{h\Delta y_i} u_{i,3} = 0, \ i = 1,2,3,4.
$$

From the given boundary conditions we have

$$
\begin{cases}
u_{0,j} = y_j^2 = \dfrac{j^2}{16}, \ j = 1,\ldots,4; \quad u_{i,0} = \dfrac{x_i^2}{4} = \dfrac{i^2}{64}; \\[3mm]
u(B_i) = 1, \quad i = 1,\ldots,7; \ u(A_i) = 1, \ i = 1,2,3.
\end{cases}
$$

In order to solve the system (8.3.14), (8.3.15), (8.3.16) of 18 unknowns $u_{i,j}$, $i = 1, \ldots, 6$, $j = 1, 2$; $u_{i,3}$, $i = 1, \ldots, 5$; $u_{7,1}$ first we evaluate the quantities Δx_i, and Δy_i:

$$\Delta y_i = \sqrt{1 - \frac{x_i^2}{4}} - 3h = \sqrt{1 - \frac{i^2}{64}} - 3h, \quad i = 1, 2, \ldots, 5,$$

$$\Delta y_i = \sqrt{1 - \frac{x_i^2}{4}} - (8 - i)h = \sqrt{1 - \frac{i^2}{64}} - (8 - i)h \quad i = 6, 7,$$

$$\Delta x_i = 2\sqrt{1 - y_i^2} - (8 - i)h = 2\sqrt{1 - \frac{i^2}{16}} - (8 - i)h, \quad i = 1, 2, 3.$$

If we enumerate the nodes (i, j) from bottom to top and left to right we obtain that the system to be solved can be written in the matrix form

(8.3.17) $$A \cdot \mathbf{u} = \mathbf{b},$$

where the 18×18 matrix has the block form

$$A = \begin{pmatrix} T_1 & d & \mathbf{O} & \mathbf{O} & \mathbf{O} & \mathbf{O} \\ A & T_2 & d & \mathbf{O} & \mathbf{O} & \mathbf{O} \\ \mathbf{O} & A & T_3 & d & \mathbf{O} & \mathbf{O} \\ \mathbf{O} & \mathbf{O} & C & T_4 & d & \mathbf{O} \\ \mathbf{O} & \mathbf{O} & \mathbf{O} & B_1 & T_5 & C \\ \mathbf{O} & \mathbf{O} & \mathbf{O} & \mathbf{O} & B_2 & T_6 \end{pmatrix},$$

where $\mathbf{O}$ is the zero 3×3 matrix, and the 3×3 matrices D, B_i, $i = 1, 2$; C and T_i, $i = 1, 2, \ldots, 6$ are given by

$$D = \begin{pmatrix} -1 & 0 & 0 \\ 0 & -1 & 0 \\ 0 & 0 & 1 \end{pmatrix}, \quad C = \begin{pmatrix} -1 & 0 & 0 \\ 0 & -1 & 0 \\ 0 & 0 & 0 \end{pmatrix},$$

$$B_1 = \begin{pmatrix} -1 & 0 & 0 \\ 0 & -1 & 0 \\ 0 & 0 & \frac{1}{h(h + \Delta x_3)} \end{pmatrix}, \quad B_2 = \begin{pmatrix} -1 & 0 & 0 \\ 0 & \frac{1}{h(h + \Delta x_2)} & 0 \\ 0 & 0 & 0 \end{pmatrix},$$

$$T_i = \begin{pmatrix} 4 & -1 & 0 \\ -1 & 4 & -1 \\ 0 & \frac{2h}{h+\Delta y_i} & -2\frac{h+\Delta y_i}{\Delta y_i} \end{pmatrix}, i = 1, 2, \ldots, 5,$$

$$T_6 = \begin{pmatrix} 4 & -1 & -1 \\ \frac{1}{h(h+\Delta y_6)} & -\frac{\Delta x_2 + \Delta y_6}{h\Delta x_2 \Delta y_6} & 0 \\ \frac{1}{h(h+\Delta x_1)} & 0 & -\frac{\Delta x_1 + \Delta y_7}{h\Delta x_1 \Delta y_7} \end{pmatrix}.$$

The matrix **b** in the system (8.3.17) is given by

$$\mathbf{b} = \left(u_{1,0} + u_{0,1}, \ u_{0,2}, \ -u_{0,3} - \frac{2h^2}{(h+\Delta y_1)\Delta y_1}, \ u_{2,0}, \ 0, \ -\frac{2h^2}{(h+\Delta y_2)\Delta y_2}, \right.$$

$$u_{3,0}, \ 0, \ -\frac{2h^2}{(h+\Delta y_3)\Delta y_3}, \ u_{4,0}, \ 0, \ -\frac{2h^2}{(h+\Delta y_4)\Delta y_4}, \ u_{5,0}, \ 0,$$

$$-\frac{1}{(\Delta x_3 + h)\Delta x_3} - \frac{1}{(\Delta y_5 + h)\Delta y_5}, \ u_{6,0}, \ -\frac{1}{(\Delta x_2 + h)\Delta x_2}$$

$$\left. -\frac{1}{(\Delta y_6 + h)\Delta y_6}, -\frac{1}{(\Delta x_1 + h)\Delta x_1} - \frac{1}{(\Delta y_7 + h)\Delta y_7} - \frac{u_{7,0}}{(h(h+\Delta y_7)}\right)^T.$$

Solving the system for $u_{i,j}$, we obtain the results given in Table 8.3.2. (the index i runs in the horizontal direction and the index j runs vertically).

Table 8.3.2

$j\backslash i$	1	2	3	4	5	6	7
1	0.34519	0.44063	0.53730	0.63861	0.73899	0.83647	0.93431
2	0.56514	0.63002	0.69497	0.77816	0.85588	0.92259	
3	0.78536	0.81932	0.83442	0.92316	0.98379		

Exercises for Section 8.3.

1. Using the finite difference method find an approximation solution of the Laplace equation $\Delta u(x,y) = 0$ on the unit square $0 < x < 1$, $0 < y < 1$, on a 9×9 uniform grid $(h = 0.9)$, subject to the Dirichlet conditions

$$u(x,0) = 20, \ u(0,y) = 80, \ u(x,1) = 180, \ u(1,y) = 0.$$

2. Using the finite difference method find an approximation solution of the Poisson equation

$$\Delta\, u(x, y) = 1$$

on the unit square $0 < x < 1$, $0 < y < 1$, on a 4×4 uniform grid ($h = 1/4$), subject to the Dirichlet conditions

$$u(x, 0) = u(0, y) = u(x, 1) = u(1, y) = 0.$$

3. Using the finite difference method find an approximation solution of the Laplace equation $\Delta\, u(x, y) = 0$ on the unit square $0 < x < 1$, $0 < y < 1$, on a 9×9 uniform grid with $h = \dfrac{1}{9}$, subject to the boundary conditions

$$u_y(x, 0) = 0, \quad u(0, y) = 80, \quad u(x, 1) = 180, \quad u(1, y) = 0.$$

4. Solve the Poisson equation

$$u_{xx} + u_{yy} = 1$$

on the rectangle $3 < x < 5$, $4 < y < 6$, with zero boundary condition, taking $h = \dfrac{1}{4}$ in the finite differences formula.

5. The function u satisfies the Laplace equation on the square $0 < x < 1$, $0 < y < 1$ and satisfies

$$u(x, y) = \cos\left(\pi x + piy\right)$$

on the boundary of the square. Take a uniform grid of side $h = \dfrac{1}{4}$ and order the nodes from bottom to top and left to right. Determine and solve the matrix equation for the 9 internal nodes.

6. Consider the Laplace equation

$$u_{xx} + u_{yy} = 0$$

on the semicircle region shown in Figure 8.3.6, subject to the Dirichlet boundary conditions

$$u(x, 0) = 1, \; u(x, y) = x^2 \; if \; x^2 + y^2 = 1.$$

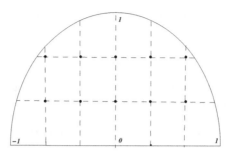

Figure 8.3.6

Approximate the solution u at the 10 interior nodes of the uniform grid with size $h = \dfrac{1}{3}$.

7. Solve the Poisson equation

$$\Delta u = -1$$

on the L-shaped region, obtained when from the unit square is removed the top left corner square of side $h = \dfrac{1}{4}$, subject to zero boundary value on the boundary of the L=shaped region (see Figure 8.3.7). Take the uniform grid for which $h = \dfrac{1}{4}$.

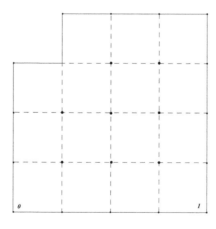

Figure 8.3.7

8.4 Finite Difference Methods for the Heat Equation.

In this section we will apply finite difference methods to solve the heat equation in one and two spatial dimensions on a rectangular domain, subject to Dirichlet boundary conditions. Parabolic partial differential equations subject to Neumann boundary conditions are treated in the same way as in the previous section by introducing new "fictitious" points.

1^0. *One Dimensional Heat Equation. Dirichlet Boundary Condition.*

Consider the heat equation on the domain $D = (0, l) \times (0, T)$, subject to Dirichlet boundary conditions

(8.4.1)
$$\begin{cases} u_t(x,t) = c^2\, u_{xx}, & (x,t) \in D = \{0 < x < a,\ 0 < t < T\} \\ u(x,0) = f_0(x), & , u(x,T) = f_T(x),\ 0 < x < a, \\ u(0,t) = g_0(t), & u(a,t) = g_a(t),\ 0 < t < T. \end{cases}$$

First we partition the rectangle D with the points

$$x_i = ih, \quad y_j = j\Delta t, \quad i = 0, 1, \ldots, m;\ j = 0, 1, 2, \ldots.$$

If $u_{i,j} \approx u(x_i, t_j)$, then we approximate the derivative $u_t(x,t)$ with the forward finite difference

$$u_t\left(x_i, y_j\right) \approx \frac{u_{i+1,j} - u_{i,j}}{\Delta t}$$

and the derivative $u_{xx}(x,t)$ with the central finite difference

$$u_{xx}\left(x_i, t_j\right) \approx \frac{u_{i+1,j} - 2u_{i,j} + u_{i-1,j}}{h^2}.$$

Then the finite difference approximation of the heat equation (8.4.1) is

$$\frac{1}{\Delta t}\left[u_{i,j+1} - u_{i,j}\right] = \frac{c^2}{h^2}\left[u_{i+1,j} - 2u_{i,j} + u_{i-1,j}\right], \ 1 \le i \le m-1;\ 1 \le j \le n-1,$$

or

(8.4.2)
$$u_{i,j+1} = \lambda u_{i-1,j} + (1 - 2\lambda)u_{i,j} + \lambda u_{i+1,j},$$

where the parameter

$$\lambda = c^2 \frac{\Delta t}{h^2}$$

is called the *grid ratio parameter*. Equation (8.4.2) is called the *explicit approximation* of the parabolic equation (8.4.1). The parameter λ plays a role in the stability of the explicit approximation scheme (8.4.2). It can be shown that we have a stable approximation scheme only if $0 < \lambda \le \frac{1}{2}$, i.e., if

$$2c^2\Delta t \le h^2.$$

Now, if we write approximations for the derivative in Equation (8.4.1) at the node $(i, j + \frac{1}{2})$, then we obtain the following implicit, so called *Crank–Nicolson approximation*.

$$(8.4.3) \quad \lambda u_{i+1,j+1} - (2+\lambda)u_{i,j+1} + \lambda u_{i-1,j+1} = -\lambda u_{i+1,j} + (2-\lambda)u_{i,j} - \lambda u_{i-1,j}.$$

The above system is solved in the following way. First, we take $j = 0$ and using the given initial values

$$u_{1,0}, u_{2,0}, \ldots, u_{i-1,0},$$

we solve the system (8.4.3) for the unknowns $u_{1,1}, u_{2,1}, \ldots, u_{i-1,1}$. Using these values, we solve the system for the next unknowns $u_{1,2}, u_{2,2}, \ldots, u_{i-1,2}$ and so on.

The advantage of the implicit Crank–Nicolson approximation over the explicit one is that its stability doesn't depend on the parameter λ.

It is important to note that for a fixed value of λ, the order of the approximation schemes (8.4.2) and (8.4.3) is $O(h^2)$.

Let us take a few examples to illustrate all of this.

Example 8.4.1. Consider the problem

$$u_t(x, y) = 10^{-2} u_{xx}(x, y), \quad 0 < x < 1, \ t > 0,$$

subject to the boundary conditions

$$u(0, t) = u(1, t) = 0, \quad t > 0$$

and the initial condition

$$u(x, 0) = \begin{cases} 10x, & 0 \le x \le \frac{1}{2} \\ 10(1 - x), & \frac{1}{2} < 1. \end{cases}$$

Using the explicit finite difference approximation find an approximation of the problem at several locations and several time instants and compare the obtained solutions with the exact solution.

Solution. Using the separation of variables method, given in Chapter 5, it can be shown (show this!) that the exact (analytical) solution of this problem is given by

$$u(x, t) = \frac{80}{\pi^2} \sum_{n=0}^{\infty} \frac{(-1)^n}{(2n + 1)^2} \sin\left((2n + 1)\pi x\right) e^{-0.01(2n+1)^2 t}.$$

Since we have homogeneous boundary conditions, system (8.4.2) can be written in matrix form

$$(8.4.4) \quad \mathbf{u}_{i,j+1} = A^j \cdot \mathbf{u}_{i,0},$$

where
$$\mathbf{u}_{i,j+1} = (u_{1,j+1} \; u_{2,j+1} \; \ldots \; u_{i-1,j+1})^T ,$$
$$u_{i,0} = (f(h) \; f(2h) \; \ldots \; f((i-1)h))^T ,$$

and the $(i-1) \times (i-1)$ matrix A is given by

$$A = \begin{pmatrix} 1-2\lambda & \lambda & 0 & \ldots & 0 & 0 & 0 \\ \lambda & 1-2\lambda & \lambda & \ldots & 0 & 0 & 0 \\ \vdots & & & & & & \\ 0 & 0 & 0 & \ldots & \lambda & 1-2\lambda & \lambda \\ 0 & 0 & 0 & \ldots & 0 & \lambda & 1-2\lambda \end{pmatrix} .$$

First, let us take $h = 0.1$, $\Delta t = 0.4$. In this case $\lambda = 0.4 < 0.5$. Solving the system (8.4.4) at $t = 1.2$, $t = 4.8$ and $t = 14.4$ we obtain approximations that are presented graphically in Figure 8.4.1 and in Table 8.4.1. From Figure 8.4.1 we notice pretty good agreement of the numerical and exact solutions.

Table 8.4.1

$t_j \backslash x_i$	0.1	0.2	0.3	0.4	0.5	0.6	0.7
0.0	2.0000	4.0000	6.0000	8.0000	9.9899	8.0000	6.0000
1.2	1.9692	3.8617	5.5262	6.7106	7.1454	6.7106	5.5262
4.8	1.5494	2.9545	4.0793	4.8075	5.0598	4.8075	4.0793
14.4	0.5587	1.0628	1.4629	1.7198	1.8083	1.7198	1.4629

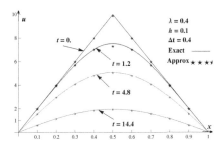

Figure 8.4.1

Next we take $h = 0.1$, $\Delta t = 0.6$. In this case $\lambda = 0.6$. Solving the system (8.4.4) at $t = 1.2$, $t = 2.4$, $t = 4.8$ and $t = 14.4$ we obtain approximations that are presented in Table 8.4.2 and graphically in Figure 8.4.2. Big oscillations appear in the solution. These oscillations grow larger and larger as time increases. This instability is due to the fact that $\lambda = 0.6 > 0.5$.

Table 8.4.2

$t_j \backslash x_i$	0.1	0.2	0.3	0.4	0.5	0.6	0.7
0.0	2.0000	4.0000	6.0000	8.0000	10.0000	8.0000	6.0000
1.2	2.0000	4.0000	6.0000	6.5600	8.0800	6.5600	6.0000
4.8	2.1244	1.8019	5.7749	2.6826	7.3010	2.6826	5.7749
14.4	73.374	-137.33	192.138	-222.257	237.528	-222.257	192.138

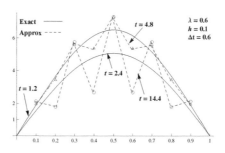

Figure 8.4.2

Example 8.4.2. Solve the problem in Example 8.4.1 by the Crank–Nicolson approximation.

Solution. Let us write the system (8.4.3) in matrix form

$$(8.4.5) \qquad\qquad A \cdot \mathbf{u}_{i,j+1} = B \cdot \mathbf{u}_{i,j},$$

where

$$\mathbf{u}_{i,j+1} = \left(u_{1,j+1} \; u_{2,j+1} \; \cdots \; u_{i-1,j+1} \right)^T,$$
$$\mathbf{u}_{i,0} = \left(f(h) \; f(2h) \; \cdots \; f((i-1)h) \right)^T,$$

and the $(i-1) \times (i-1)$ matrices A and B are given by

$$A = \begin{pmatrix} 2+2\lambda & -\lambda & 0 & \cdots & 0 & 0 & 0 \\ -\lambda & 2+2\lambda & -\lambda & \cdots & 0 & 0 & 0 \\ \vdots & & & & & & \\ 0 & 0 & 0 & \cdots & -\lambda & 2+2\lambda & -\lambda \\ 0 & 0 & 0 & \cdots & 0 & -\lambda & 2+2\lambda \end{pmatrix},$$

and

$$B = \begin{pmatrix} 2-2\lambda & \lambda & 0 & \cdots & 0 & 0 & 0 \\ \lambda & 2-2\lambda & \lambda & \cdots & 0 & 0 & 0 \\ \vdots & & & & & & \\ 0 & 0 & 0 & \cdots & \lambda & 2-2\lambda & -\lambda \\ 0 & 0 & 0 & \cdots & 0 & \lambda & 2-2\lambda \end{pmatrix}.$$

Now, if we take $\lambda = 0.6$, which can be obtained for $h = 0.1$ $(i = 10)$, $\Delta t = 0.6$, then, first we need to solve the 9×9 linear system (8.4.5) for $u_{1,1}, u_{2,1}, \ldots, u_{9,1}$. In this step we use the given initial condition $u_{i,0} = f(ih)$, $i = 1, 2, \ldots, 9$. The results obtained are used in (8.4.5) to obtain the next values $u_{i,2}$, , $i = 1, 2, \ldots, 9$, and so on. The results obtained at the first 8 nodes are summarized in Table 8.4.3.

Table 8.4.3

$t_j \backslash x_i$	0.1	0.2	0.3	0.4	0.5	0.6	0.7	0.8
0.0	2.0000	4.0000	6.0000	8.0000	10.0000	8.0000	6.0000	4.0000
1.2	1.9849	3.9346	5.7456	7.1174	7.6464	7.1174	5.7456	3.9346
4.8	1.5646	2.9861	4.1271	4.8680	5.1250	4.8680	4.1271	2.9861
14.4	0.6165	1.1728	1.6142	1.897	1.9953	1.897	1.6142	1.1728

2^0. *Two Dimensional Heat Equation on a Rectangular.*

In this section we will describe the numerical solution of the two dimensional heat equation

$$u_t(x, y, t) = c^2 [u_{xx}(x, y, t) + u_{yy}(x, y, t)], \ 0 < x < a, \ 0 < y < b, \ t > 0,$$

subject to the initial condition

$$u(x, y, 0) = f(x, y)$$

and the Dirichlet boundary conditions

$$\begin{cases} u(0, y, t) = f_1(y), & u(a, y, t) = f_2(y) \\ u(x, 0, t) = g_1(x), & u(x, b, t) = g_2(x). \end{cases}$$

With $\Delta x = 1/m$, $\Delta y = 1/n$ we partition the rectangle with the points $(i\Delta x, j\Delta y)$, $i = 0, 1, 2 \ldots, m$, $j = 0, 1, 2 \ldots, n$ and we denote the discrete time points by $t_k = k\Delta t$. If $u_{i,j}^{(k)}$ denotes an approximation of the exact solution $u(x, y, t)$ of the heat equation at the point $(i\Delta x, j\Delta y)$ at time $k\Delta t$, then using the finite difference approximations for the partial derivatives in the heat equation we obtain the explicit approximation scheme

$$u_{i,j}^{(k+1)} = \left[1 - 2c^2\Delta t\left(\frac{1}{\Delta x^2} + \frac{1}{\Delta y^2}\right)\right] u_{i,j}^{(k)} + a\Delta t\left[\frac{u_{i+1,j}^{(k)} + u_{i-1,j}^{(k)}}{\Delta x^2} + \frac{u_{i,j+1}^{(k)} + u_{i,j-1}^{(k)}}{\Delta y^2}\right].$$

In practical applications usually we take $\Delta x = \Delta y = h$, and the above explicit finite difference approximation becomes

$$(8.4.6) \qquad u_{i,j}^{(k+1)} = (1 - 4\lambda)u_{i,j}^{(k)} + \lambda(u_{i+1,j}^{(k)} + u_{i-1,j}^{(k)} + u_{i,j+1}^{(k)} + u_{i,j-1}^{(k)}),$$

where

$$\lambda = c^2 \frac{\Delta t}{h^2}.$$

In order to have a stable, and thus convergent, approximation, the increments Δt and h should be selected such that

$$\lambda = c^2 \frac{\Delta t}{h^2} \le \frac{1}{4}.$$

When solving the above system (8.4.6) for $i = 1, 2 \ldots, m-1$, $j = 1, 2 \ldots, n-1$ we use the given initial conditions

$$u_{i,j}^{(0)} = f(ih, jh), \quad i = 0, 1, 2 \ldots, m; \; j = 0, 1, 2 \ldots, n$$

and the given boundary conditions $u_{i,0}^{(k)}$, $u_{i,n}^{(k)}$, and $u_{0,j}^{(k)}$, $u_{n,j}^{(k)}$.

Example 8.4.3. Using the explicit finite difference method solve the two dimensional heat equation

$$u_t = u_{xx} + u_{yy}$$

in the rectangle $0 < x < 1.25$, $0 < y < 1$, subject to the initial and boundary conditions

$$u(x, y, 0) = 1, \quad u(x, 0, t) = u(x, 1, t) = u(0, y, t) = u(1.25, y, t) = 0.$$

Solution. If we take $h = 1/4$ ($n = 5$, $m = 4$); $\Delta t = 1/64$, then $\lambda = 1/4$ and so the stability condition is satisfied. For this choice of m, n and Δt the system (8.4.6) becomes

$$u_{i,j}^{(k+1)} = \frac{1}{4}\left(u_{i+1,j}^{(k)} + u_{i-1,j}^{(k)} + u_{i,j+1}^{(k)} + u_{i,j-1}^{(k)}\right), \; 1 \le i \le 4; \, j = 1, 2, 3; k = 0, 1, \ldots$$

The solutions of the above system taking $k = 0, 1, 2$ are presented in Table 8.4.4.

Table 8.4.4

	$u_{1,1}^{(1)}$	$u_{2,1}^{(1)}$	$u_{3,1}^{(1)}$	$u_{4,1}^{(1)}$
$u_{1,1}^{(1)}$	1/2	3/4	3/4	1/2
$u_{1,2}^{(1)}$	3/4	1	1	3/4
$u_{1,3}^{(1)}$	1/2	3/4	3/4	1/2
$j\backslash i$	$u_{1,1}^{(2)}$	$u_{2,1}^{(2)}$	$u_{3,1}^{(1)}$	$u_{4,1}^{(2)}$
$u_{1,1}^{(2)}$	3/8	9/16	9/16	3/8
$u_{1,2}^{(2)}$	1/2	13/16	13/16	1/2
$u_{1,3}^{(2)}$	3/8	9/16	9/16	3/8
$j\backslash i$	$u_{1,1}^{(3)}$	$u_{2,1}^{(3)}$	$u_{3,1}^{(1)}$	$u_{4,1}^{(3)}$
$u_{1,1}^{(3)}$	17/64	7/16	7/16	17/64
$u_{1,2}^{(3)}$	25/64	19/64	19/64	25/64
$u_{1,3}^{(3)}$	17/64	7/16	7/16	17/64

We conclude this section with a few illustrations of Mathematica's differential equation solving functions, applied to parabolic partial differential equations.

Example 8.4.4. Solve the initial boundary value problem

$$\begin{cases} u_t(x,t) = u_{xx}(x,t), & 0 < x < 10, \ 0 < t < 0.5, \\ u(x,0) = \dfrac{1}{10^5}x^2(x-10)^2 e^x, & 0 < x < 10, \\ u(0,t) = u(10,t) = 0, & 0 < t < 0.5. \end{cases}$$

Solution. The following Mathematica commands solve the problem.

$In[1]:=U = NDSolve[\{Derivative[2,0][u][x,t]==Derivative[0,1][u][x,t],$
$u[x,0] == 1/(10^5)x^2(x-10)^2 Exp[x], u[0,t] == 0, u[10,t] == 0\},$
$u, \{x,0,10\}, \{t,0,0.5\}];$

$Out[1]:= \{\{u-> InterpolatingFunction[\{\{0.,10.\},\{0,0.5\},<>]\}\}$

$In[2]:=Plot3D[Evaluate[u[x,t]/.U[[1]]], \{t,0,0.5\}, \{x,0,10\},$
Mesh $->$ None, PlotRange $->$ All]

The plot of the solution of the problem is given in Figure 8.4.3, and a plot at different time instances is given in Figure 8.4.4.

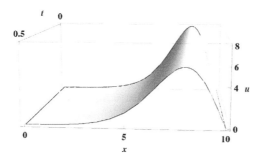

Figure 8.4.3

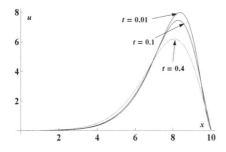

Figure 8.4.4

Example 8.4.5. Solve the heat equation

$$u_t(x,t) = u_{xx}(x,t) + x\cos 3x - 30t\cos 9x, \quad 0 < x < \pi, \, 0 < t < 1,$$

subject to the initial and Neumann boundary conditions

$$u(x,0) = \cos 5x, \quad 0 < x < \pi, \quad u_x(0,t) = u_x(\pi,t) = 0, \; 0 < t < 1.$$

Solution. The following Mathematica commands solve the problem.

$In[1]:=U = NDSolve[\{D[u[[x,t],t] == D[u[x,t],x,x] + xCos[3x]$
$-30tCos[9x], u[x,0] == Cos[5x], Devivative[1,0][u][0,t]$
$== Devivative[1,0][u][Pi,t] == 0\}, u, \{x,0,Pi\}, \{t,0,1\},$
$Method- > \{"MethodOfLines", "SpatialDiscretization"$
$> \{"TensorProductGrid", "MinPoints"- > 500\}\}];$
$Out[1]:= \{\{u- > InterpolatingFunction[\{\{0., 3.14159.\}, \{0, 1.\}\}, <>]\}\}$
$In[2]:=Plot3D[Evaluate[u[x,t]/.U[[1]]], \{t,0,1\}, \{x,0,Pi\},$
Mesh $- > None, PlotRange- > All]$
$In[3]:=Plot[Evaluate[\{u[x,.01]/.U[[1]], u[x,.4]/.U[[1]], u[x,.9]/.U[[1]]\}],$
$\{x,0,Pi\}, PlotRange- > All, Ticks- > \{\{0, Pi/10, 3Pi/10, Pi/2, 7Pi/10,$
$9Pi/10, Pi\}, \{-1, -1/2, 0, 1/2\}\}, Frame- > False]$

The plot of the solution is given in Figure 8.4.5, and plot of the solutions at different time instances is given in Figure 8.4.6.

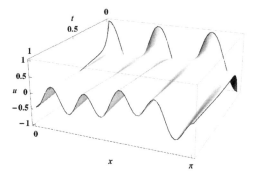

Figure 8.4.5

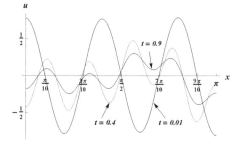

Figure 8.4.6

Example 8.4.6. Let R be the square

$$R = \{(x, y) : -5 < x < 5, \ -5 < y < 5\}$$

and ∂R its boundary.

Using Mathematica graphically display at several time instances the solution of the two dimensional heat equation

$$\begin{cases} u_t(x, y, t) = u_{xx}(x, y, t) + u_{yy}(x, y, t), & (x, y) \in R, \ 0 < t < 4, \\ u(x, 0) = 1.5e^{-(x^2+y^2)}, & (x, y) \in R, \\ u(x, y, t) = 0, & (x, y) \in \partial R, \ 0 < t < 4. \end{cases}$$

Solution. The following Mathematica commands solve the problem.

$In[1]{:=}heat = First[NDSolve[D[u[x, y, t], t] == D[u[x, y, t], x, x]$
$+D[u[x, y, t], y, y], u[x, y, 0] {==} 1.5Exp[-(x^2+y^2)]], u[-5, y, t] {==} u[5, y, t],$
$u[x, -5, t] == u[x, 5, t]\}, u, \{x, -5, 5\}, \{y, -5, 5\}, \{t, 0, 4\}]];$

$Out[1]{:=} \{\{u - > InterpolatingFunction[\{\{-5., 5.\}, \{-5., 5.\},$
$\{0., 4.\}], <>]\}\}$

To get the plots of the solution of the heat equation at $t = 0$ (the initial heat distribution) and $t = 4$ use

$In[2]{:=}Plot3D[u[0, x, y]/.heat, \{x, -5, 5\}, \{y, -5, 5\},$
$PlotRange- > All, Mesh- > 20]$

$In[3]{:=}Plot3D[u[4, x, y]/.heat, \{x, -5, 5\}, \{y, -5, 5\},$
$PlotRange- > All, Mesh- > 20]$

Their plots are displayed in Figure 8.4.7 and Figure 8.4.8, respectively.

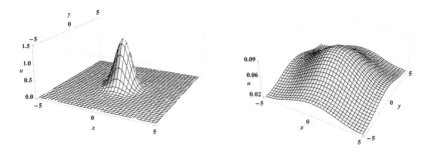

Figure 8.4.7   Figure 8.4.8

Exercises for Section 8.4.

1. Using the explicit finite difference method find a numerical solution of the heat equation

$$u_t(x,t) = u_{xx}(x,t), \quad 0 < x < 1,$$

subject to the initial and Dirichlet conditions

$$u(x,0) = \sin(2\pi x), \; u(0,t) = u(1,t) = 0$$

at the points $(x_j = 1/4, \; t_j = j \cdot 0.01), \; j = 1,2\ldots,15.$ Consider the following increments: $h = \dfrac{1}{8}, \dfrac{1}{16}$ and $\Delta t = 10^{-3}.$ Compare the obtained results with the exact solution

$$u(x,t) = e^{-4\pi^2 t} \sin(2\pi x).$$

2. Approximate the initial boundary value problem in the previous example by the Crank–Nicolson method taking $h = \dfrac{1}{8}, \; h = \dfrac{1}{16}$ and $\Delta t = 10^{-3}.$

3. Apply the explicit method to the equation

$$u_t = 0.1 u_{xx}, \; 0 < x < 20, \; t > 0,$$

subject to the initial and boundary conditions

$$u(x,0) = 2, \; u(0,t) = 0, \; u(20,t) = 10.$$

Take $h = 4.0$ and $\Delta t = 40.0$ ($\lambda = 0.25$). Perform four iterations.

4. Write down the explicit finite difference approximation for the initial boundary value problem

$$
\begin{cases}
u_t = u_x + u_{xx}, \quad 0 < x < 1, \; t > 0, \\
u(x,0) = \begin{cases} x, & 0 \le x \le \frac{1}{2} \\ 1-x, & \frac{1}{2} < x \le 1 \end{cases} \\
u(0,t) = u(1,t) = 0 \quad 0 < t < 1.
\end{cases}
$$

Perform 8 iterations taking $h = 0.1$ and $\Delta t = 0.0025.$

5. Write down the Crank–Nicolson approximation scheme for the initial boundary value problem

$$
\begin{cases}
u_t(x,t) = u_{xx})x,t), & 0 < x < 1,\ t > 0, \\
u(x,0) = x, & 0 < x < 1 \\
u(0,t) = u(i,t) = 0, & t > 0.
\end{cases}
$$

Perform the first 4 iterations taking $h = 0.25$ and $\Delta t = \dfrac{1}{16}$. Compare the obtained results with the exact solution

$$
u(x,t) = \frac{2}{\pi} \sum_{n=1}^{\infty} \frac{(-1)^n}{n} e^{-n^2\pi^2 t} \sin(n\pi x),
$$

which was derived by the separation of variables method in Chapter 5.

6. Approximate the equation

$$
u_t = u_{xx} + 1,\ 0 < x < 1,\ t > 0,
$$

subject to the initial and boundary conditions

$$
u(x,0) = 0,\ 0 < x < 1 \quad u(0,t) = u(1,t) = 0.
$$

Take $h = \dfrac{1}{4}$ and $\Delta t = \dfrac{1}{32}$.

7. Write down and solve the explicit finite difference approximation of the equation

$$
u_t = xu_{xx},\ 0 < x < \frac{1}{2},\ t > 0,
$$

subject to the initial and Neumann boundary conditions

$$
u(x,0) = x(1-x),\ u_x\left(\frac{1}{2},t\right) = -\frac{u\left(\frac{1}{2},t\right)}{2}.
$$

Take $h = 0.1$ and $\Delta t = 0.005$.

Hint: Use fictitious points.

8.5 Finite Difference Methods for the Wave Equation.

In this section we will apply the finite difference methods to hyperbolic partial differential equations of the first and second order in one and two spatial dimensions on a rectangular domain, subject to Dirichlet boundary conditions.

1^0. *Hyperbolic Equations of the First Order.* In Section 4.2 of Chapter 4, we considered the simple constant coefficient transport equation

$$(8.5.1) \qquad u_t(x,t) + au_x(x,t) = 0, \quad 0 < x < l, \, t > 0,$$

subject to the initial and boundary conditions

$$\begin{cases} u(x,0) = f(x), & 0 \le x < l, \\ u(0,t) = g(t), & t > 0. \end{cases}$$

The unique solution of this initial boundary value problem is given by

$$u(x,t) = \begin{cases} f(x - at), & \text{for } x - at > 0 \\ g(x - \frac{1}{a}t), & \text{for } x - at > 0 \end{cases}$$

and this solution is constant on any straight line of slope a. Despite the fact that we have an exact solution of the problem, a numerical approximation method of this equation will be discussed because it can serve as a model for approximation of transport equations with variable coefficients $a(x,t)$ or even more complicated nonlinear equations.

As usual, we partition the spatial and time interval with the grid points

$$x_i = ih, \quad t_j = j\Delta t, \ i = 0,1,\ldots,m; \ j = 0,1,2,\ldots,n.$$

Let $u_{i,j} \approx u(x_i,t_j)$ be the approximation to the exact solution $u(x,t)$ at the node (x_i,t_j). Now, using the forward, backward and central approximations for the partial derivatives with respect to x and t the following finite difference approximations of Equation (8.5.1) are obtained.

$$u_{i,j+1} = u_{i,j-1} - \lambda\left(u_{i+1,j} - u_{i-1,j}\right), \quad \textit{Leap-Frog}$$

$$u_{i,j+1} = \frac{1}{2}(1-\lambda)u_{i+1,j} + \frac{1}{2}(1+\lambda)u_{i-1,j}, \quad \textit{Lax--Friedrichs}$$

$$u_{i,j+1} = \frac{1}{2}(\lambda^2 - \lambda)u_{i+1,j} + (1-\lambda^2)u_{i,j} + \frac{1}{2}(\lambda^2 + \lambda)u_{i-1,j}, \quad \textit{Lax--Wendroff,}$$

where

$$\lambda = a\frac{\Delta t}{2h}.$$

It can be shown that for stability of the Lax--Friedrichs and Lax--Wendroff approximation schemes it is necessary that

$$|\lambda| = |a|\frac{\Delta t}{h} \le 1.$$

The Leap-Frog scheme is stable for every grid ratio λ.

Let us take a few examples.

Example 8.5.1. Solve the transport equation

$$u_t(x,t) + u_t(x,t) = 0, \quad 0 < x < 20, \; 0 < t < 1,$$

subject to the initial condition

$$f(x) = u(x,0) = \begin{cases} 1, & 0 \le x < 2 \\ x - 1, & 2 \le x < 4 \\ 7 - x, & 4 \le x < 6 \\ 1, & 6 \le x < 10 \\ x - 9, & 10 \le x < 11 \\ 13 - x, & 11 \le x \le 20 \end{cases}$$

and the boundary conditions

$$u(0,t) = u(1,t) = 1, \quad 0 < t < 1.$$

Use the Lax–Friedrichs scheme with

(a) $\lambda = 0.8$, $h = 0.5$ and $h = 0.25$ at $t = 0.4$.

(b) $\lambda = 1.6$, $h = 1$, $h = 0.5$ and $h = 0.25$ at $t = 1.6$.

Solution. The exact solution of the problem is given by

$$u(x,t) = f(x - t).$$

Plots of $u(x,t)$ at $t = 0$, $t = 0.4$ and $t = 1.6$ are given in Figure 8.5.1.

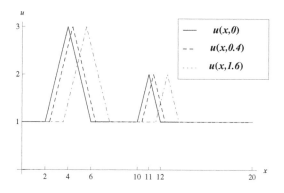

Figure 8.5.1

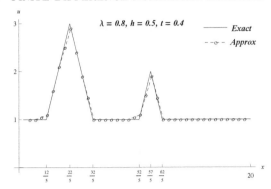

Figure 8.5.2

In order to start the Lax–Friedrichs approximation we use the initial condition $u_{i,0} = f(ih)$. For $j \geq 0$ the approximations are obtained by

$$u_{i,j+1} = \left(\frac{1}{2} - \lambda\right)u_{i+1,j} + \left(\frac{1}{2} + \lambda\right)u_{i-1,j}.$$

Using Mathematica, an approximation $u_{i,1}$ is obtained for $\lambda = 0.8$, $h = 0.5$ ($\Delta t = 0.4$) at $t = 0.4$. This solution is shown in Figure 8.5.2.

In Figure 8.5.3 the solution is shown for $\lambda = 0.8$, $h = 0.25$ ($\Delta t = 0.4$) at the time moment $t = 0.4$.

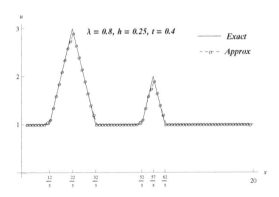

Figure 8.5.3

From these figures we see that we have reasonable approximations except at the sharp corners due to the fact that $f(x)$ is not smooth at these points.

For $\lambda = 1.6$ the situation is different, which can be seen from Figures 8.5.4, 8.5.5 and 8.5.6. Even if the grid size h is decreasing, the approximations are not getting better. The reason for this behavior is that stability condition in the Lax–Friedrich approximation scheme is not satisfied.

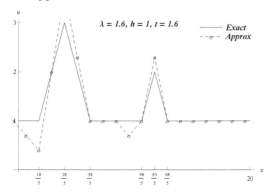

Figure 8.5.4

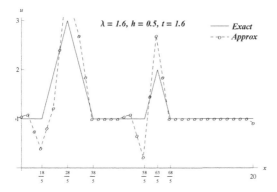

Figure 8.5.5

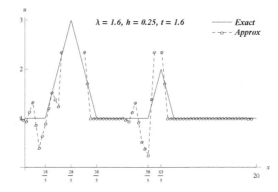

Figure 8.5.6

Example 8.5.2. Solve the transport equation in the previous example using the Leap-Frog approximation. Take $\lambda = 1.6$, $h = 0.5$ and $t = 1.6$.

Solution. The Leap-Frog approximation is given by

$$u_{i,j+1} = u_{i,j-1} - \lambda \left(u_{i+1,j} - u_{i-1,j} \right).$$

Let us note that the Leap-Frog method is a three level method. To find the values of the function at one time step, it is necessary to know the values of the function at the previous two time steps. As a result, to get started on the method we need initial values at the first two time levels. For the initial level we use the given initial condition $u_{i,0} = f(ih)$. For the next level we use the Forward-Time Central-Space (see Exercise 1 of this section).

$$u_{i,j+1} = u_{i,j} - \frac{\lambda}{2} \left(2u_{i+1,j} - u_{i-1,j} \right),$$

with $j = 0$. For $j \geq 1$ use the Leap-Frog method. Using Mathematica, the approximation $u_{i,2}$ is obtained for $\lambda = 1.6$, $h = 0.5$ ($\Delta t = 0.8$) at time $t = 1.6$. This solution along with the exact solution $f(x - 1.6)$ is shown in Figure 8.5.7.

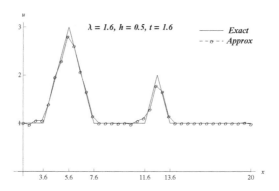

Figure 8.5.7

The above approximation schemes for the transport equation (8.5.1) with a constant coefficient a can be modified to a transport equation with a variable coefficient $a = a(x, t, u)$. For simplicity we consider the case when $a = a(x, t)$:

$$(8.5.2) \qquad u_t(x, t) + a(x, t) u_x(x, t) = 0, \quad 0 < x < l, \, t > 0,$$

subject to the initial and boundary conditions

$$u(x, 0) = f(x), \quad 0 \leq x \leq l; \; u(0, t) = g(t), \; t > 0,$$

We will discuss only the Lax–Wendroff approximation scheme for this equation. The other schemes can be derived similarly. If we assume that $u(x,t)$ is sufficiently smooth with respect to both variables, from Taylor's formula we have

$$(8.5.3) \qquad u(x, t + \Delta t) \approx u(x,t) + u_t(x,t)\Delta t + \frac{(\Delta t)^2}{2} u_{tt}(x,t).$$

If we differentiate Equation (8.5.2) with respect to t we obtain

$$\begin{aligned}
u_{tt} &= -a_t(x,t)u_x(x,t) - a(x,t)u_{tx}(x,t) = -a_t(x,t)u_x(x,t) - a(x,t)u_{xt}(x,t) \\
&= -a_t(x,t)u_x(x,t) + a(x,t)\big(a_x(x,t)u_x(x,t) + a(x,t)u_{xx}(x,t)\big) \\
&= \big(-a_t(x,t) + a(x,t)a_x(x,t)\big)u_x(x,t) + a(x,t)^2 u_{xx}(x,t),
\end{aligned}$$

from which it follows that

$$(8.5.4) \quad u_{tt}(x,t) = \big(-a_t(x,t) + a(x,t)a_x(x,t)\big)u_x(x,t) + a(x,t)^2 u_{xx}(x,t).$$

From (8.5.3) and (8.5.4) we obtain

$$(8.5.5) \quad \begin{aligned}
u(x, t + \Delta t) &\approx u(x,t) - a(x,t)u_x(x,t)\Delta t \\
&+ \frac{1}{2}\Big[\big(-a_t(x,t) + a(x,t)a_x(x,t)\big)u_x(x,t) + a(x,t)^2 u_{xx}(x,t)\Big](\Delta t)^2.
\end{aligned}$$

Now using the central approximations for the partial derivatives $u_x(x,t)$ and $u_{xx}(x,t)$, from (8.5.5) we obtain the Lax–Wendroff approximation scheme for the transport equation:

$$\begin{cases}
u_{i,j+1} = (1 - \lambda^2)u_{i,j} + \dfrac{\lambda}{2}\left(a_{i,j} + \lambda a_{i,j}^2 - \dfrac{1}{2}a_{i,j}a_{i,j}^{(x)}\Delta t + \dfrac{1}{2}a_{i,j}^{(t)}\Delta t\right)u_{i-1,j} \\
\qquad + \dfrac{\lambda}{2}\left(-a_{i,j} + \lambda a_{i,j}^2 + \dfrac{1}{2}a_{i,j}a_{i,j}^{(x)}\Delta t - \dfrac{1}{2}a_{i,j}^{(t)}\Delta t\right)u_{i+1,j},
\end{cases}$$

where $a_{i,j}^{(x)}$ and $a_{i,j}^{(t)}$ are the values of $a_x(x,t)$ and $a_t(x,t)$, respectively, at the node (i,j).

Let us take an example to illustrate this numerical method.

Example 8.5.3. Use the Lax–Wendroff approximation method with $h = 0.04$ and $\Delta t = 0.02$ to solve the problem

$$\begin{cases}
u_t(x,t) + xu_x(x,t) = 0, & 0 < x < 1,\ 0 < t < 1, \\
u(x,0) = f(x) = 1 + e^{-40(x-0.3)^2}, & 0 < x < 1.
\end{cases}$$

Display the graphs of the approximative solutions, along with the analytical solution at the time instances $t = 0.02,\ 0.10$ and $t = 0.20$.

Solution. Using the method of characteristics, explained in Chapter 4, we have that the solution $u(x,t)$ of the given initial value problem is

$$u(x,t) = f\left(xe^{-t}\right) = 1 + 1 + e^{-40(xe^{-t}-0.3)^2}.$$

For the Lax–Wendroff approximation scheme we have the following. From $a(x,t) = x$ it follows that $a_{i,j} = ih$, and $a_{i,j}^{(x)} = 1$, $a_{i,j}^{(t)} = 0$ for every i and every j. Therefore the Lax–Wendroff approximation scheme for our problem is given by

$$\begin{cases} u_{i,j+1} = (1 - \lambda^2)u_{i,j} + \dfrac{\lambda}{2}\left(ih + \lambda i^2 h^2 - \dfrac{1}{2}ih\Delta t\right)u_{i-1,j} \\[2mm] \qquad + \dfrac{\lambda}{2}\left(-ih + \lambda i^2 h^2 + \dfrac{1}{2}ih\Delta t\right)u_{i+1,j}. \end{cases}$$

The results obtained are displayed in Figure 8.5.8.

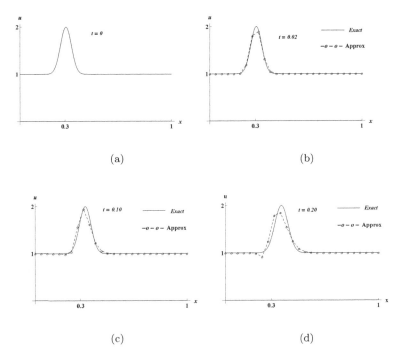

(a)   (b)

(c)   (d)

Figure 8.5.8

$2^0.$ *One Dimensional Wave Equation of the Second Order.* Now we consider the initial boundary value problem for the one dimensional wave equation

$$\begin{cases} u_{tt}(x,t) = c^2 u_{xx}(x,y,t), \quad 0 < x < a, \ t > 0, \\ u(x,0) = f(x), \quad u_t(x,0) = g(x), \ 0 < x < a \\ u(0,t) = u(l,t) = 0, \quad t > 0. \end{cases}$$

In Chapter 5 we showed that the analytical solution of the above initial boundary value problem is given by d'Alembert's formula

$$u(x,t) = \frac{f^{odd}(x-ct) + f^{odd}(x+ct)}{2} + \frac{1}{2c} \int_{x-ct}^{x+ct} g^{odd}(s)\, ds,$$

where f^{odd} and g^{odd} are the odd, $2c$ periodic extensions of f and g, respectively.

Now we will present finite difference methods for solving the wave equation. Using the central differences for both time and space partial derivatives in the wave equation we obtain that

$$(8.5.6) \qquad \frac{u_{i,j+1} - 2u_{i,j} + u_{i,j-1}}{\Delta t^2} = c^2 \frac{u_{i+1,j} - 2u_{i,j} + u_{i-1,j}}{h^2}.$$

or

$$(8.5.7) \quad u_{i,j+1} = -u_{i,j-1} + 2(1-\lambda^2)u_{i,j} + \lambda^2(u_{i+1,j} + u_{i-1,j}), \; i,j = 1,2,\ldots,$$

where λ is the *grid ratio* given by

$$\lambda = c\frac{\Delta t}{h}.$$

The question now is how to start the approximation scheme (8.5.7). We accomplish this in the following way. Using Taylor's formula for $u(x,t)$ with respect to t we have that

$$(8.5.8) \qquad u(x_i, t_1) \approx u(x_i, 0) + u_t(x_i, 0)\Delta t + u_{tt}(x_i, 0)\frac{\Delta t^2}{2}.$$

From the wave equation and the initial condition $u(x,0) = f(x)$ we have

$$u_{tt}(x_i, 0) = c^2 u_{xx}(x_i, 0) = c^2 f''(x_i).$$

Using the other initial condition $u_t(x,0) = g(x)$, Equation (8.5.8) becomes

$$(8.5.9) \qquad u(x_i, t_1) \approx u(x_i, 0) + g(x_i)\Delta t + f''(x_i)\frac{c^2 \Delta t^2}{2}.$$

Now if we use the central finite difference approximation for $f''(x)$, then Equation (8.5.9) takes the form

$$u(x_i, t_1) \approx u(x_i, 0) + g(x_i)\Delta t + \big(f(x_{i+1}) - 2f(x_i) + f(x_{i-1})\big)\frac{c^2 \Delta t^2}{2h^2}$$

and since $f(x_i) = u(x_i, 0) = u_{i,0}$ we have

$$u(x_i, t_1) \approx (1 - \lambda^2)u_{i,0} + \frac{\lambda^2}{2}\left(u_{i-1,0} + u_{i+1,0}\right) + g(x_i)\Delta t.$$

The last approximation allows us to have the required first step

(8.5.10) $$u_{i,1} = g(x_i)\Delta t + (1 - \lambda^2)u_{i,0} + \frac{\lambda^2}{2}\left(u_{i-1,0} + u_{i+1,0}\right).$$

The approximation, defined by (8.5.7) and (8.5.10), is known as the *explicit finite difference approximation* of the wave equation. The order of this approximation is $\mathcal{O}(h^2 + \Delta t^2)$.

It can be shown that this method is stable if $\lambda \leq 1$.

We can avoid the conditional stability in the above explicit scheme if we consider the following *implicit (Crank–Nicolson) scheme.*

$$\left(1 + \lambda^2\right)u_{i,j+1} - \frac{\lambda^2}{2}u_{i+1,j+1} - \frac{\lambda^2}{2}u_{i-1,j+1}$$
$$= 2u_{i,j} - \left(1 + \lambda^2\right)u_{i,j-1} + \frac{\lambda^2}{2}u_{i+1,j-1} + \frac{\lambda^2}{2}u_{i-1,j-1}.$$

The implicit Crank–Nicolson scheme is stable for every grid parameter λ.

The following example illustrates the explicit numerical scheme.

Example 8.5.4. Consider the initial boundary problem

$$\begin{cases} u_{tt}(x,t) = u_{xx}(x,t), & 0 < x < 3, \, 0 < t < 2, \\ u(x,0) = \begin{cases} 2^8 x^8 (x-1)^4, & 0 \leq x \leq 1, \\ 0, & 1 \leq x \leq 3 \end{cases} \\ u_t(x,0) = 0, & 0 < x < 3, \\ u(0,t) = u(3,t) = 0 & t > 0. \end{cases}$$

Using the explicit scheme (8.5.7) with different space and time steps we approximate the solution of the problem at several time instances. Compare the obtained numerical solutions with the analytical solution.

Solution. The following Mathematica program generates the numerical solution of the above initial boundary value problem.

```
In[1] := Wave[n_, m_] :=
Module[{i},
u_App = Table[0, {n}, {m}];
For[i = 1, i <= n, i + +,
u_App[[i, 0]] = f[i]; ];
For[i = 2, i <= n - 1, i + +,
```

$u_{App}[[i, 1]] = (1 - \lambda^2)f[i] + \lambda^2/2(f[(i + 1)] + f[(i - 1)]);];];$
$EFDS[n_-, m_-] :=$
$Module[\{i, j\},$
$For[j = 2, j <= m, j + +,$
$For[i = 2, i <= n - 1, i + +,$
$u_{App}[[i, j]] = (2 - 2\lambda^2)U_{App}[[i, j - 1]] +$
$\lambda^2(u_{App}[[i + 1, j - 1]] + u_{App}[[i - 1, j - 1]]) - uApp[[i, j - 2]];];];];$

First let us take $h = 0.02$ and $\Delta t = 0.02$. For these chosen steps we have $\lambda = 0.2 < 1$ and so we can expect a good convergence of the approximation scheme.

At the given times, the numerical and analytical solutions are displayed in Figure 8.5.9. We see that the numerical and analytical solutions almost coincide.

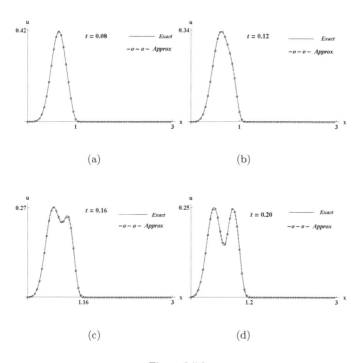

(a)   (b)

(c)   (d)

Figure 8.5.9

In Table 8.5.1 the numerical approximations are given at time instances $t = 0$, $t = 0.08$, $t = 0.16$ and $t = 0.24$ for the first 5 space points $x_1, x_2, \ldots, x_5$.

Table 8.5.1

	x_1	x_2	x_3	x_4	x_5
$t = 0.00$	$3.258/10^8$	$6.718/10^6$	$1.379/10^4$	$1.0737/10^3$	$4.943/10^3$
$t = 0.08$	0.000	$1.400/10^5$	$1.400/10^4$	$8.284/10^4$	$3.357/10^3$
$t = 0.16$	0.000	$4.922/10^4$	$2.128/10^3$	$6.857/10^3$	$1.769/10^2$
$t = 0.24$	0.000	$4.293/10^3$	$1.255/10^2$	$2.871/10^2$	$5.599/10^2$
$t = 0.50$	0.000	$8.908/10^2$	$1.657/10^1$	$2.187/10^1$	$2.382/10^1$

If we take $h = 0.05$ and $\Delta t = 0.08$, then $\lambda = 1.8 > 1$, and so instability can be expected. Indeed, this can be observed from Figure 8.5.10.

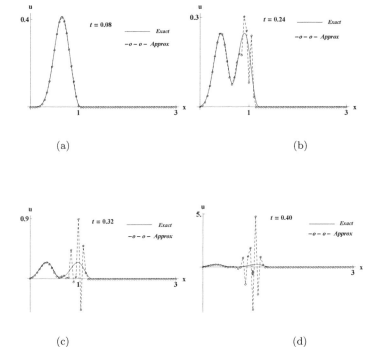

Figure 8.5.10

Example 8.5.5. Consider the initial boundary value problem

$$\begin{cases} u_{tt} = u_{xx}(x,t), & 0 < x < 1,\ 0 < t < 1, \\ u(0,t) = u(1,t) = 0, & 0 < t < 1, \\ u(x,0) = f(x) = x(1-x), & 0 < x < 1, \\ u_t(x,0) = g(x) = \sin(\pi x), & 0 < x < 1. \end{cases}$$

Using the Crank–Nicholson approximation method solve numerically the above problem. Take $h = 0.02$ and $\Delta t = 0.01$ and the following time instances: $t = 0.03$, $t = 0.10$ and $t = 0.80$. Display graphically the results obtained.

Solution. The analytical solution of the above wave is given by

$$u(x,t) = \frac{8}{\pi^3} \sum_{n=0}^{\infty} \frac{\sin(2n+1)\pi x}{(2n+1)^3} \left(\cos(2n+1)\pi t\right) + \frac{1}{\pi} \sin \pi x \sin \pi t.$$

(see the separation of variables method for the wave equation described in Section 5.3 of Chapter 5)

If we take $n = \dfrac{1}{h}$ and $m = \dfrac{1}{\Delta t}$ in the Crank–Nicholson approximation method, then the following system is obtained.

$$\begin{cases} u_{i,0} = f(ih), \ i = 0, 1, \ldots, n, \\ u_{i,1} = (1 - \lambda^2)f(ih) + \Delta t\, g(ih) + \dfrac{\lambda^2}{2}\left(f((i+1)h) + f((i-1)h)\right), \ 1 \leq i \leq n, \\ u_{i,j} = 2(1-\lambda^2)u_{i,j-1} + \lambda^2\left(u_{i+1,j-1} + u_{i-1,j-1}\right) - u_{i,j-2}, \ 1 \leq i \leq n; 2 \leq j \leq m. \end{cases}$$

The results obtained are displayed in Figure 8.5.11.

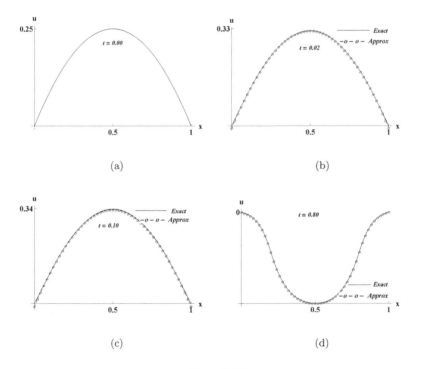

(a)

(b)

(c)

(d)

Figure 8.5.11

Exercises for Section 8.5.

1. Derive the following finite difference approximations of the transport equation (8.5.1).

$$u_{i,j+1} = u_{i,j} - \lambda\left(u_{i+1,j} - u_{i,j}\right),$$

$$u_{i,j+1} = u_{i,j} - \lambda\left(u_{i,j} - u_{i-1,j}\right),$$

$$u_{i,j+1} = u_{i,j} - \frac{\lambda}{2}\left(2u_{i+1,j} - u_{i-1,j}\right).$$

2. Solve the one dimensional wave equation

$$u_t(x,t) + u_x(x,t) = 0, \quad 0 < x < 2, \, 0 < t < 1,$$

with the initial condition

$$u(x,0) = \begin{cases} 1, & \text{if } 0 \leq x \leq \frac{1}{4}, \\ 1 + (\cos 2\pi x)^2, & \text{if } \frac{1}{4} \leq x \leq \frac{3}{4}, \\ 1 & \text{if } \frac{3}{4} \leq x \leq 2, \end{cases}$$

and the boundary conditions

$$u(0,t) = u(3,t) = 0, \quad 0 < t < 2.4.$$

Use the Lax–Friedrichs with $h = 0.04$ and $\lambda = 0.8$. Display the results at $t = 0$ and $t = 0.24$.

3. Using the Lax–Wendroff method with $h = 0.25$ and $\Delta = 0.20$ solve numerically the following problem.

$$u_t(x,t) + u_x(x,t) = 0, \quad 0 < x < \infty, \, t > 0,$$

subject to the initial condition

$$u(x,0) = f(x) = \begin{cases} 1, & 0 \leq x \leq 1, \\ 1 + (x-1)^2(x-3)^2, & 1 \leq x \leq 3, \\ 1, & 3 \leq x\infty, \end{cases}$$

and the boundary condition

$$u(0,t) = 0, \quad t > 0.$$

Compute the numerical values at the first 8 space points and the first 10 time instances. Compare the results obtained with the analytical solution.

4. Using the Leap-Frog scheme with $h = 0.1$ and $\lambda = 0.5$ approximate the solution at $t = 0.3$ of the initial boundary value problem

$$u_t(x, t) + u_x(x, t) = 0, \quad 0 < x < 1, \ t > 0,$$
$$u(x, 0) = f(x) = 1 + \sin \, pix \quad 0 < x < 1,$$
$$u(0, t) = g(t) = 1 + \sin \, \pi t, \quad t > 0.$$

Compare your results with the exact solution.

5. Consider the first order wave equation

$$u_t(x, t) + \frac{1 + x^2}{1 + 2xt + 2x^2 + x^4} u_x(x, t) = 0, \quad 0 < x < 2, \ t > 0,$$

subject to the initial and boundary conditions

$$\begin{cases} u(x, 0) = f(x) = 1 + e^{-100(x-0.5)^2}, & 0 < x < 2; \\ u(0, t) = g(t) = 0, & t > 0. \end{cases}$$

Use the Lax–Wendroff scheme to approximate its solution. Use $h = 0.005$ and $\Delta t = 0.00125$. Compare your results with the exact solution.

6. For the given initial boundary value problem

$$\begin{cases} u_{tt}(x, t) = u_{xx}(x, t), & 0 < x < 3, \ 0 < t < 3, \\ u(x, 0) = \sin(\pi x), & 0 < x < 3, \\ u_t(x, 0) = 0, & 0 < x < 3, \\ u(0, t) = u(3, t) = 0 \end{cases}$$

use the explicit scheme (8.5.7) with $h = 0.05$, $\Delta t = 0.04$ to approximate the solution and compare with the analytical solution.

7. For the initial boundary value problem

$$\begin{cases} u_{tt}(x, t) = u_{xx}(x, t), & 0 < x < 1, \ 0 < t < 1, \\ u(x, 0) = f(x) = \frac{1}{2}x(1 - x), & 0 < x < 1, \\ u_t(x, 0) = 0, \ 0 < x < 1, & u(0, t) = u(1, t) = 0 \end{cases}$$

use the explicit scheme (8.5.7) with $h = 0.1$, $\Delta t = 0.05$ to approximate the solution. Calculate the numerical solution at the first 9 space grid points and the first 5 time instances. Compare the numerical solution with the analytical solution.

8. Apply the explicit scheme to the following the initial boundary value problem.

$$\begin{cases} u_{tt}(x,t) = u_{xx}(x,t), & 0 < x < 1,\ 0 < t < 1, \\ u(x,0) = 1 + \sin(\pi x) + 3\sin(2\pi x), & 0 < x < 1, \\ u(0,t) = u(1,t) = 1, & 0 < t < 1, \\ u_t(x,0) = \sin pix, & 0 < x < 1. \end{cases}$$

Use $h = 0.05$, $\Delta t = 0.0025$ to approximate the solution. Calculate the numerical solution at the first 8 space grid points and the first 6 time instances. Compare the numerical solution with the analytical solution obtained by d'Alembert's method.

9. Derive the matrix forms of the systems of equations for the solution of the one dimensional wave equation

$$u_{tt}(x,t) = a^2(u_{xx}(x,t), \quad 0 < x < l,\ t > 0,$$

subject to the initial conditions

$$\begin{cases} u(x,0) = f(x), & 0 < x < l, \\ u_t(x,0) = g(x), & 0 < x < l, \end{cases}$$

and the boundary conditions

$$\begin{cases} u(0,t) = \alpha(t), & t > 0, \\ u(1,t) = \beta(t), & t > 0 \end{cases}$$

by the explicit approximation method.

10. Let R be the unit square

$$R = \{(x,y) : 0 < x < 1,\ 0 < y < 1\}$$

with boundary ∂R and let $\Delta_{x,y}$ be the Laplace operator

$$\Delta_{x,y}\, u = u_{xx} + u_{yy}.$$

Use the forward approximations for the second partial derivatives with respect to x, y and t to derive a finite explicit difference approximation of the following two dimensional wave boundary value problem

$$\begin{cases} u_{tt}(x, y, t) = a^2 \Delta_{x,y}\, u(x, y, t), & (x, y) \in R,\ 0 < t < T, \\ u(x, y, 0) = f(x, y), & (x, y) \in R, \\ u_t(x, y, 0) = g(x, y), & (x, y) \in R, \\ u(x, y, t) = 0, & (x, y) \in \partial R,\ 0 < t < T. \end{cases}$$

APPENDICES

A. Table of Laplace Transforms

	$f(t)$	$F(s) = \mathcal{L}\{f\}(s)$
1.	t	$\frac{1}{s}$, $s > 0$
2.	e^{at}	$\frac{1}{s-a}$, $s > a$
3.	$\sin{(at)}$	$\frac{a}{s^2+a^2}$, $s > 0$
4.	$\cos{(at)}$	$\frac{s}{s^2+a^2}$, $s > 0$
5.	$\sinh{(at)}$	$\frac{a}{s^2-a^2}$, $s > 0$
6.	$\cosh{(at)}$	$\frac{s}{s^2-a^2}$, $s > 0$
7.	$t^n e^{at}$, $n \in \mathbb{N}$	$\frac{n!}{(s-a)^{n+1}}$, $s > a$
8.	$e^{at}\sin{(bt)}$	$\frac{b}{(s-a)^2+b^2}$, $s > a$
9.	$e^{at}\cos{(bt)}$	$\frac{s}{(s-a)^2+b^2}$, $s > a$
10.	t^n, $n \in \mathbb{N}$	$\frac{n!}{s^{n+1}}$
11.	$H(t-a)$ ‡	$\frac{e^{-as}}{s}$, $s > 0$
12.	$H(t-a)f(t-a)$ ‡	$e^{-as}\mathcal{L}\{f\}(s)$
13.	$t^\nu e^{at}$, $\mathrm{Re}\,\nu > -1$	$\frac{\Gamma(\nu+1)}{(s-a)^{n+1}}$
14.	$t\cos{(at)}$	$\frac{s^2-a^2}{(s^2+a^2)^2}$, $s > 0$
15.	$t\sin{(at)}$	$\frac{2as}{(s^2+a^2)^2}$, $s > 0$
16.	$\frac{\sin{(at)}}{t}$	$\arctan\left(\frac{a}{s}\right)$, $s > 0$
17.	$J_n(at)$	$\frac{1}{\sqrt{s^2+a^2}}\left(\frac{a}{s+\sqrt{s^2+a^2}}\right)^n$, $n > -1$
18.	$I_n(at)$	$\frac{1}{\sqrt{s^2-a^2}}\left(\frac{a}{s+\sqrt{s^2-a^2}}\right)^n$, $n > -1$
19.	$t^n J_n(at)$, $n > -\frac{1}{2}$	$\frac{2^n a^n\,\Gamma\left(n+1/2\right)}{\sqrt{\pi}\left(s^2+a^2\right)^{n+1/2}}$
20.	$\frac{\sin{(2\sqrt{at})}}{\sqrt{\pi t}}$	$\frac{e^{-\frac{a}{s}}}{\sqrt{s}}$
21.	$\frac{\cos{(2\sqrt{at})}}{\sqrt{\pi t}}$	$\frac{e^{-\frac{a}{s}}}{s^{\frac{3}{2}}}$
22.	$\frac{e^{-a^2/(4t)}}{\sqrt{\pi t}}$	$\frac{e^{-a\sqrt{s}}}{\sqrt{s}}$
23.	$\delta(t)$ ‡	1
24.	$\delta(t-a)$	e^{-as}, $a > 0$

A. Table of Laplace Transforms (Continued)

	$f(t)$	$F(s) = \mathcal{L}\{f\}(s)$
25.	$t^{-1/2}e^{-\frac{a}{t}}$	$\sqrt{\frac{\pi}{s}}\,e^{-2\sqrt{as}}$
26.	$t^{-3/2}e^{-\frac{a}{t}}$	$\sqrt{\frac{\pi}{a}}\,e^{-2\sqrt{as}}$
27.	$\frac{1}{\sqrt{\pi t}}\left(1 + 2at\right)e^{at}$	$\frac{s}{(s-a)\sqrt{s-a}}$
28.	$\frac{1}{2\sqrt{\pi t^3}}\left(e^{bt} - e^{at}\right)$	$\sqrt{s-a} - \sqrt{s-b}$
29.	$e^{-a^2 t^2}$	$\frac{\sqrt{\pi}}{2a}\,e^{s^2/4a^2}\,erfc\!\left(\frac{s}{2a}\right)$
30.	$erf(at)$	$\frac{1}{s}\,e^{s^2/4a^2}\,erfc\left(\frac{s}{2a}\right)$
31.	$\frac{1}{t}\sin\sqrt{at}$	$\pi\,erf\left(\sqrt{\frac{a}{4s}}\right)$
32.	$erf(\sqrt{t})$	$\frac{1}{s\sqrt{s+1}}$
33.	$e^{at}f(t)$	$\mathcal{L}\{f\}(s-a)$
34.	$f(t+T) = f(t),\ T > 0$	$\frac{1}{1+e^{-Ts}}\int_0^T e^{-st}f(t)\,dt$
35.	$(f * g)(t)$	$\mathcal{L}\{f\}(s)\,\mathcal{L}\{g\}(s)$
36.	$f(at)$	$\frac{1}{a}\mathcal{L}\{f\}\!\left(\frac{s}{a}\right)$
37.	$\int_0^t f(\tau)\,d\tau$	$\frac{1}{s}\mathcal{L}\{f\}(s)$
38.	$f^{(n)}(t),\ n = 0,1,2,\ldots$	$s^n\mathcal{L}\{f\}(s) - s^{n-1}f(0) - s^{n-2}f'(0) -$ $\ldots - f^{(n-1)}(0)$
39.	$f'(t)$	$s\mathcal{L}\{f\}(s) - f(0)$
40.	$f''(t)$	$s^2\mathcal{L}\{f\}(s) - sf(0) - f'(0)$
41.	$t^n f(t)$	$(-1)^n\left(\mathcal{L}\{f\}\right)^{(n)}(s)$
42.	$\frac{f(t)}{t}$	$\int_s^\infty \mathcal{L}\{f\}(x)\,dx$
43.	$\frac{f(t)}{t^2}$	$\int_s^\infty \int_y^\infty \mathcal{L}\{f\}(x)\,dx\,dy$
44.	$\delta^{(n)}(t-a),\ n = 0,1,2,\ldots$	$s^n e^{-as}$

B. Table of Fourier Transforms

	$f(x)$	$\mathcal{F}\{f\}(\omega)$
1.	$sink\,(ax)$ ✠	$\frac{1}{\|a\|}\,rect\left(\frac{\omega}{2\pi a}\right)$ †
2.	e^{-ax^2}	$\sqrt{\frac{\pi}{a}}\,e^{-\frac{\omega^2}{4a}}$
3.	$e^{-a\|x\|}$	$\frac{2a}{\omega^2+a^2}$
4.	$e^{-ax}H(x)$ ‡	$\frac{1}{a+i\omega}$
5.	$sech(ax)$	$\frac{\pi}{a}\,sech\left(\frac{\pi}{2a}\omega\right)$
6.	1	$2\pi\delta_0(\omega)$
7.	$\delta(x)$ ‡	1
8.	$\cos(ax)$	$\pi\big(\delta(\omega-a)+\delta(\omega+a)\big)$
9.	$\sin(ax)$	$-\pi i\big(\delta(\omega-a)-\delta(\omega+a)\big)$
10.	$\frac{\sin(ax)}{1+x^2}$	$\frac{\pi}{2}\left(e^{-\|\omega-a\|}-e^{-\|\omega+a\|}\right)$
11.	$\frac{\cos(ax)}{1+x^2}$	$\frac{\pi}{2}\left(e^{-\|\omega-a\|}+e^{-\|\omega+a\|}\right)$
12.	$rect\,(ax)$ †	$\frac{1}{\|a\|}\,sink\left(\frac{\omega}{2\pi a}\right)$ ✠
13.	$\cos\left(ax^2\right)$	$\sqrt{\frac{\pi}{a}}\cos\left(\frac{\omega^2-a\pi}{4a}\right)$
14.	$\sin\left(ax^2\right)$	$-\sqrt{\frac{\pi}{a}}\sin\left(\frac{\omega^2-a\pi}{4a}\right)$
15.	$\|x\|^n e^{-a\|x\|}$	$\Gamma(n+1)\left(\frac{1}{(a-i\omega)^{n+1}}+\frac{1}{(a+i\omega)^{n+1}}\right)$
16.	$sgn\,(x)$ †	$\frac{2}{i\omega}$
17.	$H(x)$ ‡	$\pi\left(\frac{1}{i\pi\omega}+\delta_0(\omega)\right)$
18.	$\frac{1}{x^n}$	$-i\pi\frac{(-i\omega)^{n-1}}{(n-1)}\,sgn\,(\omega)$
19.	$J_0(x)$	$\frac{2}{\sqrt{1-\omega^2}}\,rect\left(\frac{\omega}{2}\right)$ †
20.	e^{iax}	$2\pi\delta(\omega-a)$
21.	x^n	$2\pi i^n \delta^{(n)}(\omega$ ‡$)$
22.	$\frac{1}{1+a^2x^2}$	$\frac{1}{\|a\|}e^{-\frac{\omega}{a}}$
23.	$\frac{x}{a^2+x^2}$	$-\pi i e^{-a\|\omega\|}\,sgn\,(\omega)$ †

B. Table of Fourier Transforms (Continued)

	$f(x)$	$\mathcal{F}\{f\}(\omega)$
24.	$e^{iax}f(x)$	$\mathcal{F}\{f\}(\omega - a)$
25.	$\cos(ax)f(x)$	$\sqrt{\frac{\pi}{2}}\left(\mathcal{F}\{f\}(\omega-a)+\mathcal{F}\{f\}(\omega+a)\right)$
26.	$\sin(ax)f(x)$	$-i\sqrt{\frac{\pi}{2}}\left(\mathcal{F}\{f\}(\omega-a)+\mathcal{F}\{f\}(\omega+a)\right)$
27.	$f\left(\frac{x}{a}\right)$	$|a|\mathcal{F}\{f\}(a\omega)$
28.	$f(x-a)$	$e^{-ia\omega}\mathcal{F}\{f\}(\omega)$
29.	$a\,f(x)+b\,g(x)$	$\mathcal{F}\{f\}(\omega)+\mathcal{F}\{g\}(\omega)$
30.	$f'(x)$	$i\omega\mathcal{F}\{f\}(\omega)$
31.	$f''(x)$	$-\omega^2\mathcal{F}\{f\}(\omega)$
32.	$f^{(n)}(x),\ n=0,1,\ldots$	$(i\omega)^n\mathcal{F}\{f\}(\omega)$
33.	$xf(x)$	$i\left(\mathcal{F}\{f\}\right)'(\omega)$
34.	$x^n f(x),\ n=0,1,\ldots$	$i^n\left(\mathcal{F}\{f\}\right)^{(n)}(\omega)$
35.	$(f*g)(x)$	$\left(\mathcal{F}\{f\}(\omega)\right)\left(\mathcal{F}\{g\}(\omega)\right)$
36.	$f(x)g(x)$	$\frac{1}{2\pi}\left(\mathcal{F}\{f\}*\mathcal{F}\{g\}\right)(\omega)$

Notations for the Laplace/Fourier Transform Tables

† The box function $rect\,(x)$ and the signum function $sgn\,(x)$:

$$rect\,(x) = \begin{cases} 0, & |x| > \frac{1}{2} \\ \frac{1}{2}, & x = \pm\frac{1}{2} \\ 1, & |x| < \frac{1}{2} \end{cases} \quad ; \quad sgn\,(x) = \begin{cases} -1, & x < 0 \\ 0, & x = 0 \\ 1, & x > 0 \end{cases}$$

‡ The Heaviside function $H(x)$ and the Dirac delta "function" $\delta(x)$:

$$H(x) = \begin{cases} 0, & -\infty < x < 0 \\ 1, & 0 \leq x < \infty \end{cases} \quad ; \quad \delta(x) = \begin{cases} \infty, & x = 0 \\ 0, & x \neq 0. \end{cases}$$

✠ The sinc function $sinc\,(x)$:

$$sinc\,(x) = \frac{\sin(\pi x)}{\pi x}.$$

C. Series and Uniform Convergence Facts

Suppose $S \subseteq \mathbb{R}$ and $f_n : S \to \mathbb{R}$ are real-valued functions for every natural number n. We say that the sequence (f_n) is *pointwise convergent* on S with limit $f : S \to \mathbb{R}$ if for every $\epsilon > 0$ and every $x \in S$ there exists a natural number $n_0(\epsilon, x)$ such that all $n > n_0(\epsilon)$,

$$|f_n(x) - f(x)| < \epsilon.$$

Suppose $S \subseteq \mathbb{R}$ and $f_n : S \to \mathbb{R}$ are real-valued functions for every natural number n. We say that the sequence (f_n) is *uniformly convergent* on S with limit $f : S \to \mathbb{R}$ if for every $\epsilon > 0$, there exists a natural number $n_0(\epsilon)$ such that for all $x \in S$ and all $n > n_0(\epsilon)$,

$$|f_n(x) - f(x)| < \epsilon.$$

Let us compare uniform convergence to the concept of pointwise convergence. In the case of uniform convergence, n_0 can only depend on ϵ, while in the case of pointwise convergence n_0 may depend on ϵ and x. It is clear that uniform convergence implies pointwise convergence.

The following are well known results about uniform convergence.

Theorem C.1. Cauchy Criterion for Uniform Convergence. *A sequence of functions* (f_n) *converges uniformly on a set* $S \subseteq \mathbb{R}$ *if and only if for every* $\epsilon > 0$ *there exists a natural number* $n_0(\epsilon)$ *such that for all* $x \in S$ *and all natural numbers* $n, m > n_0(\epsilon)$,

$$|f_n(x) - f_m(x)| < \epsilon.$$

Theorem C.2. *Suppose that* (f_n) *is a sequence of continuous functions on a set* $S \subseteq \mathbb{R}$ *and* f_n *converges uniformly on* S *to a function* f. *Then* f *is continuous on* S.

Theorem C.3. *Let* (f_n) *be an increasing sequence of continuous functions defined on a closed interval* $[a, b]$. *If the sequence is pointwise convergent on* $[a,b]$, *then it converges uniformly on* $[a,b]$.

Theorem C.4. *Let* (f_n) *be a sequence of integrable functions on* $[a, b]$. *Assume that* f_n *converges uniformly on* $[a, b]$ *to a function* f. *Then* f *is integrable and moreover,*

$$\int_a^b f_n(x)\, dx = \int_a^b f(x)\, dx.$$

Theorem C.5. *Let* (f_n) *be a sequence of differentiable functions on* $[a, b]$ *such that the numerical sequence* $(f_n(x_0))$ *converges at some point* $x_0 \in [a, b]$. *If the sequence of derivatives* (f'_n) *converges uniformly on* $[a, b]$, *then* (f_n) *converges uniformly on* $[a, b]$ *to a differentiable function* f *and moreover,*

$$\lim_{n \to \infty} f'_n(x) = f'(x)$$

for every $x \in [a, b]$.

Convergence properties of infinite series

(C.1)
$$\sum_{k=1}^{\infty} f_k(x)$$

of functions are identified with those of the corresponding sequence

$$S_n(x) = \sum_{k=1}^{n} f_k(x)$$

of partial sums. So, the infinite series (C.1) is pointwise, absolutely and uniformly convergent on $S \subseteq \mathbb{R}$ if the sequence $(S_n(x))$ of partial sums is pointwise, absolutely and uniformly convergent, respectively, on S.

The following criterion for uniform convergence of infinite series is useful.

Theorem C.6. Weierstrass M-Test. *Let* (f_n) *be a sequence of functions defined on a set* S *such that* $|f_n(x)| \leq m_n$ *for every* $x \in S$ *and every* $n \in \mathbb{N}$. *If*

$$\sum_{n=1}^{\infty} m_n < \infty$$

is convergent, then

$$\sum_{n=1}^{\infty} f_n$$

is uniformly convergent on S.

An important class of infinite series of functions is the power series

$$\sum_{n=0}^{\infty} a_n x^n$$

in which (a_n) is a sequence of real numbers and x a real variable. The basic convergence properties of power series are described by the following theorem.

Theorem C.7. *For the power series*

$$\sum_{n=0}^{\infty} a_n x^n$$

let R *be the largest limit point of the sequence* $\left(\sqrt[n]{|a_n|}\right)$. *Then*

a. *If* $R = 0$, *then the power series is absolutely convergent on the whole real line and it is uniformly convergent on any bounded set of the real line.*

b. *If* $R = \infty$, *then the power series is convergent only for* $x = 0$.

c. *If* $0 < R < \infty$, *then the power series is absolutely convergent for every* x *in the open interval* $\left(-\dfrac{1}{R}, \dfrac{1}{R}\right)$ *and uniformly convergent on the set* $\left\{x : |x| \le r < \dfrac{1}{R}\right\}$. *For* $|x| > \dfrac{1}{R}$ *the power series is divergent.*

A point a is called a *limit point* for a sequence (a_n) if in every open interval containing the point a there are terms of the sequence.

The following are several examples of power series:

$$\frac{1}{1-x} = \sum_{n=0}^{\infty} x^n, \ |x| < 1; \quad \arctan x = \sum_{n=0}^{\infty} (-1)^n \frac{x^{2n+}}{2n+1}, \ -1 < x \le 1.$$

$$e^x = \sum_{n=0}^{\infty} \frac{x^n}{n!}, \ x \in \mathbb{R}; \quad \ln(1+x) = \sum_{n=0}^{\infty} (-1)^n \frac{x^{n+1}}{n+1}, \ -1 < x \le 1.$$

$$\sin x = \sum_{n=0}^{\infty} (-1)^n \frac{x^{2n+1}}{(2n+1)!} \ x \in \mathbb{R}; \quad \cos x = \sum_{n=0}^{\infty} (-1)^n \frac{x^{2n}}{(2n)!}, \ x \in \mathbb{R}.$$

D. Basic Facts of Ordinary Differential Equations

1. First Order Differential Equations

We will review some basic facts of ordinary differential equations of the first and second order.

A differential equation of the first order is an equation of the form

(D.1) $$F(x, y, y') = 0,$$

where F is a given function of three variables.

A *solution* of (D.1) is a differentiable function $f(x)$ such that

$$F(x, f(x), f'(x)) = 0,$$

for all x in the domain (interval) where f is defined.

The *general solution* of (D.1) is the set of all solutions of (D.1).

An equation of the form

(D.2) $$g(y)\, dy = f(x)\, dx$$

is said to be *separable*. A separable equation can be integrated. Simply integrate both sides of equation (D.2).

An equation of the form

(D.3) $$M(x, y)\, dx + N(x, y)\, dy = 0,$$

where $M = M(x, y)$ and $N = N(x, y)$ are $C^1(R)$ functions (functions which have continuous partial derivatives in R) in a rectangle $R = [a, b] \times [c, d] \subseteq \mathbb{R}^2$ is said to be *exact* if

$$M_y(x, y) = N_x(x, y)$$

at each point $(x, y) \in R$.

If an equation is exact, then there exists a function $u = u(x, y) \in C^1(R)$ such that

$$u_x = M, \quad u_y = N \text{ in } R$$

and the general solution of (D.3) is given by $u(x, y) = C$, where C is any constant.

Example D.1. Find the general solution of the equation

$$\left(1 + y \cos xy\right) dx + \left(x \cos xy + 2y\right) dy = 0.$$

Solution. In this example

$$M = 1 + y \cos(xy), \quad N = x \cos(xy) + 2y.$$

Since $M_y = \cos xy - xy \sin xy$ and $N_x = \cos xy - xy \sin xy$ we have $M_y = N_x$, and so the equation is exact. Therefore, there exists a function $u = u(x, y)$ such that

$$u_x = M = 1 + y \cos xy \quad \text{and} \quad u_y = x \cos xy + 2y.$$

Integrating the first equation with respect to x we have

$$u(x, y) = \int (1 + y \cos xy) \, dx + f(y) = x + \sin xy + f(y).$$

Now, from $u_y = x \cos xy + f'(y) = x \cos xy + 2y$ it follows that $f'(y) = 2y$ and so $f(y) = y^2$. Thus,

$$u(x, y) = x + \sin xy + y^2$$

and so the general solution of the equation is given by

$$x + \sin xy + y^2 = C.$$

Sometimes, even though equation (D.3) is not exact, there is a function $\mu = \mu(x, y)$, not identically zero, such that the equation

$$(\mu M) \, dx + (\mu N) \, dy = 0$$

is exact. In this case the function μ is called an *integrating factor* for (D.3).

Example D.2. *(Linear Equation of the First Order).* Consider the first order homogeneous linear equation

(D.4) $$y' + p(x)y + q(x)y = 0,$$

where $p = p(x)$ and $q = q(x)$ are continuous functions in some interval. It can be easily checked that this equation has an integrating factor $e^{\int p(x) \, dx}$. Using this integrating factor, the general solution of (D.4) is obtained to be

$$y = e^{-\int p(x) \, dx} \left[\int q(x) e^{\int p(x) \, dx} \, dx + C \right],$$

where C is any constant.

A differential equation of the form

(D.5) $$y' = f(x, y),$$

subject to an initial condition

(D.6) $$y(x_0) = y_0$$

is usually called an *initial value problem*.

The following theorem is of fundamental importance in the theory of differential equations.

Theorem D.1. Existence and Uniqueness. *Suppose that $f(x,y)$ and its partial derivative $f_y(x,y)$ are continuous functions on the rectangle $R = \{(x,y) : a < x < b, c < y < d\}$ containing the point (x_0, y_0). Then in some interval $(x_0 - h, x_0 + h) \subset [a, b]$ there exists a unique solution $y = \phi(x)$ of the initial value problem (D.5), (D.6).*

2. Second Order Linear Differential Equations

A differential equation of the form

$$(D.7) \qquad\qquad a(x)y'' + b(x)y' + c(x)y = f(x)$$

is called a *second order linear equation*. The coefficients $a(x)$, $b(x)$ and $c(x)$ are assumed to be continuous on an interval I.

If $a(x) \neq 0$ for every $x \in I$, then we can divide by $a(x)$ and equation (D.7) takes the *normal form*

$$(D.8) \qquad\qquad y'' + p(x)y' + q(x)y = r(x).$$

If $a(x_0) \neq 0$ at some point $x_0 \in I$, then x_0 is called an *ordinary point* for (D.7). If $a(x_0) = 0$, then x_0 is called *singular point* for (D.7).

If $f(x) \equiv 0$, then the equation

$$(D.9) \qquad\qquad a(x)y'' + b(x)y' + c(x)y = 0$$

is called *homogeneous*.

Theorem D.2. *If the function $y_p(x)$ is any particular solution of the non-homogeneous equation (D.7) and $u_h(x)$ is the general solution of the homogeneous equation (D.8), then*

$$y(x) = y_h(x) + y_p(x)$$

is the general solution of the nonhomogeneous equation (D.7).

For homogeneous equations the following *principle of superposition* holds.

Theorem D.3. *If $y_1(x)$ and $y_2(x)$ are both solutions to the homogeneous, second order equation (D.9), then any linear combination*

$$c_1 y_1(x) + c_2 y_2(x),$$

where c_1 and c_2 are constants, is also a solution to (D.9).

For second order linear equations we have the following existence and uniqueness theorem.

Theorem D.4. Existence and Uniqueness Theorem. *Let* $p(x), q(x)$
and $r(x)$ *be continuous functions on the interval* $I = (a, b)$ *and let* $x_0 \in$
(a, b). *Then the initial value problem*

$$y'' + p(x)y' + q(x)y = r(x), \quad y(x_0) = y_0, \ y'(x_0) = y_1$$

has a unique solution on I, *for any numbers* y_0 *and* y_1.

Now we introduce linearly independent and linearly dependent functions.

Two functions $y_1(x)$ and $y_2(x)$ (both not identical to zero) are said to be
linearly independent on an interval I if the condition

$$c_1 y_1(x) + c_2 y_2(x) = 0 \text{ for every } x \in I$$

for some constants c_1 and c_2 is satisfied only if $c_1 = c_2 = 0$. In other words,
two functions are independent if neither of them can be expressed as a scalar
multiple of the other. If two functions are not linearly independent, then
we say they are *linearly dependent*. The importance of linearly independent
functions comes from the following theorem.

Theorem D.5. *Let the coefficient functions* $p(x)$ *and* $q(x)$ *in the homogeneous equation* (D.9) *be continuous on the open interval* I. *If* $y_1 = y_1(x)$
and $y_2 = y_2(x)$ *are two linearly independent solutions on the interval* I *of
the homogeneous equation* (D.9), *then any solution* $y = y(x)$ *of Equation*
(D.9) *is of the form*

$$y = c_1 y_1 + c_2 y_2,$$

for some constants c_1 *and* c_2.

In this case we say that the system $\{y_1, y_2\}$ is a *fundamental set* or *basis*
of the solutions of the homogeneous equation (D.9).

The question of linearly independence of two functions can be examined
by their *Wronskian*:

The Wronskian $W(y_1, y_2; x)$ of two differentiable functions $y_1(x)$ and
$y_2(x)$ is defined by

$$W(y_1, y_2; x) = \begin{vmatrix} y_1(x) & y_2(x) \\ y_1'(x) & y_2'(x) \end{vmatrix} = y_1(x)y_2'(x) - y_2(x)y_1'(x).$$

Theorem D.6. *If* y_1 *and* y_2 *are linearly dependent differentiable functions
on an interval* I, *then their Wronskian* $W(y_1, y_2; x)$ *is identically zero on* I.
Equivalently, if y_1 *and* y_2 *are differentiable functions on an interval* I *and
their the Wronskian* $W(y_1, y_2; x_0) \neq 0$ *for some* $x_0 \in I$, *then* y_1 *and* y_2 *are
linearly independent on this interval.*

Proof. Let y_1 and y_2 be linearly dependent differentiable functions on an
interval I. Then there are constants c_1 and c_2, not both zero, such that

$$c_1 y_1(x) + c_2 y_2(x) = 0 \text{ for every } x \in I.$$

Differentiating the last equation with respect to x we find

$$c_1 y_1'(x) + c_2 y_2'(x) = 0 \text{ for every } x \in I.$$

Since the system of the last two equations has a nontrivial solution (c_1, c_2), its determinant $W(y_1, y_2; x)$ must be zero for every $x \in I$.

The converse of Theorem D.6. is not true. In other words, two differentiable functions y_1 and y_2 can be linearly independent on an interval even if their Wronskian may be zero at some point of that interval.

Example D.3. The functions $y_1(x) = x^3$ and $y_2(x) = x^2|x|$ are linearly independent on the interval $(-1, 1)$, but their Wronskian $W(y_1, y_2; x)$ is identically zero on $(-1, 1)$.

Solution. Indeed, if $c_1 x^3 + c_2 x^2|x| = 0$ for some constants c_1 and c_2 and every $x \in (-1, 1)$, then taking $x = -1$ and $x = 1$ in this equation we obtain that $-c_1 + c_2 = 0$ and $c_1 + c_2 = 0$. From the last two equations it follows that $c_1 = c_2 = 0$. Therefore y_1 and y_2 are linearly independent on $(-1, 1)$. Also, it is easily checked that $y_1'(x) = 3x^2$ and $y_2'(x) = 3x|x|$ for every $x \in (-1, 1)$. Therefore y_1 and y_2 are differentiable on $(-1, 1)$ and

$$W(y_1, y_2; x) = y_1(x)y_2'(x) - y_2(x)y_1'(x) = 3x^4|x| - 3x^4|x| = 0$$

for every $x \in (-1, 1)$.

If one particular solution $y_1 = y_1(x)$ of the homogeneous linear differential equation (D.9) is known, then introducing a new function $u = u(x)$ by $y = y_1(x)u(x)$, Equation (D.9) takes the form

$$y_1 u'' + \left(2y_1' + p(x)y_1\right)v' = 0.$$

Now, if in the last equation we introduce a new function v by $v = u'$ we obtain the linear first order equation

$$y_1 v' + \left(2y_1' + p(x)y_1\right)v = 0.$$

The last equation can be integrated by the separation of variables and one solution of this equation is

$$v(x) = \frac{1}{y_1^2} e^{-\int p(x)\, dx}.$$

Thus,

$$u(x) = \int \frac{e^{-\int p(x)\, dx}}{y_1^2(x)}\, dx$$

and so a second linearly independent solution $y_2(x)$ of (D.9) is given by

$$(D.10) \qquad\qquad y_2(x) = y_1(x) \int \frac{e^{-\int p(x)\,dx}}{y_1^2(x)}\,dx.$$

From Theorem D.2 we have that a general solution of a nonhomogeneous linear equation of second order is the sum of a general solution $y_h(x)$ of the corresponding homogeneous equation and a particular solution $y_p(x)$ of the nonhomogeneous equation. To find a particular solution $y_p(x)$ of the nonhomogeneous equation from the general solution of the homogeneous equation we use the following theorem.

Theorem D.7. Method of Variation of Parameters. *Let the functions* $p(x)$, $q(x)$ *and* $r(x)$ *be continuous on an interval* I. *If* $\{y_1(x), y_2(x)\}$ *is a fundamental system of the homogeneous equation*

$$y'' + p(x)y' + q(x)y = 0,$$

i.e.,

$$y_h(x) = c_1 y_1(x) + c_2 y_2(x)$$

is a general solution of the homogeneous equation, then a particular solution $y_p(x)$ *of the nonhomogeneous equation*

$$y'' + p(x)y' + q(x)y = r(x)$$

is given by

$$y_p(x) = c_1(x)y_1(x) + c_2(x)y_2(x),$$

where the two differentiable functions $c_1(x)$ *and* $c_2(x)$ *are determined by solving the system*

$$\begin{cases} c_1'(x)y_1(x) + c_2'(x)y_2(x) = 0 \\ c_1'(x)y_1'(x) + c_2'(x)y_2'(x) = r(x) \end{cases}$$

Second order linear homogeneous equation

$$(D.11) \qquad\qquad ay''(x) + by'(x) + cy(x) = 0$$

with real constant coefficients a, b and c can be solved by assuming a solution of the form $y = e^{rx}$ for some values of r. We find r by substituting this solution and its first and second derivative into the differential equation (D.10) and obtain the quadratic *characteristic equation*

$$(D.12) \qquad\qquad ar^2 + br + c = 0.$$

For the roots of the characteristic equation

$$r = \frac{-b \pm \sqrt{b^2 - 4ac}}{2a}$$

we have three possibilities: two real distinct roots when $b^2 - 4ac > 0$, one real repeated root when $b^2 - 4ac = 0$, and two complex conjugate roots when $b^2 - 4ac < 0$. We consider each case separately.

Two Real Distinct Roots. Let (D.12) have two real and distinct roots r_1 and r_2. Using the Wronskian it can be easily checked that the functions $y_1(x) = e^{r_1 x}$ and $y_2(x) = e^{r_2 x}$ are linearly independent on the whole real line so the general solution of (D.11) is

$$y(x) = c_1 e^{r_1 x} + c_2 e^{r_2 x}.$$

One Real Repeated Root. Suppose that the characteristic equation (D.12) has a real repeated root $r = r_1 = r_2$. In this case we have only one solution $y_1(x) = e^{r x}$ of the equation. We use this solution and (D.10) in order to obtain a second linearly independent solution $y_2(x) = x e^{r x}$. Therefore, a general solution of (D.11) is

$$y(x) = c_1 e^{r x} + c_2 x e^{r x}.$$

Complex Conjugate Roots. Suppose that the characteristic equation (D.12) has the complex conjugate roots $r_{1,2} = \alpha \pm \beta i$. In this case we can verify that $\{e^{\alpha x} \cos(\beta x), e^{\alpha x} \sin(\beta x)\}$ is a fundamental system of the differential equation and so its general solution is given by

$$y(x) = e^{\alpha x} \big(c_1 \cos(\beta x) + c_2 \sin(\beta x)\big).$$

Example D.4. Find the general solution of the nonhomogeneous equation

$$y'' - 2y' + y = e^x \ln x, \ x > 0.$$

Solution. The corresponding homogeneous equation has the characteristic equation

$$r^2 - 2r + 1 = 0.$$

This equation has repeated root $r = 1$ and so $y_1(x) = e^x$ and $y_2(x) = x e^x$ are two linearly independent solutions of the homogeneous equation. Therefore,

$$y_h(x) = c_1 e^x + c_2 x e^x.$$

is the general solution of the homogeneous equation. To find a particular solution $y_p(x) = c_1(x)e^x + c_2(x)xe^x$ of the given nonhomogeneous equation we apply the method of variation of parameters:

$$\begin{cases} c_1'(x)e^x + c_2'(x)xe^x = 0 \\ c_1'(x)e^x + c_2'(x)(e^x + xe^x) = e^x \ln x. \end{cases}$$

Solving the last system for $c_1'(x)$ and $c_2'(x)$ we obtain

$$c_1'(x) = -x \ln x, \quad c_2'(x) = \ln x.$$

Using the integration by parts formula, from the last two equations it follows that

$$c_1(x) = \frac{1}{4}x^2 - \frac{1}{2}x^2 \ln x \text{ and } c_2(x) = x \ln x - x.$$

Therefore,

$$y_p(x) = \frac{1}{2}x^2 e^x \ln x - \frac{3}{4}x^2 e^x$$

and the general solution is

$$y(x) = y_h(x) + y_p(x) = c_1 e^x + c_2 x e^x + \frac{1}{2}x^2 e^x \ln x - \frac{3}{4}x^2 e^x.$$

The Cauchy–Euler equation is a linear, second order homogeneous equation of the form

$$\text{(D.13)} \qquad\qquad ax^2 y'' + bxy' + cy = 0,$$

where a, b and c are real constants. To solve this equation we assume that it has a solution of the form $y = x^r$. After substituting y and its first and second derivative into the equation we obtain

$$\text{(D.14)} \qquad\qquad ar(r - 1) + br + c = 0.$$

Equation (D.14) is called the *indicial equation* of (D.13).

Two Real Distinct Roots. If (D.14) has two real and distinct roots r_1 and r_2, then

$$y(x) = c_1 x^{r_1} + c_2 x^{r_2}$$

is a general solution of (D. 13).

One Real Repeated Root. If (D.14) has one real and repeated root $r = r_1 = r_2$, then

$$y(x) = c_1 x^r + c_2 x^r \ln x$$

is a general solution of (D.13).

Two Complex Conjugate Roots. If (D.14) has two, complex conjugate roots $r_1 = \alpha + \beta i$ and $r_2 = \alpha - \beta i$, then

$$y(x) = c_1 x^\alpha \cos(\beta \ln x) + c_2 x^\alpha \sin(\beta \ln x)$$

is a general solution of (D.13).

3. Series Solutions of Linear Differential Equations

Recall that a *power series* about x_0 is an infinite series of the form

$$\sum_{n=0}^{\infty} c_n (x - x_0)^n.$$

For each value of x either the series converges or it does not. The set of all x for which the series converges is an interval (open or not, bounded or not), centered at the point x_0. The largest R, $0 \le R \le \infty$ for which the series converges for every $x \in (x_0 - R, x_0 + R)$ is called the *radius of convergence* of the power series and the interval $(x_0 - R, x_0 + R)$ is called the *interval of convergence.* Within the interval of convergence of a power series, the function that it represents can be differentiated and integrated by differentiating and integrating the power series term by term.

A function f is said to be *analytic* in some open interval centered at x_0 if for each x in that interval the function can be represented by a power series

$$f(x) = \sum_{n=0}^{\infty} c_n (x - x_0)^n.$$

Our interest in series solutions is mainly in second order linear equations. The basic result about series solutions of linear differential equations of the second order is the following theorem.

Theorem D.8. *Consider the differential equation (D.7)*

$$a(x)y'' + b(x)y' + c(x)y = 0,$$

where $a(x)$, $b(x)$ and $c(x)$ are analytic functions in some open interval containing the point x_0. If x_0 is an ordinary point for (D.7) $(a(x_0) \ne 0)$, then a general solution of Equation (D.7) can be expressed in form of a power series

$$y(x) = \sum_{n=0}^{\infty} c_n (x - x_0)^n.$$

The radius of convergence of the power series is at least d, where d is the distance from x_0 to the nearest singular point of $a(x)$.

The proof of this theorem can be found in several texts, such as the book by Coddington [2].

Example D.5. Find the general solution in the form of a power series about $x_0 = 0$ of the equation

(D.15) $$y'' - 2xy' - y = 0.$$

Solution. We seek the solution of (D.15) of the form

$$y(x) = \sum_{n=0}^{\infty} c_n x^n.$$

Since $a(x) \equiv 1$ does not have any singularities, the interval of convergence of the above power series is $(-\infty, \infty)$. Differentiating twice the above power series we have that

$$y'(x) = \sum_{n=1}^{\infty} n c_n x^{n-1}, \ y''(x) = \sum_{n=2}^{\infty} n(n-1) c_n x^{n-2}.$$

We substitute the power series for $y(x)$, $y'(x)$ and $y''(x)$ into Equation (D.15) and we obtain

$$\sum_{n=2}^{\infty} n(n-1) c_n x^{n-2} - 2x \sum_{n=1}^{\infty} n c_n x^{n-1} - \sum_{n=0}^{\infty} c_n x^n = 0.$$

If we insert the term x inside the second power series, after re-indexing the first power series we obtain

$$\sum_{n=0}^{\infty} (n+2)(n+1) c_{n+2} x^n - 2 \sum_{n=1}^{\infty} n c_n x^n - \sum_{n=0}^{\infty} c_n x^n = 0.$$

If we break the first and the last power series into two parts, the first terms and the rest of these power series, we obtain

$$2 \cdot 1 c_2 + \sum_{n=1}^{\infty} (n+2)(n+1) c_{n+2} x^n - 2 \sum_{n=1}^{\infty} n c_n x^n - c_0 - \sum_{n=0}^{\infty} c_n x^n = 0.$$

Combining the three power series above into one power series, it follows that

$$2c_2 - c_0 + \sum n = 1^{\infty} \Big[(n+2)(n+1) c_{n+2} - (2n+1) c_n \Big] x^n = 0$$

Since a power series is identically zero in its interval of convergence if all the coefficients of the power series are zero, we obtain

$$2c_2 - c_0 = 0, \ (n+2)(n+1) c_{n+2} - (2n+1) c_n = 0, \ n \geq 1.$$

Therefore,

$$c_2 = \frac{1}{2}c_0 \quad \text{and} \quad c_{n+2} = \frac{2n+1}{(n+2)(n+1)}, \quad n \geq 1.$$

From this equation, recursively we obtain

$$c_2 = \frac{1}{2}c_0,$$

$$c_4 = \frac{1 \cdot 5}{4!}c_0,$$

$$c_6 = \frac{1 \cdot 5 \cdot 9}{6!}c_0,$$

$$\vdots$$

$$c_{2n-1} = \frac{1 \cdot 5 \cdot 9 \cdots (4n-3)}{(2n)!}c_0, \quad n \geq 1$$

and

$$c_3 = \frac{3}{3!}c_1,$$

$$c_5 = \frac{3 \cdot 7}{5!}c_1,$$

$$c_7 = \frac{3 \cdot 7 \cdot 11}{11!}c_1,$$

$$\vdots$$

$$c_{2n-1} = \frac{3 \cdot 7 \cdot 11 \cdots (4n-5)}{(2n-1)!}c_1, \quad n \geq 2.$$

Therefore the general solution of Equation (D.15) is

$$y(x) = c_0\left[1 + \sum_{n=0}^{\infty} \frac{1 \cdot 5 \cdots (4n-3)}{(2n)!}x^{2n}\right] + c_1\left[x + \sum_{n=2}^{\infty} \frac{3 \cdot 7 \ldots (4n-5)}{(2n-1)!}x^{2n-1}\right]$$

where c_0 and c_1 are arbitrary constants.

There are many differential equations important in mathematical physics whose solutions are in the form of power series. One such equation is the Legendre equation considered in detail in Chapter 3.

Now, consider again the differential equation (D.7)

$$a(x)y'' + b(x)y' + c(x)y = 0,$$

where $a(x)$, $b(x)$ and $c(x)$ are analytic functions in some open interval containing the point x_0. If x_0 is singular point for (D.7) $(a(x_0) = 0)$, then a general solution of Equation (D.7) cannot always be expressed in the form of a power series. In order to deal with this situation, we distinguish two types of singular points: regular and irregular singular points.

A singular point x_0 is called a *regular singular point* of Equation (D.7) if both functions

$$(x - x_0)\frac{b(x)}{a(x)} \quad \text{and} \quad (x - x_0)^2\frac{c(x)}{a(x)}$$

are analytic at $x = x_0$. A point which is not a regular singular point is called an *irregular singular point*.

Example D.6. Consider the equation

$$(x - 1)^2 xy'' + xy' + 2(x - 1)y = 0.$$

In this example, since

$$(x - 1)\frac{x}{(x - 1)^2} = \frac{x}{x - 1}$$

is not analytic at the point $x = 1$, $x = 1$ is an irregular singular point for the equation. The point $x = 0$ is a regular singular point.

Now, we explain the *method of Frobenius* for solving a second order linear equation (D.7) when x_0 is an irregular singular point. For convenience, we assume that $x_0 = 0$ is a regular singular point of (D.7) and we may consider the equation of the form

(D.16) $y'' + p(x)y' + q(x)y = 0,$

where $xp(x)$ and $x^2q(x)$ are analytic at $x = 0$.

We seek a solution $y(x)$ of (D.16) of the form

(D.17) $y(x) = x^r \sum_{n=0}^{\infty} c_n x^n = \sum_{n=0}^{\infty} c_n x^{n+r},$

where r is a constant to be determined in the following way. Since $xp(x)$ and $x^2q(x)$ are analytic at $x = 0$,

$$xp(x) = a_0 + \sum_{n=1}^{\infty} a_n x^n$$

and

$$x^2q(x) = b_0 + \sum_{n=1}^{\infty} b_n x^n.$$

If we substitute these power series and the series for y, y' and y'' in (D.17), after rearrangements we obtain

$$\sum_{n=0}^{\infty} \left[(r + n)(r + n - 1)c_n + \sum_{k=0}^{n} \left(a_{n-k}(r + k)c_k + b_{n-k}c_k \right) \right] x^n = 0.$$

For $n = 0$ we have

(D.18)
$$r(r-1) + a_0 r + b_0 = 0.$$

Equation $(D.18)$ is called the *indicial equation*.

For any $n > 1$ we obtain an equation of c_n in terms of c_0, c_1, $\cdots$, c_{n-1}. Recursively, we solve these equations and obtain a solution of Equation (D.16).

Depending on the nature of the roots r_1 and r_2 of the indicial equation (D.18) we have different forms of particular solutions.

Theorem D.9. Frobenius. *Let r_1 and r_2 be the roots of the indicial equation* (D.18). *Then*

1. *If $r_1 \neq r_2$ are real and $r_1 - r_2$ is not an integer, then there are two linearly independent solutions of* (D.18) *of the form*

$$y_1(x) = x^{r_1} \sum_{n=0}^{\infty} a_n x^n \quad and \quad y_2(x) = x^{r_2} \sum_{n=0}^{\infty} b_n x^n.$$

2. *If $r_1 \neq r_2$ are real and $r_1 - r_2$ is an integer, then there are two linearly independent solutions of* (D.18) *of the form*

$$y_1(x) = x^{r_1} \sum_{n=0}^{\infty} a_n x^n \quad and \quad y_2(x) = y_1(x \ln x + x^{r_2} \sum_{n=0}^{\infty} b_n x^n.$$

3. *If $r_1 = r_2$ is real, then there are two linearly independent solutions of* (D.18) *of the form*

$$y_1(x) = x^{r_1} \sum_{n=0}^{\infty} a_n x^n \quad and \quad y_2(x) = y_1(x) \ln x + x^{r_1} \sum_{n=0}^{\infty} b_n x^n.$$

If the roots r_1 and r_2 of the indicial equation (D.18) are complex conjugate numbers, then $r_1 - r_2$ is not an integer number. Therefore, two solutions of the forms as in Case 1 of the theorem are obtained by taking the real and imaginary parts of them.

Many important equations in mathematical physics have solutions obtained by the Frobenius method. One of them, the Bessel equation, is discussed in detail in Chapter 3.

E. Vector Calculus Facts

Vectors in $\mathbb{R}^2$ or $\mathbb{R}^3$ are physical quantities that have norm (magnitude) and direction. Examples include force, velocity and acceleration. The vectors will be denoted by boldface letters, $\mathbf{x}$, $\mathbf{y}$.

There is a convenient way to express a three dimensional vector $\mathbf{x}$ in terms of its components. If

$$\mathbf{i} = \begin{pmatrix} 1 \\ 0 \\ 0 \end{pmatrix}, \mathbf{j} = \begin{pmatrix} 0 \\ 1 \\ 0 \end{pmatrix}, \mathbf{k} = \begin{pmatrix} 0 \\ 0 \\ 1 \end{pmatrix}$$

are the three unit vectors in the Euclidean space $\mathbb{R}^3$, then any vector

$$\mathbf{x} = \begin{pmatrix} x_1 \\ x_2 \\ x_3 \end{pmatrix}$$

in $\mathbb{R}^3$ can be expressed in the form

$$\mathbf{x} = x_1\mathbf{i} + x_2\mathbf{j} + x_3\mathbf{j}.$$

The *norm, magnitude* of a vector $\mathbf{x} = x_1\mathbf{i} + x_2\mathbf{j} + x_3\mathbf{j}$ is defined to be the nonnegative number

$$\|\mathbf{x}\| = \sqrt{x_1^2 + x_2^2 + x_3^2}.$$

Addition of two vectors $\mathbf{x} = x_1\mathbf{i} + x_2\mathbf{j} + x_3\mathbf{j}$ and $\mathbf{y} = y_1\mathbf{i} + y_2\mathbf{j} + y_3\mathbf{j}$ is defined by

$$\mathbf{x} + \mathbf{y} = (x_1 + y_1)\mathbf{i} + (x_2 + y_2)\mathbf{j} + (x_3 + y_3)\mathbf{j}.$$

Multiplication of a vector $\mathbf{x} = x_1\mathbf{i} + x_2\mathbf{j} + x_3\mathbf{j}$ with a scalar c is defined by

$$c \cdot \mathbf{x} = cx_1\mathbf{i} + cx_2\mathbf{j} + cx_3\mathbf{j}.$$

The *dot product* of two vectors $\mathbf{x}$ and $\mathbf{y}$ in $\mathbb{R}^2$ or $\mathbb{R}^3$ is defined by

(E.1) $$\mathbf{x} \cdot \mathbf{y} = \|\mathbf{x}\| \|\mathbf{y}\| \cos \alpha$$

where α is the angle between the vectors $\mathbf{x}$ and $\mathbf{y}$. If $\mathbf{x} = x_1\mathbf{i} + x_2\mathbf{j} + x_3\mathbf{j}$ and $\mathbf{y} = y_1\mathbf{i} + y_2\mathbf{j} + y_3\mathbf{j}$, then

$$\mathbf{x} \cdot \mathbf{y} = x_1y_1 + x_2y_2 + x_3y_3.$$

The *cross product* of two vectors $\mathbf{x}$ and $\mathbf{y}$ $\mathbb{R}^3$ is defined to be the vector

(E.2) $$\mathbf{x} \times \mathbf{y} = \|\mathbf{x}\| \|\mathbf{y}\| (\sin \alpha) \, \mathbf{n},$$

where α is the oriented angle between $\mathbf{x}$ and $\mathbf{y}$ and $\mathbf{n}$ is the unit vector perpendicular to both vectors $\mathbf{x}$ and $\mathbf{y}$ and whose direction is given by the right hand side rule. If $\mathbf{x} = x_1\mathbf{i} + x_2\mathbf{j} + x_3\mathbf{j}$ and $\mathbf{y} = y_1\mathbf{i} + y_2\mathbf{j} + y_3\mathbf{j}$, then

$$(\text{E.3}) \qquad \mathbf{x} \times \mathbf{y} = \begin{vmatrix} \mathbf{i} & \mathbf{j} & \mathbf{k} \\ x_1 & x_2 & x_3 \\ y_1 & y_2 & y_3 \end{vmatrix} = \begin{vmatrix} x_2 & x_3 \\ y_2 & y_3 \end{vmatrix} \mathbf{i} - \begin{vmatrix} x_1 & x_3 \\ y_1 & y_3 \end{vmatrix} \mathbf{j} + \begin{vmatrix} x_1 & x_2 \\ y_1 & y_2 \end{vmatrix} \mathbf{k}.$$

A *vector-valued function* or a *vector function* is a function of one or more variables whose range is a set of two or three dimensional vectors. We can write a vector-valued function $\mathbf{r} = \mathbf{r}(t)$ of a variable t as

$$\mathbf{r}(t) = f(t)\mathbf{i} + g(t)\mathbf{j} + h(t)\mathbf{k}.$$

The *derivative* $\mathbf{r}'(t)$ is given by

$$\mathbf{r}'(t) = f'(t)\mathbf{i} + g'(t)\mathbf{j} + h'(t)\mathbf{k},$$

and it gives the tangent vector to the curve $\mathbf{r}(t)$.

We can write a vector function $\mathbf{F}(x, y, z)$ of several variables x, y and z as

$$\mathbf{F}(x, y, z) = f(x, y, z)\,\mathbf{i} + g(x, y, z)\,\mathbf{j} + h(x, y, z)\,\mathbf{k}.$$

A vector function $\mathbf{F}(x, y, z)$ usually is called a *vector field*.

The Laplace operator is one of the most important operators. For a given function $u(x, y)$, the function

$$(\text{E.4}) \qquad \nabla^2 u \equiv \Delta u = u_{xx} + u_{yy}$$

is called the *two dimensional Laplace operator* or simply two dimensional *Laplacian*.

Similarly, the three dimensional Laplacian is given by

$$(\text{E.5}) \qquad \nabla^2 u \equiv \Delta u = u_{xx} + u_{yy} + u_{zz}.$$

If $F(x, y, z)$ is a real-valued function, then vector valued function

$$(\text{E.6}) \qquad \nabla F(x, y, z) = \frac{\partial F(x, y, z)}{\partial x}\mathbf{i} + \frac{\partial F(x, y, z)}{\partial y}\mathbf{j} + \frac{\partial F(x, y, z)}{\partial z}\mathbf{k}$$

is called the *gradient* of $F(x, y, z)$ and is sometimes denoted by *grad* $F(x, y, z)$.

An important fact for the gradient is the following.

If $x = f(t)$, $y = g(t)$ and $z = h(t)$ are parametric equations of a smooth curve on a smooth surface $F(x, y, z) = c$, then for the scalar-valued function $F(x, y, z)$ and the vector-valued function $\mathbf{r}(t) = f(t)\mathbf{i} + g(t)\mathbf{j} + h(t)\mathbf{k}$ we have

$$(\text{E.7}) \qquad \nabla F(x, y, z) \cdot \mathbf{r}'(t) = 0,$$

i.e., $\nabla F(x, y, z)$ is perpendicular to the level sets of $F(x, y, z) = c$.

Other differential operations on a vector field are the *curl* and *divergence*. The curl of a vector field $\mathbf{F}(x, y, z)$ is defined by

$$\text{(E.8)} \qquad\qquad curl(\mathbf{F}) = \nabla \times \mathbf{F},$$

and is a measure of the tendency of rotation around a point in the vector field. If $curl(\mathbf{F}) = \mathbf{0}$, then the field $\mathbf{F}$ is called *irrotational*.

The *divergence* of a vector field $\mathbf{F}(x, y, z) = \big(f(x, y, z), g(x, y, z), h(x, y, z)\big)$ is defined by

$$\text{(E.9)} \qquad\qquad div(\mathbf{F}) = \nabla \cdot \mathbf{F} = f_x + g_y + h_z.$$

Some useful formulas for these differential operators are

$$\text{(E.10)} \qquad \begin{aligned} \nabla \cdot (f\mathbf{F}) &= f\nabla \cdot \mathbf{F} + \mathbf{F} \cdot \nabla f, \\ \nabla \times (f\mathbf{F}) &= f\nabla \times \mathbf{F} + \nabla f \times \mathbf{F}, \\ \nabla^2 f = \nabla \cdot \nabla f &= f_{xx} + f_{yy} + f_{zz}, \\ \nabla \times \nabla f &= \mathbf{0}, \\ \nabla \cdot (\nabla \times \mathbf{F}) &= 0, \end{aligned}$$

for every scalar function $f(x, y, z)$ and vector-function $\mathbf{F}(x, y, z)$.

If a vector field $\mathbf{F}$ is an *irrotational* field $(curl(\mathbf{F}) = \mathbf{0})$ and *nondivergent* $(div(\mathbf{F}) = 0)$ in a domain in $\mathbb{R}^3$, then there exists a real-valued function $f(x, y, z)$, defined in the domain, with the properties

$$\mathbf{F} = \nabla f \quad \text{and} \quad \nabla^2 f = 0.$$

The above function f is called *potential*.

If a vector field $\mathbf{F}$ is such that $\nabla \times \mathbf{F} = \mathbf{0}$ in the whole domain, then this vector field is called a *conservative field*. For any conservative field $\mathbf{F}$ there exists a potential function f such that $\mathbf{F} = \nabla f$.

It is often important to consider the Laplace operator in other coordinate systems, such as the polar coordinates in the plane and spherical and cylindrical coordinates in space.

Polar Coordinates. Polar coordinates (r, φ) in $\mathbb{R}^2$ and the Cartesian coordinates (x, y) in $\mathbb{R}^2$ are related by the formulas (see Figure E.1)

$$\text{(E.11)} \qquad\qquad x = r \cos \varphi, \quad y = r \sin \varphi.$$

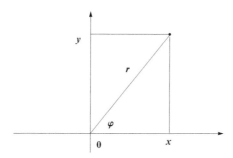

Figure E.1

If u is a function of class $C^2(\Omega)$, where Ω is a domain in $\mathbb{R}^2$, then by the chain rule we have

(E.12)
$$u_r = u_x \cos \varphi + u_y \sin \varphi,$$
$$u_\varphi = -r u_x \sin \varphi + r u_y \cos \varphi.$$

If we differentiate one more time and again use the chain rule we obtain that

(E.13)
$$u_{rr} = u_{xx} \cos^2 \varphi + 2u_{xy} \sin \varphi \cos \varphi + u_{yy} \sin^2 \varphi,$$
$$u_{\varphi\varphi} = \left(u_{xx} \sin^2 \varphi - 2u_{xy} \sin \varphi \cos \varphi + u_{yy} \cos^2 \varphi\right) - \frac{1}{r} u_r.$$

From Equations (E.10) and (E.11) we obtained

$$u_{xx} + u_{yy} = u_{rr} + r^{-1} u_r + r^{-2} u_{\varphi\varphi},$$

and therefore,

(E.14)
$$\nabla^2 u \equiv \Delta u = u_{rr} + r^{-1} u_r + r^{-2} u_{\varphi\varphi}.$$

Cylindrical Coordinates. Cylindrical coordinates (r, φ, z) in $\mathbb{R}^3$ and the Cartesian coordinates (x, y, z) are related by the formulas (see Figure E.2)

(E.15)
$$x = r \cos \varphi, \quad y = r \sin \varphi, \quad z = z.$$

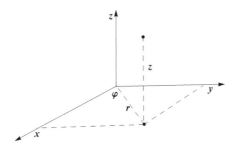

Figure E.2

If u is a function of class $C^2(\Omega)$, where Ω is a domain in $\mathbb{R}^3$, then using the result for the Laplacian in polar coordinates we have that the Laplacian in cylindrical coordinates is

(E.16) $$\nabla^2 u := \Delta u = u_{rr} + r^{-1}u_r + r^{-2}u_{\theta\varphi} + u_{zz}.$$

Spherical Coordinates. From Figure E.3 we see that the spherical coordinates (ρ, φ, θ) in $\mathbb{R}^3$ and the Cartesian coordinates (x, y, z) are related by the formulas

(E.17) $$x = \rho \cos\varphi \sin\theta, \quad y = \rho \sin\varphi \sin\theta, \quad z = \rho \cos\theta.$$

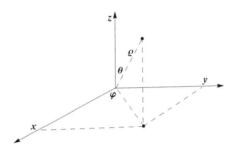

Figure E.3

The Laplacian in spherical coordinates has the form

(E.18)
$$\nabla^2 u = u_{\rho\rho} + \frac{1}{\rho}u_\rho + \frac{1}{\rho^2}\left(u_{\varphi\varphi}f_\varphi \cot\varphi + u_{\theta\theta} \cos^2\theta\right)$$
$$= \frac{1}{\rho^2}\left[\left(\rho^2 u_\rho\right)_\rho + \frac{1}{\sin\varphi}\left(u_\varphi \sin\varphi\right)_\varphi + \frac{1}{\sin^2\varphi}u_{\theta\theta}\right].$$

These equations can be obtained directly by a long and tedious calculation of the second partial derivatives, or by applying the transformation of the Laplacian from cylindrical to spherical coordinates.

Now we present several theorems that are important in the theory of partial differential equations.

Theorem E.1. Stokes's Theorem. *Let S be a bounded smooth surface in $\mathbb{R}^3$ and C a smooth closed curve on the surface S. If $\mathbf{F}(x,y,z) = f(x,y,z)\mathbf{i} + g(x,y,z)\mathbf{i} + h(x,y,z)\mathbf{k}$ is a smooth vector field on S, then*

$$\iint_S \nabla\mathbf{F} \cdot \mathbf{n}\, dS = \oint_C f\, dx + g\, dy + h\, dz$$

where $\mathbf{n}$ is the unit outward normal vector to the surface S.

Theorem E.2. Gauss/Ostrogradski Divergence Theorem. *Let* Ω *be a bounded domain in* $\mathbb{R}^3$ *with a smooth boundary surface* $S = \partial\Omega$. *If* $\mathbf{F}$ *is a smooth vector field in* Ω *and continuous on* $\Omega \cup S$, *then*

$$\iiint_\Omega \nabla \cdot \mathbf{F} \, dV = \iint_S \mathbf{F} \cdot \mathbf{n} \, dS$$

where $\mathbf{n}$ *is the unit outward normal vector to the surface* S.

Theorem E.3. Green Theorem. *Let* Ω *be a domain in* $\mathbb{R}^2$ *with a smooth boundary curve* $C = \partial\Omega$. *If* $p = p(x, y)$ *and* $q = q(x, y)$ *are smooth functions on* Ω *and continuous on* C, *then*

$$\oint_C p \, dx + q \, dy = \iint_\Omega \left(p_x - q_y \right) dx \, dy.$$

Theorem E.4. The First Green Theorem. *Let* Ω *be a domain in* $\mathbb{R}^2$ *with a smooth boundary curve* $C = \partial\Omega$. *If* $p = p(x, y)$ *and* $q = q(x, y)$ *are smooth functions on* Ω *and continuous on* $\Omega \cup C$, *then*

$$\iint_\Omega p \Delta q \, dx \, dy = \oint_C p \nabla q \cdot \mathbf{n} \, ds = \iint_\Omega \nabla p \cdot \nabla q \, dx \, dy,$$

where $\mathbf{n}$ *is the outward normal vector on the boundary* C *and* $\nabla = \left(\dfrac{\partial}{\partial x}, \dfrac{\partial}{\partial y} \right)$ *is the gradient.*

Theorem E.5. The Second Green Theorem. *Let* Ω *be a bounded domain in* $\mathbb{R}^3$ *with a smooth boundary surface* $S = \partial\Omega$. *If* $u = p(x, y, z)$ *and* $v = v(x, y, z)$ *are smooth functions on* Ω *and continuous on* $\Omega \cup S$, *then*

$$\iiint_\Omega \left(u\nabla^2 v - u\nabla^2 u \right) dx \, dy \, dz = \iint_S \left(uv_{\mathbf{n}} - vu_{\mathbf{u}} \right) dS,$$

where $\mathbf{n}$ *is the unit outward normal vector to the surface* S *and* $u_{\mathbf{n}} = \nabla u \cdot \mathbf{n}$ *and* $v_{\mathbf{n}} = \nabla v \cdot \mathbf{n}$ *are the normal derivatives of* u *and* v, *respectively.*

Multiple integrals very often should be transformed in polar or spherical coordinates. The following formulas hold.

Theorem E.6. Double Integrals in Polar Coordinates. *If $\Omega \subseteq \mathbb{R}^2$ is a bounded domain with boundary $\partial\Omega$ and $f(x,y)$ a continuous function on the closed domain $\overline{\Omega}$, $(\overline{\Omega} = \Omega \cup \partial\Omega)$, then*

$$\iint\limits_{\Omega} f(x,y)\, dx\, dy = \iint\limits_{\Omega'} f(r\cos\varphi, r\sin\varphi) r\, dr\, d\varphi,$$

where Ω' is the image of Ω under the transformation from Cartesian coordinates (x,y) to polar coordinates (r,φ).

Theorem E.7. Triple Integrals in Spherical Coordinates. *If Ω is a bounded domain in $\mathbb{R}^3$ with boundary $\partial\Omega$ and $f(x,y,z)$ a continuous function on the closed domain $\overline{\Omega} = \Omega \cup \partial\Omega$, then*

$$\iiint\limits_{\Omega} f(x,y,z)\, dV = \iiint\limits_{\Omega'} f(\rho\cos\varphi\sin\theta, \rho\sin\varphi\sin\theta, \rho\cos\theta)\rho^2 \sin\theta\, d\rho\, d\varphi\, d\theta$$

where Ω' is the image of Ω under the transformation from Cartesian coordinates (x,y,z) to spherical coordinates (ρ, φ, θ).

F. A Summary of Analytic Function Theory

A complex number z is an ordered pair (x, y) of real numbers x and y. The first number, x, is called the *real part* of z, and the second component, y, is called the *imaginary part* of z.

Addition and multiplication of two complex numbers $z_1 = (x_1, y_1)$ and $z_2 = (x_2, y_2)$ are defined by

$$z_1 + z_2 = (x_1 + x_2, y_1 + y_2), \quad z_1 \cdot z_2 = (x_1 y_1 - x_2 y_2, x_1 y_2 + x_2 y_1).$$

The complex number $(0, 1)$ is denoted by i and is called the *imaginary unit*.

Every complex number of the form $(x, 0)$ is identified by x.

Every complex number $z = (x, y)$ can be represented in the form $z = x + iy$.

The set of all complex numbers usually is denoted by $\mathbb{C}$.

For a complex number $z = x + iy$, the nonnegative number $|z|$ (called the *modulus* of z), is defined by

$$|z| = \sqrt{x^2 + y^2}.$$

For any two complex numbers z_1 and z_2 the triangle inequality holds:

$$|z_1 + z_2| \le |z_1| + |z_2|.$$

If $z = x + iy$ is a complex number, then the number $\bar{z} = x - iy$ is called the *conjugate* of z. The following is true: $\overline{z_1 + z_2} = \overline{z_1} + \overline{z_2}$; $\overline{z_1 \cdot z_2} = \overline{z_1} \cdot \overline{z_2}$ and $|z|^2 = z \cdot \bar{z}$.

If z is a nonzero complex number, then we define

$$\frac{1}{z} = \frac{1}{|z|^2} \cdot \bar{z}.$$

If z_1 and z_2 are complex numbers, then $|z_1 - z_2|$ is the distance between the numbers z_1 and z_2.

If z_0 is a given complex number and r a positive number, then the set of all complex numbers z such that $|z - z_0| = r$ represents a circle with center at z_0 and radius r.

If z_0 is a given complex number and r a positive number, then the set of all complex numbers z such that $|z - z_0| < r$ represents an open disc with center at z_0 and radius r and it is denoted by $D(z_0, r)$.

If z_0 is a given complex number and r a positive number, then the set of all complex numbers z such that $|z - z_0| \leq r$ represents a closed disc with center at z_0 and radius r and it is denoted by $\overline{D}(z_0, r)$.

A set U in the complex plane $\mathbb{R}$ is called an *open* set if for every point $a \in U$ there exists a number r such that $D(a, r) \subset U$. In other words, U is open if any point of U is a center of an open disc which is entirely in the set U.

A set U in the complex plane $\mathbb{R}$ is called a *connected* set if every two points in U can be connected by a polygonal line which lies entirely in U.

Let $z = x + iy$ be a nonzero complex number. The unique real number θ which satisfies the conditions

$$x = |z| \cos \theta, \quad y = |z| \sin \theta, \quad -\pi < \theta \leq \pi$$

is called the *argument* of z, and is denoted by $\theta = arg(z)$.

Every complex number $z \neq 0$ can be represented in the trigonometric form $z = r(\cos \theta + i \sin \theta)$, where $r = |z|$ and $\theta = arg(z)$.

If θ is a real number and r a rational number, then

$$\left(e^{i\theta}\right)^r = e^{ir\theta},$$

where

$$e^{ix} = \cos x + i \sin x.$$

If $z = r(\cos \theta + i \sin \theta)$ is a complex number and n a natural number, then there are n complex numbers z_k, $k = 1, 2, \cdots, n$ (called the nth *roots of* z) such that $z_k^n = z$, $k = 1, 2, \cdots, n$ and for each $k = 1, 2, \cdots, n$ we have

$$z_k = \sqrt[n]{r}\left(\cos \frac{\theta + 2k\pi}{n} + i \sin \frac{\theta + 2k\pi}{n}\right).$$

If $z = x + iy$ is a complex number, then e^z is defined by

$$e^z = e^x(\cos y + i \sin y).$$

The trigonometric functions $\sin z$ and $\cos z$ are defined as follows:

$$\sin z = \frac{e^{iz} - e^{-iz}}{2i}, \quad \cos z = \frac{e^{iz} + e^{-iz}}{2}.$$

If $z \neq 0$ is a complex number, then there exists unique complex number w such that $e^w = z$. The number w is called the *principal logarithm* of z and is given by

$$w = \ln |z| + iarg(z).$$

If U is an open set in the complex plane, a function $f : U \to \mathbb{C}$ is called *analytic on U* if its derivative

$$f'(z) = \lim_{\Delta z \to 0} \frac{f(z + \Delta z) - f(z)}{\Delta z}$$

exists at every point $z \in U$. f is said to be *analytic at a point z_0* if it is analytic in some open set containing z_0.

Analytic functions $f(z) = u(x, y) + iv(x, y)$, $z = x + iy$ must satisfy the *Cauchy–Riemann* equations

$$\frac{\partial u}{\partial x} = \frac{\partial v}{\partial y}, \quad \frac{\partial u}{\partial y} = -\frac{\partial v}{\partial x}.$$

Conversely, if the partial derivatives $\dfrac{\partial u}{\partial x}$, $\dfrac{\partial u}{\partial y}$, $\dfrac{\partial v}{\partial x}$ and $\dfrac{\partial v}{\partial y}$ exist and are continuous on an open set U and the Cauchy–Riemann conditions are satisfied, then f is analytic on U.

The functions e^z, z^n, $n \in \mathbb{N}$, $\sin z$, $\cos z$ are analytic in the whole complex plane. The principal logarithmic function $\ln z = \ln |z| + i\,arg(z)$, $0 < arg(z) \leq 2\pi$ is analytic in the set $\mathbb{C} \setminus \{x \in \mathbb{R} : x \geq 0\}$—the whole complex plane cut along the positive part of the Ox-$axis$.

With the principal branch of the logarithm we define complex powers by

$$z^a = e^{(a \ln z)}.$$

A point z_0 is said to be zero of *order n* of an analytic function f if

$$f(z_0) = f'(z_0) = \cdots = f^{(n-1)}(z_0) = 0, \quad f^{(n)}(z_0) \neq 0.$$

If z_0 is a zero of an analytic function f, then there is an open set U containing the point z_0 such that $f(z) \neq 0$ for every $z \in U \setminus \{z_0\}$.

If f is analytic on an open set U, then at every point $z_0 \in U$ the function f can be expanded in a power series

$$f(z) = \sum_{n=0}^{\infty} a_n (z - z_0)^n$$

which converges absolutely and uniformly in every open disc $D(z_0, R)$ whose closure $\overline{D}(z_0, r)$ lies entirely in the open set U. The coefficients a_n are uniquely determined by f and z_0 and they are given by

$$a_n = \frac{f^{(n)}(z_0)}{n!}.$$

Some power series expansions are

$$e^z = \sum_{n=0}^{\infty} \frac{z^n}{n!}, \; \sin z = \sum_{n=1}^{\infty} \frac{z^{2n-1}}{(2n-1)!}, \; \cos z = \sum_{n=0}^{\infty} \frac{z^{2n}}{(2n)!}, \; \frac{1}{1-z} = \sum_{n=0}^{\infty} z^n, \; |z| < 1.$$

If f and g are analytic functions on some open and connected set U and $f = g$ on some open disc $D \subset U$, then $f \equiv g$ on the whole U.

If f is analytic on some open and connected set U and not identically zero, then the zeros of f are isolated, i.e., for any zero a of f there exists an open disc $D(a, r) \subset U$ such that $f(z) \neq 0$ for every $z \in D(a, r) \setminus \{a\}$.

If f is analytic in a punctured disc $\tilde{D}(z_0, r) = \{z \in \mathbb{C} : 0 < |z - z_0| < r\}$ but not analytic at z_0, then the point z_0 is called an *isolated singularity* of f. In this case f can be expanded in a Laurent series about z_0:

$$f(z) = \sum_{n=-\infty}^{\infty} a_n (z - z_0)^n, \quad 0 < |z - z_0| < r.$$

If in the Laurent series of f about z_0 we have $a_n = 0$ for all $n < 0$, then z_0 is called a *removable singularity of f*.

If in the Laurent series of f about z_0 we have $a_n = 0$ for all $n < -N$ but $a_{-N} \neq 0$ for some $n \in \mathbb{N}$, then z_0 is called a *pole of order N of f*.

If in the Laurent series of f about z_0 we have $a_n \neq 0$ for infinitely many negative n, then z_0 is called an *essential singularity of f*.

A function which is analytic everywhere except at poles is called *meromorphic*

The *residue* of a function f at an isolated singularity z_0 is the coefficient a_1 in the Laurent expansion of f around the point z_0; it is denoted by $Res(f, z_0)$. If the singularity z_0 is a pole of order n then the residue is given by

$$Res(f, z_0) = \frac{1}{(n-1)!} \lim_{z \to z_0} \frac{d^{n-1}}{dz^{n-1}} \big[(z - z_0)^n f(z) \big].$$

A *contour* in $\mathbb{C}$ is defined to be a continuous map $\gamma : [a, b] \to \mathbb{C}$, $[a, b] \subset \mathbb{R}$ whose derivative $\gamma'(t)$ exists and is nonzero at all but finitely many values of t and it is piecewise continuous. If γ is a contour we say that a function f is analytic on γ if it is analytic on an open set containing the range $\{\gamma(t) : a \leq t \leq b\}$. In this case we define

$$\int_{\gamma} f(z) \, dz = \int_a^b f(\gamma(t)) \gamma'(t) \, dt.$$

If $\phi : [c,d] \rightarrow [a,b]$ is a function with a continuous first derivative and $\phi(c) = b$, $\phi(d) = a$, and ϕ is a decreasing function, then we say that the contours γ and $\gamma \circ \phi$ have *opposite orientation* and in this case we have that

$$\int_{\gamma \circ \phi} f(z)\, dz = -\int_{\gamma} f(z)\, dz.$$

A *simple closed contour* is a contour $\gamma : [a,b] \rightarrow \mathbb{C}$ such that $\gamma(a) = \gamma(b)$ but in any other case $\gamma(s) \neq \gamma(t)$.

Any simple closed contour γ divides the complex plane in two regions, the interior and the exterior region of γ. γ is said to be *positively oriented* if the interior region lies on the left with respect to the direction of motion along γ as t increases.

Cauchy's Theorem. *If f(z) is analytic inside some simple closed contour γ and continuous on the contour γ, then*

$$\int_{\gamma} f(z)\, dz = 0.$$

Suppose that Ω is a connected open set in $\mathbb{C}$ that is bounded by finitely many simple closed contours, as in Figure F.1. Ω will be in the interior of one of those contours, which will be called γ_0, and the exterior of the others, which will be called $\gamma_1, \cdots, \gamma_k$.

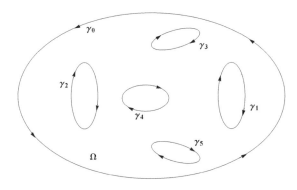

Figure F.1

Cauchy Theorem. *Let Ω and $\gamma_0, \gamma_1, \cdots, \gamma_k$ be as described above. If f is analytic on Ω and continuous on each of the γ'_k, then*

$$\int_{\gamma_0} f(z)\, dz = \sum_{j=1}^{k} \int_{\gamma_j} f(z)\, dz.$$

Cauchy Residue Theorem. *Suppose γ is a simple closed contour with interior region Ω. If f is continuous on γ and analytic on Ω except for singularities at $z_1, \cdots, z_n \in \Omega$, then*

$$\int_{\gamma} f(z)\, dz = 2\pi i \sum_{k=1}^{n} \operatorname{Res}(f, z_k).$$

Cauchy Integral Formula. *Suppose γ is a simple closed contour with interior region Ω and exterior region V. If f is analytic on Ω and continuous on γ, then for all $n = 0, 1, 2, \cdots$,*

$$\frac{n!}{2\pi i} \int_{\gamma} \frac{f(z)}{z - a}\, dz = \begin{cases} f^{(n)}(a), & \text{if } a \in \Omega, \\ 0, & \text{if } a \in V. \end{cases}$$

Let us illustrate the residue theorem by the following example.

Example. For $y \in (0, 1)$ we have

$$\int_{0}^{\infty} \frac{t^{-y}}{1 + t}\, dt = \frac{\pi}{\sin(\pi y)}.$$

Solution. We cut the complex plane along the positive real axis and consider the region bounded by the *Bromwich contour* in Figure F.2.

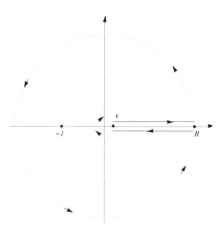

Figure F.2

On this region we define the function

$$f(z) = \frac{z^{-y}}{1+z}$$

with $arg(z^{-y}) = 0$ on the upper side of the cut. This function has a simple pole at $z = -1$ and it is easy to find that

$$Res\left(\frac{z^{-y}}{1+z}, -1\right) = e^{-\pi y}.$$

If we integrate this function along a path which goes along the upper side of the cut from $\epsilon > 0$ to R, then along the circle C_R of radius R centered at the origin, then along the lower side of the cut from R to ϵ and finally around the origin along the circle c_ϵ of radius ϵ, by the residue theorem we get

$$\int_\epsilon^R \frac{t^{-y}}{1+t}\,dt + \int_{C_R} \frac{z^{-y}}{1+z}\,dz - e^{-2\pi iy}\int_\epsilon^R \frac{t^{-y}}{1+t}\,dt - \int_{c_\epsilon} \frac{z^{-y}}{1+z}\,dz = 2\pi i e^{-2\pi iy}.$$

First let us notice that for $z \neq 0$ we have that

$$|z^{-y}| = |e^{-y\ln z}| = e^{-y Re(\ln z)} = e^{-y\ln|z|} = |z|^{-y}.$$

and therefore

$$\left|\frac{z^{-y}}{1+z}\right| = \frac{|z|^{-y}}{|1+z|} \leq \frac{|z|^{-y}}{|1-|z||}.$$

Hence, for small ϵ and large R we have

$$\left|\int_{C_R} \frac{z^{-y}}{1+z}\,dz\right| \leq 2\pi \frac{R^{y-1}}{R-1}, \qquad \left|\int_{c_\epsilon} \frac{z^{-y}}{1+z}\,dz\right| \leq 2\pi \frac{\epsilon^{y-1}}{1-\epsilon}.$$

Clearly this implies that

$$\lim_{R\to\infty} \int_{C_R} \frac{z^{-y}}{1+z}\,dz = 0 \quad \text{and} \quad \lim_{\epsilon\to 0} \int_{c_\epsilon} \frac{z^{-y}}{1+z}\,dz = 0.$$

Therefore

$$(e^{\pi iy} - e^{-\pi iy})\int_0^\infty \frac{t^{-y}}{1+t}\,dt = 2\pi i$$

and finally

$$\int_0^\infty \frac{t^{-y}}{1+t}\,dt = \frac{\pi}{\sin(\pi y)}.$$

G. Euler Gamma and Beta Functions

Among Euler's many remarkable discoveries is the *gamma function*, which for more than two centuries has become extremely important in the theory of probability and statistics and elsewhere in mathematics and applied sciences.

Historically, it was of interest to search for a function generalizing the factorial function for the natural numbers. In dealing with this problem one will come upon the well-known formula

$$\int_0^\infty t^n e^{-t}\, dt = n!, \quad n \in \mathbb{N}.$$

This formula suggests that for $x > 0$ we define the gamma function by the improper integral

$$\Gamma(x) = \int_0^\infty t^{x-1} e^{-t}\, dt.$$

This improper integral converges and it is a continuous function on $(0, \infty)$. By integration by parts, for $x > 0$ we have that

$$\Gamma(x + 1) = x\Gamma(x).$$

Since $\gamma(1) = \int_0^\infty e^{-t}\, dt = 1$, recursively its follows that

$$\Gamma(n + 1) = n!$$

for all natural numbers n. Thus $\Gamma(x)$ is a function that continuously extends the factorial function from the natural numbers to all of the positive numbers.

We can extend the domain of the gamma function to include all negative real numbers that are not integers. To begin, suppose that $-1 < x < 0$. Then $x + 1 > 0$ and so $\Gamma(x + 1)$ is defined. Now set

$$\Gamma(x + 1) = \frac{\Gamma(x)}{x}.$$

Continuing in this way we see that for every natural number n we have

(G.1) $$\Gamma(x) = \frac{\Gamma(x + n)}{x(x + 1) \cdots (x + n)}, \quad x > -n,$$

and so we can define $\Gamma(x)$ for every x in $\mathbb{R}$ except the nonpositive integers. A plot of the Gamma function $\Gamma(x)$ for real values x is given in Figure G.1.

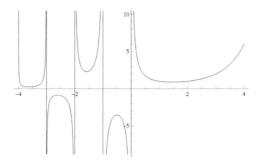

Figure G.1

Using the relation (G.1), the function Γ can be extended to a function which is meromorphic on the whole complex plane $\mathbb{C}$ and it has simple poles at the points $0, -1, 2, \cdots, -n, \cdots$, with residues at $-n$ given by

$$Res(\Gamma(z), -n) = \frac{(-1)^n}{n!}.$$

The Euler Gamma function $\Gamma(x)$ for $x > 0$ can also be defined by

$$\Gamma(x) = \lim_{n \to \infty} \Gamma_n(x)$$

where

$$\Gamma_n(x) = \frac{n! \, n^x}{x \cdot (x+1) \cdots (x+n)} = \frac{n^x}{x \cdot (1+\frac{x}{1}) \cdots (1+\frac{x}{n})}.$$

For any $z \in \mathbb{C}$ we have

$$\Gamma(z+1) = z\Gamma(z).$$

The function Γ also satisfies

$$\Gamma(z)\Gamma(1-z) = \frac{\pi}{\sin \pi z}$$

for all $z \in \mathbb{C}$.

In particular, for $z = \frac{1}{2}$ we have that

$$\Gamma(\frac{1}{2}) = \int_0^\infty t^{-\frac{1}{2}} e^{-t} \, dt = \sqrt{\pi}.$$

The previous formula together with the recursive formula for the Gamma function implies that

$$\Gamma(n + \frac{1}{2}) = \frac{1 \cdot 3 \cdot 5 \cdots (2n - 1)}{2^n} \sqrt{\pi}$$

and

$$\Gamma(-n + \frac{1}{2}) = \frac{(-1)^n 2^n}{1 \cdot 3 \cdot 5 \cdots (2n - 1)} \sqrt{\pi}$$

for all nonnegative integers n.

The function Γ does not have zeroes and $\frac{1}{\Gamma(z)}$ is analytic on the entire complex plane.

The Gamma function $\Gamma(z)$ is infinitely many differentiable on $Re(z) > 0$ and for all $x > 0$ we have that

$$\frac{\Gamma'(x)}{\Gamma(x)} = -\gamma - \frac{1}{x} + \sum_{n=1}^{\infty} \frac{x}{n(n + x)},$$

where γ is the Euler constant given by

$$\gamma = \lim_{n \to \infty} \left(1 + \frac{1}{2} + \frac{1}{3} + \cdots + \frac{1}{n} - \ln(n)\right)$$

and

$$\left(\Gamma(x)\right)^{(n)} = \int_0^{\infty} t^{x-1} e^{-t} \left(\ln t\right)^n dt.$$

For all $z \in \mathbb{C}$ we have

$$\Gamma(2z) = \frac{2^{2z-1}}{\sqrt{\pi}} \Gamma(z) \Gamma(z + \frac{1}{2}).$$

Euler discovered another function, called the *Beta function*, which is closely related to the Gamma function $\Gamma(x)$. For $x > 0$ and $y > 0$, we define the Beta function $B(x, y)$ by

$$B(x, y) = \int_0^1 t^{x-1} (1 - t)^{y-1} dt.$$

If $0 < x < 1$, the integral is improper (singularity at the left end point of integration $a = 0$). If $0 < y < 1$, the integral is improper (singularity at the right end point of integration $b = 1$).

For all $x, y > 0$ the improper integral converges and the following formula holds.

$$B(x, y) = \frac{\Gamma(x) \Gamma(y)}{\Gamma(x + y)}.$$

H. Basics of Mathematica

Mathematica is a large computer software package for performing mathematical computations. It has a huge collection of mathematical functions and commands for producing two and three dimensional graphics. This appendix is only a brief introduction to Mathematica and for more information about this software the reader is referred to *The Mathematica Book*, Version 4 by S. Wolfram [14]. There is a very convenient help system available through the "Help" menu button at the top of the notebook window. The window "Find Selected Function" gives a brief summary of what each Mathematica function does and provides references to the section of the book where the function is explained.

1.   Arithmetic Operations. Commands for expressions which should be evaluated are entered into the *input cells*, displayed in bold. The commands are evaluated by pressing the **Shift-Enter** key. The arithmetic operations addition and subtraction are as usual the keys $+$ and $-$, respectively. Multiplication and division are performed by $*$ and $/$, respectively. Multiplication of two quantities can be done by leaving a space between the quantities. Ordinary round brackets ( ) have a grouping effect as in algebra notation, but not other types of brackets. In order to perform x^y the exponential symbol $\hat{\ }$ is used: $x\hat{\ }y$.

For example, in order to evaluate

$$-5 + 3 \cdot (-2) + \frac{2^{-3} + 3 \cdot (5-7)}{-\frac{2}{3} + \frac{3}{4} \cdot (-9+3)}$$

in the cell type

In[ ]:=$-5 + 3 * (-2) + (2\hat{\ }(-3) + 3 * (5-7))/(-2/3 + (3/4) * (-9+3))$

the following result will be displayed:
Out[ ]=$-\frac{1223}{124}$.

If we want to get a numerical value for the last expression with, say, 20 digits, then in the cell type

In[ ]:=$N[\%, 20]$

Out[ ]=-9.8629032258064516129

Mathematica can do algebraic calculations.

In[ ]:=$\text{Expand}[(x-1)(2x-3)(x+9)]$

Out[ ]=$27 - 42x + 13x^2 + 2x^3$

In[ ]:=$\text{Factor}[x\hat{\ }3 + 2\,x\hat{\ }2 - 5\,x - 6]$

Out[ ]=$(-2+x)(1+x)(3+x)$

In[ ]:=Simplify$[(x^2-3\,x)\,(6\,x-7)-(3\,x^2-2\,x-1)\,(2\,x-3)]$
Out[ ]=$-3+17\,x-12\,x^2$

In[ ]:=Together$[2\,x-3+(3\,x+5)/(x\char`^2+x+1)]$
Out[ ]=$\frac{2+2\,x-x^2+2\,x^3}{1+x+x^2}$

In[ ]:=Apart$[(3\,x+1)/(x\char`^3-3\,x\char`^2-2\,x+4)]$
Out[ ]=$-\frac{4}{5\,(-1+x)}+\frac{11+4\,x}{5\,(-4-2\,x+x^2)}$

One single input cell may consist of several. A new line within the current cells is obtained by pressing the **Enter** key. Commands on the same line within a cell must be separated by semicolons. The output of any command that ends with a semicolon is not displayed.

In[ ]:=$x=5/4-3/7;y=9/5+4/9-9\,x;z=2\,x^2-3\,y^2$
Out[ ]=$-\frac{2999}{392}$

If we have to evaluate an expression with different values each time, then the substitution command symbolized by a slanted bar and a period is useful.

In[ ]:=$z=2\,x^2-3x\,y+2\,y^2/.\{x->1,y->1\}$
Out[ ]=1

When we set a value to a symbol, that value will be used for the symbol for the entire Mathematica session. Since symbols no longer in use can introduce confusion when used in new computations, clearing previous definitions is extremely important. To remove a variable from the kernel, use the Clear command.

In[ ]:=Clear$[x,y,z]$

2. Functions. All built-in mathematical functions and constants have full names that begin with a capital letter. The arguments of a function are enclosed by brackets. For example, the familiar functions from calculus $\sin x$, $\cos x$, e^x, $\ln x$ in Mathematica are Sin[x], Cos[x], Exp[x], Log[x]. The constant π in Mathematica is Pi, while the Euler number e in Mathematica is E.

In Mathematica the function $f(x)=x^2+2x$ is defined by

In[ ]:=$f[x\_]:=x^2+2\,x;$

We can evaluate $f(a+b)+f(a-b)$ by typing

In[ ]:=$f[a+b]+f[a-b]$

Out[ ]=$2a^2 + 2b^2 + 4a$.

The command for defining functions of several variables is similar:

In[ ]:=$g[x\_, t\_] := Sin[Pi\,x]\;Cos[2\,Pi\,t]$;

To define the piecewise function

$$h(x) = \begin{cases} x, & -\pi \leq x < 0 \\ x^2, & 0 \leq x < \pi \\ 0, & \text{elsewhere} \end{cases}$$

we use

In[ ]:=$h[x\_]$=Piecewise[$\{\{x, -Pi <= x < 0\}, \{0 <= x < Pi\}\}, 0$]];

If we now type

In[ ]:=$h[2]$

we obtain

Out[ ]=4

3. Graphics. Mathematica is exceptionally good at creating two and three dimensional graphs.

The plot of the function $h(x)$ above on the interval $[-2\pi, 2\pi]$ is obtained by typing

In[ ]:=Plot[$h[x], \{x, -2\,Pi, 2\,Pi\}$, PlotRange $->$ All, Ticks $->$
$\{\{-2\,Pi, -Pi, 0, Pi, 2\,Pi\}, \{-Pi, Pi^2\}\}$, AspectRatio $->$
Automatic, PlotStyle $->$ Thick, AxesLabel $-> \{x, y\}$]

The plot is displayed below.

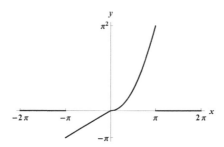

Mathematica can do multiple plots.

In[ ]:=Plot[{$Sqrt[x], x, x^2$}, {$x, 0, 2$}, PlotRange − > {0, 1.5} ,
Ticks − > {{0, 1}, {0, 1}}, PlotStyle − > {{Dashing [{0.02}]}, { Black},
{$Thickness$[0.007]}}], AspectRaito − > Automatic, AxesLabel− >{x, y}]

The plot is displayed below.

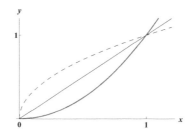

To plot many points on the plane that are on a given curve we use the
ListPlot command.

Example.

In[ ]:=list=Table[{$x, Sin[x]$}, {$x, -2\,Pi, 2\,Pi, 0.1$}];

In[ ]:=p1=ListPlot[$list$]

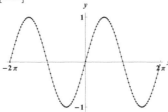

If we want to joint the points, we use ListLinePoints.

Example.

In[ ]:=$p2$=ListLinePlot[list]

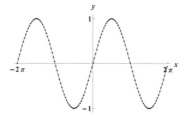

To plot both graphs $p1$ and $p2$ together we use the command Show.

Example.

In[ ]:=Show[$p1, p2$]

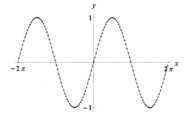

To plot the graph $p2$ next to the graph $p1$ use Show[GraphicsGrid[ ]].

Example.

In[ ]:=Show[GraphicsGrid[$\{\{p1, p2\}\}$]]

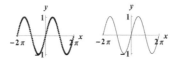

To plot implicitly given functions we use ContourPlot.

Example.

In[ ]:=ContourPlot[$\{x^3 + y^3 - 9\,x\,y == 0,$
$x + y + 3 == 0\} \{x, -6, 6\}, \{y, -6, 6\}$]

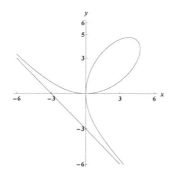

To plot in polar coordinates use PolarPlot.

Example.

Plot the following functions.

$$r = 2, \quad r = 2 + \frac{1}{3} \sin 10\varphi, \quad r = \sin 5\varphi, \ 0 \le \varphi \le 2\pi.$$

In[ ]:=PolarPlot[{2, 2 + 1/3 Sin[10 t], Sin[5 t]}, {t, 0, 2 Pi}, PlotStyle − > {Black, Dashed, Black}, Ticks − > {{−2, −1, 0, 1, 2}, {−2, −1, 0, 1, 2}}, AxesLabel − > {x, y}]

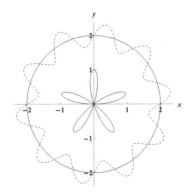

To plot curves given with parametric equations use ParametricPlot.

Plot the curve given by the parametric equations $x = \sin 2t$, $y = \sin 3t$, $0 \le t \le 2\pi$.

In[ ]:=ParametricPlot[{Sin[2 t], Sin[3 t]}, {t, 0, 2 Pi}, Ticks − > {{−1, 0, 1}, {−1, 0, 1}}, AxesLabel − > {x, y}]

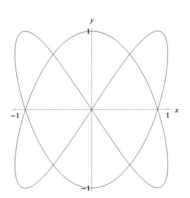

Three dimensional plots are generated using the Plot3D command.

In[ ]:=Plot3D[Sqrt[$x^2 + y^2$] Exp [$-x^2 + y^2$], {$x, -2, 2$}, {$y, -2, 2$}, Ticks
$- >$ {{$-2, 0, 2$}, {$-2, 0, 2$}, {$0, 1$}, AxesLabel $- >$ {x, y, z}]

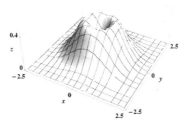

4. Differentiation and Integration.

Mathematica can perform differentiation and integration of functions of a single variable as well as multiple variables. The command for the nth derivative of a function $f(x)$ with respect to x is D[$f[x], \{x, n\}$].

To find the second derivative of the function $x^3 \cos^2 x$ type

In[ ]:=D[x^3 (Cos[x])^2, {$x, 2$}]

Out[ ]=$30x^4 \cos^2 x - 24x^5 \cos x \sin x + x^6(-2\cos^2 x + 2\sin^2 x)$

The command for partial derivatives is the same. For example, the command for the mixed partial derivative of $f(x,y)$ is D[$f[x, y], x, y$], or the command for the second partial derivative of $f(x,y)$ with respect to y is D[$f[x, y], y, 2$].

To find the mixed partial derivative of the function $x^3y^2 \cos x - x^2y^3 \cos y$ type

In[ ]:=D[$x^3\, y^2\, Cos[x] - x^2\, y^3\, Cos[y], x, y$]

Out[ ]=$6x^2y \cos x - 6xy^2 \cos y - 2x^3y \sin x + 2xy^3 \sin y$

Integrate[$f[x], x$] is the command for the indefinite integral $\int f(x)\, dx$. For the definite integral $\int_a^b f(x)\, dx$ the command is Integrate[$f[x], \{x, a, b\}$].

In[ ]:=Integrate[$1/(x^3 + 8), x$]

Out[ ]=$\dfrac{ArcTan\left(\frac{-1+x}{\sqrt{3}}\right)}{4\sqrt{3}} + \frac{1}{12} \ln (2 - x) - \frac{1}{24} \ln (4 - 2x + x^2)$

In[ ]:=Integrate$[1/(x^3+8), \{x, 0, 1\}]$

Out[ ]=$\frac{1}{72}(\pi\sqrt{3}+\ln 27)$

Integrate $[f[x, y], \{x, a, b\}, \{y, c[x], d[x]\}]$ is the command for the double
integral $\int\limits_{a}^{b}(\int\limits_{c(x)}^{d(x)} f(x, y)\, dy)\, dx$.

In[ ]:=Integrate$[Sin[x]\, Sin[y], \{x, 0, Pi\}, \{y, 0, x\}]$

Out[ ]=$\frac{\pi^2}{4}$

5. Solving Equations.
Mathematica can solve many equations: linear, quadratic, cubic and quartic equations (symbolically and numerically), any algebraic and transcendental equations (numerically), ordinary and partial differential equations (numerically and some of them symbolically). It can solve numerically systems of equations and in some cases exactly.

5a. Solving Equations Exactly.
The commands for solving equations exactly are Solve[*expr,vars*] (tries to solve the equation exactly) and Solve[*expr,vars, dom*] solves the equation (or the system) over the domain *dom* (usually real, integer or complex numbers).

Examples.

Solve a quadratic equation:

In[ ]:=Solve$[a\, x^2+x+1 == 0, x]$

Out[ ]=$\{\{x- > \frac{-1-\sqrt{1-4a}}{2a}\}, \{x- > \frac{-1+\sqrt{1-4a}}{2a}\}\}$

In[ ]:=Solve$[a\, x^2+x+1 == 0, x, Reals]$

Out[ ]=$\{\{x- > \text{ConditionalExpression}[\frac{-1-\sqrt{1-4a}}{2a}, a < \frac{1}{4}]\},$
$\{x- > \text{ConditionalExpression}[\frac{-1+\sqrt{1-4a}}{2a}, a < \frac{1}{4}]\}\}$

Solve a cubic equation (one solution):

In[ ]:=Solve$[x^3+x+1 == 0, x][[1]]$ (* the 1st solution *)

Out[ ]=$-\left(\frac{2}{3(-9+\sqrt{93})}\right)^{\frac{1}{3}}+\frac{\left(\frac{1}{2}(-9+\sqrt{93})\right)^{\frac{1}{3}}}{3^{\frac{2}{3}}}$

Solving a linear system for x and y:

In[ ]:=Solve$[a\, x+2\, y == -1\ \&\&\ b\, x-y == 1, \{x, y\}]$

$$\text{Out[ ]}=\{\{x->-\frac{1}{-a-2b},\, y->-\frac{a+b}{a+2b}\}\}$$

Solving a nonlinear system for x and y:

In[ ]:=Solve[$a\,x+2\,y==-1$ && $b\,x-y==1,\{x,y\}$]

$$\text{Out[ ]}=\{\{x->-\frac{1}{-a-2b},\, y->-\frac{a+b}{a+2b}\}\}$$

5b. Solving Equations Numerically. The Mathematica commands for solving equations numerically are NSolve[*expr,vars*] (tries to solve a given equation numerically), while NSolve[*expr,vars, Reals*] solves numerically the equation (or the system) over the domain real numbers.

Examples.

Approximate solutions to a polynomial equation:

In[ ]:=NSolve[$x^5-2\,x+1==0,x$]

Out[ ]=$\{\{x->-1.29065\},\{x->-0.114071-1.21675\,i\},$
$\{x->-0.114071+1.21675\,i\},\{x->0.51879\},\{x->1.\}\}$

In[ ]:=NSolve[$x^5-2\,x+1==0,x,Reals$]

Out[ ]=$\{\{x->-1.29065\},\{x->0.51879\},\{x->1.\}\}$

Approximate solutions to a system of polynomial equations:

In[ ]:=NSolve[$\{x^3+x\,y^2+y^3==1,3\,x+2\,y==4\},\{x,y\}$]

Out[ ]=$\{x->58.0885,\, y->-85.1328\},\{x->0.955748+0.224926\,i,$
$y->0.566378-0.337389\,i\},\,\{x->0.955748-0.224926\,i,$
$y->0.566378+0.337389\,i\}\}$

In[ ]:=NSolve[$\{x^3+x\,y^2+y^3==1,3\,x+2\,y==4\},\{x,y\},Reals|$

Out[ ]=$\{x->58.0885,\, y->-85.1328\}$

For numerical solutions of transcendental equations the Mathematica commands are

FindRoot[$f[x],\{x,x_0\}$] (searches for a numerical root of $f(x)$, starting from the point $x=x_0$);

FindRoot[$f[x]==0,\{x,x_0\}$] (searches for a numerical solution of the equation $f(x)=0$, starting from the point $x=x_0$);

FindRoot[$f[x,y]==0,\,g[x,y]==0,\{x,x_0\},\{y,y_0\}$] searches for a numerical solution of the system of equations $f(x,y)=0,\ g(x,y)=0$, starting from the point $(x=x_0,\,y=y_0)$.

Examples.

Find all roots of the function $f(x) = x - \sin x$ near $x = \pi$.

In[ ]:=$[x - 2\,Sin[x], \{x, Pi\}]$

Out[ ] = $\{x- > 1.89549\}$

Find the solution of $e^x = -x$ near $x = 0$.

In[ ]:=FindRoot$[Exp[x] == -x, \{x, 0\}]$

Out[ ] = $\{x- > -0.567143\}$

Solve the following nonlinear system of equations $\sin(x - y) + 2y - 1 = \sin x - \sin y$, $y = \sin(x + y) + \sin x + 3y$ near the point $(0, 0.5)$.

In[ ]:=FindRoot$[\{Sin[x - y] + 2\,y - 1 == Sin[x] - Sin[y], y == Sin[x +$ $y] + Sin[x] + x + 3\,y\},$
$\{\{x, 0\}, \{y, 0.5\}\}]$

Out[ ] = $\{x- > 0.0042256,\ y- > 0.500257\}$

5c. Solving Differential Equations Symbolically. The Mathematica command DSolve[eqn,y, x] solves the differential equation for the function $y(x)$. The command DSolve[$\{eqn_1, eqn_2, \ldots\}$, $\{y_1, y_2, \ldots, x\}$] solves a system of differential equations. The command DSolve[eqn, y, $\{x_1, x_2, \ldots\}$] solves a partial differential equation for the function $y = y(x_1, x_2, \ldots)$.

Examples.

Find the general solution of $y''(x) + y(x) = e^x$.

In[ ]:=DSolve[$y''[x] + y[x] == Exp[x], y[x], x$]

Out[ ] = $\{\{y[x]- > c_1 \cos x + c_2 \sin x + \frac{1}{2}e^x(\cos^2 x + \sin^2 x)\}\}$

Find the general solution of $y''(x) + y(x) = e^x$ subject to the boundary conditions $y(0) = 0$, $y'(0) = 1$.

In[ ]:=DSolve[$\{y''[x] + y[x] == Exp[x], y[0] == 1, y'[0] == 1\}, y, x$]

Out[ ] = $\{\{y- >\text{Function}[\{x\}, \frac{1}{2}(\cos x + e^x \cos^2 x + \sin x + e^x \sin^2 x)]\}\}$

Find the general solution of the partial differential equation

$$2z_x(x,t) + 5z_t(x,t) = z(x,t) - 1.$$

In[ ]:=DSolve[$2\,D[z[x,t], x] + 5\,D[z[x,t], t] == z[x,t] + 1, z, \{x,t\}$]

Out[ ] = $\{\{z- > (\{x,t\}- > e^{\frac{x}{2}}c_1(\frac{1}{2}(2t - 5x)) - 1)\}\}$

Find the general solution of the second order partial differential equation

$$3x_{xx}(x,t) - 2z_{tt}(x,t) = 1.$$

In[ ]:=DSolve[3 $D[z[x,t], \{x,2\}] - 2\,D[z[x,t], y, 2] == 1, z, \{x,t\}$]

Out[ ] = $\{\{z->(\{x,t\}->(e^{\frac{x}{2}}c_1(\frac{1}{2}(2t-5x))-1))\}\}$

In[ ] := $\{\{z->(\{x,t\}->x\,c_1(t-\sqrt{\frac{2}{3}}x)+c_2(t+\sqrt{\frac{2}{3}}x)+\frac{x^2}{6})\}\}$

5d. Solving Differential Equations Numerically. The Mathematica command NDSolve[$eqns, y, \{x, a, b\}$] finds a numerical solution to the ordinary differential equations $eqns$ for the function $y(x)$ in the interval $[a,b]$. The command NDSolve[$eqns, z, \{x, a, b\}, \{t, c, d\}$] finds a numerical solution of the partial differential equation $eqns$.

Examples.

Solve numerically the initial value problem (ordinary differential equation)

$$y'(t) = y\cos(t^2 + y^2), \quad y(0) = 1.$$

In[ ]:=nsol=NDSolve[$\{y'[t] == y[t]\,Cos[t^2 + y[t]^2], y[0] == 1\}, y, \{t, 0, 20\}$]

Out[ ] = $\{\{y->InterpolatingFunction[\{\{0., 20.\}\}, <>]\}\}$

We plot this solution with

In[ ]:=Plot[Evaluate $[y[t]/.$ nsol], $\{t, 0, 20\}$, Ticks $-> \{\{0, 5, 10, 15, 20\}$, $\{0, 1\}\}$, PlotRange $->$ All, AxesLabel $-> \{x, t\}$]

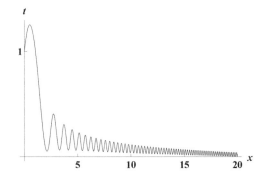

Plot the solution of the initial value problem (partial differential equation)

$$u_t(x,t) = 9u_{xx}(x,t), \quad 0 < x <, \ 0 < t < 10$$
$$u(x,0) = 0, \quad 0 < x < 5$$
$$u(0,t) = \sin 2t, \quad u(,t) = 0, \ 0 < t < 10.$$

In[ ]:=NDSolve[$\{D[u[x,t],t] == 9\,D[u[x,t],x,x], u[x,0] == x, u[0,t]$
== $Sin[2\,t], u[5,t] == 0\}, u, \{t,0,10\}, \{x,0,5\}]$

Out[ ] = $\{\{u-> \text{InterpolatingFunction}\,[\{\{0.,5.\}, \{0.10.\}\}, <>]\}\}$

We plot the solution with

In[ ]:=Plot3D[Evaluate $[u[x,t]/.\%], \{t,0,20\}, \{x,0,\}, \text{PlotRange} -> \text{All}]$

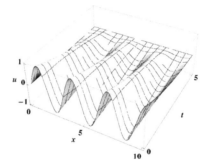

6. Matrices. In Mathematica a matrix is a list in which each component is a row of the matrix.

Examples.

In[ ] := $A = \{\{1,0,3,4\}, \{3,2,0,2\}, \{1,1,1,1\}\}$

Out[ ] = $\{\{1,0,3,4\}, \{3,2,0,2\}, \{1,1,1,1\}\}$

To print the matrix A in the traditional form use

In[ ]:=MatrixForm[A]

Out[ ]=$\begin{pmatrix} 1 & 0 & 3 & 4 \\ 3 & 2 & 0 & 2 \\ 1 & 1 & 1 & 1 \end{pmatrix}.$

We can construct matrices whose entries are defined by a function.

In[ ]:=B=Table[$1/(i+j)], \{i,2\}, \{j,3\}]$

$\text{Out}[\ ] = \{\{\frac{1}{1+1}, \frac{1}{1+2}, \frac{1}{1+3}\}, \{\frac{1}{2+1}, \frac{1}{2+2}, \frac{1}{2+3}\}\}$

$\text{In}[\ ]:=\text{MatrixForm}[B]$

$$\text{Out}[\ ]=\begin{pmatrix} \frac{1}{1+1} & \frac{1}{1+2} & \frac{1}{1+3} \\ \frac{1}{2+1} & \frac{1}{2+2} & \frac{1}{2+3} \end{pmatrix}.$$

The matrix operations addition and subtraction are represented with the usual +4 and $-$ keys. Matrix multiplication is represented using the dot key.

Examples.

Let

$\text{In}[\ ] := X = \{\{1,3,4\}, \{3,0,2\}, \{1,1,1\}\};\ Y = \{\{2,1,4\}, \{1,1,3\}, \{1,2,1\}\};$

Then

$\text{In}[\ ] := X + Y$

$\text{Out}[\ ] = \{\{3,4,8\}, \{4,1,5\}, \{2,3,2\}\}$

$\text{In}[\ ] := X.Y$

$\text{Out}[\ ] = \{\{10,12,17\}, \{8,7,14\}, \{4,4,8\}\}$

The commands $Det[A]$ and Inverse[A] are the commands for the determinant and the inverse matrix of a square matrix A. TransposeM] is the command for the transpose matrix of any matrix M.

Examples.

$\text{In}[\ ] := Det[\{\{1,3,4\}, \{3,0,2\}, \{1,1,1\}\}]$

$\text{Out}[\ ] = 7$

$\text{In}[\ ]:=\text{MatrixForm}[\text{Inverse}[\{\{1,3,4\}, \{3,0,2\}, \{1,1,1\}\}]]$

$$\text{Out}[\ ] = \begin{pmatrix} -\frac{2}{7} & \frac{1}{7} & \frac{6}{7} \\ -\frac{1}{7} & -\frac{3}{7} & \frac{10}{7} \\ \frac{3}{7} & \frac{2}{7} & -\frac{9}{7} \end{pmatrix}$$

$\text{In}[\ ]:=\text{MatrixForm}[\text{Transpose}[\{\{1,0,3,4\}, \{3,2,0,2\}, \{1,1,1,1\}\}]]$

$$\text{Out}[\ ] = \begin{pmatrix} 1 & 3 & 1 \\ 0 & 2 & 1 \\ 3 & 0 & 1 \\ 4 & 2 & 1 \end{pmatrix}$$

Using inverse matrices we can solve linear systems.

Example. Solve the system

$$3x_1 - 2x_2 + x_3 = -1,$$
$$2x_1 + x_2 - x_3 = 0,$$
$$3x_2 - 2x_2 - 3x_3 = 5.$$

Solution. We form the coefficient matrix and the right side matrix.

In[ ] := $A = \{\{3, -2, 3\}, \{2, 1, -1\}, \{3, -2, , -3\}, \ \mathbf{b} = \{-1, 0, 5\};$

Next we find the inverse matrix B of the matrix A:

In[ ]:=B=Inverse[A];

We find the solution of the system by

In[ ]:=Solution=B.**b**

Out[ ] = $\{-\frac{5}{14}, -\frac{11}{14}, -\frac{3}{2}\}$

Alternatively, we can use Mathematica command LinearSolve.

In[ ]:=Sol=LinearSolve[A, **b**]

Out[ ] = $\{-\frac{5}{14}, -\frac{11}{14}, -\frac{3}{2}\}$

Notice that Mathematica gives the exact solution of the system. If, instead of the matrix A, we use the matrix

In[ ] := $A_1 = \{\{3., -2., 3.\}, \{2., 1., -1.\}, \{3., -2., , -3.\};$

then the inverse matrix of A_1 will be the matrix

In[ ] := $B_1 = Inverse[A_1]$

Out[ ] := $\{\{0.178571, 0.285714, -0.0357143\},$
$\{-0.107143, 0.428571, -0.178571\}, \{0.25, 0., -0.25\}\}$

and the numerical solution of the system will be

In[ ]:=NumerSolution=B_1.**b**

Out[ ] := $\{-0.357143, -0.785714, -1.5\}$

To find the eigenvalues and corresponding eigenvectors of a given matrix A we use the commands Eigenvalues[A] and Eigenvectors[A], respectively.

Example. Find all eigenvalues and eigenvectors of the matrix A given by

$$A = \begin{pmatrix} 2 & -1 & -1 \\ -1 & 2 & -1 \\ -1 & -1 & 2 \end{pmatrix}$$

Solution. First define the matrix A.

In[ ] := $A = \{\{2, -1, -1\}, \{-1, 2, -1\}, \{-1, -1, 2\}\}$];

The eigenvalues of A are found by

In[ ]: = eigval=Eigenvalues[A]

Out[ ] := $\{3, 3, 0\}$

Next we find the eigenvectors

In[ ]: = eigvect=Eigenvectors[A]

Out[ ] := $\{\{-1, 0, 1\}, \{-1, 1, 0\}, \{1, 1, 1\}\}$

BIBLIOGRAPHY

[1]   W. E. Boyce, R. C. DiPrima, *Elementary Differential Equations and Boundary Value Problems, Fourth Edition*, John Wiley & Sons, New York, 1986.

[2]   E. A. Coddington, *An Introduction to Ordinary Differential Equations*, Academic Press, New York, 1966.

[3]   M. P. Coleman, *An Introduction to Partial Differential Equations with MATLAB*, CRC Applied Mathematics and Nonlinear Science Series, Chapman & Hall/CRC, Boca Raton, London, New York, 2005.

[4]   D. G. Duffy, *Advanced Engineering Mathematics*, 2nd ed., CRC Press, Boca Raton, Florida, 1998.

[5]   L. C. Evans, *Partial Differential Equations*, Graduate Studies in Mathematics, vol. 19, American Mathematical Society, Providence, Rhode Island, 1991.

[6]   G. B. Folland, *Fourier Series and Its Applications*, Mathematics Series, Wadsworth & Brooks/Cole Publishing Company, Pacific Grove, California, 1992.

[7]   P. Garabedian, *Partial Differential Equations,* 2nd *ed.*, Chelsea, New York, 1998.

[8]   G. D. Smith, *Numerical Solution of Partial Differential Equations,* 3rd *ed.*, Clarendon, Oxford, 1985.

[9]   J. W. Demmel, *Applied Numerical Linear Algebra*, SIAM, Philadelphia, 1997.

[10] E. C. Titchmarsh, *The Theory of Functions, Second Edition*, Oxford University Press, London, 1939.

[11] M. Rosenlicht, *Introduction to Analysis*, Dover Publications, Inc., New York, 1968.

[12] T. M. Apostiol, *Mathematical Analysis, Second Edition*, Addison-Wesley, New York, 1974.

[13] N. Asmar, *Partial Differential Equations and Boundary Value Problems*, Prentice-Hall, Upper Saddle River, New Jersey, 2000.

[14] S. Wolfram, *The Mathematica Book, Fourth Edition*, Cambridge University Press, New York, 1999.

[15] M. Reed, B. Simon, *Methods of Modern Mathematical Physics, I: Functional Analysis*, Academic Press, New York, 1972.

[16] G. Evans, J. Blackledge and P. Yardley, *Numerical Methods for Partial Differential Equations*, Springer-Verlag, Berlin, Heidelberg, New York, 2000.

ANSWERS TO EXERCISES

Section 1.1.

5. (a) $2 \sum\limits_{n=1}^{\infty} \frac{(-1)^{n+1}}{n} \sin nx$.

(b) $\frac{2}{\pi} - \frac{4}{\pi} \sum\limits_{n=1}^{\infty} \frac{\cos 2nx}{4n^2-1}$.

(c) $2 \sum\limits_{n=1}^{\infty} \frac{\sin nx}{n}$.

(d) $\frac{8}{\pi} \sum\limits_{n=1}^{\infty} \frac{\sin (2n-1)x}{(2n-1)^3}$.

(e) $\frac{e^{2a\pi}-1}{2a\pi} + \frac{1}{\pi} \sum\limits_{n=1}^{\infty} \left[\frac{a(e^{2a\pi}-a)}{a^2+n^2} \cos nx - \frac{n(n+e^{2a\pi})}{a^2+n^2} \sin nx \right]$.

(f) $\frac{\sinh a\pi}{a\pi} + \frac{2\sinh a\pi}{\pi} \sum\limits_{n=1}^{\infty} \left[\frac{(-1)^n)}{a^2+n^2} \cos nx - \frac{n(-1)^n}{a^2+n^2} \sin nx \right]$.

(g) $-\frac{1}{3} + \sum\limits_{n=1}^{\infty} \left[a_n \cos \frac{2n\pi}{3}x + b_n \sin \frac{2n\pi}{3}x \right]$

$a_n = -2 \frac{\sin \frac{n\pi}{3}}{n^2\pi^2} \left(n\pi + 2n\pi \cos \frac{2n\pi}{3} - 3 \sin \frac{2n\pi}{3} \right)$;

$b_n = \frac{-n\pi \cos n\pi + 3 \sin \frac{n\pi}{3}}{n^2\pi^2}$.

(h) $\frac{3}{8} + \sum\limits_{n=1}^{\infty} \left[a_n \cos \frac{n\pi}{2}x + b_n \sin \frac{n\pi}{2}x \right]$, $a_n = \frac{-2+2\cos,\frac{n\pi}{2}}{n^2\pi^2}$;

$b_n = \frac{-n\pi \cos n\pi + 2 \sin \frac{n\pi}{2}}{n^2\pi^2}$.

8. (a) $\frac{\pi}{2} - \frac{2}{\pi} \sum\limits_{\substack{n=-\infty \\ n\neq 0}}^{\infty} \frac{e^{(2n-1)ix}}{(2n-1)^2}$.

(b) $1 + \frac{i}{\pi} \sum\limits_{\substack{n=-\infty \\ n\neq 0}}^{\infty} \frac{e^{n\pi ix}}{n}$.

(c) $\frac{1}{2} - \frac{i}{\pi} \sum\limits_{\substack{n=-\infty \\ n\neq 0}}^{\infty} \frac{e^{2(2n-1)ix}}{2n-1}$.

Section 1.2.

1. (a) Not continuous; not piecewise continuous.

    (b) Continuous; not piecewise smooth.

    (c) Continuous; piecewise smooth.

    (d) Not continuous; piecewise continuous; piecewise smooth.

    (e) Continuous; piecewise smooth.

2. (a) 3 derivatives.

    (b) Derivative of any order.

    (c) None.

3. (b) The sum is 0.

    (c) The sum is $\frac{1}{2}$.

4. (b) The sum is $\frac{\pi}{2}$.

5. (b) The sums are 0.

6. Take $x = \frac{\pi}{2}$.

7. (b) The sum is $\frac{3\pi}{4}$.

9. (c) Converges at every point x.

Section 1.3.

1. (a) $\frac{x}{2} = \sum_{n=1}^{\infty} \frac{(-1)^{n+1}}{n} \sin nx.$ $\sum_{n=1}^{\infty} \frac{1}{n^2} = \frac{\pi^2}{6}.$

    (b) $f(x) = x^3 - \pi^2 x = 12 \sum_{n=1}^{\infty} \frac{(-1)^n}{n^3} \sin nx.$ $\sum_{n=1}^{\infty} \frac{1}{n^6} = \frac{\pi^6}{945}.$

2. (a) $f(x) = \frac{x}{2}$. $f(x) = \sum_{n=1}^{\infty} \frac{(-1)^{n+1}}{n} \sin nx$. $\sum_{n=1}^{\infty} \frac{1}{n^2} = \frac{\pi^2}{6}$.

(b) $f(x) = \begin{cases} -1, & -\pi < x < 0 \\ 1, & 0 < x < \pi. \end{cases}$ $f(x) = \frac{4}{\pi} \sum_{n=1}^{\infty} \frac{1}{2n-1} \sin(2n-1)\pi x$.

$\sum_{n=1}^{\infty} \frac{1}{(2n-1)^2} = \frac{\pi^2}{8}$.

(c) $f(x) = |x|$. $f(x) = \frac{\pi}{4} - \frac{2}{\pi} \sum_{n=1}^{\infty} \frac{1}{(2n-1)^2} \cos(2n-1)x$, $-\pi < x < \pi$.

$\sum_{n=1}^{\infty} \frac{1}{(2n-1)^4} = \frac{\pi^4}{96}$.

(d) $f(x) = |\sin x| = \frac{2}{\pi} - \frac{4}{\pi} \sum_{n=1}^{\infty} \frac{1}{4n^2-1} \cos 2nx$.

$\sum_{n=1}^{\infty} \frac{1}{(4n^2-1)^2} = \frac{\pi^2-8}{16}$.

4. (a) $f(x) = \sinh x = \frac{\sinh \pi}{\pi} \sum_{n=1}^{\infty} \frac{(-1)^{n+1} \sin nx}{n^2+1}$, $|x| < \pi$.

$\frac{4 \sinh^2 \pi}{\pi^2} \sum_{n=1}^{\infty} \frac{\sin^2 na}{n^2} = \frac{-\pi+\cosh pi \sinh pi}{\pi}$.

(b) $f(x) = \begin{cases} \frac{1}{2a}, & |x| < a \\ 0, & a < |x| < \pi. \end{cases}$ $f(x) = \frac{1}{2\pi} + \frac{1}{\pi} \sum_{n=1}^{\infty} \frac{\sin na}{na} \cos nx$.

$\sum_{n=1}^{\infty} \frac{\sin^2 na}{n^2} = \frac{a}{2\pi}(\pi - a)$.

5. (a) $\frac{1}{4} + 4 \sum_{n=2}^{\infty} \frac{1}{(n^2-1)^2} = \frac{1}{\pi} \int_{-\pi}^{\pi} x^2 \cos^2 x \, dx = \frac{3+2\pi^2}{6}$.

(b) $\frac{1}{2} + \frac{1}{4} + 4 \sum_{n=1}^{\infty} \frac{1}{(n^2-1)^2} = \frac{1}{\pi} \int_{-\pi}^{\pi} x^2 \sin^2 x \, dx = \frac{-3+2\pi^2}{6}$.

6. (a) By the Dirichlet test for convergence of series of complex numbers, the function f and its derivative are continuous everywhere.

7. (a) By the Dirichlet test for convergence of series of complex numbers, the series is convergent for every $x \neq 2n\pi$, $n = 0, 1, \ldots$. For $x = 2n\pi$ we obtain the harmonic series which is divergent.

9. $f(x) = \frac{1}{2\pi} + \frac{2}{\pi} \sum_{n=1}^{\infty} \frac{1-\cos na}{n^2 a^2} \cos nx$.

$\frac{1}{2\pi^2} + \frac{4}{\pi^2} \sum_{n=1}^{\infty} \frac{(1-\cos na)^2}{n^2 a^2} = \frac{1}{\pi} \int_{-\pi}^{\pi} f^2(x) \, dx$.

10. (b) (c) $f(x) = \frac{1}{2\pi} - \frac{4}{\pi} \sum\limits_{n=1}^{\infty} \frac{(-1)^n}{4n^2-1} \cos 2nx, \quad x \in \mathbb{R}.$

11. (b) $f(x) = \frac{1}{2\pi} + \frac{1}{2}\sin x - \frac{2}{\pi} \sum\limits_{n=1}^{\infty} \frac{1}{4n^2-1} \cos 2nx, \quad x \in \mathbb{R}.$

    (d) $F(x) = -\frac{1}{2}\cos x + \frac{2}{\pi} \sum\limits_{n=1}^{\infty} \frac{2n}{4n^2-1} \sin 2nx,$

13. $\alpha > 1.$

14. $0 < a < 1.$

15. No. For square integrable functions f the answer follows from the Parseval identity. For integrable functions f the answer follows by approximation with square integrable functions.

16. (a) $y(t) = C_1 \cosh t + C_2 \sinh t - \frac{1}{2} - \frac{2}{\pi} \sum\limits_{n=1}^{\infty} \frac{1}{2n-1+(2n-1)^3} \sin(2n-1)t.$

    (b) $y(t) = C_1 e^{2t} + C_2 e^t + \frac{1}{4} + \frac{6}{\pi} \sum\limits_{n=1}^{\infty} \frac{1}{\left[2-(2n-1)^2\right]^2 + 9(2n-1)^2} \cos(2n-1)t$

$+ \frac{2}{\pi} \sum\limits_{n=1}^{\infty} \frac{2-(2n-1)^2}{(2n-1)\left\{\left[2-(2n-1)^2\right]^2 + 9(2n-1)^2\right\}} \sin(2n-1)t.$

17. (a) $y(t) = C_1 \cos 3t + C_2 \sin 3t + \frac{\pi}{18} - \frac{2}{27\pi} \sin 3t$

$- \frac{4}{\pi} \sum\limits_{\substack{n=1 \\ n\neq 2}}^{\infty} \frac{1}{(2n-1)^2\left[9-(2n-1)^2\right]} \cos(2n-1)t.$

Section 1.4.

1. Fourier Cosine Series: $f(x) = \frac{\pi}{2} + \frac{4}{\pi} \sum\limits_{n=1}^{\infty} \frac{1}{2n-1} \cos(2n-1)x.$

    Fourier Sine Series: $f(x) = \frac{2}{\pi} \sum\limits_{n=1}^{\infty} \frac{1}{n} \sin nx.$

2. Fourier Cosine Series: $f(x) = \frac{2}{\pi} - \frac{4}{\pi} \sum\limits_{n=1}^{\infty} \frac{1}{4n^2-1} \cos 2nx.$

    Fourier Sine Series: $f(x) = \sin x.$

3. Fourier Cosine Series: $f(x) = \cos x.$

Fourier Sine Series: $f(x) = \frac{4}{\pi} \sum\limits_{n=1}^{\infty} \frac{n}{4n^2-1} \sin 2nx.$

4. Fourier Cosine Series: $f(x) = \frac{\pi^2}{3} + 4 \sum\limits_{n=1}^{\infty} \frac{(-1)^n}{n^2} \cos \left[(2n-1)x - \frac{\pi}{2}\right].$

Fourier Sine Series: $f(x) = 2 \sum\limits_{n=1}^{\infty} \frac{-2+(2-n^2\pi^2)(-1)^n}{n^3} \sin nx.$

5. Fourier Cosine Series: $f(x) = \frac{\pi}{4} + \frac{8}{\pi} \sum\limits_{n=1}^{\infty} \frac{\cos\frac{n\pi}{2}\sin^2\frac{n\pi}{4}}{n^2} \cos nx.$

Fourier Sine Series: $f(x) = \frac{4}{\pi} \sum\limits_{n=1}^{\infty} \frac{\sin\frac{n\pi}{2}}{n^2} \sin nx.$

6. Fourier Cosine Series: $f(x) = \frac{\pi}{2} - \frac{4}{\pi} \sum\limits_{n=1}^{\infty} \frac{1}{(2n-1)^2} \cos(2n-1)x.$

Fourier Sine Series: $f(x) = \frac{2}{\pi} \sum\limits_{n=1}^{\infty} \frac{(-1)^{n-1}}{n} \sin nx.$

7. Fourier Cosine Series: $f(x) = 3.$

Fourier Sine Series: $f(x) = \frac{12}{\pi} \sum\limits_{n-1}^{\infty} \frac{1}{2n-1} \sin(2n-1)x.$

8. Fourier Cosine Series: $f(x) = \frac{1}{2} + \frac{2}{\pi} \sum\limits_{n=1}^{\infty} \frac{(-1)^{n-1}}{2n-1} \cos \frac{(2n-1)\pi}{4} x.$

Fourier Sine Series: $f(x) = \frac{2}{\pi} \sum\limits_{n=1}^{\infty} \frac{1-\cos\frac{n\pi}{2}}{n} \sin \frac{n\pi}{4} x.$

9. Fourier Cosine Series: $f(x) = \frac{3}{4} + \frac{8}{\pi^2} \sum\limits_{n=1}^{\infty} \frac{\cos^2\frac{n\pi}{4}}{n^2} \cos \frac{n\pi}{2} x.$

Fourier Sine Series: $f(x) = \frac{4}{\pi^2} \sum\limits_{n=1}^{\infty} \frac{(-1)^{n-1}}{(2n-1)^2} \sin \frac{(2n-1)\pi}{2} x$

$-\frac{2}{\pi} \sum\limits_{n=1}^{\infty} \frac{(-1)^n}{n} \sin \frac{n\pi}{2} x.$

10. Fourier Cosine Series: $f(x) = \frac{1}{2} - \frac{2}{\pi} \sum\limits_{n=1}^{\infty} \frac{\sin\frac{n\pi}{2}}{n} \cos nx.$

Fourier Sine Series: $f(x) = \frac{2}{\pi} \sum\limits_{n=1}^{\infty} \frac{(\cos\frac{n\pi}{2}-\cos n\pi)}{n} \sin nx.$

11. Fourier Cosine Series: $f(x) = \frac{e^\pi - 1}{\pi} + \frac{2}{\pi} \sum_{n=1}^{\infty} \frac{(-1)^n e^\pi - 1}{n^2 + 1} \cos nx.$

Fourier Sine Series: $f(x) = -\frac{2}{\pi} \sum_{n=1}^{\infty} \frac{n\left[(-1)^n e^\pi - 1\right]}{n^2 + 1} \cos nx.$

Section 2.1.1.

1. (a) $F(s) = \frac{1}{s^2}.$

   (b) $F(s) = \frac{e}{-3+s}.$

   (c) $F(s) = \frac{s}{4+s^2}.$

   (d) $F(s) = \frac{2}{s(4+s^2)}.$

   (e) $F(s) = \frac{e^{-2s}(-1+e^{2s})}{s}.$

   (f) $F(s) = \frac{e^{-2s}(-1+e^{2s}+s)}{s^2}.$

   (g) $F(s) = \frac{e^{-s}}{s}.$

2. (a) $F(s) = \frac{1}{4+(s-1)^2}.$

   (b) $F(s) = \frac{1+s}{4+(s+1)^2}.$

   (c) $F(s) = \frac{s-1}{9+(s-1)^2}.$

   (d) $F(s) = \frac{s+2}{16+(s+2)^2}.$

   (e) $F(s) = \frac{5}{25+(s+1)^2}.$

   (f) $F(s) = \frac{1}{(s-1)^2}.$

3. (a) $F(s) = -18\frac{1}{(s-3)^2}.$

   (b) $F(s) = \frac{2}{(s+3)^3}.$

   (c) $F(s) = \frac{12s}{(s^2+4)^2}.$

   (d) $F(s) = \frac{s+2}{49+(s+2)^2}.$

   (e) $F(s) = \frac{1}{s} + \frac{s}{s^2+4}.$

(f) $F(s) = \frac{48s(s^2-4)}{(s^2+4)^4}$.

4. (a) $F(s) = \frac{\sqrt{\pi}}{\sqrt{s}}$.

(b) $F(s) = \frac{\sqrt{\pi}}{2s^{\frac{3}{2}}}$.

5. (a) $f(t) = t$.

(b) $f(t) = e^{2t}t$.

(c) $f(t) = \frac{1}{54}(-3t\cos 3t + t\sin 3t)$.

(d) $f(t) = e^{3t}t$.

(e) $f(t) = e^{-8t}(-1 + e^t)$.

(f) $f(t) = \frac{1}{4\sqrt{7}}\left(e^{\left(-8-2\sqrt{7}\right)t} - e^{\left(-8+2\sqrt{7}\right)t}\right)$.

(g) $f(t) = e^{\frac{7t}{2}}\cos\frac{\sqrt{61}t}{2}$.

(h) $f(t) = e^{-2t}\cos 2\sqrt{2}t$.

6. (a) $f(t) = \frac{1}{250}(-34e^t + 160e^t t - 75e^t t^2 + 34\cos 2t - 63\sin 2t)$.

(b) $f(t) = \frac{1}{216}(72t - 120t - 81\cos t + 9\cos 3t + 135\sin t - 5\sin 3t)$.

7. (a), (b), (c) - No.

Section 2.1.2.

2. (a) $F(s) = \frac{2e^{-2s}}{s^3}$.

(b) $F(s) = e^{-s}\frac{2-s+e^s s}{s^3}$.

(c) $F(s) = \frac{e^{-s}}{s}$.

(d) $F(s) = e^{-2\pi s}\frac{-1-\pi s+e^\pi s}{s^2}$.

(e) $F(s) = -6\frac{e^{-4s}}{s} + 2\frac{e^{-3s}}{s} + \frac{e^{-s}}{s}$.

(f) $F(s) = -\frac{e^{-s}}{s}$.

(g) $F(s) = \frac{1}{s^2} + \frac{e^{-s}}{s} - \frac{e^{-s}(s+1)}{s^2}$.

3. (a) $f(t) = e^{2t}t^3$.

   (b) $f(t) = \frac{1}{3}e^{-2(t-2)}\left(-1 + e^{3(t-2)}\right)H(t-2)$.

   (c) $f(t) = 2e^{t-2}\cos{(t-2)}\,H(t-2)$.

   (d) $f(t) = -\frac{1}{2}e^{-2(t-2)}\left(-1 + e^{4(t-2)}\right)H(t-2)$.

4. (a) $F(s) = 2e^{-\pi s}$.

   (b) $F(s) = e^{-3s}$.

   (c) $F(s) = 2e^{-3s} - e^{-s}$.

   (d) $F(s) = e^{-\frac{\pi s}{2}} - \frac{e^{2\pi s}}{1+s^2}$.

   (e) $F(s) = e^{-\pi s} + \frac{e^{-\pi s}s}{s^2+4}$.

Section 2.1.3.

1. (a) $f(t) = e^{-2t}\left(-1 + 3e^t\right)$.

   (b) $f(t) = \frac{1}{7}e^{-7t}\left[21 + \left(-e^{14} + e^{7t}\right)H(t-2)\right]$.

   (c) $f(t) = \frac{1}{5}e^{-7t}\left[5 + \left(-e^{14} + e^{4+5t}\right)H(t-2)\right]$.

   (d) $f(t) = \frac{1}{7}e^{-7t}\left[21 + \left(-e^{14} + e^{7t}\right)H(t-2)\right]$.

   (e) $f(t) = \frac{1}{5}e^{-7t}\left[5 + \left(-e^{14} + e^{4+5t}\right)H(t-2)\right]$.

2. (a) $f(t) = \frac{1}{13}\left(-3\cos{3t} - 2\sin{3t}\right)$
   $\qquad + \frac{e^{-\frac{t}{2}}}{39\sqrt{3}}\left(87\sqrt{3}\cos{\frac{3\sqrt{3}t}{2}} + 4\sin{\frac{3\sqrt{3}t}{2}}\right)$.

   (b) $f(t) = \frac{1}{22}\left(\cos{5(t-4)} + \cos{\sqrt{3}\,(t-4)}\right)H(t-4) - \frac{2}{\sqrt{3}}\sin{\sqrt{3}\,t}$.

   (c) $f(t) = \frac{1}{18}\left[36\cos{3t} + H(t-4)\left(-3(t-4)\cos{3(t-4)}\right.\right.$
   $\qquad \left.\left. + \sin{3(t-4)}\right)\right]H(t-2)\right]$.

   (d) $f(t) = \frac{1}{2}\left[-\left(-2 + \cos{(t-2)} + \cosh{(t-2)}\right)H(t-2)\right.$

$$+\big(-2+\cos{(t-1)}+\cosh{(t-1)}\big)H(t-1)\big].$$

(e) $f(t) = \frac{1}{9}\big[3t + H(t-1)\big(3 - 3t + \sqrt{3}\sin{\sqrt{3}}\,(ti1)\big) - 7\sqrt{3}\sin{\sqrt{3}}\,t\big].$

4. $F(s)\frac{2}{s}\tanh{\frac{s}{2}}.$

5. (a) $y(t) = te^t + 3(t-2)H(t-2).$

(b) $y(t) = 3\big(e^{-2(t-2)} - e^{-3(t-2)}\big)H(t-2) + 4\big(e^{-3(t-5)} - e^{-2(t-25)}\big)H(t-5).$

(c) $y(t) = e^{\pi - t} - \sin{(t-\pi)}\,H(t-\pi).$

(d) $y(t) = \frac{\sqrt{2}}{2}e^{-t}\sin{\sqrt{2}}\,t) + \frac{1}{4}e^{-t}\cos{\sqrt{2}}\,t + \frac{1}{4}\big(\sin{t} - \cos{t}\big)$
$+\frac{\sqrt{2}}{2}H(t-\pi)\,e^{-(t-\pi)}\sin{\sqrt{2}}\,(t-\pi).$

(e) $y(t) = \frac{1}{2}H(t-\pi)\sin{2t} - \frac{1}{2}H(t-2\pi)\,sin\,2t.$

(f) $y(t) = \big(1 + H(t-1)\big)\sin{t}.$

Section 2.1.4.

2. (a) $y(t) = t - \sin{t}.$

(b) $y(t) = e^t - t - 1.$

(c) $y(t) = \frac{t^2}{2} - \frac{1}{2}\sin^2{t}.$

(d) $y(t)\frac{1}{2}t\sin{t}.$

(e) $y(t) = \frac{t^2}{2} - \frac{1}{2}(t-2)^2\,H(t-2).$

3. (a) $F(s) = \frac{2}{s^2(s^2+4)}.$

(b) $F(s) = \frac{1}{(s+1)(s^2+1)}.$

(c) $F(s) = \frac{1}{(s-1)s^2}.$

(d) $F(s) = \frac{d}{(s^2+1)^2}.$

(e) $F(s) = \frac{2}{(s-1)s^3}.$

(f) $F(s) = \frac{2}{9(s+1)(s^2+4)}.$

4. (a) $y(t) = -1 + e^{-t} + t$.

   (b) $y(t) = \frac{1}{2}\left(t\cos t - \sin t\right)$.

   (c) $y(t) = \frac{1}{5}e^{-t} + \frac{1}{10}\left(-2\cos 2t + \sin 2t\right)$.

   (d) $y(t) = \frac{1}{256}\left(-1 + 8t^2 + \cos 4t\right)$.

   (e) $y(t) = \frac{1}{5} + \frac{1}{5}\left(\cos 2t + 2\sin 2t\right)$.

5. (a) $y(t) = t + \frac{3}{2}\sin 2t$.

   (b) $y(t) = 1 - \cos t$.

   (c) $y(t) = (1 - t)^2 e^{-t}$.

   (d) $y(t) = t^3 + \frac{1}{20}t^5$.

   (e) $y(t) = 1 + 2t$.

6. (a) $y(t) = 4 + \frac{5}{2}t^2 + \frac{1}{24}t^4$.

   (b) $y(t) = \frac{2}{\pi}\sqrt{t} - t$.

Section 2.2.1.

1. (a) $F(\omega) = \frac{2i\cos\omega}{\omega} + 2\pi\delta(\omega) - 4\frac{\sin\omega}{\omega}$.

   (b) $F(\omega) = e^{-3i\omega}\frac{1 - e^{3i\omega} + 3i\omega}{\omega^2}$.

   (c) $F(\omega) = \frac{i}{i-\omega}$.

   (d) $F(\omega) = \frac{\sqrt{2\pi}}{a}e^{-\frac{\omega^2}{2}}$.

6. $\int_0^\infty \frac{y\sin(\pi y)\cos(xy)}{1-y^2}\,dy = \cos|x|$ for $|x| < \pi$; $\int_0^\infty \frac{y\sin(\pi y)\cos(xy)}{1-y^2}\,dy = 0$ for $|x| > \pi$.

7. (a) $\mathcal{F}_c\{f\}(\omega) = \frac{2a}{\omega^2 + a^2}$.

(b) $\mathcal{F}_s\{f\}(\omega) = \frac{2\omega}{\omega^2+a^2}$.

(c) $\mathcal{F}_s\{xe^{-ax}\}(\omega) = \frac{4a\omega}{\omega^2+a^2}$.

(d) $\mathcal{F}_c\{(1+x)e^{-ax}\}(\omega) = \frac{2(-\omega^2+a^2\omega^2+a^2+a^3)}{(\omega^2+a^2)^2}$.

8. $F(\omega) = -\frac{i}{\omega} + \pi\delta(\omega)$.

Section 2.2.2.

2. (a) $F(\omega) = \frac{\pi}{|\omega|}e^{-|\frac{\omega}{a}|}$.

(b) $F(\omega) = \frac{\pi}{2}\left(e^{-|\omega-a|} + e^{-|\omega+a|}\right)$.

6. (a) $F(\omega) = -\frac{1}{2a}e^{-bt}\sin at$.

(b) $F(\omega) = \frac{1}{2}e^{-t}H(t) + \frac{1}{2}e^{t}H(-t)$.

(c) $F(\omega) = e^{-t}H(t) - e^{\frac{t}{2}}H(t) + \frac{t}{2}e^{-\frac{t}{2}}tH(t)$.

7. (a) $f(t) = \frac{i}{2}\,sgn(t)\,e^{-a|t|}$.

(b) $f(t) = \begin{cases} e^{2t}, & t > 0 \\ e^{-t}, & t < 0. \end{cases}$

(c) $f(t) = \frac{1}{4a}\left(1 - a|t|\right)e^{-a|t|}$.

8. $f(t) = \frac{1}{2a}e^{-a|t-1|}$.

9. (a) $y(t) = \frac{1}{4}e^{-|t|} + \frac{1}{2}te^{-t}H(t)$.

(b) $y(t) = \begin{cases} \frac{1}{9}e^{-t}, & t > 0 \\ \frac{1}{9}e^{2t} - \frac{1}{3}te^{2t}, & t < 0. \end{cases}$

(c) $y(t) = g(t) * f(t)$, $g(t) = \mathcal{F}^{-1}\{\frac{1}{\omega^2+1}\} = e^{-|t|}H(t)$.

(d) $y(t) = H(t)\left[e^{-t} + e^{-2t}\right]$.

Section 3.1.

2. (a) $\lambda_n = \frac{(2n-1)^2\pi^2}{4}$, $y_n(x) = \cos\frac{2n-1}{2}\pi x$.

(b) $\lambda_0 = -1$, $y_0(x) = e^{-x}$, $\lambda_n = n^2$, $y_n(x) = \sin nx - n \cos nx$.

(c) The eigenvalues are $\lambda_n = \mu_n^2$, where mu_n are the positive solutions of the equation $\cot \mu = \mu$, $y_n(x) = \sin \mu_n x$.

(d) $\lambda_n = -n^4$, $y_n(x) = \sin nx$.

4. (a) $\lambda_n = \mu_n^2$, where mu_n are the positive solutions of $\tan \mu = \mu$, $y_n(x) = \sin \mu_n x$.

(b) $\lambda_0 = -\mu_0^2$, $\mu_0 = \coth mu_0\pi$, $y_0(x) = \sinh \mu_0 x - \mu_n \cosh mu_n x$, $\lambda_n = \mu_n^2$, $y_n(x) = \sinh \mu_n x - \mu_n \cos \mu_n x$.

(c) The eigenvalues are $\lambda_n = \mu_n^2$, where mu_n are the positive solutions of $\tan 2\mu = -\mu$, $y_n(x) = \sin \mu_n x + \mu_n \cos \mu_n x$.

5. (a) $\lambda_n = n^2 + 1$, $y_n(x) = \frac{1}{x} \sin n\pi \ln x$.

(b) $\lambda_n = n^2\pi^2$, $y_n(x) = \sin n \ln x$.

(c) $\lambda_n = \frac{(2n-1)^2\pi^2}{4}$, $y_n(x) = \sin \left(\frac{2n-1}{2} \pi \ln x \right)$.

7. $\lambda_n = \frac{1}{4} + \left(\frac{n\pi}{\ln 2} \right)^2$, $y_n(x) = \frac{1}{\sqrt{x}} \sin \left(\frac{n\pi}{\ln 2} \ln x \right)$.

8. This is not a regular Sturm–Liouville problem. $\lambda_n = n^2$, $y_n(x) \in \{ \sin nx, \cos nx \}$.

Section 3.2.

1. $f(x) = \frac{4}{\pi} \sum\limits_{n=1}^{\infty} \frac{1}{2n-1} \sin \left(\frac{2n-1}{\ln 2} \pi \ln x \right)$.

2. (a) $f(x) = \frac{1}{\pi} \sum\limits_{n=1}^{\infty} \frac{1}{2n-1} \cos \frac{2n-1}{2} \pi x$.

(b) $\frac{8}{\pi^2} \sum\limits_{n=1}^{\infty} \frac{(-1)^{n-1}}{2n-1} \cos \frac{2n-1}{2} \pi x$.

(c) $f(x) = \frac{4}{\pi} \sum\limits_{n=1}^{\infty} \frac{1-\cos^2 \frac{2n-1}{4}\pi}{2n-1} \cos \frac{2n-1}{2} \pi x$.

(d) $f(x) = \frac{16}{\pi^2} \sum\limits_{n=1}^{\infty} \frac{\sin \frac{2n-1}{4} \pi \right)}{(2n-1)^2} \cos \frac{2n-1}{2} \pi x$.

3. $f(x) = \frac{l}{\pi} \sum\limits_{n=1}^{\infty} \frac{(-1)^{n-1}}{n} \sin \frac{n\pi}{l} x.$

4. $f(x) = \frac{4l}{\pi^2} \sum\limits_{n=1}^{\infty} \frac{(-1)^{n-1}}{(2n-1)^2} \sin \frac{(2n-1)\pi}{2l} x.$

5. (a) $f(x) = \sqrt{2} \sum\limits_{n=1}^{\infty} \frac{\sin\left(\sqrt{\lambda_n}\right)}{\sqrt{\lambda_n} \sqrt{1+\sin^2 \sqrt{\lambda_n}}} \cos\left(\sqrt{\lambda_n}\, x\right),$

$\cos \sqrt{\lambda_n} - \sqrt{\lambda_n} \cos \sqrt{\lambda_n} = 0.$

(b) $f(x) = \sqrt{2} \sum\limits_{n=1}^{\infty} \frac{2\cos\left(\sqrt{\lambda_n}-1\right)}{\sqrt{\lambda_n}\sqrt{1+\sin^2\sqrt{\lambda_n}}} \cos\left(\sqrt{\lambda_n}\, x\right),$

$\cos \sqrt{\lambda_n} - \sqrt{\lambda_n} \cos \sqrt{\lambda_n} = 0.$

Section 3.2.1.

5. $\frac{1}{2} \ln \frac{1+x}{1-x}.$

6. (a) $f(x) = \frac{1}{4}P_0(x) + \frac{1}{2}P_1(x) + \frac{5}{16}P_2(x) + \cdots.$

(b) $f(x) = \frac{1}{2}P_0(x) + \frac{5}{8}P_2(x) - \frac{3}{16}P_4(x) + \cdots.$

(c) $f(x) = \frac{3}{2}P_1(x) - \frac{7}{3}P_3(x) + \frac{11}{6}P_5(x).$

Section 3.2.2.

5. (a) $y = c_1 x^{-1} J_0(\mu x) + c_2 x^{-1} Y_0(\mu x).$

(b) $y = c_1 x^{-2} J_2(x) + c_2 x^{-2} Y_2(x).$

(c) $y = c_1 J_0(x^\mu) + c_2 Y_2(x^\mu).$

7. $y = -\sqrt{\frac{2}{\pi x}} \left(\sin x + \frac{\cos x}{x} \right).$

9. (a) $1 = 2 \sum\limits_{k=1}^{\infty} \frac{1}{\lambda_k J_1(\lambda_k)} J_0(\lambda_k x).$

(b) $x^2 = 2 \sum\limits_{k=1}^{\infty} \frac{\lambda_k - 4}{\lambda_k^3 J_1(\lambda_k)} J_0(\lambda_k x).$

(c) $\frac{1-x^2}{8} = \sum\limits_{k=1}^{\infty} \frac{1}{\lambda_k J_1(\lambda_k)} J_0(\lambda_k x).$

Section 4.1.

4. (a) $u(x, y) = F(x) + G(y)$, where F is any differentiable function and G is any function.

(b) $u(x, y) = xF(y) + G(y)$, where F and G are any functions of one variable.

5. (a) $u(x, y) = yF(x) + G(x)$, where F and G are any functions of one variable.

(b) $u(x, y) = F(x) + G(y)$, where F is any differentiable function and G is any function of one variable.

(c) $u(x, y) = F(y) \sin x + G(y) \cos x$, where where F and G are any functions of one variable.

(d) $u(x, y) = F(x) \sin y + G(x) \cos y$, where where F and G are any functions of one variable.

8. (a) $xu_x - yu_y = x - y$.

(b) $u = u_x u_y$. (c) $u^2 - u(u_x + yu_y) = 1$.

(d) $xu_y - yu_x = x^2 - y^2$.

(e) $xu_x - yu_y = 0$.

Section 4.2.

1. (a) $u(x, y) = f\left(\ln x + \frac{1}{y}\right)$, f is an arbitrary differentiable function of one variable.

(b) $u(x, y) = f\left(\frac{x}{y}\right) - \cos\left(\frac{xy}{2}\right)$, f is an arbitrary differentiable function of one variable.

(c) $u(x, y) = f(y - \arctan x)$, f is an arbitrary differentiable function of one variable.

(d) $u(x, y) = f\left(x^2 + \frac{1}{y}\right)$, f is an arbitrary differentiable function of one variable.

2. (a) $u(x, y) = f(bx - ay)e^{-\frac{c}{a}x}$, f is an arbitrary differentiable function of one variable.

(b) $u(x, y) = f(x^2 + y^2)e^{\frac{c}{y}}$, f is an arbitrary differentiable function of one variable.

(c) $u(x, y) = f(e^{-x(x+y+1)})e^{\frac{x^2}{2}}$, f is an arbitrary differentiable function of one variable.

(d) $u(x, y) = f(xy)e^{\frac{y^2}{2}} + 1$, f is an arbitrary differentiable function of one variable.

(e) $u(x, y) = xf\left(\frac{y}{x}\right)$, f is any differentiable function of one variable.

(f) $u(x, y) = xyf\left(\frac{x-y}{xy}\right)$, f is an arbitrary differentiable function of one variable.

(g) $u(x, y) = e^{-(x+y)^2}(x^2 + f(4x - 3y))$, f is an arbitrary differentiable function of one variable.

(h) $u(x, y) = xf\left(\frac{1+xy}{x}\right)$, f is an arbitrary an differentiable function of one variable.

3. (a) $u(x, y) = f(5x + 3y)$; $u_p(x, y) = \sin\left(\frac{5x+3y}{3}\right)$.

(b) $u(x, y) = f(3x + 2y) + \frac{1}{2}\sin x$; $u_p(x, y) = \frac{1}{2}\sin x + \frac{1}{25}(3x + 2y)^2 - \frac{1}{2}\sin\frac{3x+2y}{5}$.

(c) $u(x, y) = xy + f\left(\frac{y}{x}\right)$; $u_p(x, y) = xy + 2 - \left(\frac{y}{x}\right)^3$.

(d) $u(x, y) = x\left(y - \frac{x^2}{2}\right)^2 + f\left(y - \frac{x^2}{2}\right)$; $u_p(x, y) = x\left(y - \frac{x^2}{2}\right)^2 + e^{y - \frac{x^2}{2}}$.

(e) $u(x, y) = \frac{1}{y}f(x^2 + y)$; $u_p(x, y) = \frac{1}{y}bigl(x^2 + y)^4$.

(f) $u(x, y) = x + yf\left(\frac{y^2}{x}\right)$; $u_p(x, y) = x + \frac{x}{y}$.

4. (a) $u(x, y) = c\left(x^2 - y^2\right)^2$.

(b) $u(x, y) = \frac{1}{2}\left[2y - (x^2 - y^2)\right]$.

(c) $u(x, y) = x^2 e^{-\frac{y}{x}} + e^{\frac{y}{x}} - 1$.

(d) $u(x, y) = e^{\frac{x}{x^2 - y^2}}$.

(e) $u(x, y) = \sqrt{xy}$.

(f) $u(x, y) = e^{1 - \frac{1}{x^2 + y^2}} f\left(\frac{x}{\sqrt{x^2 + y^2}} + \frac{y}{\sqrt{x^2 + y^2}} \right)$.

(g) $u(x, y) = e^y f\left(-\ln(y + e^{-x}) \right)$.

5. (a) $u(t, x) = x e^{t - t^2}$.

   (b) $u(t, x) = f\left(x e^{-t} \right)$.

   (c) $u(t, x) = f\left(\frac{(xt - 1) + \sqrt{(xt - 1)^2 + 4x^2}}{2x} \right)$.

   (d) $u(t, x) = f(x - ct)$

   (e) $u(t, x) = \sin(x - t) + (x - t)\sin t - x + t + \cos t + t \sin t - 1$.

6. $u(t, x) = \begin{cases} t - x, & 0 < x < t, \\ (x - t)^2, & t < x < t + 2, \\ x - t, & x > t + 2. \end{cases}$

Section 4.3.

1. $x u_{xy} - y u_{yy} - u_y + 9y^2 = 0$.

2. (a) $D = b^2 - ac = 8 > 0$, hyperbolic.

   (b) $D = 0$, parabolic.

   (c) $D = -27 < 0$, elliptic.

   (d) hyperbolic.

   (e) parabolic.

   (f) elliptic.

3. (a) $D = -2y$. hyperbolic if $y > 0$, parabolic if $y = 0$, elliptic if $y > 0$.

   (b) $D = 4y^2(x^2 + x + 1)$. hyperbolic if $y < 0$, parabolic if $y = 0$, elliptic - nowhere.

(c) $D = -y^2$. elliptic if $y \neq 0$, parabolic if $y = 0$, hyperbolic - nowhere.

(d) $D = 4y^2 - x^2$. hyperbolic if $2|y| > |x|$, parabolic if $2|y| = |x|$, elliptic if $2|y| < |x|$.

(e) $D = 1 - ye^x$. hyperbolic if $y < e^{-x}$, parabolic if $y = e^{-x}$, elliptic if $y > e^{-x}$.

(f) $D = 3xy$. hyperbolic in quadrants I and III, parabolic on the coordinate axes, elliptic in quadrants II and IV.

(g) $D = x^2 - a^2$. hyperbolic if $|x| > |a|$, parabolic if $|x| = |a|$, elliptic if $|x| < |a|$.

4. (a) $D = -3/4 < 0$. The equation is elliptic. The characteristic equation is $y'^2 + y' + 1 = 0$. $\xi = y - 1/2(1 + \sqrt{3}\,i)x$, $\eta = y - 1/2(1 - \sqrt{3}\,i)x$; $\alpha = \xi + \eta$, $\beta = (\xi - \eta)i$. The canonical form is $u_{\alpha\alpha} + u_{\beta\beta} = 0$.

(b) $D = -4x$. hyperbolic if $x < 0$, parabolic if $x = 0$, elliptic if $x > 0$. The characteristic equation is $xy'^2 + 1 = 0$. If $x < 0$, then $\xi = 2 + \sqrt{-x}$, $\eta = y - \sqrt{-x}$. The canonical form in this case is

$$u_{\xi\eta} = \frac{1}{4} - \frac{1}{16^2}(\xi - \eta)^2 - \frac{u_\xi - u_\eta}{2(\xi - \eta)}.$$

If $x < 0$, $xi = y + 2\sqrt{x}\,i$, $\eta = y - 2\sqrt{x}\,i$, $\alpha = 1/2(\xi + \eta)$, $\beta = -i/2(\xi - \eta)$. The canonical form is

$$u_{\alpha\alpha} + u_{\beta\beta} = \frac{1}{\beta} + \frac{\beta^2}{4}.$$

If $x = 0$, then the equation reduces to $u_{yy} = 0$.

(c) $D = 9 > 0$. The equation is hyperbolic. The characteristic equation is $4y'^2 + 5y' + 1 = 0$. $\xi = x + y$, $\eta = 1/4x - y$. The canonical form is

$$u_{\xi\eta} = \frac{1}{3}\left(u_\eta - \frac{8}{3}\right)$$

(d) $D = 1/4$. The equation is hyperbolic. The characteristic equation is $2y'^2 - 3y' + 1 = 0$. $\xi = x - y$, $\eta = 1/2x - y$. The canonical form is

$$u_{\xi\eta} = u_\eta - u - \eta.$$

(e) $D = -y$. The equation is hyperbolic if $y < 0$, parabolic if $y = 0$ and elliptic if $y > 0$. For the hyperbolic case, $\xi = x + \sqrt{-y}$, $\eta = x - \sqrt{-y}$. The canonical form is

$$u_{\xi\eta} = \frac{u_\eta - u_\xi}{2(\xi - \eta)}.$$

For the elliptic case, $\alpha = x$, $\beta = 2\sqrt{y}$. The canonical form is

$$u_{\alpha\alpha} + u_{\beta\beta} = \frac{1}{\beta}u_\beta.$$

In the parabolic case we have $u_{xx} = 0$.

(f) $D = -x^2 y^2$. The equation is elliptic if $(x, y) \neq (0, 0)$, and it is parabolic if $y = 0$ or $y = 0$. For the elliptic case, $\xi = y^2$, $\eta = x^2$ and the canonical form is

$$u_{\xi\xi} + u_{\eta\eta} = \frac{1}{2\xi}u_\xi + \frac{1}{2\eta}u_\eta.$$

In the parabolic case we have $u_{xx} = 0$ or $u_{yy} = 0$.

(g) $D = -(a^2 + x^2)(a^2 + y^2) < 0$. The equation is elliptic in the whole plane. The canonical form is

$$u_{\xi\xi} + u_{\eta\eta} = 0.$$

5. (a) $D = 0$. The equation is parabolic. The characteristic equation is $9y'^2 + 12y' + 4 = 0$. $\xi = 2x + 3y$, $\eta = x$. The canonical form is $u_{\eta\eta} = 0$. $u(x, y) = xf(3y - 2x) + g(3y - x)$.

(b) $D = 0$. Parabolic. The characteristic equation is $y'^2 + 8y' + 16 = 0$. $\xi = 4x - y$, $\eta = x$. The canonical form is $u_{\eta\eta} = 0$. $u(x, y) = xf(4x - y) + g(4x - y)$.

(c) $D = 4 > 0$. The equation is hyperbolic. The characteristic equation is $y'^2 + 2y' - 3 = 0$. $\xi = y - 3x$, $\eta = x + y$. $u(x, y) = f(y - 3x) + g(x + y)$.

(d) $D = 1/4$. The equation is hyperbolic. The characteristic equation is $2y'^2 - 3y' + 1 = 0$. $\xi = x^2 - y^2$, $\eta = y$. The canonical form is $u_{\xi\eta} = 0$. $u(x, y) = f(x^2 - y^2) + g(y)$.

(e) $D = 0$. The equation is parabolic. The characteristic equation is $xy' = y$. $\xi = x$, $\eta = \frac{y}{x}$. The canonical form is $u_{\xi\xi} = 4$.

$u(x,y) = 2x^2 + xf\left(\frac{y}{x}\right) + g\left(\frac{y}{x}\right).$

(f) $D = 0$. The equation is parabolic. $\xi = y + \ln(1 + \sin x)$, $\eta = y + \ln(1 + \sin x)$ and the canonical form is $u_{\eta\eta} + u_\eta = 0$. $u(\xi,\eta) = f(\xi) + e^{-\frac{\eta}{4}}g(\xi)$.

6. (a) $D = \frac{9}{4} > 0$. The equation is hyperbolic. The characteristic equation is $y'^2 + y' - 2 = 0$. $\xi = x - \frac{y}{2}$, $\eta = x + y$. The canonical form is $9u_{\xi\eta} + 2 = 0$. The general solution is $u(x,y) = -\frac{2}{9}\left(x + \frac{y}{2}\right)(x + y) + f\left(x + \frac{y}{2}\right) + g(x + y)$. From the boundary conditions we obtain that $u(x,y) = x + xy + \frac{y^2}{2}$.

(b) $D = 0$. The equation is parabolic. The characteristic equation is $e^{2x}y'^2 - e^{x+y}y' + e^{2y} = 0$. $\xi = e^{-x} + e^{-y}$, $\eta = e^{-x} - e^{-y}$. The canonical form is $u_{\xi\eta} = 0$. $u(\xi,\eta) = f(\xi) + \eta g(\xi)$. $u(x,y) = f(e^{-x} + e^{-y}) + (e^{-x} - e^{-y})g(e^{-x} + e^{-y})$. Using the boundary conditions we have $f(t) = \frac{t}{2}$ and $g(t(=\frac{(2-t)^2}{2}$.

(c) $D = 0$. The equation is parabolic. $\xi = x+y$, $\eta = x$. General solution $u(x,y) = f(x+y) + xg(x+y) + \frac{4}{9}e^{3y} - \cos y$. From $u(0,y) = \frac{4}{9}e^{3y}$ we obtain that $f(y) = \cos y$. From $u(x,0) = -\frac{4}{9}$ we obtain that $g(x) = \frac{1}{x}$. Therefore $u(x,y) = \cos(x+y) + \frac{x}{x+y} + \frac{4}{9}e^{3y} - \cos y$.

7. (a) $\xi = x$, $\eta = 2x + y$, $\zeta = 2x - 2y + z$. The canonical form is

$$u_{\xi\xi} + u_{\eta\eta} + u_{\zeta\zeta} + u_\xi = 0.$$

(b) $\xi = x + \frac{1}{2}y - z$, $\eta = -\frac{1}{2}y$, $\zeta = z$. The canonical form is

$$u_{\xi\xi} = u_{\eta\eta} + u_{\zeta\zeta}.$$

(c) $t' = \frac{1}{2}t + \frac{1}{2}x - \frac{1}{2}y - \frac{1}{2}z$, $\xi = \frac{1}{2}t + \frac{1}{2}x + \frac{1}{2}y + \frac{1}{2}z$, $\eta = -\frac{1}{2\sqrt{3}}t + \frac{1}{2\sqrt{3}}x + \frac{1}{2\sqrt{3}}y - \frac{1}{2\sqrt{3}}z$, $\zeta = -\frac{1}{2\sqrt{5}}t + \frac{1}{2\sqrt{5}}x - \frac{1}{2\sqrt{5}}y + \frac{1}{2\sqrt{5}}z$. The canonical form is

$$u_{t't'} = u_{\xi\xi} + u_{\eta\eta} + u_{\zeta\zeta}.$$

(d) $t' = \frac{1}{2\sqrt{3}}t + \frac{1}{2\sqrt{3}}x - \frac{1}{2\sqrt{3}}y - \frac{1}{2\sqrt{3}}z$, $\xi = \frac{1}{\sqrt{2}}x + \frac{1}{\sqrt{2}}y$, $\eta = \frac{1}{\sqrt{2}}t + \frac{1}{2\sqrt{2}}z$, $\zeta = -\frac{1}{2}t + \frac{1}{2}x - \frac{1}{2}y - \frac{1}{2}z$. The canonical form is

$$u_{t't'} = u_{\xi\xi} + u_{\eta\eta} + u_{\zeta\zeta}.$$

Section 5.1.

3. $u(x,t) = \sin 5t \cos x + \frac{1}{15} \sin 3x \sin \frac{15t}{2}$.

4. $u(x,t) = \frac{1}{2}\left[\frac{1}{1+(x+at)^2} + \frac{1}{1+(x-at)^2}\right] + \frac{1}{a} \sin x \sin at$.

5. $u(x,t) = -\frac{1}{a}\left[e^{-(x+at)^2} - e^{-(x-at)^2}\right]$.

6. $u(x,t) = \sin 2x \cos 2at + \frac{1}{a} \cos x \sin at$.

7. $u(x,t) = \frac{1+x^2+a^2t^2}{(1+x^2+a^2t^2)^2+4a^2x^2t^2} + \frac{1}{2a}e^x\left(e^{at} - e^{-at}\right)$.

8. $u(x,t) = \cos \frac{\pi x}{2} \cos \frac{\pi at}{2} + \frac{1}{2ak}\left(e^{kx} - e^{-kx}\right)\left(e^{akt} - e^{-akt}\right)$.

9. $u(x,t) = e^{x+t} + e^{x-t} + 2\sin x \sin t$.

10. $u(x,t) = \frac{1}{a}\left[e^{-(x+at)^2} - e^{-(x-at)^2}\right] + \frac{t}{2} + \frac{1}{4a} \cos 2x \sin 2at$.

11. $u(x,t) = \begin{cases} 1, & \text{if } x - 5t < 0, \, x + 5t < 0 \\ \frac{1}{2}, & \text{if } x - 5t < 0, \, x + 5t > 0 \\ 0, & \text{if } x - 5t > 0. \end{cases}$

12. $u(x,t) = \frac{1}{2}\left[\sin(x+at) + \sin(x-at)\right] + \frac{1}{2a}\left[\arctan(x+at) + \arctan(x-at)\right]$,
if $x - 5t > 0$;

$u(x,t) = \frac{1}{2}\left[\sin(x+at) - \sin(x-at)\right] + \frac{1}{2a}\left[\arctan(x+at) - \arctan(x-at)\right]$,
if $0 < x < at$.

13. $\frac{1}{2}\left[f(x+at) + f(x-at)\right]$.

14. $\sin 2\pi x \cos 2\pi t + \frac{2}{\pi} \sin \pi x \cos \pi t$.

15. $u(x,t) = \frac{1}{2}\left[G(x+t) - G(x-t)\right]$, where G is the anti-derivative of g_p

and g_p is the 2 periodic extension of

$$g_0(x) = \begin{cases} -1, & -1 < x < 1 \\ 1, & 0 < x < 1. \end{cases}$$

The function G is 2 periodic and for $-x \le x \le 1$ we have that $G(x) = |x|$.

Let $t_r > 0$ be the times when the string returns to its initial position $u(x, 0) = 0$. This will happen only when

$$G(x + t_r) = G(x - t_r).$$

Since G is 2 periodic, from the last equation we obtain that

$$x + t_r = x - t_r + 2n, \ n = 1, 2, \ldots.$$

Therefore

$$x_r = n, \ n = 1, 2, \ldots$$

are the moments when the string will come back to its original position.

16. $u(x,t) = \frac{1}{a(1+a^2)} \sin x \left(\sin at - a \cos at + ae^{-t} \right).$

17. $E_p(t) = \frac{\pi}{4} \cos^2 t. \ E_k(t) = \frac{\pi}{4} \sin^2 t.$ The total energy is $E(t) = \frac{\pi}{4}.$

Section 5.2.

1. (a) $u(x,t) = \frac{4}{\pi} \sum_{n=1}^{\infty} \frac{1}{(2n-1)^2} \sin(2n-1)x \sin(2n-1)t.$

(b) $u(x,t) = \frac{9}{\pi} \sum_{n=1}^{\infty} \frac{1}{n^2} \sin \frac{2n\pi}{3} \sin nx \cos nt.$

(c) $u(x,t) = \sin x \sin t + \frac{4}{\pi} \sum_{n=1}^{\infty} \frac{(-1)^{n+1}}{(2n-1)^2} \sin \frac{(2n-1)\pi}{4}$
$\cdot \sin(2n-1)x \sin(2n-1)t.$

(d) $u(x,t) = \frac{4}{\pi} \sum_{n=1}^{\infty} \frac{(-1)^{n+1}}{(2n-1)^2} \sin(2n-1)x \cos(2n-1)t.$

(e) $u(x,t) = \frac{8}{\pi} \sum_{n=1}^{\infty} \frac{1}{(2n-1)^3} \sin(2n-1)x \sin(2n-1)t.$

2. (a) $u(x,t) = \sum_{n=1}^{\infty} e^{-kt} \left(a_n \cos \lambda_n t + b_n \sin \lambda_n t \right) \sin \frac{n\pi x}{l}$, where λ_n, a_n and b_n are given by

$$\lambda_n^2 = \frac{a^2 n^2 \pi^2}{l^2} - k^2, \quad a_n = \frac{2}{l} \int_0^l f(x) \sin \frac{n\pi x}{l}\, dx;$$

$$\frac{2}{l}(-ka_n + \lambda_n b_n) = \int_0^l g(x) \sin \frac{n\pi x}{l}\, dx.$$

3. $u(x,t) = \frac{4}{\pi} e^{-t} \Big[\sin x \cosh\left(\sqrt{3}t\right)$
$$+ \sum_{n=2}^{\infty} \frac{1}{2n-1} \sin(2n-1)x \cos\left(\sqrt{4n^2 - 4n - 2t}\right)\Big].$$

4. $u(x,t) = e^{-t}\Big[t \sin x + \frac{1}{\sqrt{3}} \sin 2x \sin \sqrt{3}t\Big].$

5. $u(x,t) = \sum_{n=1}^{\infty} \Big[a_n \cos\left(\frac{2n-1}{2l}\pi a t\right) + \frac{2l b_n}{(2n-1)} \sin\left(\frac{2n-1}{2l}\pi a t\right)\Big] \sin\left(\frac{2n-1}{2l}\pi x\right),$
where

$$a_n = \frac{2}{l} \int_0^l f(x) \sin\left(\frac{2n-1}{2l}\pi x\right) dx, \quad b_n = \frac{2}{l} \int_0^l g(x) \sin\left(\frac{2n-1}{2l}\pi x\right) dx.$$

6. $u(x,t) = \frac{4}{\pi^3} \sum_{n=1}^{\infty} \Big[\frac{2-2(-1)^n}{n^3} \cos 2n\pi t + \frac{1}{4n^3 - n} \sin 2n\pi t\Big] \sin n\pi x.$

7. $u(x,t) = \frac{2}{\pi} + \frac{4}{\pi} \sum_{n=1}^{\infty} \frac{1}{1-4n^2} \cos 2nx \cos 4nt.$

8. $u(x,t) = -\frac{8}{\pi^3} \sum_{n=1}^{\infty} \frac{1}{(2n-1)^3} \cos(2n-1)\pi x \sin(2n-1)\pi t.$

9. $u(x,t) = \frac{1}{2}(a_0 + a_0' t) + \sum_{n=1}^{\infty} \left(a_n \cos \frac{n\pi a}{l} t + \frac{l}{n\pi a} a_n' \sin \frac{n\pi a}{l} t\right) \sin \frac{n\pi a}{l} x.$

10. $u(x,t) = \frac{Al^2}{l^2 + a^2 \pi^2}\left(e^{-t} - \cos \frac{a\pi}{l} t + \frac{l}{a\pi} \sin \frac{a\pi}{l} t\right).$

11. $u(x,t) = \frac{2Al^3}{\pi} \sum_{n=1}^{\infty} \frac{(-1)^{n+1} \sin \frac{an\pi x}{l}}{n(l^2 + a^2 \pi^2 n^2)}\left(e^{-t} - \cos \frac{an\pi}{l} t + \frac{l}{a\pi n} \sin \frac{an\pi}{l} t\right).$

12. $u(x,t) = \frac{16 Al^2}{\pi} \sum_{n=1}^{\infty} \frac{\sin \frac{a(2n-1)\pi}{2l} x}{(2n-1)\left[a^2 \pi^2 n^2 (2n-1)^2 - 4l^2\right]} \sin t$
$$- \frac{32 Al^3}{a\pi^2} \sum_{n=1}^{\infty} \frac{\sin \frac{a(2n-1)\pi}{2l} x}{(2n-1)^2\left[a^2 \pi^2 n^2 (2n-1)^2 - 4l^2\right]} \sin \frac{a(2n-1)\pi}{2l} t.$$

13. $u(x,t) = \frac{4Al^2}{4l^2+a^2\pi^2} \sum\limits_{n=1}^{\infty} \left(e^{-t} - \cos\frac{a\pi}{2l}t + \frac{2l}{a\pi}\sin\frac{a\pi}{2l}t\right)\cos\frac{\pi}{2l}x.$

14. $u(x,t) = \left(1 - \frac{x}{\pi}\right)t^2 + \frac{x}{\pi}t^3 + \sin x \cos t$

$\qquad + \frac{4}{\pi}\sum\limits_{n=1}^{\infty}\frac{1}{n^3}\left[3(-1)^n t - 1 + \cos nt - \frac{3}{n}(-1)^n \sin nt\right]\sin nx.$

15. $u(x,t) = \left(1 - \frac{x}{\pi}\right)e^{-t} + \frac{xt}{\pi} + \frac{1}{2}\sin 2x \cos 2t$

$\qquad - \frac{2}{\pi}\sum\limits_{n=1}^{\infty}\frac{1}{n(1+n^2)}\left[e^{-t} + n^2\cos nt - \left(2n + \frac{1}{n}\right)\sin nt\right]\sin nx.$

16. $u(x,t) = x + t + \sin\frac{x}{2}\,\cos\frac{t}{2} - \frac{8}{\pi}\sum\limits_{n=1}^{\infty}\frac{(-1)^n}{(2n+1)^2}\cos\frac{2n+1}{2}t\,\sin\frac{2n+1}{2}x.$

Section 5.3.

1. $u(x,y,t) = \frac{64}{\pi^6}\sum\limits_{m=1}^{\infty}\sum\limits_{n=1}^{\infty}\frac{1}{(2m-1)^3(2n-1)^3}\sin(2m-1)\pi x\,\sin(2n-1)\pi y$

$\qquad \cos\sqrt{(2m-1)^2 + (2n-1)^2}\,t.$

2. $u(x,y,t) = \sin\pi x\,\sin\pi y\,\cos\sqrt{2}t + \frac{4}{\pi}\sum\limits_{n=1}^{\infty}\frac{\sin(2n-1)\pi y}{(2n-1)\sqrt{1+(2n-1)^2}}$

$\qquad \sin\sqrt{1+(2n-1)^2}\,t.$

3. $u(x,y,t) = v(x,y,t) + \frac{2}{\sqrt{5}}\sin\pi x\,\sin 2\pi y\,\sin\sqrt{5}t,$ where

$\quad v(x,y,t) = \frac{64}{\pi^6}\sum\limits_{m=1}^{\infty}\sum\limits_{n=1}^{\infty}\frac{1}{(2m-1)^3(2n-1)^3}\sin(2m-1)\pi x\,\sin(2n-1)\pi y$

$\quad \cos\sqrt{(2m-1)^2 + (2n-1)^2}\,t.$

4. $u(x,y,t) = \frac{16}{\pi^2}\sum\limits_{m=1}^{\infty}\sum\limits_{n=1}^{\infty}\frac{\sin(2m-1)\pi x\,\sin(2n-1)\pi y}{(2m-1)(2n-1)\sqrt{(2m-1)^2+(2n-1)^2}}$

$\qquad \sin\sqrt{(2m-1)^2 + (2n-1)^2}\,t.$

5. $u(x,y,t) = \frac{16}{\pi^6}\sum\limits_{m=1}^{\infty}\sum\limits_{n=1}^{\infty}\frac{1}{(2m-1)^3(2n-1)^3}\sin(2m-1)\pi x\,\sin(2n-1)\pi y$

$\qquad \sin\sqrt{(2m-1)^2 + (2n-1)^2}\,t.$

6. (a) $u(x,y,t) = \sum\limits_{m=1}^{\infty} \sum\limits_{n=1}^{\infty} \left[a_{mn} \cos \sqrt{\lambda_{mn}}\, t + b_{mn} \sin \sqrt{\lambda_{mn}}\, t \right] \sin \frac{m\pi x}{a}$

$\sin \frac{(2n-1)\pi y}{2b}$, where $\lambda_{mn} = \frac{m^2\pi^2}{a^2} + \frac{(2n-1)^2\pi^2}{4b^2}$.

(b) $u(x,y,t) = \sum\limits_{m=0}^{\infty} \sum\limits_{n=1}^{\infty} \left[a_{mn} \cos \sqrt{\lambda_{mn}}\, t + b_{mn} \sin \sqrt{\lambda_{mn}}\, t \right]$

$\cos \frac{m\pi x}{a} \sin \frac{n\pi y}{b}$, where $\lambda_{mn} = \frac{m^2\pi^2}{a^2} + \frac{n^2\pi^2}{b^2}$.

7. $u(x,y,t) = 3\cos 4t \sin 4y - 5\cos \sqrt{5}t \cos 2x \sin y$
$+ \frac{7}{\sqrt{10}} \sin \sqrt{10}t \cos x \sin 3y$.

8. $u(x,y,t) = e^{-k^2 t} \sum\limits_{m=1}^{\infty} \sum\limits_{n=1}^{\infty} \left(a_{mn} \cos \left(\lambda_{mn} t \right) + b_{mn} \sin \left(\lambda_{mn} t \right) \right)$
$\sin \left(\frac{m\pi x}{a} \right) \sin \left(\frac{n\pi y}{b} \right)$, where

$\lambda_{mn} = \sqrt{\frac{m^2\pi^2 c^2}{a^2} + \frac{n^2\pi^2 c^2}{b^2} - k^4}$.

9. Let $\omega_{mn} = a\pi \sqrt{\frac{m^2\pi^2 c^2}{a^2} + \frac{n^2\pi^2 c^2}{b^2}}$. You need to consider two cases:

Case 1^0. $\omega \neq \omega_{mn}$ for every $m, n = 1, 2, \ldots$. In this case

$u(x,y,t) = \sum\limits_{m=1}^{\infty} \sum\limits_{n=1}^{\infty} a_{mn} \left(\sin \omega t - \frac{\omega}{\omega_{mn}} \sin \omega_{mn} t \right) \sin \left(\frac{m\pi x}{a} \right) \sin \left(\frac{n\pi y}{b} \right)$,

where

$a_{mn} = \frac{4}{(\omega_{mn}^2 - \omega^2)ab} \int\limits_{0}^{a} \int\limits_{0}^{b} F(x,y) \sin \left(\frac{m\pi x}{a} \right) \sin \left(\frac{n\pi y}{b} \right) dy\, dx$.

Case 2^0. $\omega = \omega_{m_0 n_0}$ for some m_0 and n_0 (resonance), then

$u(x,y,t) = \sum\limits_{\substack{m=1 \\ m \neq m_0}}^{\infty} \sum\limits_{\substack{n=1 \\ n \neq n_0}}^{\infty} a_{mn} \left(\sin \omega t - \frac{\omega}{\omega_{mn}} \sin \omega_{mn} t \right) \sin \left(\frac{m\pi x}{a} \right) \sin \left(\frac{n\pi y}{b} \right)$

$+ a_{m_0 n_0} \left(\sin \omega t - \omega t \cos \omega t \right) \sin \left(\frac{m_0 \pi x}{a} \right) \sin \left(\frac{n_0 \pi y}{b} \right)$, where for $m \neq m_0$
and $n \neq n_0$ a_{mn} are determined as in Case 1^0, and

$a_{m_0 n_0} = \frac{2}{\omega ab} \int\limits_{0}^{a} \int\limits_{0}^{b} F(x,y) \sin \left(\frac{m_0 \pi x}{a} \right) \sin \left(\frac{n_0 \pi y}{b} \right) dy\, dx$.

If there are several couples (m_0, n_0) for which $\omega = \omega_{m_0 n_0}$, then instead of one resonance term in the solution there will be several resonance terms of the form specified in Case 2^0.

10. (a) $\omega = 3$ and $\omega_{mn} = \sqrt{m^2 + n^2}$. $\sqrt{m^2 + n^2} \neq 3$ for every m and n. Therefore, by Case 1^0 we have that

$$u(x,y,t) = \sum_{m=1}^{\infty} \sum_{n=1}^{\infty} a_{mn}\left(\sin 3t - \frac{3}{\sqrt{m^2+n^2}} \sin \sqrt{m^2+n^2}\, t\right) \sin mx \sin ny$$

$$= \frac{\pi^2}{4}\left(\sin 3t - \frac{3}{\sqrt{2}} \sin \sqrt{2}t\right) \sin x \sin y$$

$$+ \sum_{m=2}^{\infty} \sum_{n=2}^{\infty} a_{mn}\left(\sin 3t - \frac{3}{\sqrt{m^2+n^2}} \sin \sqrt{m^2+n^2}\, t\right) \sin mx \sin ny,$$

where for $mn \geq 2$, $a_{mn} = \frac{16mn\left(1+(-1)^m\right)\left(1+(-1)^n\right)}{\pi^2(m^2+n^2-3)(m^2-1)^2(n^2-1)^2}$.

(b) $\omega = \sqrt{5}$ and $\omega_{mn} = \sqrt{m^2+n^2}$. $\sqrt{m^2+n^2} = 5$ for $m = 1$ and $n = 2$ or $m = 1$ and $n = 3$. Therefore, by Case 2^0 we have

$$u(x,y,t) = \sum_{\substack{m,\,n=1 \\ m^2+n^2 \neq 5}}^{\infty} a_{mn}\left(\sin \sqrt{5}\,t - \frac{\sqrt{5}}{\sqrt{m^2+n^2}} \sin \sqrt{m^2+n^2}\,t\right)$$

$\sin mx \sin ny + a_{1,2}\left(\sin \sqrt{5}t - \sqrt{5}t \cos \sqrt{5}t\right) \sin x \sin 2y$

$+a_{2,1}\left(\sin \sqrt{5}t - \sqrt{5}t \cos \sqrt{5}t\right) \sin 2x \sin y$, where

$$a_{1,2} = \frac{2}{\sqrt{5}\pi^2} \int_0^\pi \int_0^\pi xy \sin^2 x \sin y \sin 2y\, dy\, dx = a_{2,1}$$

$$= \frac{2}{\sqrt{5}\pi^2} \int_0^\pi \int_0^\pi xy \sin x \sin^2 y \sin 2x\, dx\, dy = -\frac{4}{9}.$$

For $m = n = 1$ we have $a_{1,1} = \frac{2}{\sqrt{5}\pi^2} \int_0^\pi \int_0^\pi x \sin^2 x\, y \sin^2 y dy\, dx =$

$\frac{\pi^2}{8\sqrt{5}}$. For every other m and n we have

$$a_{mn} = \frac{4}{(m^2+n^2-5)\pi^2} \int_0^\pi \int_0^\pi x \sin x \sin mx\, y \sin y \sin ny\, dy\, dx$$

$$= \frac{16mn(1+(-1)^m)(1+(-1)^n)}{(m^2+n^2-5)(m^2-1)^2(n^2-1)^2\pi^2}.$$

11. $u(x,y,t) = \sum_{m=1}^{\infty} \sum_{n=1}^{\infty} \frac{(\omega_{mn}^2+\omega^2)\sin \omega t + 2k\omega t \cos \omega}{(2m-1)(2n-1)\left[(\omega_{mn}-\omega)^2+4k^2\omega^2\right]} \sin \frac{m\pi x}{a} \sin \frac{n\pi y}{b}$,

where $\omega_{mn} = \frac{m^2\pi^2}{a^2} + \frac{n^2\pi^2}{b^2}$.

Section 5.4.

1. $u(r,\varphi,t) = \sum_{n=1}^{\infty} \frac{8}{z_{0n}^3 J_1(z_{0n})} J_0(z_{0n}r) \cos(z_{0n}t)$.

2. $u(r,\varphi,t) = \sum_{n=1}^{\infty} \frac{J_1\left(\frac{z_{0n}}{2}\right)}{z_{0n}^2 (J_1(z_{0n}))^2} J_0(z_{0n}r) \sin(z_{0n}t)$.

3. $u(r,\varphi,t) = 16\sin\varphi \sum\limits_{n=1}^{\infty} \frac{1}{z_{1n}^3 J_2(z_{1n})} \, J_1(z_{1n}r) \, \cos(z_{1n}t)$.

4. $u(r,\varphi,t) = 16\sin\varphi \sum\limits_{n=1}^{\infty} \frac{1}{z_{1n}^3 J_2(z_{1n})} \, J_1(z_{1n}r) \, \cos(z_{1n}t)$

$\quad +24\sin 2\varphi \sum\limits_{n=1}^{\infty} \frac{1}{z_{2n}^4 J_3(z_{2n})} \, J_2(z_{2n}r) \, \sin(z_{2n}t)$.

5. $u(r,\varphi,t) = 5J_4(z_{4,1}r)\cos 4\varphi \, \cos(z_{4,1}t) - J_2(z_{2,3}r)\sin 2\varphi \, \cos(z_{2,3}t)$.

6. $u(r,\varphi,t) = J_0(z_{03}r)\cos(z_{03}t) + 8 \sum\limits_{n=1}^{\infty} \frac{1}{z_{0n}^4 J_1(z_{1n})} \, J_0(z_{0n}r)\sin(z_{0n}t)$.

7. $u(r,\varphi,t) = 4 \sum\limits_{n=1}^{\infty} \frac{1}{z_{0n}^2 J_1(z_{0n})} \, J_0\left(\frac{z_{0n}}{2}r\right) \sin\left(\frac{z_{0n}}{2}t\right)$.

8. $u(r,t) = \sum\limits_{n=1}^{\infty}\left[a_n\cos\left(c\frac{z_{0n}}{a}t\right) + b_n\sin\left(c\frac{z_{0n}}{a}t\right)\right] J_0\left(\frac{z_{0n}}{a}r\right)$, where

$\quad a_n = \dfrac{2}{a^2\left(J_1\left(z_{0n}\right)\right)^2} \int\limits_0^a rf(r)J_0\left(\frac{z_{0n}}{a}r\right) dr$,

$\quad b_n = \dfrac{2}{a\,c\,z_{0n}\left(J_1\left(z_{0n}\right)\right)^2} \int\limits_0^a rg(r)J_0\left(\frac{z_{0n}}{a}r\right) dr$.

9. $u(r,t) = \frac{2}{a^2} \int\limits_0^a (f(r)+tg(r))r\,dr$

$\quad + \sum\limits_{n=1}^{\infty}\left[a_n\cos\left(c\frac{z_{1n}}{a}t\right) + b_n\sin\left(c\frac{z_{1n}}{a}t\right)\right] J_0\left(\frac{z_{1n}}{a}r\right)$, where

$\quad a_n = \dfrac{2}{a^2\left(J_0\left(z_{1n}\right)\right)^2} \int\limits_0^a rf(r)J_0\left(\frac{z_{1n}}{a}r\right) dr$,

$\quad b_n = \dfrac{2}{a\,c\,z_{1n}\left(J_0\left(z_{1n}\right)\right)^2} \int\limits_0^a rg(r)J_0\left(\frac{z_{1n}}{a}r\right) dr$.

10. $u(r,t) = \frac{A}{c^2}\left[\frac{a^2-r^2}{4} - 2a^2 \sum\limits_{n=1}^{\infty} \frac{1}{z_{0n}^2 J_1\left(z_{0n}\right)} J_0\left(\frac{z_{0n}}{a}r\right)\cos\left(c\frac{z_{0n}}{a}t\right)\right]$.

11. $u(r,t) = \sum\limits_{n=1}^{\infty} a_n(t)J_0\left(\frac{z_{0n}}{a}r\right)$,

$\quad a_n(t) = \frac{1}{\lambda_n} \int\limits_0^t \left(\int\limits_0^a f(\xi,\eta)J_0\left(\frac{z_{0n}}{a}\xi\right)\sin\left(\lambda_n(t-\eta)\right) d\xi\right) d\eta, \quad \lambda_n = \frac{c z_{0n}}{a}$.

12. $u(r,t) = \frac{1}{r} \sum\limits_{n=1}^{\infty} a_n\left[\cos\left(\lambda_n t\right) + \mu_n\sin\left(\lambda_n t\right)\right]$, where $\lambda_n = \frac{2n-1}{2(r_2-r_1)}\pi$,

$$\mu_n = \frac{\lambda_n \sin(\lambda_n r_2) + r_2^{-1} \cos(\lambda_n r_2)}{\lambda_n \cos(\lambda_n r_2) - r_2^{-1} \sin(\lambda_n r_2)},$$

$$a_n = \int_{r_1}^{r_2} \left(\cos(\lambda_n r) + \mu_n \sin(\lambda_n r) \right) r f(r) \, dr.$$

13. $u(r, \varphi, t) = A \cos \varphi \sum_{n=1}^{\infty} a_n J_1\left(\frac{\mu_{1n}}{a} r\right) \cos\left(\frac{a\mu_n}{a} t\right)$, where μ_n are the positive roots of $J_1'(x) = 0$,

and $a_n = \dfrac{2\mu_n^2}{a(\mu_n^2 - 1)\left(J_1(\mu_n)\right)^2} \displaystyle\int_0^a J_1\left(\frac{\mu_n}{a} r\right) dr.$

Section 5.5.

1. $u(x,t) = 3 + 2H\left(t - \frac{x}{2}\right)$, $H(\cdot)$ is the unit step Heaviside function.

2. $u(x,t) = \sin(xit) - H(t-x)\sin(x-t)$, $H(\cdot)$ is the unit step Heaviside function.

3. $u(x,t) = \left[\sin(x-t) - H(t-x)\sin(x-t)\right]e^{-t}$, $H(\cdot)$ is the unit step Heaviside function.

4. $u(x,t) = \left[\sin(x-t) - H(t-x)\sin(x-t)\right]e^{t}$, $H(\cdot)$ is the unit step Heaviside function.

5. $u(x,t) = t^2 e^{-x} - t e^{-x} + t.$

6. $u(x,t) = f\left(t - \frac{x}{c}\right)H\left(t - \frac{x}{c}\right)$, $H(\cdot)$ is the unit step Heaviside function.

7. $u(x,t) = (t-x)\sinh(t-x)H(t-x) + xe^{-x}\cosh t - te^{-t}\sinh t.$

8. $u(x,t) = \frac{t^3}{6} - \frac{1}{6}(t-x)^3 H(t-x)$, $H(\cdot)$ is the unit step Heaviside function.

9. $u(x,t) = t + \sin(t-x)H(t-x) - (t-x)H(t-x)$, $H(\cdot)$ is the unit step Heaviside function.

10. $u(x,t) = \frac{k}{c^2\pi^2}\left[1 - \cos\left(\frac{\pi c}{a}t\right)\right]\sin\left(\frac{\pi}{a}x\right)$.

11. $u(x,t) = \frac{1}{\pi}\sin\pi x \sin\pi t$.

12. $u(x,t) = \sum\limits_{n=1}^{\infty}(-1)^n\left[t - (2n+1-x)\right]H\left(t - (2n+1-x)\right)$.

13. $u(x,t) = \frac{4}{\pi^2}\sum\limits_{n=1}^{\infty}\frac{1}{(2n-1)^2}\sin\left((2n-1)\pi x\right)\sin\left((2n-1)\pi t\right)$.

14. $u(x,t) = \sin\pi x \cos\pi t - \frac{1}{\pi}\sin\pi x \sin\pi t$.

15. $u(x,t) = f\left(x - \frac{t^2}{2}\right)$.

16. $u(x,t) = f\left(x - \frac{t}{3}\right)$.

17. $u(x,t) = 3\cos bigl(x + \frac{t^3}{3}\right)$.

18. $u(x,t) = \left(e^{-t} + te^{-t}\right)f(x) + te^{-t}g(x)$.

19. $u(x,t) = \sqrt{\frac{\pi}{2}}\,e^{-|x+t|}$.

20. $u(x,t) = \frac{1}{2}\int\limits_{-\infty}^{\infty}e^{-|\omega|}\cos\omega t\, e^{i\omega x}$.

21. $u(x,t) = -c\int\limits_{0}^{t-\frac{x}{c}}f(s)\,ds$.

22. $u(x,t) = \frac{1}{2c}\int\limits_{0}^{t}\left(\int\limits_{|x-c(t-\tau)|}^{x+c(t-\tau)}f(s,\tau)\,ds\right)d\tau$.

23. $u(x,t) = \frac{1}{2c}\int\limits_{0}^{t}\left(\int\limits_{0}^{x+c(t-\tau)}f(s,\tau)\,ds\right)d\tau$

$$-\frac{1}{2c} \int\limits_0^t \left(\int\limits_0^{|x-c(t-\tau)|} f(s,\tau)\,ds \right) sign\big[x - c(t-\tau)\big]\,d\tau.$$

24. $u(x,t) = e^{-cmx} f(t - cx) H(t - cx)$, where $m = \frac{a}{c^2}$.

25. $u(x,t) = -ce^{k(x-ct)} \left(\int\limits_0^{t-\frac{x}{c}} e^{ck\tau} f(\tau)\,d\tau \right) H(ct - x)$.

Section 6.1.

3. $u(x,t) = \frac{100}{\sqrt{4a\pi t}} \int\limits_0^\infty e^{-\frac{(\xi-x)^2}{4at}}\,d\xi = 50\big(1 + er\,f(\frac{x}{\sqrt{4at}})\big)$, where

$er\,f(x) = \frac{2}{\sqrt{\pi}} \int\limits_0^x e^{-s^2}\,ds.$

4. $u(x,t) = e^t \frac{1}{\sqrt{4a\pi t}} \int\limits_{-\infty}^\infty e^{-\frac{(\xi-x)^2}{4at}}\,d\xi.$

Section 6.2.

1. $u(x,t) = \frac{8}{\pi} \sum\limits_{n=1}^\infty \frac{1}{(2n-1)^2} e^{-(2n-1)^2 t} \sin(2n-1)x.$

2. $u(x,t) = 3 - \frac{12}{\pi^2} \sum\limits_{n=1}^\infty \frac{1}{(2n-1)^2} e^{-\frac{(2n-1)^2 \pi^2}{9} t} \cos \frac{n\pi}{3} x.$

3. $u(x,t) = \frac{4}{\pi^2} \sum\limits_{n=1}^\infty \frac{(2n-1)\pi - 8(-1)^n}{(2n-1)^2} e^{-\frac{(2n-1)^2 \pi^2}{64} t} \sin \frac{(2n-1)\pi}{8} x.$

4. $u(x,t) = \frac{80}{\pi} \sum\limits_{n=1}^\infty \frac{1}{2n-1} e^{-2(2n-1)^2 t} \sin(2n-1)x.$

5. $u(x,t) = \frac{40}{\pi} \sum\limits_{n=1}^\infty \frac{1-\cos \frac{n\pi}{2}}{n} e^{-\frac{n^2\pi^2}{2} t} \sin \frac{n\pi}{2} x.$

6. $u(x,t) = \frac{\pi^2}{3} + 4 \sum\limits_{n=1}^\infty \frac{(-1)^n}{n} e^{-3n^2 t} \cos nx.$

7. $u(x,t) = \frac{400}{\pi} \sum\limits_{n=1}^\infty \frac{1}{2n-1} e^{-\frac{(2n-1)^2}{4} t} \sin \frac{2n-1}{2} x.$

8. $u(x,t) = \frac{4}{\pi} \sum\limits_{n=1}^{\infty} \frac{(-1)^{n-1}}{(2n-1)^2} e^{-a^2(2n-1)^2 t} \sin(2n-1)x.$

9. $u(x,t) = \sum\limits_{n=1}^{\infty} \left[\frac{4}{2n-1} - \frac{8(-1)^{n+1}}{(2n-1)^2 \pi^2} \right] e^{-a^2 \frac{(2n-1)^2}{4} t} \sin \frac{2n-1}{2}x.$

10. $u(x,t) = \frac{T_0}{\pi} x + \frac{2T_0}{\pi} \sum\limits_{n=1}^{\infty} \frac{1}{n} e^{-n^2 t} \sin nx.$

11. $u(x,t) = \frac{\pi}{2} e^{-t} \sin x - \frac{16}{\pi} \sum\limits_{n=1}^{\infty} \frac{1}{(4n^2-1)^2} e^{-4n^2 t} \sin 2nx.$

12. $u(x,t) = \frac{1}{2} - \frac{1}{2} e^{-4t} \cos 2x.$

13. $u(x,t) = \frac{4rl^2}{a^2 \pi^3} \sum\limits_{n=1}^{\infty} \frac{1}{(2n-1)^3} \left(1 - e^{-\frac{(2n-1)^2 a^2 \pi^2}{l^2} t} \right) \sin \frac{(2n-1)\pi}{l}x.$

14. $u(x,t) = \frac{1}{3} - t - \frac{2}{\pi^2} \sum\limits_{n=1}^{\infty} \frac{(-1)^n}{n^2} e^{-a^2 n^2 \pi^2 t} \cos n\pi x.$

15. $u(x,t) = \frac{4}{\pi} \sum\limits_{n=1}^{\infty} \frac{(-1)^{n-1}}{(2n-1)^4} \left[1 - e^{-a^2(2n-1)^2 \pi^2 t} \right] \sin(2n-1)x.$

16. $u(x,t) = \frac{1-e^{-t}}{2} + e^{-4t} \cos 2x.$

17. $u(x,t) = \sum\limits_{n=0}^{\infty} e^{-n^2 t} \left(a_n \cos nx + b_n \sin nx \right),$ where

$a_n = \frac{1}{\pi} \int\limits_{-\pi}^{\pi} f(x) \cos nx\, dx, \quad b_n = \frac{1}{\pi} \int\limits_{-\pi}^{\pi} f(x) \sin nx\, dx, \quad n = 1, 2, \dots$

18. $u(x,t) = \sum\limits_{n=1}^{\infty} u_n(t) \sin \frac{n\pi}{l}x,$ where $u_n(t) = \int\limits_{0}^{t} e^{-\frac{a^2 n^2 \pi^2 (t-s)}{l^2}} f_n(\tau)\, d\tau,$

$f_n(\tau) = \frac{2}{l} \int\limits_{0}^{l} F(s,\tau) \sin \frac{n\pi}{l} s\, ds.$

20. $u(x,t) = \sum\limits_{n=1}^{\infty} a_n e^{-z_n^2 t} \sin z_n x,$ where z_n is the nth positive solution
of the equation $x + \tan x = 0,$
and $c_n = 2 \int\limits_{0}^{1} f(x) \sin z_n x\, dx, \quad n = 1, 2, \dots$

21. $u(x,t) = \sum\limits_{n=1}^{\infty} c_n e^{-\left(\frac{a^2 n^2 \pi^2}{l^2}+b\right)t} \sin \frac{n\pi}{l}x, \quad c_n = \frac{2}{l} \int\limits_0^l f(x) \sin \frac{n\pi}{l}x \, dx.$

22. $u(x,t) = e^{-\left(\frac{a^2 n^2 \pi^2}{l^2}+b\right)t} \sin \frac{\pi}{2l}x.$

23. $u(x,t) = \sum\limits_{n=0}^{\infty} c_n e^{-\left(\frac{a^2 n^2 \pi^2}{l^2}+b\right)t} \cos \frac{n\pi}{l}x, \quad c_n = \frac{2}{l} \int\limits_0^l f(x) \cos \frac{n\pi}{l}x \, dx.$

24. $u(x,t) = \frac{aA}{\cos \frac{l}{a}} e^{-t} \sin \frac{x}{a} + \frac{2}{l} \sum\limits_{n=0}^{\infty} \left[\frac{T}{\omega_n} + (-1)^n \frac{Aa^2}{1-a^2\omega_n^2}\right] e^{-a^2 n^2 \omega_n t} \sin \omega_n x,$

where $\omega_n = \frac{(2n+1)\pi}{a}, \quad \omega_n \neq \frac{1}{a}.$

Section 6.3.

3. $u(x,y,z,t) = \sum\limits_{m=1}^{\infty} \sum\limits_{n=1}^{\infty} \sum\limits_{j=1}^{\infty} A_{mnj} e^{-k^2 \lambda_{mnj} t}$

$\sin\left(\frac{m\pi}{a}x\right) \sin\left(\frac{m\pi}{b}y\right) \sin\left(\frac{j\pi}{c}z\right)$, where

$A_{mnj} = \frac{8}{abc} \int\limits_0^a \int\limits_0^b \int\limits_0^c f(x,y,z) \sin\left(\frac{m\pi}{a}x\right) \sin\left(\frac{m\pi}{b}y\right) \sin\left(\frac{j\pi}{c}z\right) dz \, dy \, dx.$

4. $u(x,y,t) = \sum\limits_{n=1}^{\infty} a_n e^{-\frac{n^2 \pi^2}{a^2}t} \sin \frac{n\pi}{a}x$

$+ \sum\limits_{m=1}^{\infty} \sum\limits_{n=1}^{\infty} a_{mn} e^{-\pi^2\left(\frac{m^2}{a^2} + \frac{n^2}{b^2}\right)t} \sin \frac{m\pi}{a}x \cos \frac{n\pi}{b}y.$

5. $u(x,y,t) = 3e^{-44\pi^2 t} \sin 6\pi x \sin 2\pi y - 7e^{-\frac{11}{2}\pi^2 t} \sin \pi x \sin \frac{3\pi}{2}y.$

6. $u(x,y,t) = \frac{8}{\pi^2} \sum\limits_{m=1}^{\infty} \sum\limits_{n=1}^{\infty} \frac{1}{(2m-1)(2n-1)} e^{-\left((2m-1)^2+(2n-1)^2\right)t}.$

$\sin(2m-1)x \sin(2m-1)y.$

7. $u(x,y,t) = e^{-2t} \sin x \sin y + 3e^{-5t} \sin 2x \sin y + e^{-13t} \sin 3x \sin 2y.$

8. $u(x,y,t) = \frac{16}{\pi^2} \sum\limits_{m=1}^{\infty} \sum\limits_{n=1}^{\infty} \left[\frac{\sin mx \sin ny}{m^2 n^2} \sin \frac{m\pi}{2} \sin \frac{n\pi}{2}\right] e^{-\left(m^2+n^2\right)t}$

9. $u(x,y,t) = \frac{1}{13}\left(1 - e^{-13t}\right) \sin 2x \sin 3y + e^{-65t} \sin 4x \sin 7y.$

10. $u(x, y, t) = e^{-\left(k_1 x + k_2 y\right)} \sum\limits_{m=1}^{\infty} \sum\limits_{n=1}^{\infty} A_{mn} e^{-\left(m^2 + n^2 + k_1^2 + k_2^2 + k_3\right)t}$
$\sin mx \sin my$, where

$$A_{mn} = 4 \int\limits_0^\pi \int\limits_0^\pi f(x, y) e^{k_1 x + k_2 y} \sin mx \sin my \, dy \, dx.$$

11. $u(x, y, t) = \sum\limits_{m=1}^{\infty} \sum\limits_{n=1}^{\infty} e^{-\left(m^2 + n^2\right)t} \left(\int\limits_0^t F_{mn}(s) e^{-(m^2 + n^2)s} \, ds \right)$
$\sin mx \sin ny$, where

$$F_{mn}(t) = \tfrac{4}{\pi^2} \int\limits_0^\pi \int\limits_0^\pi F(x, y, t) \sin mx \sin ny \, dy \, dx.$$

12. $u(r, t) = 2T_0 \sum\limits_{n=1}^{\infty} \dfrac{1}{z_{0n} J_1\left(z_{0n}\right)} J_0\left(\tfrac{z_{0n}}{a} r\right) e^{-\frac{z_{0n} c^2}{a^2} t}$, where z_{0n} is the nth
positive zero of the Bessel function $J_0(\cdot)$.

13. $u(r, t) = 16 \sum\limits_{n=1}^{\infty} \dfrac{1}{z_{1n}^3 J_2\left(z_{1n}\right)} J_1\left(z_{1n} r\right) e^{-z_{1n}^2 t}$; z_{1n} is the nth positive
zero of the Bessel function $J_1(\cdot)$.

14. $u(r, t) = r^3 \sin 3\varphi - 2 \sin 3\varphi \sum\limits_{n=1}^{\infty} \dfrac{1}{z_{3n} J_4\left(z_{3n}\right)} J_3\left(z_{3n} r\right) e^{-z_{3n}^2 t}$, where
z_{3n} is the nth positive zero of the Bessel function $J_3(\cdot)$.

15. $u(r, z, t) = \tfrac{4}{\pi} \sum\limits_{m=1}^{\infty} \sum\limits_{n=1}^{\infty} \dfrac{1 - (-1)^m}{m \, z_{0n} J_1(z_{0n})} e^{-\left(z_{0n}^2 + m^2 \pi^2\right) t} J_0\left(z_{0n} r\right) \sin m\pi z$,
where z_{0n} is the nth positive zero of the Bessel function $J_0(\cdot)$.

16. $u(r, t) = \tfrac{2}{\pi r} \sum\limits_{n=1}^{\infty} \dfrac{(-1)^{n-1}}{n} e^{-c^2 n^2 \pi^2 t} \sin n\pi r.$

17. $u(r, t) = \sum\limits_{n=1}^{\infty} A_n e^{-c^2 n^2 \pi^2 t} \tfrac{1}{r} \sin n\pi r$, $A_n = 2 \int\limits_0^1 r f(r) \sin n\pi r \, dr.$

18. $u(r, t) = \sum\limits_{m=1}^{\infty} \sum\limits_{n=1}^{\infty} A_{mn} \dfrac{1}{\sqrt{z_{nm} r}} e^{-z_{mn}^2 t} J_{n+\frac{1}{2}}\left(z_{mn} r\right) P_n\left(\cos \theta\right)$,
where z_{nm} is the the mth positive zero of the Bessel function $J_n(\cdot)$,

and $P_n(\cdot)$ is the Legendre polynomial of order n. The coefficients A_{mn} are evaluated using the formula

$$A_{mn} = \frac{(2n+1)P_n'(0)\sqrt{z_{nm}}}{n(n+1)\,J_{n-\frac{1}{2}}^2(z_{nm})} \int_0^1 r^{\frac{3}{2}} J_{n+\frac{1}{2}}^2(z_{nm}\,r)\,dr.$$

Section 6.4.

1. $u(x,t) = u_0 \sum\limits_{n=0}^{\infty} \left[erf\left(\frac{2n+1+x}{2\sqrt{t}}\right) - erf\left(\frac{2n+1-x}{2\sqrt{t}}\right) \right]$, where

$erfc(x) = \frac{2}{\sqrt{\pi}} \int\limits_x^{\infty} e^{-u^2}\,du$ is the complementary error function.

2. $u(x,t) = u_1 + (u_0 - u_1)\,erfc\left(\frac{x}{2\sqrt{t}}\right).$

3. $u(x,t) = u_0 \left[1 - erfc\left(\frac{x}{2\sqrt{t}}\right) + e^{x+1}\,erfc\left(\sqrt{t} + \frac{x}{2\sqrt{t}}\right) \right].$

4. $u(x,t) = \frac{x}{2\sqrt{\pi}} \int\limits_0^t \frac{f(t-\tau)}{t^{\frac{3}{2}}} e^{-\frac{x^2}{4\tau}}\,d\tau.$

5. $u(x,t) = 60 + 40\,erfc\left(\frac{x}{2\sqrt{t-2}}\right) \mathcal{U}_2(t).$

6. $u(x,t) = 100 \left[-e^{1-x+t}\,erfc\left(\sqrt{t} + \frac{1-x}{2\sqrt{t}}\right) + erfc\left(\frac{1-x}{2\sqrt{t}}\right) \right].$

7. $u(x,t) = u + 0 + u_0\,e^{-\pi^2 t}\,\sin \pi x.$

8. $u(x,t) = \frac{u_0}{2\sqrt{\pi}}\,x \int\limits_0^t \frac{e^{-h\tau - \frac{x^2}{4\tau}}}{\tau^{\frac{3}{2}}}\,d\tau.$

9. $u(x,t) = 10^{t-x} + 10 \int\limits_0^t \left(1 + t - \tau + e^{t-\tau}\right) erfc\left(\frac{x}{2\sqrt{\tau}}\right)\,d\tau.$

10. $u(x,t) = \frac{1}{2}x(1-x) - \frac{4}{\pi^3} \sum\limits_{n=1}^{\infty} \frac{1}{(2n-1)^3}\,e^{-(2n-1)^2\pi^2 t}\,\sin(2n-1)\pi x.$

11. $u(x,t) = \frac{u_0}{2}\,e^{-kx}\,erfc\left(\frac{x}{2c\sqrt{t}} + \frac{c(1-k)}{2}\sqrt{t}\right) + erfc\left(\frac{1-x}{2\sqrt{t}}\right)$

$$+ \tfrac{u_0}{2} e^{-x} erfc\left(\tfrac{x}{2c\sqrt{t}} - \tfrac{c(1-k)}{2}\sqrt{t} \right) + erfc\left(\tfrac{1-x}{2\sqrt{t}} \right).$$

12. $u(r,t) = \tfrac{r^2}{2} + 3t - \tfrac{3}{10} - \tfrac{2}{r} \sum\limits_{n=1}^{\infty} \tfrac{\sin(z_n r)}{z_n^2 \sin z_n} e^{-z_n^2 t}, \ \tan z_n = z_n.$

13. $u(x,t) = Be^{-3t} \cos 2x - \tfrac{A}{2}x \sin x.$

14. (a) $u(x,t) = \dfrac{x}{2c\sqrt{\pi}\,t^{\frac{3}{2}}} e^{-\frac{x^2}{4c^2 t}}.$

(b) $u(x,t) = \dfrac{x}{2c\sqrt{\pi}\,t^{\frac{3}{2}}} \int\limits_0^t f(t-\tau)\, \dfrac{e^{-\frac{x^2}{4c^2\tau}}}{\tau^{\frac{3}{2}}}\, d\tau.$

15. $u(x,t) = \dfrac{x}{2c\sqrt{\pi}} \int\limits_{-\infty}^{\infty} \mu(t-\tau)\, e^{-\frac{x^2}{4c^2\tau}}\, d\tau.$

16. $u(x,t) = \dfrac{x}{2c\sqrt{\pi}} \int\limits_0^t \int\limits_{-\infty}^{\infty} f(\eta,\tau)\, \dfrac{e^{-\frac{(x-\eta)^2}{4c^2(t-\tau)}}}{\sqrt{t-\tau}}\, d\eta\, d\tau.$

17. $u(x,t) = \tfrac{1}{2} erf\left(\tfrac{1-x}{2c\sqrt{t}} \right) + \tfrac{1}{2} erf\left(\tfrac{1+x}{2c\sqrt{t}} \right).$

18. $u(x,t) = \tfrac{1}{\pi} \int\limits_{-\infty}^{\infty} \tfrac{\cos \omega x}{1+\omega^2} e^{-c^2\omega^2 t}.$

19. $u(x,t) = \dfrac{1}{t\sqrt{2\pi}} \int\limits_{-\infty}^{\infty} f(\tau)\, e^{-\frac{(x-\tau)^2}{2t^2}}\, d\tau.$

20. $u(x,t) = -\dfrac{c}{\sqrt{\pi}} \int\limits_0^t \dfrac{\mu(\tau)}{\sqrt{t-\tau}}\, e^{-\frac{x^2}{4c^2(t-\tau)}}\, d\tau.$

21. $u(x,t) = \dfrac{1}{2c\sqrt{\pi}} \int\limits_0^t \left(\dfrac{1}{\sqrt{t-\tau}} \int\limits_0^{\infty} \left[e^{-\frac{(x-\xi)^2}{4c^2(t-\tau)}} - e^{-\frac{(x+\xi)^2}{4c^2(t-\tau)}} \right] f(\xi,\tau)\, d\xi \right) d\tau.$

22. $u(x,t) = \tfrac{2}{\pi} \int\limits_0^{\infty} \tfrac{\sin \omega}{\omega} e^{-\omega^2 t} \cos x\omega\, d\omega.$

23. $u(x,t) = \tfrac{2}{\pi} \int\limits_0^{\infty} \tfrac{1-e^{-\omega^2 t}}{\omega} \sin x\omega\, d\omega.$

24. $u(x,t) = \frac{1}{2c\sqrt{\pi}} \int\limits_0^t \int\limits_0^\infty f(\eta,\tau) \, \frac{e^{-\frac{(x-\eta)^2}{4c^2(t-\tau)}} - e^{-\frac{(x-\eta)^2}{4c^2(t+\tau)}}}{\sqrt{t-\tau}} \, d\eta \, d\tau.$

25. $u(x,t) = \frac{1}{2c\sqrt{\pi}} \int\limits_0^t \int\limits_0^\infty f(\eta,\tau) \, \frac{e^{-\frac{(x-\eta)^2}{4c^2(t-\tau)}} + e^{-\frac{(x-\eta)^2}{4c^2(t+\tau)}}}{\sqrt{t-\tau}} \, d\eta \, d\tau.$

Section 7.1.

3. The maximum value of $u(x,y)$ is 1, attained at the boundary points $(1,0)$ and $(1,0)$.

4. $u(x,y) = \frac{x}{x^2+(y+1)^2}.$

5. $u(x,y,z) = \frac{z-1}{(x^2+y^2+(z-1)^2)^{\frac{3}{2}}}.$

6. For $\mathbf{x} = (x,y)$ and $\mathbf{y} = (x',y')$ in the right half plane

$$G(\mathbf{x},\mathbf{y}) = -\frac{1}{2\pi} \left[\ln | \mathbf{x} - \mathbf{y} | - \ln | \mathbf{x} - \mathbf{y}^* | \right],$$

where $\mathbf{y}^* = (-x', y')$.

The solution of the Poisson equation is given by

$$u(\mathbf{x}) = \iint\limits_{RP} f(\mathbf{y}) G(\mathbf{x},\mathbf{y}) \, d\mathbf{y} - \int\limits_{\partial(RP)} g(\mathbf{y}) \, G_{\mathbf{n}}(\mathbf{x},\mathbf{y}) \, dS(\mathbf{y}).$$

7. For $\mathbf{x} = (x,y)$ and $\mathbf{y} = (x',y')$ in the upper half plane UP

$$G(\mathbf{x},\mathbf{y}) = -\frac{1}{2\pi} \left[\ln | \mathbf{x} - \mathbf{y} | - \ln | \mathbf{x} - \mathbf{y}^* | \right],$$

where $\mathbf{y}^* = (x', -y')$.

The solution of the Poisson equation is given by

$$u(\mathbf{x}) = \iint\limits_{RP} f(\mathbf{y}) G(\mathbf{x},\mathbf{y}) \, d\mathbf{y} - \int\limits_{\partial(RP)} g(\mathbf{y}) \, G(\mathbf{x},\mathbf{y}) \, dS(\mathbf{y}).$$

9. $u(\mathbf{x}) = -\frac{1}{\pi} \int\limits_{|\mathbf{y}|=R} \ln \sqrt{| \mathbf{x} - \mathbf{y} |^2} \, g(\mathbf{y}) \, dS(\mathbf{y}) + C.$

10. $u(x,y) = 1 - \frac{1}{\pi} \left[\arctan\left(\frac{1-x}{y}\right) - \arctan\left(\frac{x^2+y^2-x}{y}\right) \right].$

Section 7.2.

1. (a) $u(x,y) = \frac{200}{\pi} \sum_{n=1}^{\infty} \frac{1-(-1)^n}{n} \cosh n\pi y \sin n\pi x$

$\qquad + \frac{200}{\pi} \sum_{n=1}^{\infty} \frac{1-(-1)^n}{n} \frac{2-\cosh n\pi}{\sinh n\pi} \sinh n\pi y \sin n\pi x.$

(b) $u(x,y) = \frac{2}{\pi} \sum_{n=1}^{\infty} \frac{1-(-1)^n}{n} \frac{\sinh n\pi}{} \sinh ny \sin nx$

$\qquad + \frac{2}{\pi} \sum_{n=1}^{\infty} \frac{1-(-1)^n}{n} \sinh n\pi \left(\sinh ny + \sinh n(\pi - x) \right) \sin ny.$

(c) $u(x,y) = \frac{2}{\pi} \sum_{n=1}^{\infty} \frac{(-1)^{n-1}}{n \sinh 2n\pi} \sin n\pi x \sinh n\pi y.$

(d) $u(x,y) = \frac{400}{\pi} \sum_{n=1}^{\infty} \frac{\sin \frac{(2n-1)\pi}{2} x}{(2n-1) \sinh \frac{(2n-1)\pi}{2}} \sinh \left(\frac{(2n-1)\pi}{2}(1-y) \right)$

$\qquad + \frac{200}{\pi} \sum_{n=1}^{\infty} \frac{1}{n \sinh 2n\pi} \sinh n\pi x \sin n\pi y.$

(e) $u(x,y) = \frac{\sin 7\pi x \sinh \left(7\pi(1-y) \right)}{\sinh 7\pi} + \frac{\sin \pi x \sinh \pi y}{\sinh \pi}$

$\qquad + \frac{\sin 3\pi y \sinh \left(3\pi(1-x) \right)}{\sinh 3\pi} + \frac{\sinh 6\pi x \sin 6\pi y}{\sinh 6\pi}.$

(f) $u(x,y) = \frac{400}{\pi} \sum_{n=1}^{\infty} \frac{\sin (2n-1)\pi x \sinh (2n-1)\pi y}{(2n-1) \sinh (2n-1)\pi}.$

(g) $u(x,y) = \frac{2}{a} \sum_{n=1}^{\infty} A_n \frac{\sin \frac{n\pi}{a} x \sinh \frac{n\pi}{a} y}{\sinh \frac{n\pi}{a} b}, \quad A_n = \int_0^a f(x) \sin \frac{n\pi}{a} x \, dx.$

2. $u(x,y) = \frac{2}{a} \sum_{n=1}^{\infty} A_n \frac{\sin \frac{n\pi}{a} x \sinh \left(\frac{n\pi}{a}(b-y) \right)}{\sinh \frac{n\pi}{a} b}, \quad A_n = \int_0^a f(x) \sin \frac{n\pi}{a} x \, dx.$

3. $u(x,y) = \frac{1}{2}x + \frac{2}{\pi^2} \sum_{n=1}^{\infty} \frac{1-(-1)^n}{n^2 \sinh n\pi} \sinh n\pi x \cos n\pi y.$

4. $u(x,y) = \frac{2}{\pi} \sum_{n=1}^{\infty} \frac{1-(-1)^n}{n(n \cosh n\pi + \sinh n\pi)} \left(n \cosh nx + \sinh nx \right) \sin ny.$

5. $u(x,y) = A_0 y + \sum_{n=1}^{\infty} A_n \sinh \left(\frac{n\pi}{a} y \right) \cos \frac{n\pi}{a} y; \quad A_0 = \frac{1}{ab} \int_0^a f(x) \, dx,$

$\qquad A_n = \frac{1}{a \sin \frac{n\pi}{a} b} \int_0^a f(x) \cos \frac{n\pi}{a} x \, dx.$

6. $u(x,y) = 1.$

7. $u(x,y) = 1 - \frac{4}{\pi} \sum_{n=1}^{\infty} \frac{\sin \left(\frac{(2n-1)\pi x}{a} \right)}{(2n-1) \cosh \frac{(2n-1)\pi b}{a}} \cosh \left(\frac{(2n-1)\pi}{a} y \right).$

8. $u = \frac{4}{\pi} \sum_{n=1}^{\infty} \left[\frac{\sin nx}{2n-1} \cosh (2n-1)y + \frac{\left(4-\cosh (2n-1) \right) \sin nx}{(2m-1) \sinh (2n-1)} sinh (2n-1)y \right].$

9. $u(x,y) = \sum_{n=1}^{\infty} \left[A_n \cosh \left(n - \frac{1}{2} \right)\pi x + B_n \sinh \left(n - \frac{1}{2} \right)\pi x \right] \sin \left(n - \frac{1}{2} \right)\pi y,$

where $A_n = 4\dfrac{(-1)^{n-1}}{(2n-1)^2}$, $B_n = -A_n \dfrac{\cosh\left(n-\frac12\right)\pi}{\sinh\left(n-\frac12\right)\pi}$.

10. $\displaystyle\int_0^1 f(y)\,dy = 0$.

11. $u(x,y) = \dfrac{2}{\pi}\displaystyle\sum_{n=1}^{\infty}\left(\int_0^\pi f(t)\sin nt\,dt\right)e^{-ny}\sin nx$.

12. $u(x,y) = (\sin x + \cos x)\,e^{-y}$.

13. $u(x,y) = \displaystyle\sum_{n=1}^{\infty} A_n e^{-\frac{(2n-1)\pi}{2b}x}\sin\frac{(2n-1)\pi}{2b}y$,

$A_n = \dfrac{2}{b}\displaystyle\int_0^b f(y)\sin\frac{(2n-1)\pi}{2b}y\,dy$.

14. $u(x,y) = 2\displaystyle\sum_{n=1}^{\infty} A_n\left(\int_0^b \cos\lambda_n\xi\,d\xi\right)e^{-\lambda_n x}\cos\lambda_n y$, $A_n = \dfrac{\lambda_n^2+h^2}{b(\lambda_n^2+h^2)+h}$,

λ_n are the positive solutions of the equation $\lambda\tan b\lambda = h$.

15. $u(x,y) = -\dfrac{16}{\pi^2}\displaystyle\sum_{m=1}^{\infty}\sum_{n=1}^{\infty}\dfrac{\sin(2m-1)\pi x\,\sin(2n-1)\pi y}{(2m-1)(2n-1)\left[(2m-1)^2+(2n-1)^2\right]}$.

16. $u(x,y) = \dfrac{8}{\pi^4}\displaystyle\sum_{m=1}^{\infty}\sum_{n=1}^{\infty}\dfrac{(-1)^m\sin m\pi x\,\sin(2n-1)\pi y}{m(2n-1)\left[(m^2+(2n-1)^2\right]}$.

17. $u(x,y) = -\dfrac{4\sin\pi x}{\pi^3}\displaystyle\sum_{n=1}^{\infty}\dfrac{\sin(2n-1)\pi y}{(2n-1)\left[1+(2n-1)^2\right]}$

$+\dfrac{2}{\pi}\displaystyle\sum_{n=1}^{\infty}\dfrac{(-1)^{n-1}\sin n\pi x\,\sinh n\pi y}{n\sinh n\pi}$.

18. $u(x,y) = \dfrac{4}{\pi^4}\displaystyle\sum_{m=1}^{\infty}\sum_{n=1}^{\infty}\dfrac{(-1)^{m+n-1}\sin m\pi x\,\sin n\pi y}{mn\left[(m^2+n^2)\right]}$

$+\dfrac{2}{\pi}\displaystyle\sum_{n=1}^{\infty}\dfrac{(-1)^{n-1}\sin n\pi x\,\sinh n\pi y}{n\sinh n\pi}$.

19. $u(x,y) = \left(a\sinh y - \dfrac13\cosh y + \dfrac13 e^{2\pi}\right)\sin x + \displaystyle\sum_{n=2}^{\infty} A_n\dfrac{\sinh ny\sin ny}{\sinh n\pi}$,

where $a = \dfrac{A_1 + \frac13\sinh\pi - \frac13 e^{2\pi}}{\sinh pi}$, $A_n = \displaystyle\int_0^\pi f(x)\sin nx\,dx$.

Section 7.3.

1. $u(r,\varphi) = \displaystyle\sum_{n=1}^{\infty}\dfrac{r^n}{na^{n-1}}\left(a_n\cos n\varphi + b_n\sin\varphi\right) + C$, where C is any con-

stant; $a_n = \dfrac{1}{2\pi}\displaystyle\int_0^{2\pi} f(\varphi)\cos n\varphi\,d\varphi$, $b_n = \dfrac{1}{2\pi}\displaystyle\int_0^{2\pi} f(\varphi)\sin n\varphi\,d\varphi$.

2. $u(r,\varphi) = -\displaystyle\sum_{n=1}^{\infty}\dfrac{r^{n+1}}{na^n}\left(a_n\cos n\varphi + b_n\sin\varphi\right) + C$, where C is any con-

stant; $a_n = \dfrac{1}{2\pi}\displaystyle\int_0^{2\pi} f(\varphi)\cos n\varphi\,d\varphi$, $b_n = \dfrac{1}{2\pi}\displaystyle\int_0^{2\pi} f(\varphi)\sin n\varphi\,d\varphi$.

3. $u(r, \varphi) = Ar \sin \varphi$.

4. $u(r, \varphi) = B + 3Ar \sin \varphi - 4Ar^3 \sin 3\varphi$.

5. $u(r, \varphi) = Ar \sin \varphi - \frac{8A}{\pi} \sum\limits_{n=1}^{\infty} r^{2n} \frac{1}{4n^2 - 9} \sin 2n\varphi$.

6. $u(r, \varphi) = \frac{1}{2} + \frac{2}{\pi} \sum\limits_{n=1}^{\infty} \frac{(-1)^{n-1}}{2n-1} r^{2n-1} \sin(2n-1)\varphi$.

7. $u(r, \varphi) = \sum\limits_{n=1}^{\infty} \frac{1}{n} r^n \sin n\varphi$.

8. $u(r, \varphi) = B_0 \ln r + A_0$

$$+ \sum\limits_{n=1}^{\infty} \left[\left(A_n r^n + \frac{b_n}{r^n}\right) \cos n\varphi + \left(C_n r^n + \frac{D_n}{r^n}\right) \sin n\varphi \right], \text{ where}$$

$A_n = \frac{b^n g_n^{(c)} - a^n f_n^{(c)}}{b^{2n} - a^{2n}}$, $\quad B_n = a^n b^n \frac{b^n f_n^{(c)} - a^n g_n^{(c)}}{b^{2n} - a^{2n}}$, $\quad A_0 = \frac{f_0^{(c)} - g_0^{(c)}}{\ln a - \ln b}$,

$C_n = \frac{b^n g_n^{(s)} - a^n f_n^{(s)}}{b^{2n} - a^{2n}}$, $\quad D_n = a^n b^n \frac{b^n f_n^{(s)} - a^n g_n^{(s)}}{b^{2n} - a^{2n}}$, $\quad B_0 = \frac{g_0^{(c)} \ln a - f_0^{(c)} \ln b}{\ln a - \ln b}$;

$f_n^{(c)}$, $f_n^{(s)}$, $g_n^{(c)}$, $g_n^{(s)}$ are the Fourier coefficients of $f(\varphi)$ and $g(\varphi)$,

$f_n^{(c)} = \frac{1}{2\pi} \int\limits_0^{2\pi} f(\varphi) \cos n\varphi \, d\varphi$, $\quad f_n^{(s)} = \frac{1}{2\pi} \int\limits_0^{2\pi} f(\varphi) \sin \varphi \, d\varphi$,

$g_n^{(c)} = \frac{1}{2\pi} \int\limits_0^{2\pi} g(\varphi) \cos n\varphi \, d\varphi$, $\quad g_n^{(s)} = \frac{1}{2\pi} \int\limits_0^{2\pi} g(\varphi) \sin \varphi \, d\varphi$.

If $b = 1, f(\varphi) = 0$ and $g(\varphi) = 1 + 2\sin \varphi$, then $u(r, \varphi) = 1 + 2\frac{a^2 + r^2}{(1+a^2)r} \sin \varphi$.

9. $u(r, \varphi) = \frac{1}{2} + \frac{4}{\pi} \sum\limits_{n=1}^{\infty} \frac{(-1)^{n-1}}{2n-1} \left(\frac{r}{a}\right)^{2n-1} \cos 2(2n-1)\varphi$.

10. $u(r, \varphi) = \frac{c}{2} + \frac{4c}{\pi} \sum\limits_{n=1}^{\infty} \frac{(-1)^{n-1}}{2n-1} \left(\frac{r}{2}\right)^{2n-1} \cos(2n-1)\varphi$.

11. $u(r, \varphi) = -2 \sum\limits_{n=1}^{\infty} \frac{1}{\lambda_{0n}^3 J_1(\lambda_{0n})} J_0(\lambda_{0n} r)$, where λ_{0n} is the nth positive zero of $J_0(\cdot)$.

12. $u(r, \varphi) = -2 \sum\limits_{n=1}^{\infty} \left[\frac{J_0(\lambda_{0n} r)}{\lambda_{0n}^3 J_1(\lambda_{0n})} + \frac{J_3(\lambda_{3n} r)}{\lambda_{3n}^3 J_4(\lambda_{3n})} \cos 3\varphi \right]$, where λ_{0n} is the nth positive zero of $J_0(\cdot)$ and λ_{3n} is the nth positive zero of $J_3(\cdot)$.

13. $u(r, \varphi) = r^2 \sin 2\varphi - 2 \sum\limits_{n=1}^{\infty} \frac{1}{\lambda_{0n}^3 J_1(\lambda_{0n})} J_0(\lambda_{0n} r)$, where λ_{0n} is the nth positive zero of $J_0(\cdot)$.

14. $u(r, \varphi, z) = u(r, z) = 2 \sum\limits_{n=1}^{\infty} \frac{1}{\lambda_{0n} J_1(\lambda_{0n})} J_0(\lambda_{0n} r) \frac{\sinh(\lambda_{0n} z)}{\sinh(\lambda_{0n})}$, where λ_{0n} is the nth positive zero of $J_0(\cdot)$.

15. $u(r, z) = \sum\limits_{n=1}^{\infty} \dfrac{J_0\left(\lambda_{0n} r\right)}{\sinh\left(\lambda_{0n}\right)} \left[A_n \sinh\left(\lambda_{0n}\left(1-z\right)\right) + B_n \sinh\left(\lambda_{0n} z\right)\right]$, where
λ_{0n} is the nth positive zero of $J_0(\cdot)$. The coefficients are determined by the formula
$$A_n = \dfrac{2}{J_1^2\left(\lambda_{0n}\right)} \int\limits_0^1 G(r) J_0\left(\lambda_{0n} r\right) r \, dr, \quad B_n = \dfrac{2}{J_1^2\left(\lambda_{0n}\right)} \int\limits_0^1 H(r) J_0\left(\lambda_{0n} r\right) r \, dr.$$

16. $u(r, \theta) = 2 + \sum\limits_{n=1}^{\infty} (4n - 1) \dfrac{(-1)^{n-1}(2n-2)!}{n 2^{2n-2}\left(n-1)!\right)^2} r^{2n-1} P_{2n-1}(\cos \theta)$.

17. $u(r, \theta) = 1 - r \cos \theta$.

18. $u(r, \theta) = \frac{7}{3} + \frac{1}{3} r^2 \left(3 \cos^2 \theta - 1\right)$.

19. $u(r, \theta) = 1 + 2r\cos \theta + 4 \sum\limits_{n=1}^{\infty} (-1)^{n-1} \dfrac{4n+1}{n+1} \dfrac{n(2n-2)!}{2^{2n+1}\left(n)!\right)^2} r^{2n} P_{2n}(\cos \theta)$.

20. $u(r, \theta, \varphi) = \sum\limits_{m=0}^{\infty} \sum\limits_{n=0}^{m} r^{-m-1} \left[\left(A_{mn} \cos n \varphi + B_{mn} \sin \varphi\right) P_m^{(n)}(\cos \theta)\right]$,
$$A_{mn} = \dfrac{(2m+1)!\,((m-n)!}{2\pi(m+n)!} \int\limits_0^{2\pi} \int\limits_0^{\pi} f(\theta, \varphi) \, P_m^{(n)}(\cos \theta) \cos n\varphi \sin \theta \, d\theta \, d\varphi,$$
$$B_{mn} = \dfrac{(2m+1)!\,((m-n)!}{2\pi(m+n)!} \int\limits_0^{2\pi} \int\limits_0^{\pi} f(\theta, \varphi) \, P_m^{(n)}(\cos \theta) \sin n\varphi \sin \theta \, d\theta \, d\varphi.$$

21. $u(r, \theta, \varphi) = \sum\limits_{m=1}^{\infty} \sum\limits_{n=0}^{m} \dfrac{r^m}{m} \left[\left(A_{mn} \cos n \varphi + B_{mn} \sin \varphi\right) P_m^{(n)}(\cos \theta)\right] + C$,
where C is any constant and A_{mn} and B_{mn} are as in Exercise 18.

Section 7.4.

1. $u(x, y) = \frac{2}{\pi} \int\limits_0^{\infty} \dfrac{\sinh \omega x}{(1+\omega^2) \sinh \omega \pi} \, d\omega$.

2. $u(x, y) = \frac{2}{\pi} \int\limits_0^{\infty} F(\omega) \dfrac{\sinh \omega(2-y)}{(1+\omega^2) \sinh 2\omega} \sin \omega x \, d\omega$.

3. $u(x, y) = \frac{2}{\pi} \int\limits_0^{\infty} \dfrac{\omega}{1+\omega^2} \left[e^{-\omega x} \sin \omega y + e^{-\omega y} \sin \omega x\right] d\omega$.

4. $u(x, y) = \frac{y}{\pi} \int\limits_0^{\infty} \dfrac{f(\omega)}{y^2+(x-\omega)^2} \, d\omega$.

5. $u(x, y) = \frac{1}{\pi} \left[\arctan\left(\dfrac{1+x}{y}\right) + \arctan\left(\dfrac{1-x}{y}\right)\right]$.

6. $u(x, y) = \frac{1}{2} \dfrac{2+y}{x^2+(2+y)^2}$.

7. $u(x, y) = \frac{y}{\pi} \int\limits_{-\infty}^{\infty} \dfrac{\cos \omega}{(x-\omega)^2+y^2} \, d\omega$.

8. $u(x,y) = \frac{2}{\pi} \int\limits_0^\infty \frac{\sinh \omega x}{(1+\omega^2)\sinh \omega} \cos \omega y \, d\omega$.

9. $u(x,y) = \frac{2}{\pi} \int\limits_0^\infty \frac{1-\cos \omega}{\omega} e^{-\omega x} \sin \omega y \, d\omega$.

12. $u(r,z) = a \int\limits_0^\infty J_0(\omega r) J_1(\omega) e^{-\omega z} \, d\omega$.

13. $u(r,z) = \int\limits_0^\infty J_0(\omega r) e^{-\omega(z+a)} \, d\omega = \frac{1}{\sqrt{(z+a)^2+r^2}}$.

Section 8.1.

1. $\lambda_1 = 5$, $\mathbf{x}_1 = \begin{pmatrix} -4 \\ 0 \\ -3 \\ 5 \end{pmatrix}$; $\lambda_2 = 9$, $\mathbf{x}_2 = \begin{pmatrix} 0 \\ 0 \\ -3 \\ 0 \end{pmatrix}$.

$\lambda_3 = -\sqrt{10}$, $\mathbf{x}_3 = \begin{pmatrix} -\frac{1+\sqrt{10}}{3} \\ 0 \\ 1 \\ 0 \end{pmatrix}$; $\lambda_4 = \sqrt{10}$, $\mathbf{x}_4 = \begin{pmatrix} -\frac{1-\sqrt{10}}{3} \\ 0 \\ 1 \\ 0 \end{pmatrix}$.

2. $\|A\|_1 = 5$, $\|A\|_\infty = 7$, $\|A\|_E = 6$. The characteristic equation of

$$A^T A = \begin{pmatrix} 2 & 2 & 6 \\ 2 & 13 & 7 \\ 6 & 7 & 21 \end{pmatrix}$$

is $\lambda^3 - 36\lambda^2 + 252\lambda - 64 = 0$. Numerically we find that $\lambda_{max} \approx 26.626$.

3. We find that the eigenvalues of A and A^{-1} are

$$\lambda_1(A) = 1.99999, \quad \lambda_2(A) = 0.00001$$
$$\lambda_1(A^{-1}) = 0.500003, \quad \lambda_2(A^{-1}) = 100000.$$

Therefore $\|A\|_2 = 1.9999$, $\|A^{-1}\|_2 = 100000$ and hence

$$\kappa(A) = 199999 \approx 200000.$$

The exact solution of the system

$$\begin{pmatrix} 1 & 0.99999 \\ 0.99999 & 1 \end{pmatrix} \begin{pmatrix} x_1 \\ x_2 \end{pmatrix} = \begin{pmatrix} 2.99999 \\ 2.99998 \end{pmatrix}$$

is

$$x_1 = 2 \quad \text{and} \quad x_2 = 1,$$

but the exact solution of the system

$$\begin{pmatrix} 0.99999 & 0.99999 \\ 0.99999 & 1 \end{pmatrix} \begin{pmatrix} y_1 \\ y_2 \end{pmatrix} = \begin{pmatrix} 2.99999 \\ 2.99998 \end{pmatrix}$$

is

$$y_1 = 4.00002 \quad \text{and} \quad y_2 = -1.$$

We observe that the solution of the second system differs considerably from the solution of the original system despite a very small change in the coefficients of the original system. This is due to the large conditional number of the matrix A.

4. Let $\mathbf{x} = \begin{pmatrix} x_1 \\ x_2 \\ x_3 \end{pmatrix}$ be any nonzero vector. Then

$$\mathbf{x}^T \cdot A \cdot \mathbf{x} = (x_1 \; x_2 \; x_3) \cdot \begin{pmatrix} 2 & -1 & 0 \\ -1 & 2 & -1 \\ 0 & -1 & 2 \end{pmatrix} \cdot \begin{pmatrix} x_1 \\ x_2 \\ x_3 \end{pmatrix}$$
$$= x_1^2 + (x_1 - x_2)^2 + (x_2 - x_3)^2 + x_3^2 \geq 0.$$

The above expression is zero only if $x_1 = x_2 = x_3 = 0$ and so the matrix is indeed positive definite.

5. First notice that the matrix of the system is strictly diagonally dominant (check this). Therefore the Jacobi and Gauss–Seidel iterative methods converge for every initial point $\mathbf{x}_0$. Now if we take $\epsilon = 10^{-6}$ and

$In[1] := A = \{\{4, 1, -1, 1\}, \{1, 4, -1, -1\}, \{-1, -1, 5, 1\}$
$, \{1, -1, 1, 3\}\};$

$In[2] := b = \{5, -2, 6, -4\};$

in the modules "ImprovedJacobi" and "ImprovedGaussSeidel"

$In[3] := ImprovedJacobi[A, b, \{0, 0, 0, 0\}, 0.000001]$

$In[4] := ImprovedGaussSeidel[A, b, \{0, 0, 0, 0\}, 0.000001]$

we obtain

(a) $Out[4] := k = 23 \; \mathbf{x} = \{-0.75341, 0.041092, -0.28081, 0.69177\}$

(b) $Out[5] := k = 12 \; \mathbf{x} = \{-0.75341, 0.04109, -0.28081, 0.69177\}$.

6. It is obvious that the matrix

$$A = \begin{pmatrix} 10 & -1 & -1 & -1 \\ -1 & 10 & -1 & -1 \\ -1 & -1 & 10 & -1 \\ -1 & -1 & -1 & 10 \end{pmatrix}$$

is symmetric. Next we show that this matrix is strictly positive defi-

nite. Indeed, if $\mathbf{x} = \begin{pmatrix} x_1 \\ x_2 \\ x_3 \\ x_4 \end{pmatrix}$ is any vector, then

$$\mathbf{x}^T \cdot A \cdot \mathbf{x} = (x_1 - x_2)^2 + (x_1 - x_3)^2 + (x_1 - x_4)^2 + (x_2 - x_3)^2$$
$$+ (x_2 - x_4)^2 + (x_3 - x_4)^2 + 7(x_1^2 + x_2^2 + x_3^2 + x_4^2) > 0.$$

We work as in Example 8.1.5. If we take
$In[1] := A = \{\{10, -1, -1, -1\}, \{-1, 10, -1, -1\}, \{-1, -1, 10, -1\}$
$, \{-1, -1, -1, 10\}\};$
$In[2] := b = Transpose[\{\{34, 23, 12, 1\}\}];$
$In[3] := x[0] = \{0, 0, 0, 0\};$
then we obtain the following table.

k	x_1	x_2	x_3	x_4	d_k
0	0	0	0	0	
1	4.0854	2.7636	1.4419	0.1202	5.3606
2	4.0000	3.0000	2.0000	1.0000	0.0112

The exact solution is obtained by
$In[4] := XExact = Inverse[A]b;$
$Out[4] := \{4.0, 3, 0, 2.0, 1.0\}.$

7. The matrix A is not a strictly definite positive since, for example, for
its first row we have $|2| + |-2| = 4 = |4|$. To show that the Jacobi
and Gauss–Seidel iterative methods are convergent we use Theorem
8.1.3.

For the Jacobi iterative method, the matrix $B = B_J$ in (8.1.17) is
given by

$$B_J = D^{-1}(L + U) = \begin{pmatrix} 4 & 0 & 0 \\ 0 & -4 & 0 \\ 0 & 0 & 4 \end{pmatrix}^{-1} \cdot \left[\begin{pmatrix} 0 & 0 & 0 \\ 0 & -4 & 0 \\ 0 & 0 & 0 \end{pmatrix} \right.$$

$$\left. + \begin{pmatrix} 0 & 2 & -2 \\ 0 & 0 & -1 \\ 0 & 0 & 0 \end{pmatrix} \right] = \begin{pmatrix} 0 & \frac{1}{2} & -\frac{1}{2} \\ -\frac{1}{3} & 0 & \frac{1}{3} \\ \frac{3}{4} & -\frac{1}{4} & 0 \end{pmatrix}.$$

Using Mathematica to simplify the calculations we find that the characteristic polynomial $p_{B_J}(\lambda)$ is given by

$$p_{B_J}(\lambda) = -\lambda^3 - \frac{8}{8}\lambda + \frac{1}{12}\lambda.$$

Again using Mathematica we find that the eigenvalues of B_J (the roots of $p_{B_J}(\lambda)$) are

$$\lambda_1 = 0.129832, \quad \lambda_2 = -0.0649159 + 0.798525\, i,$$
$$\lambda_3 = -0.0649159 - 0.798525\, i.$$

Their modules are

$$|\lambda_1| = 0.129832, \quad |\lambda_2| = 0.801159, \quad |\lambda_3| = 0.801159.$$

Therefore the spectral radius $\rho(B_J) = 0.801159 < 1$ and so by Theorem 8.1.3 the Jacobi iterative method converges.

For the Gauss–Seidel iterative method, the matrix $B = B_{GS}$ in (8.1.18) is given by

$$B_{GS} = (L + D)^{-1} \cdot U = \left[\begin{pmatrix} 0 & 0 & 0 \\ 1 & 0 & 0 \\ 3 & -1 & 0 \end{pmatrix} + \begin{pmatrix} 4 & 0 & 0 \\ 0 & -3 & 0 \\ 0 & 0 & 4 \end{pmatrix} \right]^{-1}$$
$$\cdot \begin{pmatrix} 0 & 2 & -2 \\ 0 & 0 & -1 \\ 0 & 0 & 0 \end{pmatrix} = \begin{pmatrix} 0 & \frac{1}{2} & -\frac{1}{2} \\ 0 & \frac{1}{6} & \frac{1}{6} \\ 0 & \frac{1}{3} & \frac{5}{12} \end{pmatrix}.$$

Using Mathematica (to simplify the calculations) we find that the characteristic polynomial $p_{B_{GS}}(\lambda)$ is given by

$$p_{B_{GS}}(\lambda) = -\lambda^3 - \frac{7}{12}\lambda^2 - \frac{1}{8}\lambda.$$

We find that the eigenvalues of B_{GS} (the roots of $p_{B_{GS}}(\lambda)$) are

$$\lambda_1 = 0, \quad \lambda_2 = 0.291667 + 0.199826\, i, \quad \lambda_3 = 0.291667 - 0.199826\, i.$$

Their modules are

$$|\lambda_1| = 0, \quad |\lambda_2| = 0.353553, \quad |\lambda_3| = 0.353553.$$

Therefore the spectral radius $\rho(B_{GS}) = 0.353553 < 1$ and so by Theorem 8.1.3 the Gauss–Seidel iterative method converges.

Section 8.2.

1. (a) We know from calculus that $f'(x) = \cos x$ and so the exact value of $f'(\pi/4)$ is $\sqrt{2}/2 \approx 0.707$.

   With the forward finite difference approximation for $h = 0.1$ we have

   $$f'\left(\frac{\pi}{4}\right) \approx \frac{f\left(\frac{\pi}{4} + 0.1\right) - f\left(\frac{\pi}{4}\right)}{0.1} = \frac{\cos\left(\frac{\pi}{4} + 0.1\right) - \cos\left(\frac{\pi}{4}\right)}{0.1} \approx 0.63.$$

   With the backward finite difference approximation for $h = 0.1$ we have

   $$f'\left(\frac{\pi}{4}\right) \approx \frac{f\left(\frac{\pi}{4}\right) - f\left(\frac{\pi}{4} + 0.1\right)}{0.1} = \frac{\cos\left(\frac{\pi}{4}\right) - \cos\left(\frac{\pi}{4} + 0.1\right)}{0.1} \approx 0.773.$$

   With the central finite difference approximation for $h = 0.1$ we have

   $$f'\left(\frac{\pi}{4}\right) \approx \frac{\cos\left(\frac{\pi}{4} + 0.1\right) - 2\cos\left(\frac{\pi}{4}\right) + \cos\left(\frac{\pi}{4} - 0.1\right)}{0.1} \approx 0.702.$$

   (b) $f'(x) = \sec^2 x$ so the exact value of $f'(x)$ is 2.

   With the forward approximation $f'(x) \approx 2.23049$.

   With the backward approximation $f'(x) \approx 1.82371$.

   With the central approximation $f'(x) \approx 2.0271$.

2. For the mesh points $x = 0.5$ and $x = 0.7$ we use the forward and backward finite differences approximations, respectively:

   $$f''(0.5) \approx \frac{0.56464 - 0.47943}{0.1} = 0.8521,$$

   $$f''(0.7) \approx \frac{0.56464 - 0.64422}{0.1} = -0.7958.$$

   For the mesh point $x = 0.6$ we can use either finite difference approximation. The central approximation gives

   $$f'(0.6) \approx \frac{0.64422 - 0.47943}{2 * 0.1} = 0.82395.$$

   For the second derivative with the central approximation we obtain

   $$f''(0.6) \approx \frac{0.64422 - 2 * 0.56464 + 0.47943}{0.1^2} = -0.563.$$

3. Apply Taylor's formula to $f(x+h)$, $f(x-h)$, $f(x+2h)$ and $f(x-2h)$.

4. We find from calculus that $f'(x) = 2xe^{x^2}$ and $f''(x) = 2e^{x^2} + 4x^2 e^{x^2}$.
Therefore the exact value of $f''(1) = 6e \approx 16.30969097$.
With the central finite difference approximation taking $h = 0.001$ we have

$$f''(1) \approx \frac{e^{(1+0.001)^2} - 2e^{1^2} + e^{(1-0.001)^2}}{0.001^2} \approx 13.30970817.$$

5. Partition $[0, 1]$ into n equal subintervals.

$$x_i = ih, \quad i = 0, 1, 2, \ldots, n; \ h = \frac{1}{n}.$$

If we discretize the given boundary value problem we obtain the linear system

$$y_{i+1} + (h^2 x_i - 2)y_i + y_{i-1} = 0, \quad i = 1, 2, \ldots, n-1,$$

where $y_0 = y(0) = 0$ and $y_n = y(1) = 1$.

Taking $n = 6$, using Mathematica (as in Example 8.2.3) you should obtain the following results.

x	0.1	0.2	0.3	0.4	0.5	0.6
$y(x)$	0.013382	0.046762	0.100124	0.173396	0.260390	0.378719

Section 8.3.

1. The 81 unknown $u_{i,j}, i, j = 1, 2, \ldots, 9$ are given in Table E. 8.3.1.

Table E. 8.3.1

$j \backslash i$	1	2	3	4	5	6	7
1	125.8	141.2	145.4	144.0	137.5	122.6	88.61
2	102.10	113.50	116.50	113.10	103.30	84.48	51.79
3	89.17	94.05	93.92	88.76	77.97	60.24	34.05
4	80.53	79.65	76.40	70.00	59.63	44.47	24.17
5	73.30	67.62	62.03	55.22	46.08	33.82	18.18
6	65.05	55.52	48.87	42.76	36.65	26.55	14.73
7	51.39	40.52	35.17	31.29	27.23	21.99	14.18

2. The following $3\times$ system in the $u_{i,j}, i, j = 1, 2, \ldots, 3$ unknown should be solved

$$4u_{i,j} - u_{i-1,j} - u_{i,j+1} + u_{i,j-1} = -\frac{1}{64}, \quad i, j = 1, 2, 3.$$

From the boundary conditions use $u_{i,0} = u_{0,j} = 0$, $i, j = 1, 2, 3$. The results are given in Table E. 8.3.2.

Table E. 8.3.2.

$j\backslash i$	1	2	3
1	−0.043	−0.055	−0.043
2	−0.055	−0.070	−0.055
3	−0.043	−0.055	−0.043

3. The 81 unknown $u_{i,j}, i, j = 1, 2, \ldots, 9$ are givn in Table E. 8.3.3.

Table E. 8.3.3.

$j\backslash i$	1	2	3	4	5	6	7
1	126.5	142.3	146.8	145.5	138.8	123.6	89.1
2	103.5	116.0	119.6	116.3	106.0	86.47	52.81
3	91.66	98.41	99.21	94.05	82.49	63.47	35.71
4	84.72	86.79	84.83	78.21	66.46	49.21	26.55
5	80.44	79.21	75.12	67.49	55.92	40.37	21.29
6	77.84	74.47	68.97	60.69	49.36	35.04	18.25
7	76.42	71.88	65.58	56.96	45.70	32.20	16.65

4. Let $u(3+0.25i, 4+0.25j) \approx u_{i,j}$, $i = 1, 2, 3$, $j = 1, 2, 3, 4$. The results are given in Table E. 8.3.4.

Table E. 8.3.4.

$j\backslash i$	1	2	3
1	−0.017779	−0.044663	−0.065229
2	−0.0277746	−0.053768	−0.068876
3	−0.032916	−0.056641	−0.072783
4	−0.034524		

5. The matrix equation is $A\mathbf{u} = \mathbf{b}$, where

$$
A = \begin{pmatrix}
4 & -1 & 0 & -1 & 0 & 0 & 0 & 0 & 0 \\
-1 & 4 & -1 & 0 & -1 & 0 & 0 & 0 & 0 \\
0 & -1 & 4 & 0 & 0 & -1 & 0 & 0 & 0 \\
-1 & 0 & 0 & 4 & -1 & 0 & -1 & 0 & 0 \\
0 & -1 & 0 & -1 & 4 & -1 & 0 & -1 & 0 \\
0 & 0 & -1 & 0 & -1 & 4 & 0 & 0 & -1 \\
0 & 0 & 0 & -1 & 0 & 0 & 4 & -1 & 0 \\
0 & 0 & 0 & 0 & -1 & 0 & -1 & 4 & -1 \\
0 & 0 & 0 & 0 & 0 & -1 & 0 & -1 & 4
\end{pmatrix},
$$

and

$$
\mathbf{u} = \begin{pmatrix} u_{1,1} & u_{1,2} & u_{1,3} & u_{2,1} & u_{2,2} & u_{2,3} & u_{3,1} & u_{3,2} & u_{3,3} \end{pmatrix}^T
$$

is the unknown matrix, and

$$
\mathbf{b} = \begin{pmatrix} 2\cos\dfrac{\pi}{4} & 0 & 2\cos\dfrac{3\pi}{4} & 0 & 0 & 0 & 2\cos\dfrac{5\pi}{4} & 0 & 2\cos\dfrac{7\pi}{4} \end{pmatrix}^T.
$$

Solving the system we obtain the results given in Table E. 8.3.5.

Table E. 8.3.5.

$j\backslash i$	1	2	3
1	0.35355339	0	−0.35355339
2	0	0	0
3	−0.35355339	0	0.35355339

6. The solution is given in Table E. 8.3.6.

Table E. 8.3.6.

$j\backslash i$	1	2	3	4	5
1	0.79142	0.73105	0.71259	0.73105	0.79142
2	0.52022	0.42019	0.38823	0.42019	0.52022

7. Let $u(0.25i, 0.25j) \approx u_{i,j}$, $i = 1, 2, 3$, $j = 1, 2$, $u(0.25i, 0.75) \approx u_{i,3}$, $i = 2, 3$. The results are given in Table E. 8.3.7.

Table E. 8.3.7.

$j\backslash i$	1	2	3
1	165/268	218/268	176/268
2	174/268	263/268	218/268
3		174/268	165/268

Section 8.4.

1. The results are given in Table E.8.4.1.

Table E.8.4.1

t_j	$h = 1/8$	$h = 1/16$	$u(1/4, t_j)$
0.01	0.68	0.62	0.67
0.02	0.47	0.42	0.45
0.03	0.32	0.28	0.31
0.04	0.22	0.19	0.21
0.05	0.15	0.13	0.14
0.06	0.10	0.09	0.09
0.07	0.07	0.06	0.06
0.08	0.05	0.04	0.04
0.09	0.03	0.03	0.03
0.10	0.02	0.02	0.02
0.11	0.01	0.01	0.01
0.12	0.01	0.01	0.01
0.13	0.01	0.01	0.01
0.14	0.00	0.00	0.00
0.15	0.00	0.00	0.00

2. The results are given in Table E.8.4.2.

Table E.8.4.2.

t_j	$h = 1/8$	$h = 1/16$	$u(1/4, t_j)$
0.01	0.68	0.63	0.67
0.02	0.47	0.42	0.45
0.03	0.32	0.29	0.31
0.04	0.22	0.19	0.21
0.05	0.15	0.13	0.14
0.06	0.10	0.09	0.09
0.07	0.07	0.06	0.06
0.08	0.05	0.04	0.04
0.09	0.03	0.03	0.03
0.10	0.02	0.02	0.02
0.11	0.02	0.01	0.01
0.12	0.01	0.01	0.01
0.13	0.01	0.01	0.01
0.14	0.01	0.00	0.00
0.15	0.00	0.00	0.00

3. The results are given in Table E.8.4.3.

Table E.8.4.3.

$t_j \backslash x_i$	4.0	8.0	12.0
40.0	1.5	2.0	4.0
80.0	1.25	2.375	5.0
120.0	1.2188	2.75	5.5939
160.0	1.2970	3.0780	5.9843

4. Using the central finite difference approximations for u_x and u_{xx} the explicit finite difference approximation of the given initial boundary value problem is

$$u_{i,j+1} = \left(1 - \frac{2\Delta t}{h^2}\right)u_{i,j} + \left(\frac{\Delta t}{2h} + \frac{\Delta t}{h^2}\right)u_{i+1,j}$$
$$+ \left(-\frac{\Delta t}{2h} + \frac{\Delta t}{h^2}\right)u_{i-1,j}.$$

With the given $h = 0.1$ and $\Delta t = 0.0025$ the results are presented in Table E.8.4.4.

Table E.8.4.4.

$t \backslash x$	x_1	x_2	x_3	x_4	x_5	x_6	x_7
0.0025	0.1025	0.2025	0.3025	0.5338	0.7000	0.5162	0.2975
0.0025	0.1044	0.2050	0.3395	0.5225	0.6123	0.5025	0.3232
0.0075	0.1060	0.2163	0.3556	0.5026	0.5621	0.4815	0.3321
0.0100	0.1098	0.2267	0.3611	0.4832	0.5268	0.4614	0.3328
0.0125	0.1144	0.2342	0.3612	0.4657	0.4993	0.4432	0.3293
0.0150	0.1187	0.2391	0.3585	0.4497	0.4766	0.4266	0.3239
0.0175	0.1221	0.2419	0.3541	0.4351	0.4571	0.4114	0.3172
0.2000	0.1245	0.2429	0.3487	0.4216	0.4399	0.3976	0.3103

5. Since $\lambda = 1/2$, the Crank–Nicolson finite difference scheme becomes

$$u_{i+1,j+1} - 5u_{i,j+1} + u_{i-1,j+1} = -u_{i+1,j} + 3u_{i,j} - u_{i-1,j},$$

for $i = 1, 2, 3; j = 0, 1, 2, \ldots$.

The boundary conditions imply that $u_{0,j} = u_{4,j} = 0$ for every j. The initial condition implies that $u_{i,0} = i/4$. Solving the system we obtain the results presented in Table E.8.4.5.

Table E.8.4.5.

$t \backslash x$	0.00	0.25	0.50	0.75	1.00
1/32	0	0.25	0.50	0.75	0
2/32	0	0.25	0.50	0.25	0
3/32	0	0.25	0.25	0.25	0
4/32	0.125	0.25	0.25	0.125	0

6. The results are presented in Table E.8.4.6.

Table E.8.4.6.

$t\backslash x$	0.00	0.25	0.50	0.75	1.00
1/32	0	0.0312500	0.0312500	0.0312500	0
2/32	0	0.0468750	0.0625000	0.0468750	0
3/32	0	0.0625000	0.0782500	0.0625000	0
4/32	0.125	0.0703125	0.0937500	0.0703125	0

Section 8.5.

1. Use finite difference approximations for the partial derivatives.

2. The analytical solution of the problem is given by $u(x,t) = f(x-t)$.

    The graphical results are presented in Figure E.8.5.1. (a), (b).

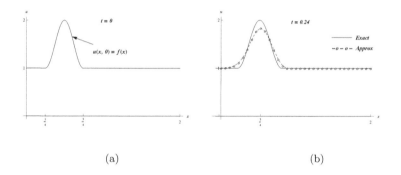

(a)   (b)

Figure E.8.5.1

3. The numerical results are given in Table E.8.5.1. The numbers are rounded to 4 digits.

Table E.8.5.1.

$t\backslash x$	0.25	0.50	0.75	1.00	1.25	1.50	1.75	2.00
0.2	1.0	1.0	1.0	0.985	1.024	1.270	1.641	1.922
0.4	1.000	1.000	1.000	0.993	0.976	1.063	1.352	1.715
0.6	1.000	0.999	1.000	1.000	0.981	0.977	1.115	1.434
0.8	1.000	1.000	1.000	1.000	0.995	0.969	0.991	1.177
1.0	1.000	1.000	1.000	1.000	1.000	0.986	0.960	1.015
1.2	1.000	1.000	1.000	1.000	1.000	1.000	0.974	0.957

The values of the analytical solution are presented in Table E.8.5.2.

Table E.8.5.2.

$t\backslash x$	0.25	0.50	0.75	1.00	1.25	1.50	1.75	2.00
0.2	1.000	1.000	1.000	1.191	1.562	1.979	2.00	2.00
0.4	1.000	1.000	1.000	1.000	1.036	1.038	1.039	1.706
0.6	1.000	1.000	1.000	1.000	1.000	1.000	1.077	1.410
0.8	1.000	1.000	1.000	1.000	1.000	1.000	1.000	1.130
1.0	1.000	1.000	1.000	1.000	1.000	1.000	1.000	1.000
1.2	1.0000	1.000	1.000	1.000	1.000	1.000	1.000	1.000

A graphical solution of the problem at the time instances $t = 0$ and $t = 0.24$ is displayed in Figure E.8.5.2 (a), (b).

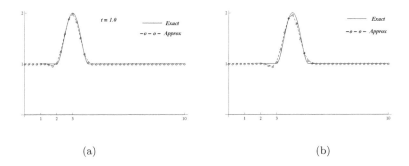

(a)   (b)

Figure E.8.5.2

4. For the initial level use the given initial condition $u_{i,0} = f(ih)$. For the next level use the Forward-Time Central-Space

$$u_{i,1} = u_{i,0} - \frac{\lambda}{2}\left(2u_{i+1,0} - u_{i-1,0}\right).$$

For $j \geq 1$ use the Leap-Frog method

$$u_{i,j+1} = u_{i,j-1} - \lambda\left(u_{i+1,j} - u_{i-1,j}\right).$$

Using Mathematica, the approximation $u_{i,6}$ is obtained for $\lambda = 0.5$, $h = 0.1$ ($\Delta t = 0.05$) at time $t = 0.3$. This solution along with the solution $f(x - 0.3)$ is shown in Figure E.8.5.3 (b). In Figure E.8.5.3 (a) the initial condition $u(x,0)$ is displayed.

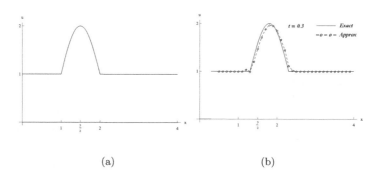

(a)   (b)

Figure E.8.5.3

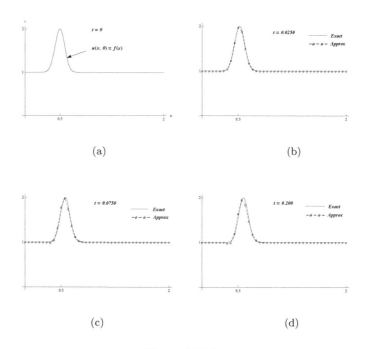

(a)   (b)

(c)   (d)

Figure E.8.5.4

5. The analytical solution of given initial value problem is

$$u(x,t) = f\left(x - \frac{t}{1+x^2}\right) = 1 + e^{-100\left(x - \frac{t}{1+x^2} - 0.5\right)^2}.$$

To obtain the numerical results, use $h = 0.05$ and $\Delta t = 0.0125$. The obtained numerical results at several time instances t are displayed in Figure E.8.5.4 (a), (b), (c) and (d).

6. The numerical solution is displayed in Figure E.8.5.5 and the analytical solution is presented in Figure E.8.5.6. It is important to point out that the x and t axes are scaled: value 60 on the x-axis corresponds to position $x = 3$, and the value 75 on the t-axis corresponds to time $t = 3$.

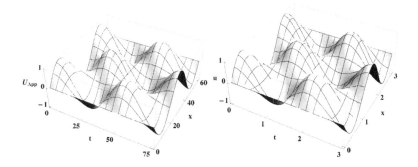

Figure E.8.5.5   Figure E.8.5.6

7. The analytical solution $u_{an}(x, t)$ is given by

$$u_{an}(x, t) = \frac{1}{2}\left(F_{ext}(x + t) + F_{ext}(x - t)\right),$$

where $F_{ext}(x)$ is the 2 periodic extension of the odd function

$$f_{odd}(x) = \begin{cases} -f(-x), & -1 \le x \le 0, \\ f(x), & 0 \le x \le 1. \end{cases}$$

The numerical results are given in Table E.8.5.3 and the results obtained from the analytical solution are given in Table E.8.5.4. The numbers are rounded to 4 digits.

Table E.8.5.3.

$t\backslash x$	0.1	0.2	0.3	0.4	0.5	0.6	0.7
0.05	0.079	0.104	0.119	0.124	0.119	0.104	0.079
0.10	0.074	0.100	0.115	0.120	0.115	0.100	0.075
0.15	0.068	0.091	0.109	0.113	0.109	0.094	0.069
0.20	0.062	0.084	0.099	0.105	0.100	0.085	0.061
0.25	0.04	0.070	0.085	0.093	0.089	0.075	0.054

Table E.8.5.4.

$t\backslash x$	0.1	0.2	0.3	0.4	0.5	0.6	0.7
0.05	0.079	0.104	0.119	0.124	0.119	0.104	0.079
0.10	0.075	0.100	0.115	0.120	0.115	0.100	0.075
0.15	0.0687	0.094	0.109	0.113	0.109	0.094	0.069
0.20	0.006	0.085	0.100	0.105	0.100	0.085	0.060
0.25	0.05	0.074	0.089	0.094	0.089	0.074	0.050

8. The analytical solution $u_a(x,t)$ is given by

$$u_a(x,t) = \frac{1}{2}F_{ext}(x+t) + \frac{1}{2}F_{ext}(x-t)$$

$$+ \frac{1}{2}G_{ext}(x+t) - \frac{1}{2}G_{ext}(x-t),$$

where $F_{ext}(x)$ and $G_{ext}(x)$ are the 2 periodic extension of the odd functions

$$F_{odd}(x) = \begin{cases} -f(-x), & -1 \leq x \leq 0, \\ f(x), & 0 \leq x \leq 1; \end{cases}$$

$$G_{odd}(x) = \begin{cases} -g(-x), & -1 \leq x \leq 0, \\ g(x), & 0 \leq x \leq 1. \end{cases}$$

To find G_{ext} first find an antiderivative $G(x)$ for $-1 < x < 1$:

$$G(x) = \int_{-1}^{x} g(x)\,dx = \int_{-1}^{x} \sin \pi x\,dx = -\frac{1}{\pi}(1 + \cos \pi x).$$

After that extend periodically $G(x)$.

The numerical solution of the problem along with the analytical solution at the specified time instances is presented in Figure E.8.5.7.

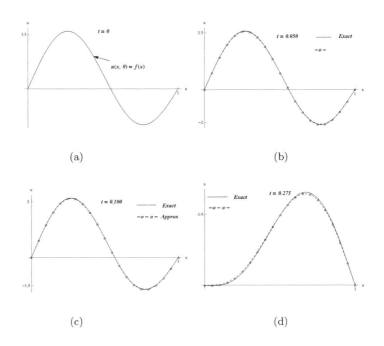

Figure E.8.5.7

9. The system has the following matrix form

$$\mathbf{U_{i,j+1}} = L \cdot \mathbf{U_{i,j}} - \mathbf{U_{i,j-1}} + \lambda^2 \, \mathbf{b_{j-1}},$$

where

$$A = \begin{pmatrix} 2(1-\lambda^2) & \lambda^2 & \cdots & 0 & 0 \\ \lambda^2 & 2(1-\lambda^2) & \cdots & 0 & 0 \\ \vdots & & & & \\ 0 & 0 & \cdots & 2(1-\lambda^2) & \lambda^2 \\ 0 & 0 & \cdots & \lambda^2 & 2(1-\lambda^2), \end{pmatrix}$$

$$\mathbf{U_{i,j}} = \begin{pmatrix} u_{1,j} \\ u_{2,j} \\ \vdots \\ u_{i,j} \end{pmatrix}, \qquad \mathbf{b_{j-1}} = \begin{pmatrix} \alpha(t_{j-1}) \\ 0 \\ \vdots \\ 0 \\ \beta(t_{j-1}) \end{pmatrix}.$$

10. Divide the unit square by the grid points $x_i = ih$, $i = 0, 1, \ldots, n$, $y_i = jh$, $j = 0, 1, \ldots, n$, for some positive integer n where $h = \Delta x = \Delta y = 1/n$ and divide the time interval $[0, T]$ with the points $t_k = k\Delta t$, $k = 0, 1, \ldots, m$ for some positive integer m, where $\Delta t = T/m$. Let $u_{i,j}^{(k)} \approx u(x_i, y_j, t_k)$ be the approximation to the exact solution $u(x, y, t)$ at the point (x_i, y_j, t_k). Substituting the forward approximations for the second partial derivatives into the wave equation, the following approximation is obtained.

$$u_{i,j}^{(k+1)} = 2\left(1 - 2\lambda^2\right)u_{i,j}^{(k)} + \lambda^2\left(u_{i+1,j}^{(k)} + u_{i-1,j}^{(k)}\right.$$
$$\left. + u_{i,j+1}^{(k)} + u_{i,j-1}^{(k)}\right) - u_{i,j}^{(k-1)},$$

where

$$\lambda = a\frac{\Delta t}{h}.$$

INDEX OF SYMBOLS

$\mathbb{R} = (-\infty, \infty)$ (the set of all real numbers), 2

$S_N(f, x)$ (Fourier partial sum), 6

$S(f, x)$ (Fourier series), 6

$\mathbb{N} = \{1, 2, \ldots\}$ (the set of all natural numbers), 8

$\in$ (belongs to), 16

$f(a^+)$, $f(a^-)$ (right hand side limit, left hand side limit), 20

(f, g) (inner product), 20

f', $f^{(k)}$ (derivative, k^{th} derivative), 46

Γ (Euler's Gamma function), 55

$\mathbb{Z} = \{0, \pm 1, \pm 2, \ldots\}$ (the set of all integer numbers), 57

$\mathcal{L}$ (Laplace transform), 83

$\mathcal{F}$ (Fourier transform), 83

$\mathcal{F}_c$, $\mathcal{F}_s$ (Fourier cosine transform, Fourier sine transform), 83

$\mathcal{L}^{-1}$ (inverse Laplace transform), 90

H (Heaviside step function), 96

δ (Dirac delta function), 99

$f * g$ (convolution), 109

$\mathcal{F}^{-1}$ (inverse Fourier transform), 117

$\mathcal{F}_c^{-1}$ (inverse Fourier cosine transform), 117

$\mathcal{F}_s^{-1}$ (inverse Fourier sine transform), 117

sgn   (sign function), 119

P_n (Legendre's polynomial), 172

$P_n^{(m)}$ (associated Legendre's polynomial), 181

J_μ (Bessel function of the first kind), 184

Y_μ (Bessel function of the second kind), 185

$\mathbb{R}^n$ (n-real space), 204

$\mathbf{x} = (x_2, x_2, \ldots, x_n)$ (row vector in $\mathbb{R}^n$), 204

u_x, $\dfrac{\partial u}{\partial x}$, u_{xx}, $\dfrac{\partial^2 u}{\partial x^2}$ (partial derivatives), 204

Ω (domain), 204

∂ (boundary), 204

$C^k(\Omega)$, $C^\infty(\Omega)$ (classes of differentiable functions), 204

$\subseteq$ (subset), 210

Δ (Laplace operator, Laplacian), 224

∇ (divergence), 408

$\mathcal{H}_n$ (Hankel transform), 438

$\mathcal{H}_n^{-1}$ (inverse Hankel transform), 438

$\| \ \|$ (norm of a vector, norm of a matrix), 454

O ("big Oh"), 473

$\approx$ (approximately equal to), 473

Sinc   sink function), 524

Res (residue), 551

erf   (error function), 368

630

INDEX

For Product Safety Concerns and Information please contact our EU representative GPSR@taylorandfrancis.com Taylor & Francis Verlag GmbH, Kaufingerstraße 24, 80331 München, Germany

Printed and bound by CPI Group (UK) Ltd, Croydon, CR0 4YY

01/05/2025

01858518-0006